TECTONIC GEOMORPHOLOGY

COMPANION WEBSITE

This book has a companion website:

www.wiley.com/go/burbank/geomorphology

with Figures and Tables from the book for downloading

Tectonic Geomorphology

Second Edition
Douglas W. Burbank
and Robert S. Anderson

WILEY-BLACKWELL

A John Wiley & Sons, Ltd., Publication

This edition first published 2012 © 2012 by Douglas W. Burbank and Robert S. Anderson
Previous edition: 2001 © Douglas W. Burbank and Robert S. Anderson

Blackwell Publishing was acquired by John Wiley & Sons in February 2007. Blackwell's publishing program
has been merged with Wiley's global Scientific, Technical and Medical business to form Wiley-Blackwell.

Registered Office
John Wiley & Sons, Ltd, The Atrium, Southern Gate, Chichester, West Sussex, PO19 8SQ, UK

Editorial Offices
9600 Garsington Road, Oxford, OX4 2DQ, UK
The Atrium, Southern Gate, Chichester, West Sussex, PO19 8SQ, UK
111 River Street, Hoboken, NJ 07030-5774, USA

For details of our global editorial offices, for customer services and for information about how to apply for
permission to reuse the copyright material in this book please see our website at
www.wiley.com/wiley-blackwell.

Library of Congress Cataloging-in-Publication Data

Burbank, Douglas West.
Tectonic geomorphology / Douglas W. Burbank and Robert S. Anderson. – 2nd ed.
      p.   cm.
   Includes bibliographical references and index.
   ISBN 978-1-4443-3886-7 (cloth) – ISBN 978-1-4443-3887-4 (pbk.)
1. Geomorphology.   2. Geology, Structural.   I. Anderson, Robert S. (Robert Stewart), 1952–
II. Anderson, Robert S.   III. Title.
   GB401.5.B86 2012
   551.41–dc22

                                                2011012030

A catalogue record for this book is available from the British Library.

This book is published in the following electronic formats: ePDF 9781444345032;
Wiley Online Library 9781444345063; ePub 9781444345049; Mobi 9781444345056

Set in 9.5/12pt Garamond by SPi Publisher Services, Pondicherry, India

1   2012

DEDICATED

to

Rachel and Suzanne, Helen and Jack, Flodie and Andy.

Contents

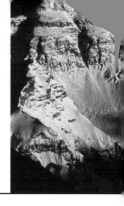

Colour plate section appears between pages 226 and 227

COMPANION WEBSITE

This book has a companion website:

www.wiley.com/go/burbank/geomorphology

with Figures and Tables from the book for downloading

Preface to First Edition

No one can gaze at satellite images of the earth's surface without being struck by the crumpled zones of rocks that delineate regions of ongoing and past tectonic deformation. Landscapes in tectonically active areas result from a complex integration of the effects of vertical and horizontal motions of crustal rocks and erosion or deposition by surface processes. In a sense, many landscapes can be thought of as resulting from a competition among those processes acting to elevate the earth's surface and those that tend to lower it. The study of this competition and the interpretation of the geodynamic and geomorphic implications of such landscapes is the focus of tectonic geomorphology.

Tectonic geomorphology is a wonderfully integrative field that presents stimulating challenges to anyone trying to extract information from deforming landscapes. Consider, for example, a typical subject for a tectonic geomorphic study: a growing fold in the hangingwall of a thrust fault. The fault cuts the surface along much of its length, but diminishes in displacement towards its tip, where no surface break is present. How is the shape of the fold related to the mechanical properties of the faulted material? How much slip occurs in an individual earthquake on that thrust fault and what is the detailed pattern of deformation? How far does the tip of the thrust propagate laterally and how much does the hangingwall grow with each rupture? Is a similar pattern of deformation repeated in successive earthquakes? What is the interval between earthquakes? Can the preserved structural form of the fold be explained as the sum of deformation during earthquakes alone, or is there significant interseismic deformation? How is the surface of the fold modified by erosion and deposition as it grows in length and height, and are there systematic changes along the fold that define a predictable evolutionary pattern of the landscape above such faults? Are the patterns and rates of erosion influenced more by tectonic or climatic forcing? What are the relative and absolute rates of rock uplift and erosion within the hangingwall and how is any imbalance reflected by the resulting landscape? In order to answer these questions, one has to draw on an understanding of paleoseismology, geodesy, structural geology, geomorphology, geochronology, paleoclimatology, stratigraphy, and rock mechanics, because each underpins some aspects of the interplay of tectonics and erosion. Key parts of the answer to each question are embedded in the landscape and are the subject of tectonic geomorphology.

Although efforts to interpret the processes that generate dynamic landscapes are not new, several key advances in the past few decades have provided a more robust foundation for rigorous analysis, and have opened up entirely new avenues of research.

- The development of new chronologic techniques permits us to inject time control into many facets of a landscape that were formerly unconstrained. The plethora of new dating approaches underpins most of the success in defining rates of processes.

- Process-oriented geomorphic studies have served to improve our understanding of the physical basis of surface processes. Insights into the key physical phenomena that control erosion and sediment transport have led to the development of both conceptual and quantitative models that can be applied in landscape analysis.
- Understanding of the record of past climate change has deepened immeasurably, such that the magnitude and rate of past climatic variability is now better calibrated than ever before. The impact of climate change on surface processes and landform development can now be better addressed in this detailed framework.
- New geodetic tools are providing an incomparable overview of rates of deformation of the earth's surface. Not only can changes in distances between specific points be documented with unprecedented accuracy, but the spatial pattern of deformation following an earthquake can now be documented across vast areas.
- Paleoseismology, the study of the record of past earthquakes, has matured into a diverse discipline and has provided a basis for documenting the behavior of faults, their impact on adjacent landscapes, and the societal hazards that they pose.
- Instead of conducting regional tectonic analyses, many structural geologists have focussed increasingly on the physical characterization of faulting and folding, on scaling relationships, and on the interactions among faults. Such studies show how fundamental building blocks like faults evolve through time and influence patterns of deformation.
- The widespread availability of digital topography and the development of new ways to obtain accurate topographic data through forest canopies, clouds, and oceans has facilitated a time-efficient quantification of topographic attributes that was previously unattainable.
- The accessibility to high-speed computing has underpinned the development of numerical models that exploit the results of many of the fields enumerated above. More importantly than providing a means to numerically re-create some observation, these models facilitate exploration of the potential interactions, sensitivities, and response times among changing processes.

The list could be much longer, but the point is clear: this is a time of unusual opportunity in tectonic geomorphology. The confluence of new approaches, tools, and data bases is leading to rapid changes in our understanding of how the surface of the earth evolves in regions of active tectonism. These topics are the focus of the present book.

This book is organized somewhat differently than most other books on this topic. The initial chapters are devoted to fundamental building blocks and individual disciplines in tectonic geomorphic studies: geochronology, structural characteristics of faults and folds, the nature of geomorphic "markers" that can be used to keep track of deformation, geodesy, paleoseismology, and strategies for calculating rock uplift and erosion rates. Rather than following these chapters with discussions of tectonic geomorphology in specific geomorphic settings, such as fluvial or coastal environments, we have chosen to view the landscape within different time frames, ranging from <10^4 yr to more than 1 million years. The choice of each time window is not arbitrary, but instead reflects our understanding of how the tectonic and geomorphic information that can be extracted from a landscape varies as a function of time. In our view, these time scales reflect the relative dominance of surface or tectonic processes in shaping a landscape, the degree to which a landscape preserves the record of individual or small groups of events, and the time scale of interactions between climate, erosion, and tectonics. Contrasting types of tectonic geomorphic information are embedded in landscapes at these different time scales. The appropriate questions to ask and the proper techniques for extracting answers also differ as a function of time.

The reasoning behind this subdivision of our text may become more clear if you again consider the fold growing above a thrust fault. Individual earthquake ruptures occur at Holocene time scales. Essentially instantaneous, co-seismic deformation changes geomorphic gradients and forces adjustments within the geomorphic system. Just after the earthquake, the entire displacement field will be almost perfectly recorded by the deformed ground surface and consequently provides an excellent opportunity to delineate relationships among rock properties, displacement gradients, and seismic characteristics. Over the ensuing centuries, fault scarps will degrade, and geomorphic processes will modify the hillslopes and the channels of the landscape toward a new equilibrium with the altered geometry of the surface – until the next earthquake induces another cycle of adjustment. Throughout this interval, the average geomorphic surface is likely to mimic the aggregate strain accumulated over several seismic cycles.

At times scales of 100,000 years or more, major swings in the climate will have occurred, dozens of earthquakes will have progressively deformed the fold, and geomorphic processes will have incised uplifted surfaces of the fold. The geomorphic surface is less likely to mimic the strain field, and climatic changes are likely to have created a distinctive imprint on the fold's surface. The steepest portions of a river crossing the fold may now occur over the fold crest, rather than on the forelimb of the fold. At these time scales, the geomorphic system associated with the fold will evolve as the structural relief increases and the fold propagates laterally.

At intervals of a million years, it is likely that the blind thrust beneath the fold has linked up with other faults, or has died, or is now embedded in a series of growing structures. The topography and the regional pattern of accumulated strain may be largely decoupled, such that the highest topography may not match the structural crest of the fold. But the gradients of rivers, the presence of water gaps, and the spatial array of geomorphic elements, such as catchment geometries or slope distributions, will reflect the long-term evolution of the fold-and-thrust belt.

The final chapter of this book is devoted to numerical modeling of landscapes. These models have now evolved to be able to handle the complexity of the deformation fields, and the temporal and spatial complexity of the geomorphic processes that modify a landscape. They can be used to test old conceptual models of landscape evolution, to generate new strategies for field documentation of the relevant processes and their rates, and as educational tools to allow visualization of landscape evolution. We do not attempt to provide a complete guide or a cookbook for such modeling. Rather, this chapter represents an introduction to different types and scales of numerical models, and provides what could be termed "building blocks" for landscapes: the displacement fields associated with individual faults, and with orogen-scale deformation fields at depth, and the "rules" commonly used for movement of regolith on hillslopes, the transport of sediment in channels, and bedrock incision of streams. We illustrate the linkage of tectonic and geomorphic process models with examples ranging from simple models of evolving fault scarps to more complex models of entire tectonically active landscapes.

The field of tectonic geomorphology has been advancing and expanding so rapidly that it has been difficult to keep this text abreast of the latest developments. Despite its obvious shortfalls, it is our hope that this text will serve as a useful springboard for those geologists wanting an introduction to tectonic geomorphology and as a helpful reference for those of us who are pursuing research in aspects of this broad, highly integrative, and rapidly evolving field. It is our hope that upper level undergraduates and students early in their graduate careers will find this to be a useful and accessible introduction to tectonic geomorphology.

Many people played a role in the production of this book. John Grotzinger invited one of us (DWB) to come to MIT as a Crosby Lecturer and to develop a course on tectonic geomorphology. The idea of this book sprang from those lectures and the realization that there appeared to be a need for such a text. Doug Hammond and Bob Douglas at USC and Dick Walcott at Victoria University provided logistical support during various stages of writing, while Bill Bull, Steve Wells, Ed Keller, and Paul Hoffman provided early suggestions regarding content and format. We greatly appreciate the wisdom and guidance of our own teachers and research advisors, who set us on a path that has allowed us to range widely through the disciplines of the earth sciences. These include Steve Porter, Tom Dunne, Noye Johnson, Bernard Hallet, Jim Smith, Peter Haff, and Gary Johnson. Our research collaborators, colleagues, and many former students provoked us with their ideas and engaged us in discussions that helped shape this book. In particular, we would like to thank Nic Brozovic, Andrew Meigs, Julio Friedmann, Jaume Vergés, Richard Beck, Fritz Schlunegger, Ian Brewer, Mike Ellis, Jeff Marshall, Alex Densmore, Greg Hancock, Eric Small, Dan Orange, Kelin Whipple, Liz Safran, Eric Fielding, Jerôme Lavé, Niels Hovius, Rudy Slingerland, Hugh Sinclair, Marith Reheis, Peter Molnar, Euan Smith, Ann Blythe, Merri Lisa Formento-Trigilio, and Jamie Shulmeister. The technical production of this book has benefited from assistance of Doug Myers and the capable staff at Blackwells, particularly Jane Humphreys and Jill Connors. Many ideas and concepts in this book were developed during research supported by the National Science Foundation (EAR 92-2056, 96-14765, 96-27865, 97-06269, 99-09647) and NASA (NAG W-3762, 5-2191, 5-7781). Finally, we would like to thank our families and particularly our spouses for their support of our efforts and their tolerance of our preoccupations.

Doug Burbank
Bob Anderson

Preface to Second Edition

This past decade has seen an explosion in research across the entire spectrum of tectonic geomorphology. Such a blossoming of creative approaches and novel results has begun to generate a new perspective on the interactions of tectonics with surface processes: a perspective that has spurred us to produce a second edition of our 2001 book.

Several key elements have underpinned significant advances over the past decade. The abundance of readily available, high-resolution digital topography for much of the globe, such as that made publicly available at increasingly higher resolution by NASA, has enabled a new quantification and comparison of landscapes. New imagery around the globe that is accessible through portals like Google Earth™ has ushered in a new era of armchair exploration and analysis of distant and remote landscapes by geomorphologists, whereas meter-scale topography, such as that derived from radar altimetry, such as Lidar, has provided unprecedented spatial resolution and a striking view of the "bare Earth" in which trees and shrubs have been removed. Soon, we expect that terrestrial laser scanning will represent a common tool exploited by geodesists and geomorphologists for cm-scale (and smaller) measurements, whereas Google Earth may enable meter-scale measurements of rivers, glaciers, and landslides.

This improved quantification of landscapes has proceeded hand-in-hand with an enhanced view of how the Earth deforms. Not only has InSAR analysis matured, but the number of continuous GPS sites has increased by an order

of magnitude since 2000. Currently, some tectonically active terrains, such as Taiwan or the western US, host dozens of geodetic arrays that provide mm-scale precision at sub-annual time scales. Similarly, dense seismological arrays are yielding better reconstructions than ever before of earthquake ruptures, fault geometries, and seismogenic structures in the crust.

New syntheses of space-based observations of the earth, oceans, and atmospheres are providing both a more holistic Earth view, as well as a enhanced ability to draw quantitative comparisons. Satellite measurements of rainfall on an annual, seasonal, and even daily basis are now available at high spatial resolution for mid-to-low latitudes. The topography of the sea floor and height of the ocean surface is better defined than ever before. Such data provide a valuable starting point for exploring climate-tectonic linkages, topographic gradients, and subsurface loads. No longer do we need to merely speculate about either the magnitude of orographic precipitation gradients or the relationship of submarine and subaerial topography: these can be quantitatively examined.

Our ability to date landscape features and geomorphic or tectonic events continues to improve. New, low-temperature thermochronometers continue to be developed. Cheaper, faster, and smaller sample processing with many chronologic methods enables far more ages to be measured. As a consequence, the timing and rates of diverse events and processes are better known than ever before. Long- and short-term erosion rates, recurrence intervals, and innumerable deformed geomorphic features are now much better calibrated.

Our process-based understanding of phenomena ranging from rupture dynamics and fault creep to landslide triggering and fluvial erosion is rapidly improving. Geomorphologists seem to have become increasingly clever at finding those atypical field sites where enough variables can be held constant that the controls on some process can be reasonably explored. In addition, experimental facilities and approaches have permitted assessment of conceptual models, calibration of controlling parameters, and examination of the interplay and trade-offs within simple systems. As a consequence, an improved physics-based understanding of both tectonic and geomorphic processes has emerged.

The use of numerical models to explore the tectonic and geomorphic world has grown rapidly in the past decade. Increasingly powerful computers now enable modeling of landscapes and deformation at spatial and temporal scales that are appropriately scaled for geomorphic and tectonic processes. Models with foci ranging from lithospheric dynamics to atmospheric interactions with topography can now be used to predict dynamic topography, rainfall distributions, river profiles, glacial flow, and spatial variations in erosion. Moreover, our expanded quantification of rates of processes, as well as their geomorphic and tectonic signatures, enables better testing and improvement of model predictions.

Finally, a new generation of tectonic geomorphologists has emerged in the past decade. Armed with broad array of analytical skills, the ability to quantify processes better than anytime in the past, and a suite of numerical tools, modeling skills, and global data sets, these new scientists have sustained and accelerated the rejuvenation of the field that is the subject of this book.

We have retained the organization of our original book in which we begin with some basics on useful geomorphic features for tracking deformation, on approaches for dating such features, and on crustal structures that result from faulting and folding. We then examine both geodetic approaches to quantifying ongoing deformation, as well as paleoseismic techniques for reconstructing the history of past earthquakes. After an exploration of approaches to define variations in erosion or uplift (in its various forms), we turn to a time-based framework to examine interactions between landscapes and tectonics with three chapters that focus on timescales encompassing the Holocene, late to mid Pleistocene, and late Cenozoic. These temporal divisions are chosen because contrasting balances between tectonics and erosion prevail within each era. At Holocene time scales, tectonic deformation can act as a first-order control on landscapes, which are commonly in disequilibrium. At increasingly longer time scales, erosion and rock uplift may approach an equilibrium, whereas at still-longer time scales, tectonics and associated landscapes may wax and wane.

Our final chapter is devoted to numerical modeling of landscapes, orographic precipitation, and the thermal structure of a dynamic crust. As in the previous edition, we initially introduce different types and scales of numerical models, and then focus on "building blocks" for deforming and eroding landscapes. We include a new emphasis on interactions of topography with climate, the thermochronologic signal of differential erosion, dynamic topography, and long-term landscape evolution.

Throughout this edition, we incorporate research results from the past decade. Over 500 new references and more than 150 new figures have been added: most of these have been newly published since 2000. We bring to the fore a selection of the new techniques and results that appear to open inspiring vistas on interactions between geomorphology and tectonics.

We have not, however, attempted a thorough coverage of this burgeoning field. For example, despite the amazing new views of the seafloor that are continuing to emerge, we have largely ignored submarine topography and the unique perspective it offers of tectonic-geomorphic interactions. Similarly, we largely ignore volcanic terranes, and we cover only a modest selection of the remarkable array of chronologic and geochemical techniques that are now being applied to tectonic geomorphic problems. Nonetheless, we hope that this edition will provide an accessible and effective foundation for upper-level undergraduates and graduate students with interests in tectonic geomorphology and that it will serve as a helpful reference for more skilled professionals.

Many people played a role in the second edition of this book. Macquarie University and the University of Canterbury hosted one of us (DWB) on a sabbatical during which much of the writing and drafting occurred. Our research collaborators, colleagues, and many former students provoked us with their ideas and engaged us in discussions that helped shape this book. In particular, we would like to thank Bodo Bookhagen, Eric Kirby, Mike Oskin, Jerôme Lavé, Nathan Niemi, Colin Amos, Brian Clarke, Alison Duvall, Beth Pratt-Sitaula, Ian Brewer, Mike Ellis, Jeff Marshall, Alex Densmore, Greg Hancock, Eric Small, Dan Orange, Liz Safran, Niels Hovius, Kelin Whipple, Manfred Strecker, Neil Humphrey, Ann Blythe, Todd Ehlers, Jaakko Putkonen, Peter Molnar, Mark Quigley, Kip Hodges, Ben Crosby, George Hilley, Jean-Phillipe Avouac, Dork Sahagian, Ramon Arrowsmith, Olaf Zielke, and Ray Weldon.

The technical production of this book has benefited from attentive editing of Kelvin Matthews, the responsiveness of Pascal Raj, and the patience of Ian Francis at Blackwell and Wiley. Many ideas and concepts in this edition were developed during research supported by the National Science Foundation (EAR 00-01044, 01-17242, 02-30403, 02-29911, 04-08675, 05-07431, 08-19874, 08-38265) and by NASA (NAG5-9039, 13758, NNX08AG05G), National Geographic Research, the Petroleum Research Fund, and the National Earthquake Hazards Reduction Program. Lastly, we would like to thank our spouses for their persistent encouragement and their much appreciated forebearance over the past decade.

Doug Burbank
Bob Anderson
April 2011

1 Introduction to tectonic geomorphology

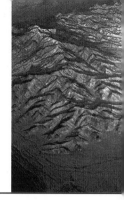

Introduction

The unrelenting competition between tectonic processes that tend to build topography and surface processes that tend to tear them down represents the core of tectonic geomorphology. Anyone interested in the Earth's surface has wondered why it has the shape it does and what forces are responsible for that shape. For more than a century, this natural curiosity has inspired numerous conceptual models of landscape evolution under varied tectonic and climatic regimes. In the past, our ability to assign reliable ages to geomorphic and tectonic features was commonly very limited. In the absence of a chronological framework, testing competing concepts of landscape evolution was nearly impossible. As a consequence, these unquantified models were often viewed skeptically and treated as speculative notions.

During the past few decades, innovative applications of new techniques for determining the ages of landscape features, for assessing the mechanisms and rates of geomorphic processes, and for defining rates of crustal movement, have helped revitalize the field of tectonic geomorphology. It is now possible to measure at the scale of millimeters how rapidly a given site is moving with respect to another and how those rates of relative convergence or divergence are partitioned among various faults and folds. Similarly, we can now quantify how rapidly

rivers and glaciers are incising into bedrock and the rates at which landslides are stripping mountain slopes. Clearly, the merger of such data sets can underpin a new understanding of the balance between the rates at which crustal material is added at a given site and the rates at which this material is eroded away. Defining this balance and interpreting the landscape that results from this competition represents a major component of modern tectonic geomorphology.

One of the remarkable attributes of tectonic geomorphology is the breadth of fields it often encompasses. Few other geological fields easily blend as disparate topics as seismology, Quaternary climate change, geochronology, structure, geodesy, and geomorphology. In fact, such breadth makes this field particularly exciting, as new data and ideas emerge from fields as diverse as paleobotany and fault mechanics. This diversity also presents a formidable challenge, because successful studies commonly require blending of appropriate data from specialized fields that were traditionally considered to be unrelated. Certainly, specialists in well-established disciplines like structural geology or stratigraphy often make important contributions to tectonic geomorphology. Fundamental leaps in our understanding, however, more commonly emerge from an integration across several disciplines.

This book is intended primarily for readers who are familiar with basic geomorphological and structural concepts and terms. Although we define many terms and review basic concepts and models throughout the book, we assume that our audience understands many commonly used geomorphological and structural terms, such as *base level* or *conjugate faults*. For those readers without such a background, occasional reference to a basic geomorphological or structural text will probably be helpful.

It is beyond the scope of this book to cover the contributions of each of the subdisciplines of tectonic geomorphology. Instead, we focus on some of the key tools, approaches, and concepts that have served to advance tectonic geomorphological studies during the past few decades. Several building blocks underpin many such studies, including a knowledge of how the Earth deforms both during earthquakes and between them, what sorts of features can be used to track deformation in the past, and what types of techniques are useful for dating features of interest.

Initially, we introduce geomorphic markers: landscape features, such as marine or fluvial terraces, that can be used to track deformation. Prior to deformation, the surfaces of these markers represent planar or linear features with a known geometry that can be predictably tracked across the landscape. Faulting or folding can subsequently deform such markers. Documentation of any departure from their unperturbed shape can serve to define the magnitude of deformation. Consequently, the recognition and measurement of such displaced or deformed markers is critical to many tectonic-geomorphic studies. These deformed markers are the raw data that cry out for interpretation.

In order to delineate rates of deformation, both the timing and amount of deformation must be defined. Fortunately, dozens of techniques have been used successfully to define the age of displaced features. In a chapter on dating, we present examples of some of the more commonly used techniques, and we try to convey a sense of the situations in which use of each technique would be appropriate. For each technique, we provide its conceptual underpinnings, discuss some of its limitations, and describe what data are actually collected in the field.

The oft-used term "tectonic processes" is a grab-bag expression that encompasses all types of deformation, including the motion of tectonic plates, slip on individual faults, ductile deformation, and isostatic processes. We concern ourselves here primarily with those processes that are most relevant to relatively localized deformation of the Earth's surface. In this book, we have generally chosen to ignore deformation and surface processes related to volcanism. Many of the concepts that are developed here are descriptive of landscape responses to volcanic processes, but the way in which material is added to the system (sometimes from above or only from below), as well as the time and spatial scales, are commonly different between volcanic and non-volcanic settings.

Because the accumulated movements of individual faults and folds have built many landscapes, we describe some current concepts from seismology and structural geology concerning coseismic rupture, the scaling of fault slip, fault-related folding, and geometries of deformation in different tectonic settings (compressional, extensional, and strike-slip environments). This deformation generates the fundamental topographic morphology on which erosive forces act. Much has been learned about ongoing deformation of the Earth's surface through geodetic studies: detailed surveys that delineate regional to local crustal displacements, often at time scales of a few years. During the past two decades, the increasingly widespread application of new techniques, such as the Global Positioning System (GPS) and radar interferometry, has yielded spectacular new images of current crustal motions. In a chapter on geodesy, we describe these techniques and some of the insights gained from them. In many ways, it is in the context of this greatly improved understanding of modern deformation that we now can make the largest strides in interpreting past deformation and landscape evolution.

Knowledge of current rates of crustal deformation leads to a host of provocative (and socially relevant) questions: Have these rates been steady over time? Do individual faults rupture at regular intervals and produce earthquakes of similar magnitude through time? Do groups of faults accelerate synchronously at the expense of other faults? How large are the earthquake ruptures that accommodate the stress that is built by the incessant deformation of the crust? In a chapter on paleoseismology, we discuss some of the techniques for delineating the past behavior of faults, defining recurrence intervals and paleoslip rates, and assessing whether earthquakes occur with temporal regularity or randomness or in clusters.

Naturally, discovering ways to measure the amount and rate of erosional losses from an area and to define changes in the height of the surface of that same area through time lies at the heart of many tectonic-geomorphic studies. Armed with new tools for dating and with digital topographic databases, we can now address these topics more accurately than in the past. In a chapter on erosion and uplift, we describe several strategies for quantification of rates of erosion, surface uplift, and rock uplift, and we illustrate how these can be synthesized to examine the balance between building and tearing down of topography.

In the latter half of this book, we step away from a focus on the key building blocks and instead introduce concepts and examples of landscape evolution and tectonic interpretations at different time scales. Moreover, rather than examine specific geomorphic environments or topics, such as the ways in which rivers and fluvial features can provide insights on tectonic processes, we focus on more integrated studies. The key issues can be encompassed by two questions: "What information from the Earth's surface improves our understanding of the nature of the interactions between tectonics and geomorphological processes?" and "How do we interpret preserved geomorphic features in order to reveal rates and patterns of tectonic deformation in the past?" The more recent record is often best suited to answer the first question, whereas great value comes from studies that are able to answer the second question for intervals in the distant past.

There is a natural, time-dependent progression in the nature of tectonic–geomorphic landscapes and the insights that can be gained from them. Perhaps surprisingly, part of this progression is dictated by past climate changes (Fig. 1.1).

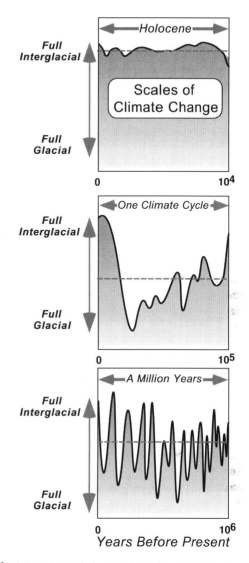

Fig. 1.1 Schematic illustration of climate change scaled at orders of magnitude (Holocene, one glacial cycle, Middle–Late Pleistocene).
The relative stability of the Holocene climate is atypical of a full glacial–interglacial, 100-kyr cycle. Similarly, the sawtooth pattern of change during past glacial–interglacial cycles does not show oscillations of similar magnitude, as is typical of the past million years.

At Holocene time scales, we are confined to a post-glacial era in which climatic conditions have varied relatively little. At time scales of 100000 years or more, complete or multiple glacial–interglacial cycles have occurred.

Large-scale changes in sea level, the size of glaciers, and the discharge of rivers at these time scales leave strong imprints on the landscape. These climate cycles are often the "drivers" that generate the geomorphic markers, in the form of moraines or marine and fluvial terraces, that are so important in the field of tectonic geomorphology. At time scales of a million years or more, numerous climatic cycles can create a more "time-averaged" landscape.

These different time scales also often create a natural segregation of the focus of tectonic-geomorphic studies. At Holocene time scales, cause and effect can be examined rather directly. For example, if faulting causes the land surface to tilt in a given direction, the timing and rate of diversion of rivers in response to that tilt can be measured. Present rates of crustal deformation can be compared against the sediment load of rivers. Commonly, geomorphic markers, such as river terraces, are almost untouched by erosion and can function as pristine recorders of Holocene deformation. The strength of Holocene studies is that the record is often most complete, the dating is most precise, rates of many competing processes can be directly measured, and their interactions can be examined. At least three potential disadvantages can limit Holocene studies. First, the rates of tectonic processes may be sufficiently slow or the occurrence of deformational events, such as earthquakes, may be sufficiently rare that the tectonic signal is obscure. Second, rates of geomorphic processes during Holocene (interglacial) times may not be very representative of long-term rates. Third, the time it takes for geomorphic systems to respond to a change imposed by tectonic forcing can be longer than the Holocene. Thus, the geomorphic system may be in a state of transition with respect to tectonic perturbations imposed upon it.

As described in subsequent chapters, the limited precision of most dating techniques means that, as one delves beyond the Holocene and farther into the past, it becomes progressively more challenging to associate specific events in time. Thus, it can be difficult to define direct responses to individual forcing events. On the other hand, major climate changes often have created robust geomorphic markers, such as river terraces, that persist as recognizable features and provide a lengthy time framework within which to document patterns of deformation. At short time scales, tectonic forcing is commonly unsteady, because it results from discrete events, such as earthquakes, that are widely separated in time. At time scales of more than 10^4 yr, this unsteadiness is commonly smoothed out, and average rates of deformation can be defined. Because many geomorphic markers persist in the landscape at time scales of single glacial–interglacial cycles, this is an ideal interval within which to document past rates of tectonic forcing. At still longer time scales, erosion has typically removed most markers that are more than about 10^5 yr old. Yet, it is at this time scale that large-scale landscape responses to sustained tectonic forcing can be clearly seen. Typically, landscapes must be treated at a coarse spatial scale in order to examine the products of tectonic–geomorphic interactions, such as the topographic characteristics of a collisional mountain belt or the broad swath of deformation that occurs as a continent passes over a hot spot.

We conclude with a chapter on numerical modeling of tectonically active landscapes. In contrast to efforts aimed at directly measuring tectonic–geomorphic processes, the interactions between deformation and surface processes can also be studied theoretically. If we could write numerical rules that represent phenomena such as the displacement of the crust due to faulting or the erosion and redistribution of mass due to surface processes, we could proceed to investigate interactions among these processes. Consider, for example, displacement on a normal fault that bounds the front of a mountain range. The topographic offset will change the local gradient of any river crossing the fault, and a numerical rule for river incision could predict how that reach of the river would respond and how that response would be propagated upstream.

Rather complex models for landscape evolution in different tectonic environments have recently been formulated. It is not our intent to describe or compare these in detail. Instead, we describe several of the basic building blocks that could go into a numerical model, and we illustrate some of the predictions of these models.

In our view, the usefulness of numerical models is not in creating a reproduction of some actual landscape. Instead, models provide a means to explore potential interactions within a landscape, so that we have a firmer basis for understanding how variables as diverse as crustal rigidity, susceptibility of bedrock to landslides, and the distribution of precipitation may interact with each other, perhaps in unexpected ways. Such models are meant to develop our insight into the complex interactions among the processes, and they can serve to point toward measurements one might make in the field that would most efficiently constrain the process rates.

Energetics

Energy drives the interactions between tectonics and surface processes. In order to build topography, work must be done against gravity. The energy needed to accomplish this work comes ultimately from the conversion of a small fraction of the energy involved in the horizontal motions of the lithospheric plates that constitute the more rigid exterior of the planet. The energy driving plate tectonics comes from primordial heat associated with building of the planet, from the decay of radioisotopes, and from phase changes in the interior of the Earth.

It is perhaps surprising to cast the energy expenditure represented by plate motion in everyday terms. Consider, for example, a simple calculation of the kinetic energy of that part of the Indian Plate lying south of where it is colliding with mainland Asia and building the Himalaya. We may specify its approximate dimensions as 3000 km wide by 7000 km long by 50 km thick. If we assume a mean density of 3000 kg/m^3, this yields a mass of approximately 3×10^{21} kg. Taking a mean plate velocity of 5 cm/yr yields a kinetic energy ($\frac{1}{2}mv^2$) for the plate of about 4×10^3 J (joules), equivalent to about 1/200 of a "Snickers" bar (10^6 J)! On the other hand, the rate of energy expenditure required to move the Indian Plate (force × distance/second) is equivalent to approximately 10^{13} J/s or 10^7 "Snickers" bars per second! Over the past 50 million years, some tiny fraction (~1/100 of 1%) of that energy

has been expended (or converted, really, to potential energy) in elevating the crust of the Himalaya and Tibet far above the geoid and thereby creating the world's largest topographic anomaly, averaging about 5 km above sea level.

The energy for driving surface processes derives from a combination of gravitational potential energy and solar energy. All rocks that have been elevated above the geoid have a potential energy equivalent to the product of their mass, gravitational acceleration, and their height above the geoid (PE = mgh). At the top of the atmosphere, the Earth receives solar energy equivalent to about 1300 W/m^2. The solar energy received at the Earth's surface is several orders of magnitude greater than the energy that leaks out from the Earth's internal heat engine (~40 mW/m^2). Solar energy evaporates water and heats the air that carries the moisture with it as the air rises, expands, and cools. The water attains its maximum potential energy at the top of its atmospheric trajectory. When the vapor condenses and falls as precipitation, it converts some of its potential energy to kinetic energy that it delivers to the surface of the Earth with the force of its impact. The potential energy that is represented by the mass of water at some elevation above the geoid is then available to be expended doing geomorphic work or to be lost through heat dissipation or frictional processes. Solar energy is also an important factor in chemical weathering processes, especially in any temperature-dependent reactions, as well as in mechanical weathering processes, such as freeze–thaw cycles.

Active tectonics and models of landscape development

Sharp contrasts in the appearances of landscapes in a given climatic or tectonic regime inspired geologists in the past to devise schemes to explain those contrasts. One of the most prominent such geologists was William Morris Davis, who in the late 1800s and early 1900s developed the well-known geomorphic models (Davis, 1899) showing a progression from "youth" to "maturity" to "old age" (Fig. 1.2). Living in the wake of the revolutionary ideas of

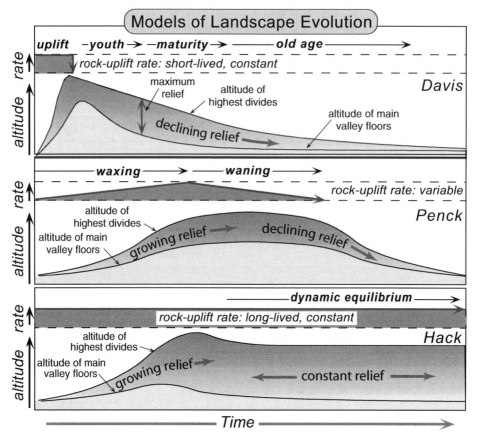

Fig. 1.2   Classical models of tectonic forcing and landscape responses.
Based on the landscape response theories of (top) Davis (1899), (middle) Penck (1953), and (bottom) Hack (1975), each panel is characterized by differences in the duration and rate of tectonic forcing (rock uplift) and by the topographic response engendered by these different styles of rock uplift. Modified after Summerfield (1991).

Charles Darwin, it was perhaps natural to conceive of landscapes as evolving from one stage to the next. In Davis's models, tectonic forcing is an impulsive phenomenon that occurs at the beginning of the "geomorphic cycle." All of the building of topography occurs at the start of the cycle (Fig. 1.2). Subsequently, geomorphic processes attack and degrade the topography, with the end-result predicted to be a peneplain. This view of landscape evolution prevailed through much of the 1900s and is still widely cited by introductory geology textbooks.

A new theory of landscape development that opposed Davis's models was promoted by Walther Penck in the 1950s (Penck, 1953). Rather than having all of the tectonic deformation at the beginning of a cycle, Penck suggested a more wave-like pattern of tectonic forcing through time. In this scheme, the magnitude of deformation gradually increases toward a climax and then slowly wanes away. Instead of calling upon an impulsive building of topography followed by long intervals of erosion, Penck's model invoked steadily increasing rates of deformation that would accelerate rock uplift and gradually build the topography toward a maximum state of topographic relief. Geomorphic processes were conceived as attacking the uplifting region throughout the period of mountain building, so that the resulting landscape could be interpreted as a product of this competition between deformation and erosion. As the rate of mountain building waned, erosion was proposed to overtake deformation rates, thereby causing a gradual reduction in the residual topography (see Fig. 1.2).

A third approach to landscape development was proposed by Hack (1975), who suggested that, when rates of deformation and rates of erosion are sustained for long intervals, landscapes will come into a sort of balance or dynamic equilibrium. Given the finite strength of rocks, Hack realized that topography could not increase without limit, even if rates of tectonic forcing persisted for very long periods. At some point, as topography grew and grew, the relief on hillslopes would create forces exceeding the rock strength, and they would collapse. With continued uplift of the bedrock, additional slope failures would limit the height that the topography could attain. Eventually, the topography would enter into a rough steady state or a dynamic equilibrium. Consequently, in this model, rates of mountain building and rates of erosion would come into a long-term balance (see Fig. 1.2). Unlike the models of Davis or Penck, there is no need for rates of deformation to become negligible, or even to wane at all. Similarly, after attaining a maximum sustained topography, there is no need for the landscape to "evolve." Instead, it simply fluctuates around an equilibrium topography until such time as the rates of tectonic forcing change.

One might wonder whether any of these theories are applicable to real landscapes. One way to consider this problem is to compare the time scales of deformation and of geomorphic responses. For example, an earthquake is a very impulsive event that "instantaneously" creates topographic change, say in the form of a fault scarp. If the time between earthquakes is long relative to the time it takes geomorphic processes to cause significant topographic evolution, the scarp will be degraded before the next earthquake. This topographic change could be thought of as a small-scale version of a Davisian scheme for landscape evolution. On the other hand, if the recurrence interval between earthquakes is short, then the scarp will be continuously refreshed and will not degrade in a predictable fashion toward a

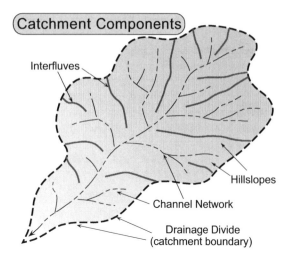

Fig. 1.3 Geomorphic elements of a drainage basin. Diverse geomorphic components of a catchment have a hierarchy of inertia to change, with the channel network as most responsive, and the drainage divide as least responsive.

low-relief surface. In fact, some balance might be expected to develop between the renewal of the scarp by faulting and erosion of it during interseismic periods. Thus, even though the total displacement grows through time, the scarp itself might attain a "steady-state" shape.

It is also useful to consider the response times or inertia of different elements in a geomorphic system. Consider, for example, a drainage basin in a mountain belt (Fig. 1.3). The basin consists of river channels, hillslopes, the crests of interfluves, and the drainage divide that defines the shape of the catchment. Some of these elements will respond more rapidly to changes imposed on them than will others. Suppose that, near its outlet from the mountains, this basin is bounded by a normal fault. How will the various basin elements respond to displacement on that fault? The channel that crosses the fault will "see" the displacement first. The local base level will have fallen abruptly, causing the river gradient to steepen locally and erosion of the channel to intensify. At this point, all other positions in the catchment will have been undisturbed by the faulting event, so the faulting elicits no initial geomorphic response. Subsequently, a wave of erosion of the channel bed can be expected to propagate up the fluvial system. As this channel erosion passes the base of an adjacent hillslope, it causes the hillslope gradient to increase and will initiate a change in the rate of hillslope processes. These hillslope processes could eventually affect the interfluves and finally the shape of the entire catchment. Large geomorphic elements, such as the catchment shape, are unlikely to be affected by individual seismic events. Thus, a clear hierarchy of response times exists (rivers respond sooner than hillslopes, etc.), and an analogous hierarchy of topographic inertia is evident (catchment shape is resistant to change, whereas river gradients are susceptible to small perturbations).

Consideration of the scaling of elements in a given geomorphic system, of response times or inertia of those elements with respect to imposed changes, and of the rates, magnitude, and duration of different styles of tectonic forcing suggests a way in which the apparently incompatible landscape evolution concepts of Davis, Penck, and Hack can be reconciled. In fact, recent numerical models of tectonically perturbed landscapes have explicitly addressed this problem. Using a surface-process model that links channel incision, sediment transport, and hillslope erosion, Kooi and Beaumont (1996) developed a model that predicts a lag between the onset of deformation and the response of the geomorphic system to that deformation (Fig. 1.4). The overall response of the geomorphic system and the magnitude of the lag depend strongly on the nature of the tectonic forcing. If the forcing is impulsive (*à la* Davis), then the topography is rapidly created and simply degrades through time. If the deformation increases to a maximum through time and then wanes (*à la* Penck), topography gradually builds in the face of progressively increasing rates of erosion. The maximum topographic expression occurs slightly after the rate of deformation begins to wane, because the rock uplift still outpaces the rate of erosion. Finally, in the latter half of the cycle, the topography wanes, despite gradually diminishing rates of erosion. If the tectonic forcing is continuously sustained, then the Kooi and Beaumont model predicts that, after an initial interval of building of topography, rates of rock uplift and erosion will

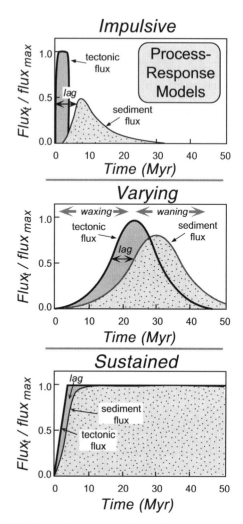

Fig. 1.4   Tectonic versus sediment flux process–response models.
Duration and magnitude of rock uplift (the vertical tectonic flux) are compared with the erosional sediment flux from uplifted mountains. The three scenarios (impulsive, varying, sustained) are analogous to the models of Davis, Penck, and Hack (Fig. 1.2). Note that the time lag between tectonic forcing and sediment response is variable. Modified after Kooi and Beaumont (1996).

become balanced (*à la* Hack), and the topography will attain a persistent dynamic equilibrium (Fig. 1.4). A change in the rate of tectonic forcing would push the system toward a new equilibrium, whereas cessation of deformation would return the system to an almost Davisian state in which the topography is systematically degraded.

These predictions are the output of a numerical model that represents a great simplification of both tectonic and geomorphic processes. In fact, the physics of many of these processes are still poorly understood. Nonetheless, it is satisfying to think that the thoughtful, yet apparently contradictory, landscape development models put forth during the past 100 years are indeed reconcilable and that simple numerical models can help enhance our understanding of tectonically active landscapes.

The new world

Over the past two decades, technological advances have dramatically improved our view of the Earth. As the diversity and availability of digital topography has expanded, our ability to visualize the shape of the Earth's surface has become much easier and more accurate. For 11 days in February 2000, NASA's Space Shuttle used an active radar system to map the topography of the world between 60°N and 60°S (Farr et al., 2007). The digital elevation model (DEM) derived from this mission now provides almost complete elevation coverage with a spatial resolution of 90 m. Still more recently, a higher-resolution, 30-m DEM that covers the world between 83°N and 83°S has been developed from satellite stereoimages and is freely available (http://www.gdem.aster.ersdac.or.jp/). The success of Google Earth in merging such topographic data with remote sensing images has provided tectonic geomorphologists with an unprecedented opportunity to explore the Earth's surface. As high-resolution (≤1 m) imagery is increasingly incorporated into Google Earth, individual fault scarps, uplifted marine terraces, and channels on actively growing folds in previously inaccessible areas can now be visually explored from a computer almost anywhere in the world!

In the temperate mid-latitudes, however, where forests blanket much of the landscape, even high-resolution imagery does not commonly permit a clear view of the actual land surface. But, this restriction is also changing. A new technology, lidar (light detection and ranging), which is also known as laser scanning (Carter et al., 2007), uses concentrated pulses of light emitted from an airborne instrument to penetrate through openings in a forest canopy. By measuring the return time and direction of the light, the vector to the ground can be calculated from the "last returns" (those that were not reflected by above-ground vegetation). Via an integration of all of the last returns and with precise knowledge of the position of the aircraft, a *bare-Earth* topographic image of the Earth's surface can be created. The resolution of the image depends on how high the aircraft is flying and the nature of the light beam, but commonly DEMs are created with 1-m spatial resolution and topographic uncertainties of a few centimeters. The stunning success of lidar in revealing previously unknown fault scarps beneath the canopy of dense northwestern US forests (Fig. 1.5) made converts of skeptics and launched widespread efforts to acquire lidar data over diverse tectonic and geomorphic targets. Tripod-mounted laser scanners are also gaining increasing popularity. Based on the same physical principles, these instruments permit topographic reconstructions with millimeter- to centimeter-scale uncertainties to be created of individual hillslopes, fault scarps, or river beds; and, with a series of topographic scenes over time, the details of landscape changes can be recorded and quantified. The vastly improved accuracy of these new topographic views, ranging from global topography to individual hillslopes, has underpinned a new view of the Earth and the processes that mold its surface.

Some modern controversies

At present, many lively controversies animate studies in tectonic geomorphology. Some of these provide an interesting backdrop for reading subsequent chapters of this book. For example, the history of Cenozoic cooling of the rocks within many mountain ranges has been interpreted to suggest that the ranges experienced accelerated rates of uplift during late Cenozoic times. With increased recognition that this apparent increase was approximately coeval with the onset of the "Ice Ages," it was commonly

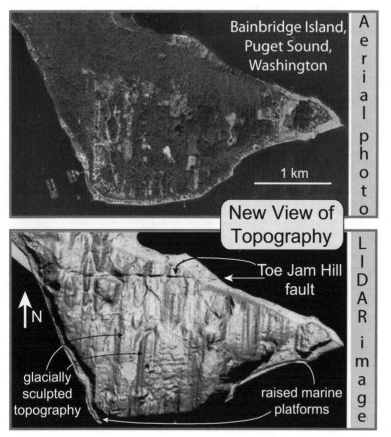

Fig. 1.5   Comparison of aerial photograph with "bare-Earth" lidar image of an active fault. Dense forest obscures geological structure and geomorphology in an aerial photograph of the southern tip of Bainbridge Island in Washington's Puget Sound (top). Removal of the forest canopy using lidar data (bottom) yields a shaded-relief, bare-Earth model whose topography reflects the climatic and tectonic history of the site. Glacial grooves and ridges reveal the former ice flow direction to the south, whereas the abrupt linear escarpment in the north shows the surface expression of the Toe Jam Hill Fault, a fault that was not previously known to break the surface. This fault is a north-dipping backthrust of the Seattle Fault, and the lidar imagery provided the foundation for a trenching campaign that revealed the earthquake history of this hazard-producing fault zone (Nelson *et al.*, 2003). Images courtesy of Samuel Johnson and the USGS.

believed that mountain uplift led to cooling and helped to precipitate the Ice Ages. In 1990, Molnar and England challenged this entrenched idea and suggested nearly the opposite: as climate changed to more glacial conditions in late Cenozoic times, enhanced rates of erosion within mountain belts caused increased rates of valley incision, which in turn incited isostatic uplift of the residual peaks (Molnar and England, 1990). Although isostatic uplift can produce ranges with higher summits, it does not implicate tectonic uplift as the cause of climate change.

So, how do we tell whether the climate caused uplift of the summits or whether surface uplift of the ranges caused changes in climate? Potential resolutions to this quandary require many ingredients, including the nature and magnitude of changes in mean elevation of mountains, in rates of erosion, in climate, and in elevations of summits and valley bottoms. Moreover, we would

like to know when and how rapidly changes took place. If you think carefully about any one of these ingredients, you quickly realize the reason why tectonic geomorphological studies are often interdisciplinary in nature. Consider, for example, the concept of mean elevation of a range. Using digital topographic data, the current mean elevation is straightforward to calculate. But how do you determine mean elevations in the past? As described in subsequent chapters, approaches to estimating former altitudes range from paleobotanical to isotopic studies. Or, consider the effects of changes in climate. During Ice Age times, was there more precipitation or less? Did the expansion of glaciers lead to enhanced rates of erosion? There is intense interest in, and considerable argument about, whether glaciers are effective agents of erosion in comparison to rivers. The controversy has spawned a flood of recent research into the physical and

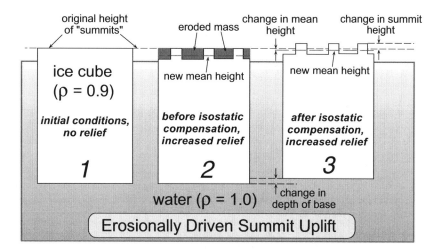

Fig. 1.6   Isostatic uplift of mountain summits due to enhanced erosion.
The density contrasts of ice and water are analogous to crust and mantle contrasts, respectively. Erosion of the top of the ice cube decreases its mass and its mean height. More rapid erosion of valley bottoms than of summits leads to increased relief and uplift of summits, at the same time that the mean elevation decreases. Note that stage 2 will never occur unless some force restrains the ice cube from rebounding due to melting at the surface.

chemical processes involved in erosion of bedrock by both fluvial and glacial processes.

Can enhanced erosion drive uplift of mountain summits? Yes, but only if there is an increase in topographic relief associated with the erosion. In essence, there has to be less erosion of the summits than there is of the valley bottoms. Think about an ice cube floating in a drink (Fig. 1.6). Its upper surface is at a mean elevation equivalent to ~1/10th of its total thickness, equivalent to $(1 - \rho_{ice}/\rho_{water})$, where ρ = density. If you were to cut canyons into the upper surface of the cube, the mean elevation of the cube's surface would decrease and the base of the cube would bob upward in the water in order to maintain the isostatic balance (maintaining 1/10th of the total mass above the water surface). If you could cut the canyons without "eroding" other parts of the upper surface of the cube, these remnants would actually rise higher than their original height in response to the lowering of the mean elevation.

Although melting of an ice cube demonstrates that peak uplift can occur due to enhanced erosion, has this commonly occurred in the past, is it related to a more "erosive" climate, and what are the magnitudes of the uplift of peaks involved? The common view has been that increased rates of erosion tend to increase topographic relief and, therefore, would promote the uplift of summits. On the other hand, theoretical studies of river profiles suggest that increased erosive power causes river gradients to decrease (Whipple, 2004). If this is true, then, in order to generate an increase in topographic relief in the landscape, hillslopes would have to be concurrently lengthened and steepened. Tectonic-geomorphic studies that document the temporal evolution of both the valley bottoms and the adjacent hillslopes are needed to resolve these issues.

Another modern controversy in tectonic geomorphology revolves around earthquake prediction. Over the past two decades, a substantial research effort has focused on answering such questions as: "Which fault is most likely to rupture in the next large earthquake?", "When is that event likely to occur?", and "How large an earthquake can we expect to occur?" Many scientists would maintain that the best way to make such predictions is to understand the past history of faulting. Such topics fall into the realm of paleoseismology with its focus on the reconstruction of past earthquakes in terms of their distribution in space and time, coseismic displacements, and interactions among faults.

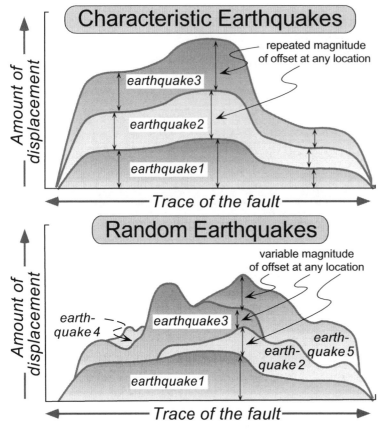

Fig. 1.7 Scenario of characteristic earthquake displacement versus unpredictable, random displacement.
Characteristic earthquakes repeatedly rupture the same part of a fault and show similar along-strike variations in coseismic displacements from one event to the next. Random earthquakes vary in both the location of the rupture and patterns of displacement.

If paleoseismological studies were to indicate that successive ruptures of a particular fault occurred at regularly spaced intervals in time, it would then be possible to define a "recurrence interval." If we also know when the last rupture occurred, we could predict when the next earthquake was likely to happen. But, do faults display regularity in their rupture histories? The known paleoseismological record commonly extends back through only a few earthquakes, making it difficult to define a statistically significant recurrence interval.

One popular model among paleoseismologists suggests that many faults can be said to experience "characteristic earthquakes," implying that each successive rupture mimics the displacement of the previous event. If true, this implies that, along any particular section of a fault, the amount of displacement that results

from a single earthquake will be duplicated in subsequent ruptures (Fig. 1.7). Imagine the predictive power that would result if faults were shown to generate both characteristic earthquakes and to have predictable recurrence intervals! City planners and paleoseismologists would all be delighted. For most faults, however, this seductively attractive concept has yet to be demonstrated. Consequently, much current debate revolves around whether or not certain faults display characteristic earthquakes and predictable recurrence intervals. Or, could some faults display a very irregular behavior in both the time between ruptures and displacement variations in successive events? Far from being resolved, this debate lies at the core of paleoseismology, and has spawned much recent work both in the field and in numerical models of the faulting process.

Many paleoseismological studies rely on trenches that are dug across faults to reveal the slip in past earthquakes. Measurements of displaced strata serve to define the magnitude of displacement in an earthquake, whereas dating of those strata can determine when the rupture occurred. But do such measurements capture the full slip history? Might they overestimate the slip? Several recent discoveries complicate the interpretation of trenches. As faults approach the Earth's surface, slip commonly decreases, so trench exposures at the surface may tend to underestimate the total coseismic slip. This underestimate is especially likely for "blind" thrust faults that do not rupture the surface. Even where faults do break the surface, is it possible that significant strain occurs as diffuse deformation between faults? Although difficult to quantify, such broadly distributed strain may represent a significant fraction of the total deformation. In such conditions, displacement in trenches would be underestimates of the integrated deformation across an area. The advent of extensive geodetic networks (mostly relying on GPS) has provided new insights on slip during and between earthquakes. Recent recognition of "slow earthquakes" that occur over hours to days (Heki *et al.*, 1997) indicates that slip during an earthquake may be equaled by slow slip of nearly equal magnitude. If such slip were propagated to the surface, it would be impossible to recognize as a slow slip event in an exposure within a paleoseismic trench.

The rates of convergence between tectonic plates, as well as measured rates of local deformation, indicate that, in many active mountain ranges, rocks are moving upward with respect to sea level at rates of several millimeters per year. Recall that a rate of 1 mm/yr is equivalent to 1 km in a million years. Thus, in the absence of erosion, vertical rock uplift rates of several mm/yr would build very high mountains in only a few million years. Clearly, mountains do not grow indefinitely: only 14 peaks poke more than 8 km above sea level. But what controls their ultimate height? Is there a limit to the energy available to lift the mass of rock and increase its potential energy? Does rock strength set the height limits, or do

changing rates of erosion determine the topography of ranges? Such questions lie at the core of another current controversy in tectonic geomorphology: Can the concept of "dynamic equilibrium" be applied in active orogens at the mountain range scale? Dynamic equilibrium implies that, on average over time, the land-scape maintains a steady-state form, whereby the height of the summits, the steepness of the valley walls, and the topographic relief fluctuate around long-term mean values. If the mean height of the mountains stays the same through time, this persistence implies that rates of rock uplift (vertical movement of rocks with respect to sea level or the geoid) are balanced by rates of erosion. How is this accomplished? Are surface processes capable of eroding at several mm/yr? Which processes are responsible (river erosion, landsliding, glacial erosion, conversion of rock to soil) and do these processes operate in different ways in different mountain belts? Or, is the traditional idea correct that rapid rates of rock uplift are commonly compensated, not by geomorphic agents at all, but by events of tectonic denudation (extensional faulting) that efficiently lower the regional height of the landscape?

In order to answer such questions, we have to be able to document rates of modern and past erosion, to quantify rates of rock uplift and changes in the mean elevation and topo-graphic relief of mountain ranges (Fig. 1.8), and to document the role of extension within orogens. Also, because climate undergoes large glacial–interglacial fluctuations at 100 000-year intervals, rates of erosion that are responsive to climate (most erosional processes are) will also vary strongly through time. Consequently, when thinking about the problem of dynamic equilibrium and steady-state topography, it is most useful to consider time spans that exceed a full glacial–interglacial cycle, so that average rates can be determined.

Such constraints present some great challenges to researchers in tectonic geomorphology. It is inadequate to document only modern rates (which are difficult enough to measure accurately!). Rates from intervals throughout a climate cycle or rates that integrate an entire

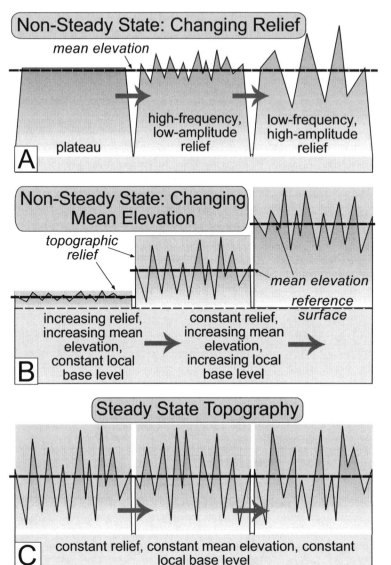

Fig. 1.8   Steady-state versus non-steady-state topographic characteristics.
Non-steady-state topography can have (A) constant mean elevation, but changing topographic relief, or (B) constant relief, but changing mean elevation. In steady-state conditions, (C) relief, mean elevation, and base level remain constant, although the elevation of an individual point can vary through time.

cycle are needed. This integration requires clever ways to measure quantities of material removed from or added to a landscape. Moreover, researchers must somehow inject a reliable "clock" into the rock record, because rate calculations can be no more precise than the time interval across which they are measured. Such measurements are difficult to make. An alternative approach would be to determine the operational "rules" by which various surface processes erode the landscape, and to incorporate them into a theoretical model of landscape evolution. For example, how does the rate of bedrock erosion at the base of a glacier relate to the speed at which the glacier is sliding, its thickness, the steepness of its bed, freezing and thawing at the ice–rock interface, and the resistance of the bedrock beneath it? Only if we can both define these rate relationships and determine how glaciers have extended and retreated in the past, will we be able to model the mean rates of erosion within this portion of the landscape.

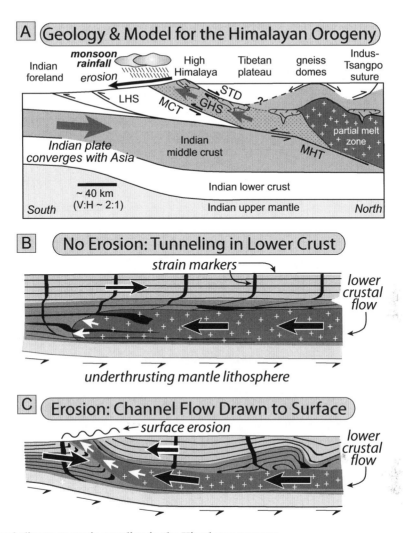

Fig. 1.9 Proposed climate–tectonic coupling in the Himalayan orogeny.
A. Geological observations and conceptual model that connect partial melting in Tibet and lower crustal flow to climatically driven erosion and major faults in the Himalaya. Monsoon-driven erosion along the southern flank of the Himalaya is proposed to weaken the underlying crust and to draw the channel flow toward the surface. GHS: Greater Himalayan Sequence; LHS: Lesser Himalayan Sequence; MCT: Main Central Thrust; MHT: Main Himalayan Thrust; STD: South Tibetan Detachment. B. In the absence of erosion, numerical modeling suggests simple outward lower crustal flow within a confined channel (tunneling). C. With intense erosion localized in a specific region, modeling suggests lower crustal flow is drawn upwards toward the zone of high erosion. Note that this geometry predicts relative motion along the channel's margins that is consistent with the slip on the major Himalayan faults (MCT and STD in A). Modified after Beaumont *et al.* (2001).

In recent years, provocative linkages between climate, erosion, and tectonics have been proposed. Even at the scale of orogens, the large-scale organization of major faults and rock uplift has been attributed to patterns of differential erosion that are in turn controlled by climate.

Underpinning these interpretations is the observation that rapid erosion both removes rock mass, thereby perturbing the stress field on the underlying rocks, and compresses the geothermal gradient, thereby heating and weakening the underlying rocks. Together, these effects are interpreted to guide patterns of deformation and metamorphism.

Consider, for example, four attributes of the Himalayan–Tibetan orogen (Fig. 1.9A) and how they might be linked. First, thickening of the crust beneath the Tibetan Plateau appears to have caused partial melting of at least some of the lower crust (Nelson *et al.*, 1996), which then becomes weak and tends to flow outward as a channel confined above and below by stronger rock. Second, the rocks of the Greater Himalaya are bounded by two large faults that have been active at the same time: below the Greater Himalayan rocks, the Main Central Thrust is an up-to-the-south thrust, whereas above them, the South Tibetan Detachment is a down-to-the-north normal fault. Third, every summer the Indian monsoon dumps about 4 m of rainfall on the southern flank of the Greater Himalaya, but less than 10% of that amount penetrates into Tibet. Across this decreasing rainfall gradient, current erosion rates also diminish by three- to four-fold (Gabet *et al.*, 2008). Fourth, combining these observations, some geologists suggest that the intense, monsoon-driven erosion on the southern flank of the Himalaya weakened the underlying crust and "drew" the partial-melt channel toward the surface, thereby determining where the big faults bounding the channel are located. Numerical modeling of lower crustal flow tends to reinforce the importance of erosion (Beaumont *et al.*, 2001, 2004). In the absence of erosion, lateral tunneling is predicted as channel flow remains confined within the lower crust (Fig. 1.9B), whereas when erosion is introduced, the flow migrates upward toward the region of high erosion (Fig. 1.9C). A striking aspect of these numerical models is that, if the pattern of rainfall were reversed so that rain was focused on the opposite side of the orogen, the models predict that the orientation of the large faults could also flip direction! Whereas more questions than answers remain about potential linkages between climate, erosion, and tectonics, their proposed interactions set an exciting stage for future exploration of the role of climate in mountain building.

Looking ahead

Resolutions to these controversies are beyond the scope of this book, but they provide a framework for thinking about many of the topics discussed in the subsequent chapters. These controversies illustrate some of the breadth of modern tectonic geomorphological studies. Any serious consideration of potential solutions to these controversies quickly reveals the interdisciplinary nature of the research required to address them. Although certainly not unique in its demands for interdisciplinary work, tectonic geomorphology attains much of its current vibrancy from the cross-pollination that is occurring between specialists of many disciplines who are coming together to address major unresolved issues. It is our intent that the following chapters provide some insight into the tools, approaches, and interpretational techniques that are currently used in tectonic geomorphological studies. We hope to convey the striking innovation and creativity of past researchers, upon whose shoulders future advances will be made.

2 Geomorphic markers

In order to measure the amount of deformation that has occurred due to tectonic processes, it is typically necessary to have an identifiable feature that has been displaced. Unique rock types or structures that formerly extended across a fault in an unbroken pattern provide a datum or "piercing point" from which the magnitude of subsequent displacements can be determined. In order to calculate the displacement reliably, the pre-deformational geometry of a presently offset feature has to be reconstructed accurately. The better that geometry is known, the more reliably the offset can be calculated.

In tectonic geomorphology, we are often concerned with offset geomorphic *markers*, by which we mean identifiable geomorphic features or surfaces that provide a reference frame against which to gauge differential or absolute deformation. The best geomorphic markers are readily recognizable landforms, surfaces, or linear trends that display these three characteristics: (i) a known initial, undeformed geometry; (ii) a known age; and (iii) high preservation potential with respect to the time scale of the tectonic process being studied. Oftentimes, only some of these characteristics may have been determined for a displaced marker. Because of the usefulness of geomorphic markers, considerable effort and care are warranted in defining their geometry and age. In this chapter, we examine the pristine shape of many useful markers, such as river or coastal terraces, and

discuss the conditions under which they are likely to form.

The geometry of a pristine geomorphic feature is a crucial attribute of a marker, because the deformation of the surface of such a feature records the tectonic signal. Modifications of an original undisturbed feature by subsequent erosion or deposition, however, may make it difficult to define the original geometry of a presently offset geomorphic surface. In such cases, the probable geometry of the feature can sometimes be predicted through comparison with undeformed modern analogs. For example, river terraces are frequently used as geomorphic markers to document fault offsets or folds (Molnar *et al.*, 1994; Rockwell *et al.*, 1984; Thompson *et al.*, 2002). Unfortunately, the preservation of river terraces becomes increasingly fragmentary as they get older and more extensively dissected. In such a situation, the profile of the modern river or of young, nearly undissected terraces along the same or analogous reaches of the river provides models for the smooth downstream changes in the longitudinal profiles of rivers. Such profiles can then be used to predict the geometry of older terraces prior to deformation.

The age of a geomorphic marker is a critical ingredient when calculating rates of deformation. Most markers form in response to either climatic or tectonic controls. Some markers form due to autocyclic processes, such as landsliding, which

Tectonic Geomorphology, Second Edition. Douglas W. Burbank and Robert S. Anderson.
© 2012 Douglas W. Burbank and Robert S. Anderson. Published 2012 by Blackwell Publishing Ltd.

creates lakes within mountains, or avulsion of river channels, which occurs due to aggradation on floodplains. If the relationship of a suite of markers to known and dated climatic variations can be established, the markers themselves can be indirectly dated through correlation to the climatic record. For example, climate change often affects the water discharge in a river and its sediment load. Consider the situation in which, owing to a climatically induced increase in discharge, a river begins to incise its bed. After sufficient incision, the river may be flanked by fluvial terraces (geomorphic markers) that now represent a geomorphic response to the climate change. If we know the age of the climate change that caused the change in discharge and sediment load, then the age of the terrace itself can be deduced from the climatic record. With these terraces as time markers in the landscape, it becomes possible to measure both the amount and rate of deformation since they were created. Therefore, it is very useful to understand those aspects of the climatic record that are most likely to be relevant to the creation of geomorphic markers. In the absence of climatic calibration, or for markers whose origin may not reasonably be attributed to climatic cycles, direct dating of a marker is required. A variety of dating approaches is discussed in Chapter 3.

Ephemeral features, such as small levees or even tire tracks, that could wash away in the next storm can provide markers that are adequate to calibrate coseismic offsets of a recent earthquake (McGill and Rubin, 1999). Long-lived geomorphic features, however, are required in order to document deformation over many thousands of years. But, erosion is continually modifying the geometry of such markers, and ongoing tectonic and climatic changes are overprinting the landscape with new features. Consequently, a tectonic geomorphologist benefits both from a knowledge of a marker's potential for long-term preservation in the landscape and from an ability to recognize useful fragments of older markers within the complexly intertwined geomorphic elements of most natural landscapes.

The sensitivity (amplitude of the response) and response time of geomorphic systems to changes in the variables that control them vary markedly. The discharge of rivers, for example, will respond almost instantaneously to changes in precipitation, whereas glaciers often take several years to translate increases in snowfall into an advance of the snout of the glacier. In general, the response time of a geomorphic system to changes in the climatic forcing of the system increases dramatically with the scale of the system, and inversely with respect to the efficiency of the processes involved. When the variables that control a geomorphic system change, the highly sensitive components of an integrated system, such as the cross-sectional area of a river or the elevation of its bed, can usually achieve a new equilibrium rapidly. In contrast, larger and less sensitive components, such as an entire drainage basin (Fig. 1.3), can take many millennia to come into equilibrium with any new controls on the system.

Planar geomorphic markers

To be useful as a reference frame against which to measure displacement, the initial, pre-deformational geometry of a geomorphic feature must be defined. Because erosion and/or deposition continually modify old geomorphic markers, a map of a presently exposed marker often reveals only fragments of formerly continuous surfaces or features. The task of anyone wishing to use such markers to calibrate deformation is to create a reliable reconstruction of the undeformed geometry of the marker. Because modern analogs commonly provide the best model to use in reconstructing the shape of older, presently fragmented features, we will describe coastal, lacustrine, fluvial, and several terrestrial markers that have been extensively used to define tectonic deformation.

Marine terraces, beaches, and shorelines

Along coastal regions of many parts of the world, bench-like features, or *marine terraces*, have been created by the interaction of the ocean with the adjacent landmass. Marine terraces fall into two classes: constructional terraces associated with coral reefs, and destructional or erosional terraces. Terraces ultimately result from the wide variation in sea level with time, this variation being driven by a combination of the histories of the local movement of the landmass with respect to the geoid and the global variations in sea level. The global or *eustatic* sea-level changes within the Pleistocene are caused primarily by changes in the volumes of the continental ice sheets. The associated swings in sea level are huge: 120–150 m. Only 20 kyr ago, sea level was ~120 m below its present level (Fairbanks, 1989; Lambeck and Chappell, 2001). As both physical and biological processes are strongly focused at the intersection of land and sea, it is not surprising that the variations in the location of this interface through time have left a pronounced and, therefore, useful geomorphic record.

The constant attack on the shoreline by waves creates erosional marine terraces. As waves crash against a coast, fluid turbulence and hydraulic pressure are capable of loosening and plucking fragments of bedrock, whereas sediment entrained in the water can impact forcefully against the cliffs. In places, this persistent attack can form a *wave-cut notch* at the base of a sea cliff. As plucking and abrasion continue, the notch deepens, the base of the cliff retreats, and the shore platform widens. In response to undercutting, the slope of the sea cliff is effectively steepened and can be expected to collapse when stable slope angles are exceeded. If the material that is loosened at the cliff face is not removed, it simply mantles the bedrock and prevents subsequent erosion. More commonly, however, a seaward-sloping shoreline platform or *abrasion ramp* permits wave energy to transport the eroded debris into deeper water offshore. This ramp typically slopes at about 1° and extends sufficiently seaward to allow sediment that is transported along its surface to be dumped into deeper water, dissolved, or transported away in suspension following further comminution. Because wave energy is needed to transport sediment on the abrasion ramp, these ramps do not extend below the typical wave base (~10 m in most coastal areas). Given

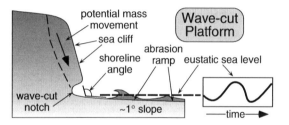

Fig. 2.1 Abrasion platform, sea cliff, wave-cut notch, and shoreline angle.
The vertical position of the abrasion ramp varies through time as a function of sea-level variations.

the 1° slope of the ramp, this sets a maximum width of about 500 m for most abrasion ramps. Wider ramps can result from a deeper wave base, gentler seaward slopes, or multiple occupations of approximately the same or slightly higher relative sea-level positions.

A newly created erosional marine terrace typically consists of two distinct surfaces: an abrasion ramp dipping seaward at ~1°; and a sea cliff dipping seaward at the stable angle of repose, given the strength and cohesion of the bedrock that it comprises (Fig. 2.1 and Plate 1A). The junction of these bedrock surfaces is called the *shoreline angle* (also called the *inner edge* of the platform), and it is here that a wave-cut notch may be preserved. Because this wave-cut notch and/or the junction between the abrasion ramp and the sea cliff occurs at the shoreline, it represents a marker that closely approximates local sea level at the time of its formation. Formally, this feature records the intersection of sea level (which, when averaged over many tidal cycles, closely approximates the geoid) and the landmass. Hence, it becomes a paleo-horizontal indicator that permits assessment of both tilting parallel to the coast and spatial variations in vertical motions of the crust. When reconstructing the former inner edge of an uplifted abrasion platform, it is important to try to pinpoint the position of the shoreline angle. Commonly, however, the shoreline angle will have been obscured by downslope motion of eroded material from topographically higher slopes. In these circumstances, sometimes the surface of the abrasion platform well away from

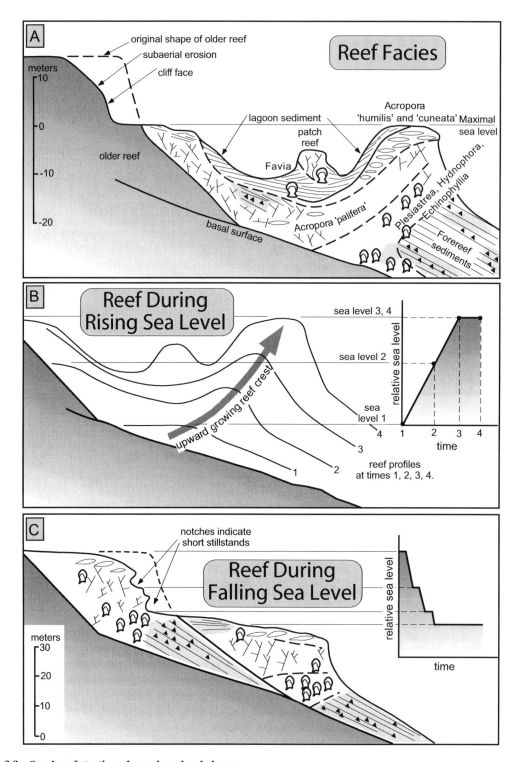

Fig. 2.2   Coral reef stratigraphy and sea-level change.
A. Idealized model of facies zonation and geometry of a coral reef that grew in response to rising sea level. Note that the coral species *Acropora humilis* and *A. cuneata* occupy the reef crest and most closely approximate sea level.

the paleo-sea cliff and that of the sea cliff itself can be projected to an intersection that approximates the former shoreline elevation. In contrast, the depth below sea level of the outer edge of the abrasion ramp can vary by 10 m or more. Moreover, the position of the preserved outer edge of the platform depends on how much it has retreated due to erosion and smoothing since it became emergent. Consequently, the position of this outer edge provides a much less reliable marker with which to measure deformation, even though it may be more readily preserved and observed in the landscape.

Constructional marine terraces can form when marine conditions are favorable for the growth of corals and coralline algae. Typically, these conditions occur where water temperatures in winter remain above about 18°C, where clear water permits penetration of sunlight to support photosynthesis, and where the salinity is normal. Under conditions conducive for growth, coral reefs can be rapidly built upward toward the sea surface (at rates exceeding 10 cm/yr) and outward toward available space. When sea level is stable for sufficiently long periods, corals will build platforms that are closely tied to sea level and, therefore, provide useful geomorphic markers. During a relative rise in sea level, a predictable upward and outward coral growth occurs (Fig. 2.2A and B). If the rise is particularly rapid, the rate of vertical growth of the crest may lag behind the relative sea-level rise. Only when the sea level stabilizes will the reef crest attain its maximum height near the sea surface. During a subsequent fall in relative sea level, wave-cut notches may by cut into the forereef during brief stillstands (Fig. 2.2C). If sea level stabilizes at a new local level, corals will build a reefal bench.

Because different coral species are adapted to different marine conditions, a marked zonation of species occurs within a reef. For example, wave-resistant corals will be found in the nutrient-rich waters on the leading edge of a reef, whereas less robust forms occupy the backreef (Chappell, 1974). Knowledge of this zonation (Fig. 2.2A) and an ability to recognize various coral species, some of which are more faithful recorders of sea level, permits reconstruction of the geometry of presently emerged reefs and their relationship to former sea levels at the time of their growth. For constructional terraces comprising coral reefs, the reef crest and buttress represent the leading edge of the reef and yield the best approximation of sea level. Because this edge is most subject to erosion if the reef is uplifted above sea level, knowledge of the depth below the sea surface at which various coral species are found must often be combined with a recognition of the preserved species zonation within a reef (Fig. 2.2A) to estimate the position of the former reef crest.

Along many tectonically rising coasts, flights of marine terraces provide direct evidence for multiple decreases in relative sea level. But when do these terraces form, and how are they related to sea-level changes? In order for a coastal terrace to be generated, sea level must remain at approximately the same relative position with respect to the land so that corals can build outward to form a reefal terrace or so that wave attack is focused along the same abrasion platform through time. Owing to rapid rates of lateral erosion and of coral growth, it often takes only a few thousand years to create a broad terrace (Anderson *et al.*, 1999). Consequently, we expect terraces to form during intervals when rates of vertical movement of the land and rates of sea-level change are nearly equivalent. If we knew well the history of sea level and could correlate between individual terraces and former sea-level positions, we would have a basis for calculating vertical deformation rates.

Fig. 2.2 (*cont'd*) These near-surface corals can be eroded, however, and commonly species that reside slightly below the mean surface are better indicators of former sea levels. B. Idealized cross-section of the reef in A showing how the topography of the reef is related to changing relative sea level. During the intervals of rapid sea-level rise (from sea level 1 to 3, inset), the crest of the reef grows upward, but remains below the sea surface. As sea level stabilizes, the reef reaches sea level and grows seaward. C. Wave-cut notches are a response to falling sea level. A new reefal platform can be built once sea level stabilizes. Modified after Chappell (1974).

Box 2.1   Isotopic changes in the ocean due to glaciation.

The history of climatic change in Quaternary times is primarily a function of the growth and decay of ice sheets. The key record of changing volumes of glaciers through time comes from the isotopic record of the oceans. During glaciations, water evaporated from the ocean is stored in ice sheets (see figure). The ratio of two common isotopes of oxygen, ^{16}O and ^{18}O, is different in ice sheets than in the ocean as the result of a fractionation process. When water evaporates (primarily near the equator), the lighter isotope (^{16}O) is preferentially evaporated, causing the ocean to become enriched in ^{18}O. During condensation and precipitation, ^{18}O is preferentially removed from the water vapor, so that during its poleward transport, the remaining vapor becomes increasingly enriched in ^{16}O. Precipitation at high latitudes is therefore strongly depleted in ^{18}O compared to "standard mean ocean water". (SMOW). This causes the ice sheets to be isotopically "light" compared to SMOW (lower $^{18}O/^{16}O$ ratio). Conversely, the oceans become isotopically heavier (higher $^{18}O/^{16}O$ ratio) due to the storage of "extra" ^{16}O in the ice sheets. The more ice stored on land, the isotopically heavier the ocean becomes. Foraminifera that grow in equilibrium with the sea water record its isotopic composition. Consequently, the stratigraphic record of isotopic changes displayed by foraminifera during the Quaternary can be interpreted as a record of changing ice volumes.

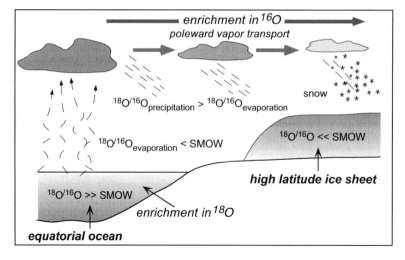

Isotopic changes due to sequestration of water in ice sheets.

Fortunately, within the past few decades, the history of Quaternary sea level has been reconstructed with increasing accuracy (Bloom *et al.*, 1974; Chappell, 1974; Chappell *et al.*, 1996; Lambeck and Chappell, 2001). During maximum glacial conditions, the volume of water stored in ice sheets on land caused average sea level to be lowered by more than 100 m. On the other hand, during the peak of the previous interglaciation at about 125 ka, there was apparently less ice on

Earth than there is today, and mean sea level was approximately 6 m higher than it is today. The most detailed reconstructions of the growth and decay of ice sheets have been derived from variations in the oxygen isotopic composition of seawater (Box 2.1). The removal of water from the ocean via evaporation and the sequestering of this isotopically lighter (lower $^{18}O/^{16}O$ ratio) water in ice sheets caused the oceans to become isotopically heavier (higher $^{18}O/^{16}O$ ratio) during

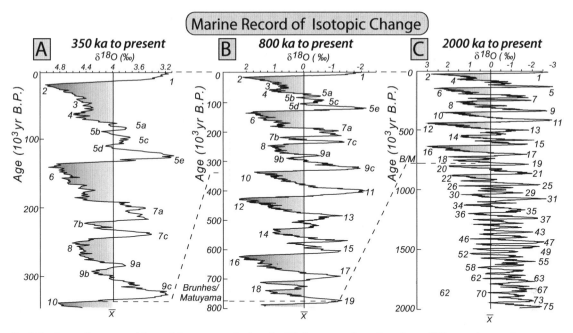

Fig. 2.3 Isotopic composition of the oceans during the Pleistocene viewed at three different time scales. Heavier oxygen isotopic compositions (increasing to the left) correlate with greater ice-sheet volumes and lower mean sea level. Interglacial intervals correspond with high sea-level stands. A. 0–350 ka; B. 0–800 ka; and C. 0–2 Ma. Note the prominent, 100 kyr periodicity during the past 800 kyr. Prior to that, periodicities are dominated by a 40 kyr cycle; 20 kyr cycles are superimposed on both 100 kyr and 40 kyr cycles. Labels next to the peaks and troughs refer to isotopic stages. Stage 5e, for example, represents the last interglacial maximum. Modified after Porter (1989) and Lisiecki and Raymo (2005).

glaciations. Thus, the pattern of isotopic fluctuations derived from deep-sea cores provides a proxy record for both climate and sea-level variations (Fig. 2.3).

Unfortunately, a one-to-one correspondence does not exist between seawater isotopic variations and sea-level changes. This non-equivalence occurs for several reasons. The world's oceans are not simple bathtubs, meaning that an equal volume of water does not translate into a uniform increment in sea-level change because, as sea level rises, the surface area of the ocean also increases. In addition, withdrawal of water from the ocean and sequestration on land rearranges the water load on the Earth's crust and drives isostatic rearrangements of deep crustal and mantle materials, which differ from place to place (Clark et al., 1978; Lambeck et al., 2002). Therefore, the best estimates for past sea-level variations require calibration and have been largely derived from

studies of radiometrically dated coral terraces on tectonically rising coasts. Some key calibration studies have been conducted on the striking successions of coral terraces preserved on the Huon Peninsula of New Guinea (Bloom et al., 1974; Chappell, 1974; Chappell et al., 1996), a coastline responding to rapid collision of an island-arc terrain against the edge of the Australian Plate. These terraces get older with increasing elevation, and they record the *relative* sea-level change through time. This relative change results from the sum of the *real* changes in sea level and the *apparent* changes in sea level (Fig. 2.4):

relative = real + apparent

Real sea-level changes are due to absolute vertical changes of the ocean surface (due primarily to changing volumes of water in the ocean as a result of glaciation; also called *eustatic*) and can be global in extent. Apparent sea-level

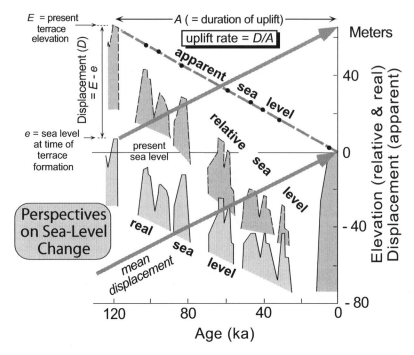

Fig. 2.4 Relationship between relative, real, and apparent sea levels.
Apparent sea level is defined as the inverse of the vertical tectonic displacement (D). Displacement is determined as the difference between the present elevation of a terrace (E) and the original elevation of the terrace (e); thus $D = E - e$. The rate of tectonic displacement is assumed constant here and is equal to the displacement divided by the age (rate = D/A). Relative sea level is represented by the altitudinal position of preserved terraces with respect to present sea level (dashed line). Real sea level represents actual changes in the height of the ocean surface. Because the apparent and real sea levels add to yield the relative sea level, if two of these quantities are known, the third can be derived. Modified from Lajoie (1986).

changes result from and represent the inverse of the actual vertical displacement of the land, that is, tectonic uplift or subsidence. For example, if you are standing at present sea level, and you see an ancient abrasion platform high above you, either the sea level has fallen from that elevated height to its present position (real sea-level fall) or the terrace has been tectonically raised above the position at which it formed (causing apparent sea-level fall), or some combination of both effects has occurred. Such deformation is clearly local or at most regional in extent. If tectonic rates are assumed to be steady during the time interval encompassed by a flight of terraces, then the apparent sea-level change will be linear (Fig. 2.4). In any case, apparent sea level represents the difference between the relative and the real sea-level change:

apparent = relative − real

So, how does one use these relationships to calculate real sea-level changes in the past? Consider a flight of emergent marine terraces for which the age of each terrace and its elevation above modern sea level are known. This suite of terraces represents the relative sea-level change through time. To determine the apparent sea-level change, the tectonic uplift rate has to be determined. If data from other areas have revealed the position of sea level at some time in the past with respect to the present sea level, and if a terrace of that age is present within the local succession, then the difference between the present elevation of the terrace and the elevation of the sea level at the time the terrace formed defines the amount of tectonic uplift. For example, if a 125-ka terrace

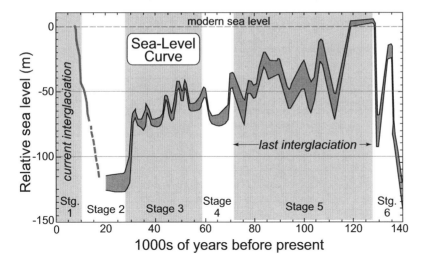

Fig. 2.5 Variations in real sea level since 140 ka. Sea-level variations as reconstructed from coastal terraces and sediments on continental shelves. Note the 1–20 m changes at frequencies of a few thousand years. Modified from Lambeck and Chappell (2001).

is presently 131 m above sea level (this is its relative sea-level position), it can be assumed to have formed during the last interglacial maximum sea level of +6 m. The tectonic uplift is consequently 125 m (131 m − 6 m), and the average uplift rate is 1 m/kyr (125 m/125 kyr). The rate of apparent sea-level change is, therefore, −1 m/kyr. Either the graphical or the arithmetic subtraction of the magnitude of the apparent sea level from the elevation of the correlative, dated terrace yields the real sea-level change through time (Fig. 2.4):

real = relative − apparent

Over the past few decades, a major effort has been made to create a reliable curve of "real" sea-level change. Field studies have focused primarily on coral terraces and deposits on shallow continental shelves, and have yielded a curve that is considered to be quite well known for the past 135 ka, especially for the highstands (Fig. 2.5), but is less confidently defined before that time. The uncertainty arises in part from imprecision in dates of older terraces and in part from the paucity of reliable sea-level calibration points that are older than 135 kyr. It is questionable whether tectonic uplift rates defined for more recent intervals can be confidently extrapolated into the past, so that apparent sea-level changes are difficult to define at longer time scales. As a consequence, the sea-level curve prior to 135 ka is often closely modeled on the variations shown by

the oxygen isotopic record. Comparison of the sea-level curve for the past 135 ka (Fig. 2.5) with the oxygen isotopic curve (see Fig. 2.3) shows that they are clearly similar, but the magnitude of successive peaks varies considerably. Hence, our knowledge of sea levels prior to 135 ka should be regarded as only approximate. Moreover, recent studies show that high-resolution records of sea-level change are commonly valid only at local or regional scales (Milne and Mitrovica, 2008). Although the overall pattern, timing, and general magnitude of sea-level change are likely to be correct, complex deformation of the geoid by changing loads and mantle processes dictates that few places on Earth will faithfully record past eustastic changes at better than ±5 m. Despite these caveats, along many tectonically active coastlines, flights of marine terraces provide a very powerful tool for calibrating absolute and differential uplift and tilting all along the coast.

Eustatic sea level has not varied much during the past several thousand years. As a consequence, well-developed abrasion platforms (Plate 1A) and coral terraces exist along many coastlines. In addition, several features with a much more limited preservation potential can provide useful markers for delineating Holocene deformation. Along stony beaches during major storms, gravelly beach ridges and berms are formed a few meters above the typical high-tide line (Stevens, 1974). Because the crests of these features are parallel to the sea surface, they

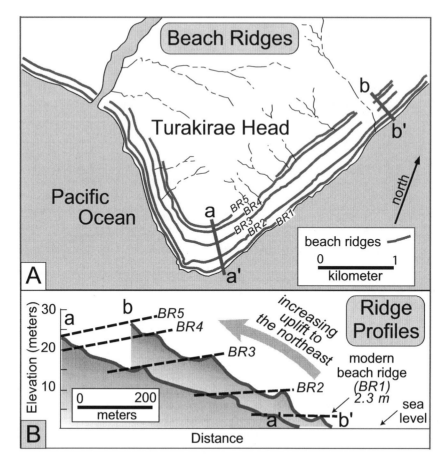

Fig. 2.6   Modern and uplifted beach ridges on a deforming coast.
A. Map of beach ridges at Turakirae Head, southern tip of the North Island of New Zealand. The Wairarapa Fault is situated a few kilometers to the east of the beach and has caused coseismic uplift of the ridges. B. Cross-sections of topography of beach ridges showing modern storm beach and successive coseismically uplifted ridges. The crests of correlative ridges rise progressively higher above sea level toward the northeast (from "a" to "b"), indicating increasing coseismic displacements to the northeast in each event. Modified after Stevens (1974) and McSaveney *et al.* (2006).

provide a horizontal reference surface (Fig. 2.6 and Plate 1B). If the height of the present storm ridge above sea level is known, then uplifted and deformed older ridges can be used to define deformation (e.g., Marshall and Anderson, 1995; McSaveney *et al.*, 2006).

Although not geomorphological in nature, zonations of marine plants or animals can sometimes provide useful biological markers for defining deformation. For example, those creatures that are attached to rocks and grow just below the high-tide line can be used to define the local sea surface just as easily as can abrasion platforms. If this coastal community were uplifted coseismically, the upper elevational limit of these organisms would record the vertical displacement at that spot on the coast (Plafker and Ward, 1992). Although such a record can be quite precise, it is clearly more transient than most geomorphic markers: attached organisms exposed continuously to the air soon die and eventually fall off the rocks to which they were attached, implying that the record of such instantaneous uplift events must be obtained within a few months to years of the event (Carver *et al.*, 1994).

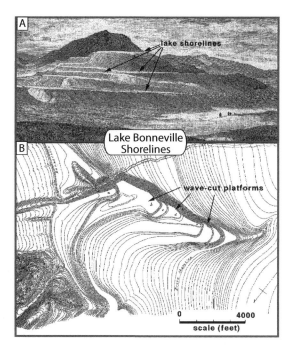

Fig. 2.7 Lacustrine shorelines formed by Pleistocene Lake Bonneville.
Wave-cut platforms are incised into headlands, the flanks of islands, and across spits. A. Perspective view of shorelines. B. Topographic map of wave-cut platforms incised into an elongate spit.
Modified after Gilbert (1890).

Lacustrine shorelines

Like marine terraces, lacustrine shoreline features are almost perfectly horizontal at the time of their formation. As waves impinge on the shore, they create *wave-cut benches* or *lacustrine strand lines*. The width of these benches depends on the erosional resistance of the bedrock at the shoreline, the length of time during which the lake level remained constant, and the strength of wave attack, which is a function of fetch, storm winds, and local shoreline geometry. Along the margins of Pleistocene Lake Bonneville in Utah, for instance, benches as much as 100 m wide have been formed. In Gilbert's (1890) pioneering study on Lake Bonneville, he identified many of these shorelines and used them as markers to document crustal rebound that followed the shrinkage of the lake (Fig. 2.7).

Lake levels in closed tectonic depressions fluctuate significantly in response to the water balance in the catchment. Unfortunately, unlike major changes in eustatic sea level, which are essentially synchronous on a global basis, lake-level changes (Fig. 2.8A) commonly vary between adjacent basins (Benson *et al.*, 1990). Compilations of lake-level records from the southwestern United States (Smith and Street-Perrott, 1983) show that, even in a limited geographic region, the timing of highstands can be quite variable (Fig. 2.8B). Such variability can result from sustained sweeps of climatic systems across a region in response to, for example, latitudinal shifts of the Intertropical Convergence Zone or to retreat of ice sheets. Temporal variability also results at least in part from the complex routing of water through the landscape, such that one basin may begin to fill only after an adjacent basin overflows (Adams *et al.*, 1999). Once the water in a lacustrine basin fills to the height of an outlet, the water level cannot get significantly higher, even if the discharge into the lake continues to increase. Conversely, bedrock thresholds that control the level of a lake outlet and, therefore, its maximum height may erode and can cause the lake level to drop, irrespective of climatic variations. These regional and local factors dictate that the age of a displaced shoreline must be determined within the particular basin being examined, rather than inferred through correlation with dated shorelines elsewhere or with global climate records.

Deltas

Both marine and lacustrine deltas provide clear geomorphic evidence for former water levels. Deltas have an advantage over terraces or wave-cut benches in that they are often larger geomorphic features that are therefore more likely to be preserved. The disadvantage of deltas is that they only form where rivers enter bodies of water, and thus they typically offer less extensive spatial coverage than do shoreline features: they provide control on changes in elevation at a point, rather than along a line.

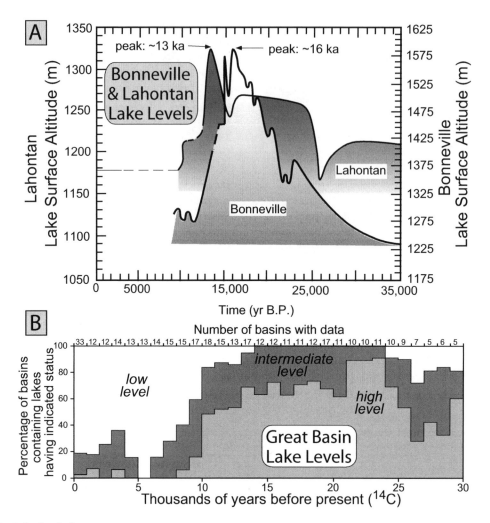

Fig. 2.8   Lake-level changes.
A. Lake-level history for Lake Bonneville and Lake Lahontan. Despite the proximity of these two large lakes in the Great Basin, they have significantly different histories due both to contrasts in discharge and to changing topographic thresholds within each basin. Modified after Benson *et al.* (1990). B. Late Quaternary lake-level records in southwestern United States. These data indicate that the majority of lakes fluctuate synchronously at the time scale of a few thousand years, but that lake-level variations in any individual lake may differ markedly from the mean. Modified from Smith and Street-Perrott (1983).

Gilbert (1890) described the internal bedding geometries of lacustrine deltas long ago, and showed that the contact of the topset and foreset beds closely approximates the level of the body of water into which the delta is prograding (Fig. 2.9). If a lake only existed during one particular interval, mapping of the foreset–topset contact for many deltas formed along its margin would define a paleo-horizontal surface that

may have been subsequently displaced. As with lake shorelines, the magnitude of isostatic rebound or fault displacements can be deduced from crustal warping of the paleo-horizontal surface defined by deltas (Thorson, 1989). If, however, multiple high lake levels occurred within the basin, then not only must the height of a topset–foreset contact be determined, but the age of the delta must also be ascertained to

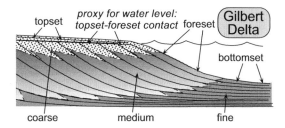

Fig. 2.9 Internal bedding geometries in a simple delta. The contact between the foreset and topset beds closely approximates the lake level or sea level at the time of delta growth.

permit reliable correlation of deltas formed in the same interval.

River terraces

All of the previously discussed markers provide a horizontal reference frame for assessing deformation. Even if the geomorphic evidence for a displaced shoreline feature is discontinuously preserved across the area of interest, the former geometry of the pristine feature is known to be horizontal, such that vertical displacements can be confidently calculated. If the timing of the surface cannot be determined, one can still have considerable confidence in the relative displacements of points, and hence determine vertical displacement field (but not the displacement *rate*). When geomorphic features that were not originally horizontal are used as markers, care must be taken to ascertain the gradient and geometry of the feature prior to offset. For example, if a change in gradient is used to define tectonic warping, one needs to be confident that these gradient changes are not natural ones resulting from some non-tectonic cause, such as variable resistance of bedrock to erosion or the normal downstream gradient of a river.

The term *stream power* refers to the rate of expenditure of potential energy per unit length of stream (Box 2.2) and is proportional to the slope of the water surface and to the river discharge. Analysis of stream power, including spatial and temporal changes in power, provides one perspective on the causes of river

erosion or aggradation. As stream power increases, the energy supply at the channel bed is also interpreted to increase, such that more energy becomes available to overcome friction, erode the bed, or transport sediments. The term *unit stream power* or *specific stream power* represents the stream power per unit area of the bed and is numerically related to bed shear stress and mean velocity. It can be easily imagined that changes in stream power will affect the behavior of a river, such as whether it is aggrading or degrading (Whipple and Tucker, 1999). However, other variables, such as the sediment load, the caliber of the sediment, and the roughness of the bed, also influence the aggradational or degradational state of the river (Fig. 2.10). Increases in load, caliber, or roughness are commonly interpreted to increase the resistance of the river bed to erosion, because larger loads require a higher expenditure to transport and greater roughness dissipates more energy through turbulence. A river that is neither aggrading nor degrading can be considered to be in equilibrium (Bull, 1991) and to be at the *threshold of critical power* (Fig. 2.10). At this threshold, the stream power is just sufficient to transport the sediment load that is being supplied from upstream, and the height of the bed remains constant. In general, if other factors are held steady, increases in river slope or in discharge, or decreases in bed roughness, sediment load, or sediment caliber, will cause the river to cross the threshold of critical power and begin to erode its bed. In contrast, changes in the opposite sense will push the river into an aggradational mode. The concept of a threshold of critical power has been usefully applied to the interpretation of the genesis of river terraces, because it indicates the potential linkages among different variables and suggests how changes in climate or tectonics could cause the river to switch from aggradation to degradation, or vice versa.

River terraces are common examples of preserved, sloping geomorphic features. Two classes of river terraces are typically defined: *aggradational* (or constructional or fill), and degradational (or erosional or cut or *strath*).

Box 2.2   Stream power.

Consider a "packet" of water that flows along a river. As it loses altitude along its course, it loses potential energy (PE) (see figure). This energy loss (ΔPE) occurs over some increment of time (Δt), during which the altitude of the packet of water is lowered by some amount (Δh) along some length of stream bed (Δx). Power is the rate of doing work, or energy expenditure. Therefore, stream power is defined as the rate of change of potential energy; and stream power/unit length defines the amount of energy that is available to do work over a give length of stream bed during a given interval. Thus,

$$\frac{\text{stream power}}{\text{unit length}} = \Omega = \frac{\Delta\text{PE}}{\Delta t \, \Delta x}$$

Recall that $\Delta\text{PE} = mg\Delta h$, where m is mass and g is gravitational acceleration, and that $m/\Delta t = \rho_w Q$, where ρ_w is the density of water and Q is discharge. So

$$\Omega = \frac{\Delta\text{PE}}{\Delta t \, \Delta x} = \frac{mg\Delta h}{\Delta t \, \Delta x} = \frac{m}{\Delta t} g \frac{\Delta h}{\Delta x}$$

$$= \rho_w QgS = kQS$$

where slope $S = \Delta h/\Delta x$ and the constant $k = \rho_w g$.

Specific stream power is defined as the power available per unit area of the bed. Thus,

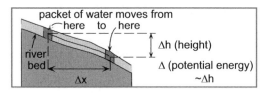

Downstream loss of potential energy in streams provides power to do work on the bed and banks.

$$\text{specific stream power} = \frac{\Omega}{\text{width}} = \frac{\rho_w gQS}{w}$$

where w is the width of the stream bed. But discharge $Q = wd\Delta x/\Delta t$, so

$$\text{specific stream power} = \frac{\Omega}{w} = \frac{\rho_w gS}{w} \frac{wd\Delta x}{\Delta t}$$

$$= \rho_w gSd \frac{\Delta x}{\Delta t} = \tau\bar{v}$$

where τ is the bed shear stress, and $\bar{v}$ is the mean velocity.

Variations in stream power define the changes in the amount of energy available to do work on the bed of a stream. Overcoming frictional forces, transport of sediment, and erosion of the bed are all dependent on the energetics of the river system. In the context of stream power, increases in discharge or slope will have a strong effect on the energy available to do work by the river.

The former result from aggradation of river-transported alluvium along a river's course, followed by downcutting, which leaves the former aggradational surface abandoned as a terrace. This incision could be regarded as a consequence of crossing the threshold of critical power, such that the river moves from an aggradational or equilibrium mode to a degradational mode (Fig. 2.11). As the water discharge of a river increases downstream, its surface slope generally decreases. Thus, at the regional scale, the longitudinal profile of the terrace should typically be represented by a smoothly decreasing, concave gradient in the downstream direction. Tectonic perturbations to such a gradient often are straightforward to determine.

Degradational terraces can form in several ways, some of which provide more predictable downstream profiles than others. For example, if a river is incising into alluvium and reaches an equilibrium during which incision ceases and the river bevels sideways into the valley fill, or even

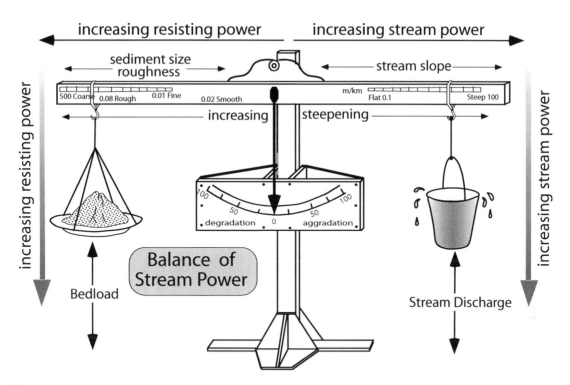

Fig. 2.10   Schematic representation of the threshold of critical power as a balance between eroding and resisting forces.

An increase in stream slope or discharge, or a decrease in sediment load, sediment caliber, or bed roughness, will move the system toward erosion of its bed. Modified after Bull (1991).

aggrades a bit, paired terraces (correlative terraces preserved on both sides of the river) with considerable downstream continuity can be formed (Fig. 2.11). If a river is degrading through alluvium and is also switching its course back and forth within a valley during degradation, it can create terraces that are unpaired (no correlative terrace on the opposite side of the river). The downstream geometry of such terraces may be hard to reconstruct (Merritts *et al.*, 1994), because unpredictable successions of terraces are preserved at any location and an age equivalency would need to be demonstrated prior to confident correlation among terrace remnants. Such terraces can provide useful markers for deformation that is contained within an individual terrace remnant, but they are much less practical for examining broader patterns of tectonic deformation.

A river incising into bedrock can create a bedrock terrace or *strath terrace* (Plate 1C). Such bedrock incision typically occurs within or immediately adjacent to mountains, where variations in bedrock resistance to erosion are common along the river's course. Across less resistant bedrock, stream power might tend to be lower due to some combination of river widening and gradient decrease. Even in an equilibrium condition without tectonism, the river gradient in a bedrock river will be more variable than in an alluvial river. Thus, as with unpaired degradational terraces, strath terraces provide useful local geomorphic markers, but are typically not as useful in the documentation of regional deformation patterns. This restriction is all the more true due to the limited downstream extent of such strath surfaces, many of which are only 100 m or so in extent when cut into resistant bedrock. Clear exceptions occur both when the underlying bedrock is weak, so that rivers can more readily bevel regionally extensive terraces,

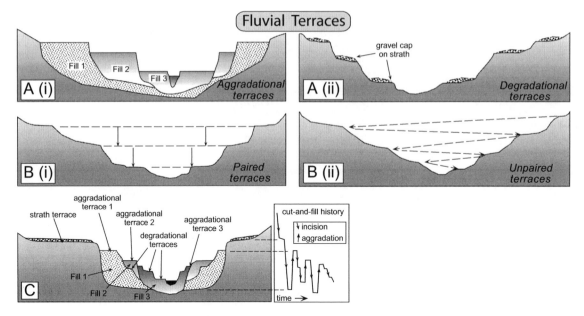

Fig. 2.11   Schematic configurations of river terraces.
A. Cross-sectional sketches of (i) aggradational and (ii) degradational fluvial terraces. B. (i) Paired and (ii) unpaired river terraces. C. Cross-section showing complex sequence of aggradational and degradational surfaces. Multiple cut-and-fill events are outlined in the right-hand box.

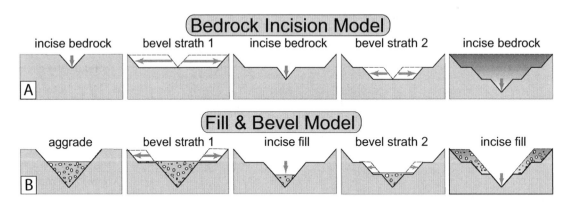

Fig. 2.12   Models for formation of strath terraces.
A. Traditional bedrock incision model for strath terrace formation results from extensive lateral beveling by a river, causing a broadening of the valley floor and retreat of the valley walls. Intervals of strath cutting are separated by intervals of river downcutting through the bedrock. Height of a strath above the bedrock valley floor indicates how much bedrock incision has occurred since strath formation. B. Fill-and-bevel model (e.g., Hancock and Anderson, 2002) occurs within a valley already incised into the bedrock. Aggradation within the valley protects the bedrock valley floor from erosion, but enables the river to attack the valley walls above the bedrock floor, where new straths are then cut. If the river subsequently incises partway through its fill, new straths at a lower level can be cut. Note that the height of the strath above the bedrock valley floor is unrelated to the amount of bedrock incision since the strath formed. Although the geometry of the straths for the two models is identical, strong contrasts exist in the volume of bedrock that must be removed during strath formation and in the strath's relationship to the history of bedrock incision.

and when particularly large and turbulent rivers transport heavy sediment loads that enable them to bevel laterally into bedrock.

The controls on the formation of bedrock straths are still debated. The traditional model for the formation of strath terraces involves gradual fluvial widening of the valley floor to create a broad, flat, bedrock surface into which the river later incises, leaving behind a strath (Fig. 2.12). This process was inferred to occur during intervals of tectonic quiescence, and sequences of straths were interpreted as records of episodic tectonic uplift. More recently, straths have been suggested to form when valleys have aggraded due to increased sediment loads (Hancock and Anderson, 2002). Not only does the aggraded sediment shield the bedrock beneath the river from further erosion, it also provides *tools* that collide with the valley walls and can bevel them laterally (Hartshorn *et al.*, 2002; Sklar and Dietrich, 2004; Turowski *et al.*, 2008). Subsequent changes in sediment or water discharge cause the river to incise down through its alluvial fill (which is readily

removed), leaving behind strath terraces that are elevated above the bedrock floor (Fig. 2.12). In this scenario, climate changes are also likely to modulate strath formation and abandonment.

In order to use deformed terraces to calculate tectonic rates, their ages must be known. Although obtaining reliable ages is commonly difficult, we know that many paired aggradational terraces form in response to climatic cycles. In many areas of alpine glaciation, for example, fluvial terraces can be demonstrated to correlate with moraines associated with glacial advances or stillstands (Penck and Brückner, 1909). Changes in sediment and water fluxes during these climatic intervals can lead to river aggradation that is followed by incision. The ages of such climatically controlled surfaces are often similar across a region that has experienced similar climatic conditions. Notably, when straths are beveled across weak bedrock (commonly Cenozoic sedimentary strata), their formation appears to be modulated primarily by climate, and hence their ages (Fig. 2.13) are likely to be regionally

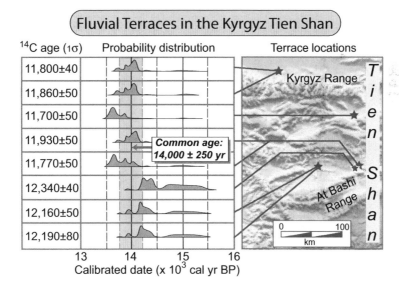

Fig. 2.13 Late Pleistocene terrace ages in the Tien Shan, Kyrgyzstan.
Radiocarbon ages from eight fluvial terraces are each depicted as probability density plots of the calendar ages corresponding to the radiocarbon age of each site and its uncertainty. These dates were collected from terraces in three tectonically distinct basins as much as 200 km apart. Each terrace surface is underlain by a strath that was cut on Tertiary sedimentary rocks and covered by 2–10 m of fluvial gravels. The terrace ages derive from organic matter preserved in the gravels. Their consistency argues that their formation was climatically controlled. Note that the probability distribution for each age is a function of the radiocarbon date, its uncertainty, and variations in atmospheric ^{14}C through time, as explained in Chapter 3. Modified after Thompson *et al.* (2002).

Distance upstream from the basin mouth (km)

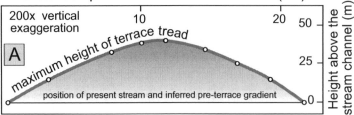

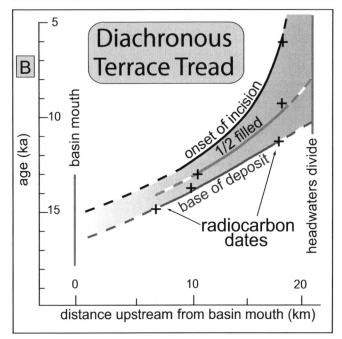

Distance upstream from the basin mouth (km)

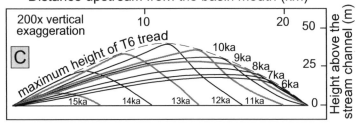

Fig. 2.14   Terrace formation.
A. Height of the top surface of terrace T6 at Cajon Creek, California, compared with the modern river. The gradient of the modern river is removed, in order to emphasize the apparent upward bowing of the surface. The highest part of the terrace occurs where it crosses the San Andreas Fault, but there is no apparent offset here. B. Variations in the timing of initiation of aggradation and incision along Cajon Creek, both of which are highly diachronous. Radiocarbon dates (crosses) on the fill are used to calibrate the aggradation–degradation cycle. Incision is assumed to begin shortly after the maximum thickness of fill is attained. C. Interpretation of the evolving topographic profile of the terrace due to migration of the locus of deposition and erosion. Modified after Weldon (1986).

coeval (Pan *et al.*, 2003; Pazzaglia and Brandon, 2001; Thompson *et al.*, 2002). Thus, correlation of undated terrace surfaces with ones dated at another locality or with known climatic changes has often been used to estimate the age of the unknown surface (Avouac and Peltzer, 1993).

Even where the physical downstream continuity of a terrace is clear, it is not always safe to assume that the upper surface represents an essentially isochronous horizon. The potentially complex nature of terrace formation is strikingly illustrated by a terrace that crosses the San Andreas Fault in southern California near Cajon Pass (Weldon, 1986). When the terrace surface is compared with the modern stream gradient, the terrace appears to have been warped upwards more than 30 m (Fig. 2.14A), although it is not physically disrupted across the fault.

Dating of the aggradational fill underlying the terrace, however, was facilitated by the discovery of organic debris within the sediments. The resultant radiocarbon ages indicate that both the initiation of aggradation and the initiation of incision are remarkably diachronous along the terrace surface (Fig. 2.14B). Compared to the upstream reaches, deposition commenced ~4 kyr earlier in the downstream reaches, and incision of the aggraded surface began there ~7 kyr earlier. Thus, while upstream parts of the terrace were just beginning to aggrade, the dissection of the downstream areas had already begun! Even though the entire terrace is less than ~15 kyr old, the age and time of abandonment of the upper surface of the terrace varies by ~7 kyr across a distance of ~10 km (Weldon, 1986).

The cause of this diachronous response is unknown. Alluvial terraces form within coupled, complex systems that include erosion, transport, and deposition of sediments. Such systems do not respond instantaneously or uniformly to changes in controlling parameters, such as discharge, sediment supply, rock uplift rate, or base-level lowering (Humphrey and Heller, 1995). Instead, changes in aggradation or degradation propagate up stream and down through both alluvial and bedrock systems at rates which may be related to the diffusivity of each system and to the velocity of kinematic waves that move along alluvial and bedrock reaches (Humphrey and Heller, 1995). The duration of these waves of deposition or erosion often greatly exceeds the duration of the perturbation that initiated the wave. Many drainage basins have equilibrium response times that are considerably greater than 10^5 yr. Given that many changes imposed on river systems, such as discharge variations due to climate, occur on much shorter times scales, it seems likely that depositional and/or erosional systems will be constantly integrating the responses to successive variations in controlling parameters. Thus, equilibrium may be rarely attained in fluvial systems. The diachrony displayed by the Cajon terrace (Weldon, 1986) might typify many fluvial terraces. In the common absence of data

to delineate chronological differences along a terrace, however, most workers assume the upper terrace surface is essentially isochronous along its length. Diachrony at the scale of a few thousand years becomes less important when older terraces are considered, but for post-glacial and Holocene terraces, such variability would significantly distort rates that were calculated assuming an isochronous terrace surface.

Alluvial fans

In many respects, the surfaces of alluvial fans are similar to fluvial terraces. Slopes generally decrease down-fan and any longitudinal section displays a slightly concave upward profile. Deposition on most fans is considerably more episodic than in rivers, and, in addition to channelized fluvial processes, various types of mass flows and unchannelized flows commonly dominate deposition. Although most fans are characterized by steady, down-fan decreases in gradient, some fans appear to be segmented (Bull, 1964), such that in cross-section they comprise a suite of fairly straight slopes that abruptly change where the linear slopes intersect. Such segmentation can result from shifting of the boundary between deposition and erosion up or down the fan. For example, if deposition is focused on the fan apex (also termed the fanhead), alluvium will accumulate and steepen this area with respect to the rest of the fan. If the fanhead is subsequently entrenched, remnants of the abandoned steeper surface will remain, but the apex of deposition will shift down the fan. As aggradation on this gentler, down-fan surface proceeds, the upper surface of aggradation may eventually intersect and begin to overlap the steeper, up-fan surface, such that an abrupt slope change occurs at the intersection. In this situation, the upper and lower parts of the fan surface will have different ages. In such circumstances, before using fans as geomorphic markers, the age of each surface should be separately determined. Even in the absence of fan segmentation, most fans are composed of a mosaic of surfaces of different ages, ranging from the modern channels to dissected remnants of long-abandoned surfaces. Weathering of clasts on the surface of a fan can permit the relative ages of different segments of a fan to be defined. Following isolation from deposition, weathering processes, such as fracturing, rind and varnish development, reddening of the underside of clasts, and granular disintegration, begin to modify the appearance and character of the fan surface (McFadden et al., 1982; Ritter et al., 1993). Commonly, in arid regions, readily visible contrasts in surface color permit classification of the relative ages of different fan surfaces (Bull, 1991). In more humid fans, more time-consuming quantification of changes in the depositional surface may be required to divide the fan into surfaces of differing ages. In order to use a fan surface as a reference against which to measure deformation, only similarly aged portions of the fan should be used to define a reference plane.

Many arid alluvial fans are dominated by deposition from debris flows (Whipple and Dunne, 1992). The surface topography of debris-flow deposits (up to several meters) and the roughening of the fan surface that results from them can determine the scale at which the local surface of the fan can be considered to have a predictable geometry and, therefore, to be useful as a geomorphic marker. Even surfaces with considerable roughness, however, can often be used successfully as markers, because, despite its irregularities, the average surface gradient can be defined (Avouac et al., 1993), and offsets of this gradient by faulting can be readily recognized (Fig. 2.15).

Lava flows, debris flows, and landslides

Despite the meter-scale roughness of the upper surface of most lava flows, the surfaces of elongate flows can provide excellent geomorphic markers. Topographic gradients along the flow surface can be measured directly, and numerous surface features can be recognized for cross-fault correlation. The highly resistant flows have a high preservation potential, and the flow itself is often directly datable with radiometric methods.

The surface topography of mass movement deposits, such as debris flows, mudflows, and landslides, often depends in part on their water content and viscosity at the time of deposition (Whipple and Dunne, 1992). The higher the water content, the smoother the upper surface of the deposit and the greater the downstream travel distance are likely to be. These mass movement deposits have the advantage of having formed as the result of an instantaneous event, thereby avoiding the problem of diachrony in the surface seen, for example, in Cajon Creek fluvial terraces (Fig. 2.14). Moreover, it is not uncommon that mass movements overrun or contain within their deposits organic debris that can be radiocarbon dated. As a consequence, a date from anywhere within them can often pinpoint the age of the deposit closely. In contrast, a ^{14}C date within the strata of an aggradational terrace simply provides a lower limit on the time of abandonment of the depositional surface. Multiple dates are commonly needed on such surfaces in order to judge the rate of aggradation and the timing of abandonment (see Fig. 2.14).

Erosional surfaces

It has long been recognized that, during long intervals of tectonic quiescence, topography can be beveled off by erosional processes and a low-relief landscape can be produced. Isolated erosional remnants (monadnocks or inselbergs) may be surrounded by pediments: erosional surfaces of low relief carved into bedrock. Across broader geographic regions, peneplains characterized by slightly undulating and generally featureless topography can result from long-continued erosion and deposition in the absence of active deformation. When such erosional surfaces formed in the distant geological past, they were commonly buried by subsequent deposition, such that a regionally extensive unconformity is preserved. When such low-relief surfaces (pediments, peneplains, regional unconformities) are uplifted tectonically, they can form prominent markers in the landscape. Typically these

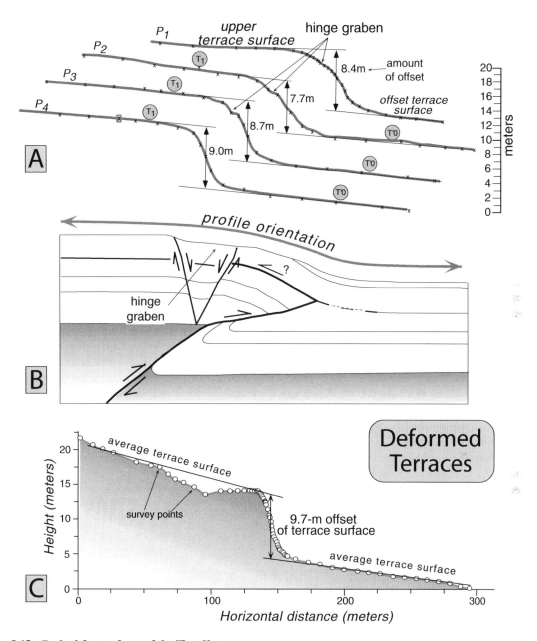

Fig. 2.15   Faulted fan surfaces of the Tien Shan.
A. Surveyed offsets of the surface of alluvial fans where displaced by a thrust fault. Fresh faults scarps have been degraded. B. Interpretation of the schematic structure beneath the faulted terraces. Note that even small irregularities in the surface, such as the "hinge graben," can sometimes be interpreted in terms of the underlying structure. C. On surfaces displaying several meters of local relief, displacements of the "average" down-fan gradient can be derived from surveyed profiles. Modified after Avouac *et al.* (1993).

erosional surfaces are identified by the low relief of the uplifted surface, deep weathering of an undulating regional surface, an accordance among summit heights, and/or the smoothly varying topography of broadly folded or faulted terrains. Most commonly, the actual erosion surface is no longer preserved because slow bedrock weathering has exhumed through the surface since it was created. As a consequence, multiple lines of evidence commonly need to be assembled to justify the reconstruction of a formerly contiguous surface. In places, a low-relief surface stands in contrast to a nearby tectonically active front characterized by high relief, such that the uplifted surface provides a useful marker for estimating tectonic displacement at long time scales. For example, in the San Bernardino Mountains of southern California, the Big Bear plateau is distinguished by an undulating, deeply weathered surface that covers ~1500 km² and averages about 2 km in elevation. The plateau sits atop a 600- to 1000-m-high, north-facing escarpment that is underlain by a north-vergent, active thrust fault (Spotila and Sieh, 2000). The erosion surface is interpreted to have developed since late Cretaceous times and to have been at much lower elevations in the Miocene when basalt flows were erupted across parts of it. Where preserved, these basalts fossilize the former erosion surface. Elsewhere, however, deep weathering (up to 30 m) has obliterated the actual Miocene erosion surface across most of the plateau. Nonetheless, the lateral continuity of this weathered surface, its clear spatial relationship to sites where the erosion surface is preserved, and consistent cooling ages across the surface all support its interpretation as an uplifted erosion surface (Spotila and Sieh, 2000).

Similarly, in the Tien Shan of Central Asia, a regionally extensive erosion surface at least 100 000 km² in extent was beveled across Paleozoic and Mesozoic rocks and buried by Cenozoic sedimentary rocks (Chediya, 1986; Sadybakasov, 1990). Wherever basal Cenozoic strata are exposed above this unconformity, their bedding parallels the dip of the unconformity at kilometer scales, implying very low relief on the unconformity surface when it was buried. This surface has been recently exhumed due to rock uplift. The striking contrast in erodibility of the rocks above and below the unconformity has caused the Cenozoic sediment to be rapidly eroded, revealing the unconformity surface (Plate 1D), which provides an excellent marker for

recording folding and faulting of ranges that rise as much as 2 km above the surrounding terrain (Burbank *et al.*, 1999). Not only do erosion surfaces like those in the San Bernardino Mountains or the Tien Shan serve to define the three-dimensional pattern of differential rock and/or surface uplift, but the unconformity surface itself forms a reference for calibrating the amount of erosion that has occurred beneath it and for assessing the processes by which such uplifted bedrock surfaces are dissected (Oskin and Burbank, 2005, 2007; Goode and Burbank, 2011).

Linear geomorphic markers

Whereas the previously described geomorphic markers represent areally extensive surfaces, it is also possible to use linear geomorphic and man-made features to determine deformation. Although displaced planar features are more suitable for defining regional tilting, linear features, such as glacial moraines (Plate 1E), can provide ideal piercing points from which an offset can often be unambiguously measured. Unlike many two-dimensional surfaces, such as marine or fluvial terraces, many linear geomorphic features can be formed by individual events, some of which may have occurred instantaneously from a geological perspective, for example, the levees that form on the margins of a debris flow (Plate 1F). Such features often have no direct relation to climatic variations, so that ages need to be determined for each event in order to determine rates of deformation.

Rivers and ridge crests

The courses of rivers and ridge crests that are displaced across strike-slip faults can clearly record lateral offsets (Fig. 2.16A). It is important to ascertain, however, that the deflection of a stream is due directly to differential displacement of its course by faulting and is not the result of the intersection between a regionally sloping surface and a fault scarp. If streams are offset in directions that oppose the regional

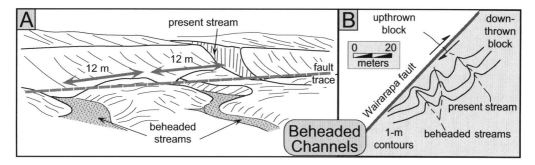

Fig. 2.16   Beheaded channels along a strike-slip fault.
A. Sketch of beheaded streams along a strike-slip fault, the Wairarapa Fault, North Island, New Zealand. Spacing between streams indicates two previous earthquakes with about 12 m of displacement in each event. B. Map with 1 m contours of offset and beheaded stream channels along the Wairarapa Fault. Modified after Grapes and Wellman (1993).

slope, the cause of the offset is more likely to be tectonic than when the deflection occurs in the direction of the regional slope. Owing to strike-slip motions, streams can be *beheaded*, by which it is meant that an abandoned stream channel abruptly terminates as it crosses a fault (Fig. 2.16B). The difficulty in assessing offsets of rivers and ridge crests lies in making reliable correlations from one side of a fault to the other. Commonly, multiple ridges and streams cut across faults, so that specific correlations can be ambiguous (Frankel *et al.*, 2007).

Because rivers are capable of incising and modifying any displaced profile, vertical movements often are underestimated by the apparent displacement of the river channel at the location of the fault. If the upstream part of a stream bed is elevated by faulting with respect to its downstream continuation, the stream will tend to incise through the scarp. Remnants of the former valley floor may be preserved as small terraces on either side of the channel, and their height above the downstream, but offset, continuation of the channel can be used to assess the amount of vertical displacement (Beanland and Clark, 1994).

Glacial moraines

The elongate ridges of ice-transported debris that form glacial moraines provide linear geomorphic markers (Plate 1E) that have an obvious direct climatic cause. Lateral displacements can

often be readily measured in map view from the offset of the linear trend of the moraine crest, whereas vertical offsets can be assessed by comparing the topographic trend along the length of the moraine crest on either side of a fault.

If advances attributed to surging glaciers are excluded, then most major glacial advances are responses to large-scale climatic changes. Thus, one might expect to be able to correlate the record of successive glacial advances with the record of Quaternary climatic fluctuations (see Box 2.1). For the most recent advances, this is commonly true, although the timing of the maximum extent of alpine glaciers in any particular mountain range often differs by thousands of years from the time of maximum ice-sheet extent (Gillespie and Molnar, 1995). Therefore, whereas an absolute date on a moraine is always preferred, an approximate age can be assigned to undated moraines, and tectonic rates (with appropriate uncertainties) can be calculated based on observed offsets.

When moraines other than those associated with the most recent advances are considered, the one-to-one correlation with the climatic record typically breaks down due to incomplete moraine preservation (Box 2.3). In such circumstances, a local glacial chronology associated with specific preserved moraines (Owen *et al.*, 2008) needs to be established in order to have reliable control on long-term rates

Box 2.3   The problem of moraine survival.

Ever since the first synthesis of the record of alpine glaciation (Penck and Brückner, 1909), it has been common to recognize three or four major moraines in glaciated valleys. At the same time, the global climate record indicates that there have been at least 10 major glaciations in the past 1 Myr and many more in the previous 1 Myr (Fig. 2.3). Why is there such a mismatch between the number of glaciations and the preserved morainal record of those glaciations? One answer comes from a statistical analysis of the probability of moraine preservation (Gibbons *et al.*, 1984).

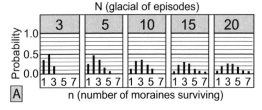

A. Examples of the predicted number of moraines surviving after 3, 5, 10, 15, and 20 glacial advances of random length. Modified after Gibbons *et al.* (1984).

Let us assume that there has been a succession of 15 glaciations and that, with respect to other glaciations, the relative magnitude of each glaciation and its associated advance is randomly distributed. What would happen if the most recent glacial advance were also the largest? It would wipe out most or all of the geomorphic record of all previous advances. If, on the other hand, the glaciations happened to fall sequentially from the most extensive at the beginning to the least extensive at the end, then every single glaciation would be represented. The question of how many moraines will survive can be posed statistically as follows. The probability (P) that n moraines will survive, if there were N glaciations, is

$$P(n/N) = \frac{1}{N} \sum_{N=n-1}^{N-1} P((n-1)/N)$$

For example, the probability of two moraines surviving, if there were four glaciations, is

$$P(2/4) = \frac{1}{4}[P(1/1) + P(1/2) + P(1/3)]$$

Because $P(1/N) = 1/N$, we get

$$P(2/4) = \frac{1}{4}\left[1 + \frac{1}{2} + \frac{1}{3}\right] = \frac{1}{4}\left[\frac{11}{6}\right] = 0.46$$

The probabilities for differing numbers of preserved moraines can be quite readily computed (see figure A). Perhaps surprisingly, they indicate that, for 8–20 glaciations with randomly distributed magnitudes, the most likely number of moraines to survive is only three!

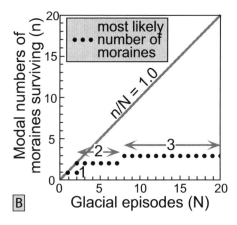

B. Plot of most likely number of surviving moraines as a function of the number of glacial episodes. Modified after Gibbons *et al.* (1984).

Overall, it is clear that a succession of glacial moraines will typically provide only a fragmentary record of climate change (see figure B). Similarly, to the extent that aggradational terraces are correlated with the magnitude of glacial advances (as is often supposed), it is likely that the preserved aggradational terraces have buried older, smaller terraces beneath them.

of deformation. It is commonly assumed that, within any area experiencing a regionally consistent climate, most of the glaciers would be expected to advance and retreat approximately synchronously, but there are few well-dated tests of this contention (Gillespie and Molnar, 1995). In the absence of dates, it is, therefore, not uncommon to try to correlate with the global climatic record (Fig. 2.3), but such correlations should be made with considerable skepticism.

Other linear features

Numerous other linear features, both naturally occurring and man-made, can be used as markers for gauging displacement. Fairly viscous debris flows can create raised levees of coarse debris along their channel margins (Beanland and Clark, 1994). Fence lines, railroad tracks, curbs, sidewalks, lines painted on streets, and even tracks created by cars, motorcycles, or bikes, can provide linear markers that are readily measured and are useful for documenting either coseismic offsets of recent earthquakes (e.g., the scarp of the 1992 Landers earthquake in California displaced alluvial fans that sported hundreds of motorcycle paths) or slower rates of creep. These man-made markers are not geomorphic features in a strict sense, but when trying to generate a catalog of displacements along a recently ruptured fault zone, any displaced linear feature with a known pre-faulting geometry should be evaluated.

Commonly encountered problems with markers

The most typical difficulty in extracting the maximum information from a displaced geomorphic surface is the absence of a well-documented age for the surface. Considerable effort often is warranted to try to uncover datable material that can constrain the age of the feature. A knowledge of the array of available dating tools, experience with the appropriate field procedures for collecting suitable material for a specific dating method, and a thorough and

innovative approach to the problem of defining geomorphic ages is a boon to anyone attempting to obtain reliable age constraints and hence deformation rates.

Correlation of partially preserved geomorphic features presents another challenge. As older and older markers are examined, their preservation typically becomes increasingly fragmentary. Unless distinctive characteristics permit discrimination among features of differing ages, the correct correlation among the remnants may be difficult to achieve. A traditional field technique for terrace remnants, for example, has been to survey the height of terraces along local reaches and to correlate the terraces from one site to another based on their relative height above the river. This approach can be misleading, because it assumes a consistency in the longitudinal profile of the terrace through time. A better technique is to employ continuous, geodetic surveying along terrace surfaces (Merritts et al., 1994) and to document the connection from one reach to the next whenever possible (Fig. 2.17). Even with such data, reconstruction of the long profiles of multiple terraces can be ambiguous. Distinctions need to be drawn between strath and aggradational terraces, because, when aggradational terraces primarily result from a downstream rise in local base level, their gradients are likely to be gentler than those of most upstream strath terraces. Recognition of different terrace types, continuous geodetic surveying, dating of terraces, and analysis of how and why the terrace sequence developed all assist in creating a reliable reconstruction.

New tools continue to emerge that can assist geomorphologists in reconstructing landscapes. Even where features, such as terraces, are quite continuous, dense vegetation can often obscure the correlations among them. With the availability of lidar bare-Earth DEMs, the vegetation can be removed and a much clearer view of the landscape typically emerges (Fig. 2.18). Especially in heavily forested landscapes, lidar can revealed the detailed geomorphic structure of the landscape. Unfortunately, lidar acquisition is not cheap: costs for surveys are commonly 200–400 US

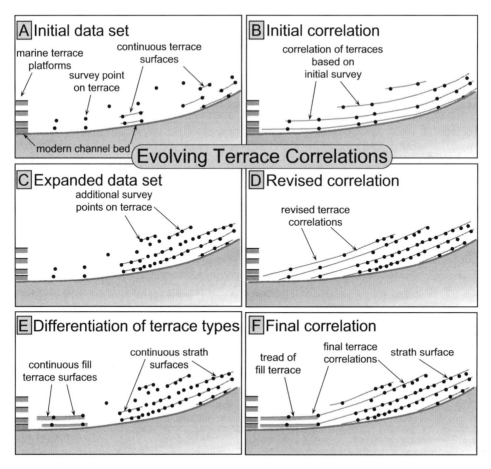

Fig. 2.17   Correlation of remnants of river terraces based on an example from the Mattole River, northern California.

A. Traditional technique of surveying the altitude of terraces in local reaches provides isolated points that must be correlated. B. If it is assumed that the past and present river gradients were parallel, correlations are then based on the height of a terrace above the modern river bed. C. More spatially continuous surveying of the upstream extent of terraces can define their gradients more reliably than can be done solely with correlated spot heights. D. When compared with the modern gradient, the reconstructed terraces are clearly not parallel to the modern profile. E. When the surfaces of aggradational terraces are distinguished from those of strath terraces, contrasts in gradients between the two terrace types become apparent. F. Terrace reconstruction based on survey, dating, and geomorphic field evidence shows strath terraces with steep gradients merging into the present river profile. Aggradational terraces with gentle gradients result from deposition forced by rising sea level (rising base level). Modified after Merritts *et al.* (1994).

dollars per square kilometer. But, if the problem is sufficiently important, then an investment in very high-resolution lidar topography can be well worth it.

Other problems can arise when a regional chronology of geomorphic markers has been developed and is subsequently applied to the deformational analysis of local sites. Suites of aggradational terraces or of glacial moraines are commonly dated by assembling ages from individual outcrops spread over a broad area. Such amalgamations can be reasonable and may be the only practical approach. Nonetheless, it is important to be aware of the possible presence

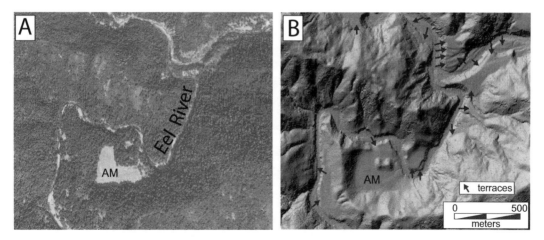

Fig. 2.18 Lidar imaging of terraces beneath a dense forest canopy.
Paired images of about 2.5 km² of the Eel River catchment in northern California. A. The high-altitude photograph depicts the dense vegetation that dominates this humid, temperate landscape and obscures the smaller-scale topography. B. The shaded-relief DEM (1-m pixels) reveals dozens of fluvial terraces (arrows) flanking the river banks. Flights of terraces a few meters across are clearly visible, such as at the confluence in the upper right corner. AM: abandoned meander. Modified after images from the National Center for Airborne Laser Mapping (NCALM) and Google Earth.

of terraces or moraines that are unrelated to the regional sequence. Surging glaciers or ones covered by rockfalls, for example, often advance largely independently of climatic controls (Kamb and Engelhardt, 1987; Santamaria Tovar *et al.*, 2008). Yet, they create moraines that are not readily distinguishable from those of glaciers intimately controlled by climate. Ages assigned to those moraines based on correlation to a climatic record would be erroneous. Similarly, a very large landslide in a catchment may overwhelm the transport capacity of a river (Ouimet *et al.*, 2007). This excess sediment would cause the river bed to aggrade, and an aggradational terrace could be left behind that is unrelated to terraces in nearby drainages, where terraces reflect instead the regional climatic controls.

Summary

Geomorphic markers abound within many landscapes. Common geomorphic markers include wave-cut benches (marine and lacustrine), fluvial surfaces (terraces and fans),

and linear features, such as moraines. Clever reading of the landscape and innovative adaptations of the general principles and approaches discussed here provide a basis for utilizing geomorphic markers in almost any geomorphic setting. One must attempt to understand the pristine, undeformed shape of the marker (formally, the initial conditions of the problem), because this shape forms the basis for all interpretations of deformation. Dating of geomorphic features takes persistence, familiarity with available techniques, innovation, and some luck. The search for datable material is often tedious, but as one geologist said, "If you haven't found datable material, it's because you haven't looked carefully enough!" All calculations of deformation rates depend on assigning ages to displaced features. Therefore, those long searches can often pay off by yielding new insights into how rapidly deformation has occurred in the past. In the next chapter, we examine several approaches to dating of geomorphic, stratigraphic, and structural features that record deformation. Just as geomorphic markers, such as moraines, terraces, and alluvial fans, have some predictable geometries, different kinds of faults and folds display contrasting styles of surface deformation, as discussed in an ensuing chapter. When fault-zone displacements are combined with rates derived from dated and deformed geomorphic markers, a three-dimensional reconstruction of fault-zone evolution, rupture history, landscape perturbations, and geomorphic responses can be attained.

3 Establishing timing in the landscape: dating methods

The science of tectonic geomorphology relies strongly on placing time controls on the landscape. In order to determine the rate at which a fault moves or a surface deforms, we must establish the age of offset features. The precision needed depends on the questions being asked, and on the recurrence time scales of the events themselves. Until the advent of radioactive dating methods, the principal means of establishing timing in the landscape was through the use of relative dating methods. The distinction between *relative dating* and *absolute dating* methods is that relative methods yield relational information only (surface X is older than surface Y, which is in turn older than surface Z), whereas absolute dating allows us to place a number on the age (terrace X was created 3500 years before present). The latter is absolute in the sense that it does not require attention to any other surface, but one must be aware of the potential errors inherent in each of the absolute dating methods. When calibrated against some absolute ages, several relative dating techniques lend themselves to some degree of quantification, making them "semi-quantitative" methods. In Table 3.1, we list several of the frequently applied methods, the age ranges for which they can be used, and a primary reference to the literature where they are better discussed. This chapter does not present an exhaustive review of the available techniques, but rather illustrates a few of the

techniques, focusing on the newer and more quantitative methods. For a rather comprehensive review of dating techniques, see Noller *et al.* (2000).

Relative dating methods

Clink versus thump, and its quantification: the clast seismic velocity method

For years, geomorphologists have determined the relative age of a surface by walking up to the surface to be dated and hammering on the boulders protruding from it. The resulting sound, which is a sharp bell-like clink if the boulder is fresh, and a dull and boring thump if the boulder is old and decrepit, is then used to assign a relative age. Many boulders are pounded and a statistical sense is developed about the age of this surface relative to some other. About 25 years ago, this method was modified to allow quantification (Crook, 1986). The clast seismic velocity method is based on the principle that a rock exposed at the Earth's surface will develop microcracks through a variety of weathering processes, and that the number of microcracks controls the propagation speed of seismic waves (compressional waves, or sound) through the rock. Although it is difficult to measure microcrack density, it is

---

Table 3.1 Relative dating methods.

Method	Age range	Materials needed	References
Clast seismic velocity	1–100 ka	Boulders	Crook (1986), Gillespie (1982)
Obsidian hydration	1–500 ka	Obsidian-bearing lavas	Pierce et al. (1976)
Soils	1–500 ka	Soils	Harden (1982)
Mineral weathering	10 ka–1 Ma	Boulders	Colman and Dethier (1986)
Landform modification	10 ka–1 Ma	—	Davis (1899), Cotton (1922)

quite easy to measure the seismic velocity by using a micro-seismic timer (Fig. 3.1A). For each of several spacings between the accelerometer and the hammer, many measurements of travel time are made. The travel time is recorded and is converted to velocity by dividing the separation distance between hammer and sensor by the travel time. The clast seismic velocity of a particular boulder is then calculated, and the next boulder is selected. Many boulders on each surface to be dated, e.g., moraine, alluvial fan, debris-flow terrace, are measured (Fig. 3.1B). Where the conditions are appropriate, the resulting age progressions have made sense (Crook and Gillespie, 1986), such that surfaces expected to be older than others have clast seismic velocities that are slower, indicating higher microcrack densities. The applicability of the technique is limited to sites where one may assume that the production rate of microcracks is uniform among the boulders to be sampled. This restriction forces us to focus on sites with uniform lithologies, or at least ones where many boulders of the same rock type are available, because the rate of production of microcracks, from any of many possible production mechanisms, is likely to be strongly dependent upon rock type. As with many of the other relative dating methods, however, the assignment of absolute ages from such data is difficult, given that we have no theoretical basis for predicting the rate of decline of clast seismic velocity through time. We do not know how long it will take for the clast seismic velocity to decline by 50%. This limitation forces us to calibrate the technique locally against surfaces of known age or to consider the resulting differences in clast seismic velocity in only a semi-quantitative sense (Gillespie, 1982).

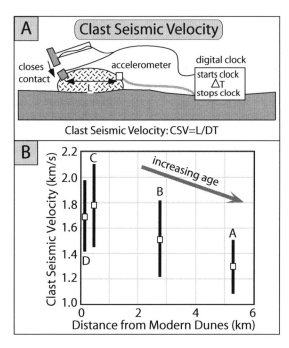

Fig. 3.1 Clast seismic velocity measurements. A. Cartoon of clast seismic velocity (CSV) methodology. B. Results of application of CSV method to the dating of debris-flow benches in Panamint Valley, California. Inferred age increases as the CSV declines, reflecting growth of microcrack population within the boulders sampled. Modified after Anderson and Anderson (1990).

Weathering rinds

Rocks exposed to temperature and wetness at the surface of the Earth weather by a variety of mechanisms. In general, this weathering proceeds inward from the surface of the rock, where moisture is most accessible and where temperature changes most rapidly and with greatest magnitude on both daily and storm cycles. This variability can result in the production of a weathering rind: an identifiable layer that has experienced more weathering than the material beneath it. If you crack open most clasts from an old surface, you will see that a thin layer near the surface of the rock is discolored, revealing that the minerals near the surface have been altered in some way. The thickness of this layer is thought to be a proxy for the time the rock has spent in near-surface conditions. Like the clast seismic velocity technique, this technique is fraught with problems associated with variability of the rind thickness among the surface clasts, the lithologic dependence of the rate of rind growth, and inheritance of a weathering rind from previous exposure. Nonetheless, one may control for many of these problems at selected sites. Moreover, the method is inexpensive, low-tech, and readily applied to a variety of deposits. It has been argued that the growth of the weathering rind ought to proceed as the square root of time (Colman, 1986). Presumably, this inference reflects the fact that the mineralogical changes required to create a visible rind are mostly chemical and require diffusion of species into and out of the rock. Because diffusive processes always result in thicknesses, L, that vary as the square root of the diffusivity, k, multiplied by the time, t, since the process was initiated ($L \sim \sqrt{\kappa t}$), one might expect a square-root relation of rind thickness to age. Only rarely have sufficient data been arrayed to test this model (Fig. 3.2B), because absolute ages on a variety of surfaces are required for the test.

In a review of chronologies derived for alpine glaciations in the western United States, Colman and Pierce (1992) noted that this decline in the rate of rind growth through time (no matter what the exact nature of the nonlinearity) allows one to place limits on the relative ages of deposits based on the ratios of the rind thicknesses. In other words, the ratio of the ages (old/young) should always be greater than the ratio of the rind thicknesses (thick/thin).

The primary problems with the technique are the need for calibration of the effects of local climate, the dependence of weathering rates on the lithology of the clasts, and the possibility of inheritance of rinds from prior exposure. This latter complexity is especially important if the clasts are

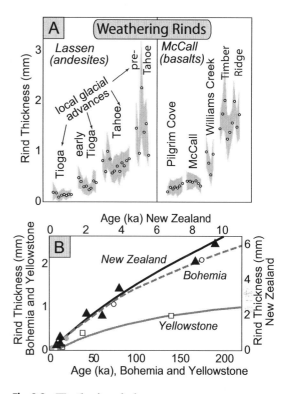

Fig. 3.2 Weathering rinds.
A. Weathering rind thicknesses are used to distinguish glacial deposits in Mt Lassen and Mt McCall, Idaho. Absolute age control is lacking on these moraines; the bottom axis is arbitrary. B. Growth histories of rinds in Bohemia and Yellowstone (left and bottom axes) and New Zealand (right and top axes). Note decline in growth rate with time. Modified after Colman and Pierce (1992).

highly resistant to abrasion within the transport system that delivers the clasts to the site to be dated. For instance, quartzite clasts weathering out of Jurassic conglomerates in central Utah appear on the surface with both weathering rinds and percussion marks from transport during the Jurassic (and perhaps even earlier episodes of transport). Most studies have therefore focused on weathering rinds developed in easily weathered, more abradable clasts such as andesites and basalts (Colman and Pierce, 1992) (Fig. 3.2).

Obsidian hydration rinds

A technique used to great effect in archeology is based on the growth of a hydration rind as a glassy surface weathers. Obsidian is a natural

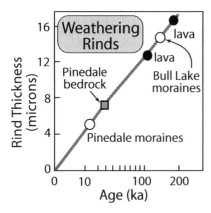

Fig. 3.3 Hydration rind thickness as a function of age. Semi-log plot shows logarithmic fall-off in growth rate with time. Calibration line from rinds in cracks on surfaces of independently dated rhyolite lava flows and from subglacially produced cracks in Last Glacial Maximum (Pinedale) bedrock. Open circles indicate data from cracks on surfaces of boulders in undated glacial moraines. Modified after Pierce *et al.* (1976).

glass, generally having a high silica content. The glass hydrates once a surface is exposed to the air. The thickness of the weathered rind is measured in thin section normal to the exposed surface, and is identified by an abrupt roll-off in the refractive index of the glass. The rind is on the order of several microns thick. Because the rate of hydration-rind growth is dependent at least slightly on the composition of the obsidian, and surely on the temperature to which the surface has been subjected, the technique requires calibration against surfaces of known age to yield quantitative dates.

Working in a field area containing dated lava flows, Pierce *et al.* (1976) demonstrated the usefulness of the technique in dating both glaciated volcanic bedrock and glacial moraines that incorporated clasts that had been glacially abraded. The glaciation is presumed to have generated new cracks in the rocks, which then hydrated (Fig. 3.3). Again, this technique suffers from the likelihood that the rind growth rate is dependent on both rock type and climate. Given that climate varied significantly over the Late Pleistocene, during which most of the surfaces with which we are concerned have evolved, this variability can be a significant drawback.

Soil development

It is common to be faced with the need to date a depositional surface in which a soil has developed. Several techniques have evolved, most of them quite qualitative, to attempt to place ages or relative ages on these surfaces based on the degree of soil development. Soil color is often the easiest indicator of soil age, although, as in the identification of minerals, there are many ways to go wrong using such a simple criterion. More robust indicators include the accumulation of carbonate, clays, and iron within the subsurface. In formal pedogenic studies that attempt to quantify the soil development through time on a series of geomorphic surfaces of similar parent material (a soil chronosequence), many indicators are documented. One integrated measure that synthesizes diverse indicators of pedogenic maturity, including soil clay content, soil color, and soil structure, is called the Harden index (Harden, 1982). Although it is not our purpose here to review all such techniques (see, e.g., Birkeland (1990) for such a review), we will illustrate a couple of methods that have been used to estimate ages.

Carbonate coatings and other pedogenic indicators

In arid regions, soils accumulate calcium carbonate in the near-surface. Rainwater delivers some calcium directly and dissolves calcium-bearing minerals (mostly carbonates) both from the parent material and from airborne dust that has accumulated on the soil surface. This dissolved load can be re-precipitated at depth as the water is wicked back up to the surface during evaporation or uptake by plants such that the remaining water becomes supersaturated with respect to calcite. Precipitation is also favored if the CO_2 content decreases or if the temperature increases in the soil, thereby lowering the solubility of calcite. The total mass of carbonate in the soil, and that part of it that occurs as coatings on the bases of soil clasts, have both been used in documenting relative ages of surfaces. Studies in Idaho (Vincent *et al.*, 1994) have shown that the rate of growth of carbonate coatings has varied by at least a factor

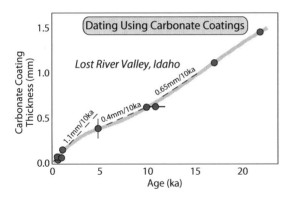

Fig. 3.4   **Carbonate coatings as a function of deposit age, from soils in the Lost River Valley, Idaho.** Although rates vary by a factor of more than 2, a long-term average rate of about 0.6 mm/10 kyr appears relevant for this setting. Modified after Vincent *et al.* (1994).

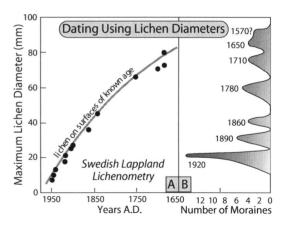

Fig. 3.5   **Lichen diameter as a function of age.** A. Lichen growth curve calibrated using surfaces of known age in the Swedish Lappland. Note nonlinear growth, with the most rapid growth for youngest surfaces. B. Histogram of the number of moraines with a given maximum lichen diameter and, hence, of a given age inferred from the calibration curve. Modified after Denton and Karlen (1973).

of 2 and perhaps 3 over the last few tens of thousands of years (Fig. 3.4). The mean rate over the last 20 kyr has been about 65 μm/kyr. Those clasts from surfaces well over 100 ka commonly have coatings on the order of 10 mm thick. We note that, whereas in this particular region of Idaho one might be able to determine the age of a surface by documenting the thickness of these coatings, the technique requires a local calibration, which in turn necessitates having some surfaces of known absolute age. Uncertainties, therefore, creep in, not only from the variability of the coating thicknesses themselves, but from the errors in the dating methods used to obtain the absolute ages of the surfaces used in the calibration.

Lichenometry

A technique that has long enjoyed use in geomorphic studies is lichenometry, in which the diameter of a specific type of lichen is used as a proxy for the time a rock has been exposed at the surface (e.g., Benedict, 1967; Innes, 1984, 1985; Locke *et al.*, 1979; Porter, 1981). The lichen most commonly employed is *Rhizocarpon* (one or another subgenus), which is crudely circular in shape and whose diameter is thought to increase linearly with time. Until recently, the practice was

to record the largest lichen found from the surfaces of commonly hundreds of boulders on the surface to be dated, or to take the mean of the five largest lichens. Again, local calibration is necessary, as climate dictates the rate of lichen growth. Typically, the calibration of the growth-rate curve comes from surfaces that have been exposed for a known period of time, most commonly from man-made structures. One clever approach uses lichens on tombstones, which of course have the age stamped on them! Lichenometry has been used effectively to date moraines in glacial settings (Denton and Karlen, 1973; Porter, 1981) (Fig. 3.5) and major rockfalls in rugged landscapes (Porter and Orombelli, 1981).

This technique has seen a resurgence recently through the work of Bill Bull (Bull, 1996; Bull and Brandon, 1998; Bull *et al.*, 1994). He argues that the lichenometric method is much more robust if one records the maximum diameters of lichens on many hundreds to thousands of rocks on the feature to be dated (he measures using a digital micrometer linked to a minicomputer in his backpack). A major problem with the previous practice was that one relied heavily upon the regularity of the statistics in the population of lichen diameters, i.e., that after sampling *n* random boulders one would record the same maximum diameter, even though *n* is relatively small. Bull's (1996) revision of the technique retains all the population statistics and builds up a dense histogram (probability density function) of the lichen diameters. This approach is facilitated by the availability of low-cost digital micrometers whose resolution is high, and from which the digital data can easily be exported as an electronic file. Bull has applied his technique in several settings, recording thousands of measurements, and developing impressive statistics. He calibrates his growth-rate curve using surfaces that are independently dated, often on the basis of tree-ring counts. From the resulting distributions of maximum lichen diameters, he argues convincingly that one can resolve discrete events in the record that are separated by a few decades or even less. Most interestingly, and most pertinent to our topic, Bull argues convincingly that he can resolve discrete pulses of rockfall material associated with seismic activity, at sites in both California (Bull, 1996) and New Zealand (Bull and Brandon, 1998). At the rockfall sites in California, the composite probability density plots of lichen diameters (Fig. 3.6) revealed several peaks, which, when calibrated using the tree-ring dated sites, correspond to historical Sierra Nevada and San Andreas earthquakes within a few years. Importantly, the breadths of the distributions allow assignment of error estimates, which in this case are ±3–10 years: a remarkably small error for ages on the order of hundreds of years, given that the clock is a biological one.

As in all techniques, limitations and uncertainties exist. Here the proper climate is required in which a well-behaved lichen thrives, and one must be able to obtain independent ages on surfaces for use in calibrating the growth-rate curve. The biology behind the method has recently been addressed by Loso and Doak (2005). They remind us that lichens in fact die. The components of a biologically sound model include the colonization of the surface we wish to date, the growth of individual thalli, and the probability of death. Combination of these essential ingredients gives rise to a

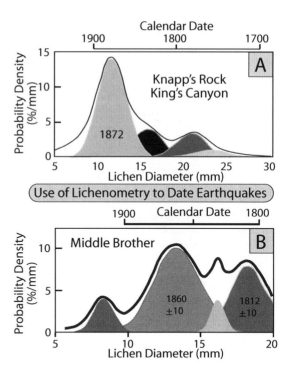

Fig. 3.6 *Rhizocarpon* lichen diameters on talus boulders in the Sierra Nevada, California.
Histograms (probability density functions) of diameters record times of generation of new talus blocks, corresponding to major historical earthquakes in California. A. Lichens in King's Canyon. B. Lichens at Middle Brother. Modified after Bull (1996).

modeled evolution of the population statistics (essentially the demographics of the population). They show that the demographic evolution depends upon the individual species, in that they will differ in one or another of the biological components (Fig. 3.7). They also demonstrate that other population statistics are more robust as a measure of surface age than is the maximum diameter – although, of course, such information is much more difficult to document.

Some studies (e.g., Porter, 1981) have suggested that lichen growth rates may be higher on fine-grained volcanic substrates than on coarser-grained intrusive rocks, whereas other studies conclude that lithology, smoothness of the substrate, local mean annual temperature, precipitation, and length of the growing season apparently do not exert a strong influence on

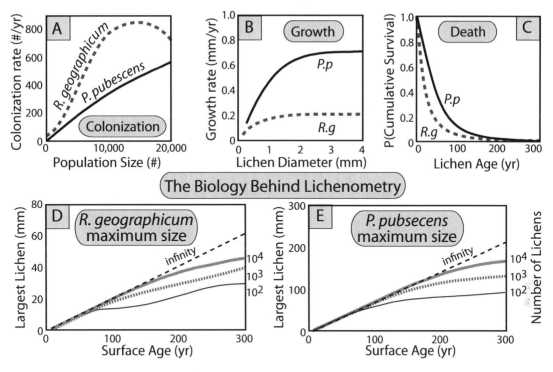

Fig. 3.7 The biology behind lichenometry.
A–C. Rates of colonization, growth, and cumulative survival estimated for *Rhizocarpon geographicum* (dashed lines) and *Phyllostachys pubescens* (solid lines) populations at Iceberg Lake, Alaska. The growth rates of both species B are lowest in the smallest lichens, as biologists predict. D, E. Results for *R. geographicum* and *P. pubescens* from many simulations of an evolving lichen population that predicts the maximum lichen diameter to be measured by randomly sampling the population on surfaces of increasing age. Dashed black line shows the largest possible thallus in an infinitely large sample. Mortality ensures that large lichens are extremely rare on old surfaces, so in practice the largest lichen increasingly falls off from this theoretical maximum as sample size decreases, giving rise to the "great growth" pattern displayed in most lichenometric dating curves. Modified after Loso and Doak (2005).

the growth rate. Protection from the sun and wind does promote more rapid growth, such that Bull (1996), for instance, measured only exposed lichens. Harking back to the biological model, this dependence translates into spatial variation in colonization rates, growth rates, and probabilities of death that are likely functions of exposure of the site to one or another environmental variable. Whereas further testing of the population techniques for lichenometry is warranted, this approach holds remarkable promise for developing rather detailed chronologies on geomorphic surfaces less than 500 years old. This age range is of great interest to paleoseismology, but one in which absolute age control is often difficult to achieve.

Absolute dating methods

Most absolute dating methods rely on some process that occurs at a regular rate – in essence, a clock. In some instances, these clocks leave a physical record that may be biological, as typified by tree rings, or geological, as represented by annual lake beds or varves. The remaining techniques rely on either atomic clocks or cosmic clocks or both. Here we review the fundamental concepts of the atomic clocks we will use.

Some (parent) atoms spontaneously decay through fission to other (daughter) atoms plus associated nuclear fragments plus energy. These events are called radioactive decays. Whereas any particular parent atom may decay at any

Table 3.2 Absolute dating methods.

Method	Useful range	Materials needed	References
Radioisotopic			
^{14}C	35 ka	Wood, shell	Libby (1955), Stuiver (1970)
U–Th	10–350 ka	Carbonate (corals, speleothems)	Ku (1976)
Thermoluminescence (TL)	30–300 ka	Quartz or feldspar silt	Berger (1988)
Optically stimulated luminescence (OSL)	30–300 ka	Quartz silt	Aitken (1998)
Cosmogenic			
In situ ^{10}Be, ^{26}Al	0–4 Ma	Quartz	Lal (1988), Nishiizumi *et al.* (1991)
He, Ne	unlimited	Olivine, quartz	Cerling and Craig (1994)
^{36}Cl	0–4 Ma		Phillips *et al.* (1986)
Chemical			
Tephrochronology	0–several Ma	Volcanic ash	Westgate and Gorton (1981), Sarna-Wojcicki *et al.* (1991)
Amino acid racemization	0–300 ka, temperature dependent		
Paleomagnetic			
Identification of reversals	>700 ka	Fine sediments, volcanic flows	Cox *et al.* (1964)
Secular variation	0–several Ma	Fine sediments	Creer (1962, 1967), Lund (1996)
Biological			
Dendrochronology	0–10 ka, depending upon existence of a local master chronology	Wood	Fritts (1976), Jacoby *et al.* (1988), Yamaguchi and Hoblitt (1995)
Sclerochronology	0–1000 yr	Coral	Buddemeier and Taylor (2000)

random instant, the probability of such decay depends on the parent–daughter pair. The lower the probability of decay at any instant, the longer it will take for a population of parent atoms to decay to half its size, and vice versa. The time required to reduce the parent population by half defines the *half-life*. Mathematically, the process is captured in the differential equation

$$dN/dt = -\lambda N \tag{3.1}$$

This equation is a commonly used example of a first-order linear differential equation found in the front of many introductory differential equation textbooks. Here N is the number of parent

atoms, dN/dt is the rate of change of this number, and the decay constant λ expresses the probability of decay of any parent atom at any instant. The solution to this equation is an exponential function:

$$N = N_0 e^{-\lambda \tau} \tag{3.2}$$

where N_0 is the initial abundance of the parent atom at time $t = 0$. You can easily see that the time, τ, it takes for the decay of the population from N_0 to N_0/e (recalling that e = 2.718..., this is roughly $N_0/3$) is $1/\tau$. The half-life, found by solving for the time at which $N = N_0/2$, is $t_{1/2} = -\ln(1/2)/\lambda$, or $0.693/\lambda$. In Table 3.2, we list a few of the commonly used radioactive isotopes, their decay constants, and their associated half-lives. We will see that in some cases it is either more instructive or easier to measure the accumulation of the daughter isotopes, or the ratio of the parent to the daughter, rather than the parent.

Tree rings

An entire science of dendrochronology has evolved around the dating of geomorphic surfaces using tree rings and the exploration of climatic change using the tree-ring width as a surrogate for climatic stresses. In climates with a distinct growing season, trees put on new wood in discrete layers that may be counted to reveal the age of the tree. Growth is rapid in the wet season and the summer, then slows in the winter, resulting in relatively low-density wood with large cells in the growth season, separated by layers of denser, finer wood. Whereas the easiest application is obviously to date a surface with the trees still growing (by cutting down the tree or by coring the tree with a boring device), the technique has been augmented greatly by extending the time series derived from living trees through the addition of time series from dead trees. Tree-ring thickness and the density of the wood in the rings are documented using any of several methods, including X-ray images and now even blue-light intensity (e.g., Campbell *et al.*, 2007). As the thickness or width of the annual growth ring is dictated by the climatic conditions of water availability and temperature, tree-ring width time series can be viewed as a proxy for a climatic time series. That this climate time series reflects at least a regional, if not a global, phenomenon means that trees far separated should record the same temporal pattern of ring widths. We note that one must take care to account both for the species of the tree and for the long-term trend (decrease) in the rate of ring growth through time as any tree matures (Fig. 3.8). A ring-matching technique is, therefore, based on a master tree-ring time series that is compiled from a set of trees whose ring-width series overlap. For example,

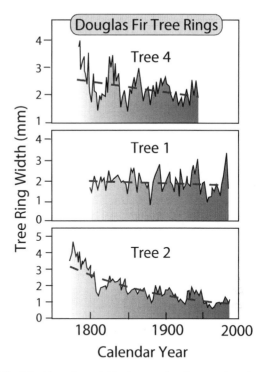

Fig. 3.8   Tree-ring widths in Douglas firs as a function of time in the Pacific Northwest of the United States. The tree-ring width shows decline in growth rate with age, best seen in Tree 2 (dashed line). Superposed on this trend are the modulations in growth rate associated with variations of local climate. Modified after Yamaguchi and Hoblitt (1995).

Yamaguchi and Hoblitt (1995) demonstrated the time of burial of trees on Mount St. Helens by debris from prehistoric volcanic eruptions by matching the ring-width time series from trees embedded in or buried by lahars with that of the master tree-ring chronology. This methodology entailed choosing a date for the volcanic event (= the age of the outermost ring), calculating the correlation coefficient between the sample tree-ring width series and that of the master tree-ring series, choosing another sample age, recalculating the correlation, and so on. The resulting plot of correlation versus outermost ring age reveals a strong spike in a particular year, which is interpreted to be the correct match (Fig. 3.9). The technique relies strongly on a trustworthy master tree-ring time series and works best for dating large trees with

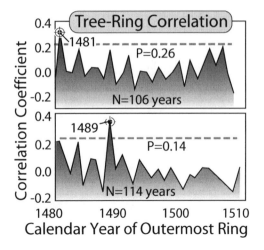

Fig. 3.9 Correlation of tree-width time series with the master tree-ring time series as a function of chosen start year.
Top series shows strongest correlation for a start date (outermost ring) of calendar year 1481, whereas bottom series is strongest for start date of calendar year 1489. Correlation coefficients for all other start years have probabilities of random occurrence that exceed the labeled probability, *P*. Modified after Yamaguchi and Hoblitt (1995).

longer ring-width time series to test against the master chronology.

Tree-ring studies have also provided the basis for calibration of the radiocarbon method (see following section). The carbon from a particular tree ring of known age (from counting backward from the bark) can be ^{14}C dated using recently developed techniques that permit dating of very small samples. Such calibration is feasible for about the past 10 000 years, the longest compiled tree-ring record.

Detailed studies of tree rings have allowed geoscientists to place very precise dates on important short-lived prehistoric geological events, including rockslides, volcanic flows, and even earthquakes. We will revisit this technique in our discussion of prehistoric Cascadia earthquakes.

Radiocarbon dating

The most commonly used technique to date geomorphic features and surfaces that are less than about 30–40 ka employs carbon-14 (^{14}C), known also as radiocarbon because it radioactively decays. Both ^{14}C and ^{13}C are formed in the atmosphere by cosmic radiation interactions with atmospheric nitrogen (N), resulting in a mixture of atmospheric carbon that is roughly 98.9% ^{12}C, 1.1% ^{13}C, and $1.17 \times 10^{-10}\%$ ^{14}C, most of it in the form of CO_2. Because the atmosphere is well mixed on short time scales, this relative abundance is uniform around the globe – although we note that, in the very short term, the plumes from coal-fired power plants result in low ^{14}C content in the air, as all CO_2 from the power plant is ^{14}C-dead (see Turnbull *et al.*, 2007). From this atmospheric pool, the carbon in CO_2 is fixed through photosynthesis in plants. Plants should, therefore, be made of organic carbon that mimics the isotopic ratio in the atmosphere during growth, although the plant's ratio is offset from that of the atmosphere due to fractionation during photosynthesis. Once a plant or animal cell dies, it no longer incorporates new carbon into its organic material. Radioactive decay takes over and begins to modify the ratio of the carbon isotopes in the cells. So, it is upon a cell's death that the clock starts. We will see that the ^{14}C dating technique is intertwined with at least two other methods, U–Th and tree-ring dating, demonstrating how interdependent some of these techniques have become.

Two principal means exist for measuring the ratio of ^{14}C to ^{12}C in a sample. *Conventional* dating is done by counting the decays of ^{14}C to its daughter ^{14}N through emission of an electron: a beta (β) particle. The rate of decay, as represented by the beta particles, is commonly measured in decays per minute per gram of carbon and requires samples on the order of a few grams of carbon in size. For old samples, it is necessary to count decays for long periods of time to record enough decays for statistical reliability, and the facilities in which such measurements are made must be shielded from other radiation to avoid miscounting. Even so, one can see easily from the graph in Fig. 3.10 that the inevitable finite errors associated with the isotopic measurement grow in importance with sample age. For example, we cannot distinguish between a sample of, say, about 40 ka and one of infinite age. This

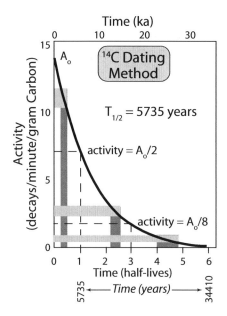

Fig. 3.10   Decay of ^{14}C concentration with time follows classic exponential curve.
Shaded bars show transformation of uncertainty in activities to uncertainties in age. Uncertainties become very large for low activities. Ages greater than about seven half-lives cannot be distinguished from infinite ages. Modified after Olsson (1968).

limitation exists because, if the sample is older than about seven half-lives, a little less than 1% $((1/2)^7 = 0.8\%)$ of the original ^{14}C is left in it – so with the half-life of ^{14}C being 5735 years, this is about 40 ka. For example, a gram of modern carbon experiences about 15 decays per minute. If this sample were 40 kyr old, it would have spanned about seven half-lives and would generate about $15/2^7 = 15/128$ decays per minute or only about 7 decays per hour. Although such low rates of decay *can* be counted, incoming cosmic radiation can also trigger the detectors, and consequently causes large uncertainties in the detector counts attributable to decay of ^{14}C. (This is why ^{14}C labs are often either deep underground or lead-shielded or both.)

Although measuring the decay rate remains the dominant (and far cheaper) technique, it has been augmented tremendously within the last decade or so by the development of *accelerator mass spectrometer* (AMS) techniques. In AMS dating, the carbon is measured according to its isotopic weight, not according to its radioactivity, or rate of decay. This approach means that one counts all the carbon atoms in a sample, not just those that decay, so that the sample size can be dramatically reduced to a milligram. When AMS dating was first introduced, it was hoped that AMS would also allow extension of the age range of the technique to many more half-lives. In practice, the technique is still not routinely capable of extending beyond about 50 ka. Both commercial and academic labs will run samples using either the conventional or AMS method.

Although ^{14}C dating revolutionized archeology and allowed scientists to establish the absolute timing from organic remains for the first time in the 1950s, several pitfalls must be acknowledged. In order to estimate the age of an object using ^{14}C, one must know: (i) the starting ratio of ^{14}C/^{12}C; (ii) the decay rate; and (iii) the final ratio. We know the decay rate well, characterized by a half-life of 5735 years, and can measure the present ratio well using AMS or can estimate it based on measured rates of decay. However, because ^{14}C is created in the atmosphere by cosmic radiation, any variations in the production rate of ^{14}C result in a different starting ratio in the organic material. Variations in production rate result from several sources, chiefly variations in the cosmic-ray flux, which are associated with fluctuations of the Earth's and the Sun's magnetic fields. Because our measurements of these fluctuations have lasted for only a few decades, we must rely on calibration. Happily, as mentioned previously, tree rings provide an organic record that can be pieced together back several thousand years – in the case of bristlecone pines, 10 kyr. It is from this record that we know the ^{14}C production-rate record, which is itself of interest to those studying the magnetic fields of the Sun and Earth. With a ^{14}C ratio in hand, from which a radiocarbon age is calculated using the half-life, one may turn to published computer codes (e.g., Stuiver and Reimer, 1993) that allow translation to calendar years (generating a "calibrated radiocarbon age"). The problems are largest at young ages, where the conversion between radiocarbon ages and calendar ages is typically ambiguous (Fig. 3.11): a given sample could be either x or y years old, owing to a large fluctuation in the ^{14}C production

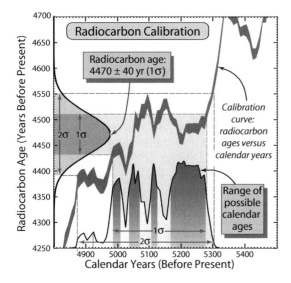

Fig. 3.11   Radiocarbon calibration and ambiguity of ^{14}C ages.
The calibration curve of radiocarbon age versus calendar age is used to estimate the calendar age represented by a radiocarbon date of 4470±40 years. Despite a 1σ uncertainty of only 40 yr in the radiocarbon date, the range of calendar ages consistent with that date is ~600 yr older and extends across ~300 yr (or up to 400 yr for a 2σ uncertainty). This large range of ages can be narrowed somewhat if either smaller analytical uncertainties or multiple dates for the same event can be obtained. Data derived from the CALIB radiocarbon calibration program of Stuiver *et al.* (2009).

rate within the last couple of millennia. This tree-ring calibration method is valid to about 10 ka, beyond which another calibration method is needed. Here we turn to the precise dating of calcium carbonate deposits (in particular, corals) by both U–Th dating and AMS with ^{14}C.

Uranium–thorium (U–Th) dating

The uranium decay series represents a very trustworthy set of clocks. The series consists of a set of several isotopes that are generated from their parents at varying rates and that decay to their own daughters at yet other rates (Fig. 3.12). The ultimate parents of the chains are ^{238}U, ^{235}U, and ^{232}Th. Decays take place at a statistically steady rate that is independent of temperature, of the magnetic fields of the Earth and of the Sun, and of all other environmental factors that

are the bane of many geological and biological clocks, including radiocarbon. Fortunately, the decay constants for some of these pairs include those that correspond to half-lives that are very useful in neotectonic and climatic investigations.

Because the U–Th series is a more robust clock than is ^{14}C, it has been used as a means of calibrating the ^{14}C clock beyond the end of the tree-ring record (Fig. 3.13). In a clever sampling scheme, Bard *et al.* (1990, 1998) used the calcium carbonate from submerged corals off Barbados as a means of obtaining sample pairs from the same material. The corals contained both ^{14}C as some small fraction of the carbon, and U substituting for Ca in the carbonate lattice. Accelerator mass spectrometric (AMS) measurements of the ^{14}C and accurate measurements of the U–Th revealed a consistent offset between the two clocks for samples older than the Holocene. This important result indicates that ages calculated using ^{14}C are as much as a few thousand years too young for samples dating to about 25 ka.

The substitution of U for Ca in carbonate lattices makes shells and carbonate coatings in soils prime targets for U–Th dating. Unfortunately, not all carbonate materials retain the parents and the daughters in their lattice through time – that is, they are not "closed systems." In particular, most shells (for instance, typical clams) leak uranium, making them essentially worthless as clocks. Luckily, most unrecrystallized corals work well, both the colonial and solitary types. In Barbados, Fairbanks (1989) reconstructed past sea levels using *Acropora palmata*, a coral that grows within a couple of meters of sea level (Fig. 2.2). The west coast of North America sports numerous marine terraces whose ages are needed in order to determine deformation rates of the coastline. Because most of the coastline is at higher latitudes than those at which colonial corals grow, the target has been instead the solitary coral *Balanophyllia elegans*. These corals are difficult to find, as they are only a centimeter in diameter and look like wagon-wheel spaghettis. Nonetheless, they have been used by numerous workers (see review in Muhs, 1992) to identify which marine terrace was formed during which sea-level highstand. U–Th has also been

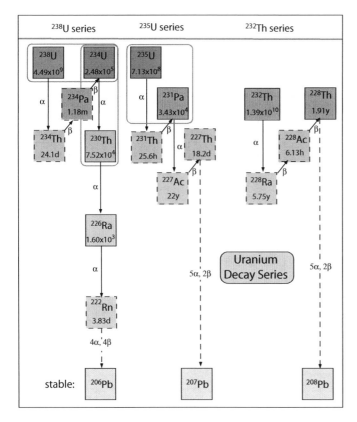

Fig. 3.12 Uranium and thorium decay chains.
Radioactive parents [238]U, [235]U, and [232]Th decay through α and β decay steps to stable daughter isotopes of Pb. Half-lives shown in boxes are in years except where noted.

used to date the carbonate coatings on pebbles, whose thickness has already been discussed as a surrogate for soil age. The method entails scraping off all but the innermost coating, adjacent to the clast upon which it is growing, in order to analyze this innermost rind. Although this method works in some desert settings (e.g., Ku *et al.*, 1979), it fails in many, owing either to leakage of one or another product from the carbonate or to severe impurity of the coatings. Such unpredictable success renders this approach as a tool of low priority, to be used only if no other can be identified. Thankfully, one can determine if significant leakage has occurred and, hence, whether the age obtained is worthy of interpretation – see, for example, Muhs *et al.* (1994) for discussion of these tests.

Whereas both societal concerns and geological efforts to understand well the most recent tectonic events often require that we focus on the past thousand years, this interval is difficult to date accurately using [14]C (Fig. 3.11). Although

the measurement precision in the best labs may be as little as 10–20 years, the calendar uncertainty is commonly many decades due to the fluctuations of atmospheric ^{14}C during the past 1000 yr – see Atwater *et al.* (1991) and Sieh *et al.* (1989) for examples of high-precision ^{14}C dating. High-precision U–Th dating of corals that grew during this same interval, however, can have calendar uncertainties of <10 yr, and thus can provide a detailed time resolution that was previously unattainable with most other techniques (Edwards *et al.*, 1988, 1993).

The production rate of ^{14}C has been even further constrained by yet another method, which employs the high-resolution sediment core from the Cariaco Basin, South America (Hughen *et al.*, 2004). High-precision ^{14}C dating of the sediment core now extends more than 50 000 years (Fig. 3.14) and reveals significant departure of the ^{14}C clock from calender years that is greatest during the last major glacial. The offset history is best explained by appeal to variations in the

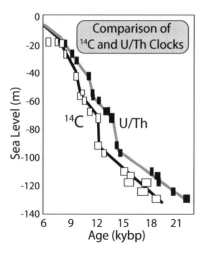

Fig. 3.13 Paired U–Th and radiocarbon ages of corals.
The sampled corals from Barbados grew during rise
of sea level from its Last Glacial Maximum of
roughly −150 m to within 10 m of modern levels. Dates
obtained using both radiocarbon and U–Th series
methods are shown. Two strong spikes in sea-level rise
rates deduced from high slopes on the plot (glacial
meltwater pulses) correspond only if the radiocarbon
ages are systematically shifted to older ages. The U–Th
clock is more trustworthy given its independence from
fluctuations in the rate of cosmic-ray bombardment.
Modified after Bard *et al.* (1990).

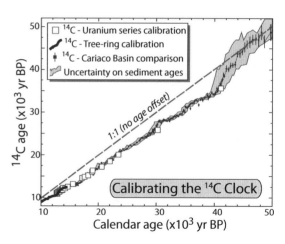

Fig. 3.14 Calibration of ¹⁴C with ocean sediment cores.
High-precision ¹⁴C age of sediment versus age model
derived from correlation of climate proxies from Cariaco
Basin sediment and GISP2 ice core, Greenland. Tree-ring
and U–Th calibration points are also shown, and agree
with the sediment core-deduced offset history. Maximum
offset of roughly 5000 years at about 40 ka is significant
throughout the last glacial interval and declines to the 1 : 1
line in the Holocene. Modified after Hughen *et al.* (2004).

Earth's magnetic field, which modulates the production rate of ¹⁴C in the atmosphere.

Amino acid racemization

All fossils contain at least trace amounts of
organic matter than can be retained for long
periods of time. The proteins that constitute this
organic material in both skeletal and shell material are themselves composed of large numbers
of amino acids. After death, the amino acids in
these proteins are altered by a set of chemical
and physical processes. The degree to which
this alteration has taken place can be used as a
clock (Bada *et al.*, 1970). Although transformations represent a very complex set of processes,
some transformations appear to be reliable
enough to provide both a relative and an absolute dating method (Kaufman *et al.*, 1992). Many
types of organic materials have been used, including bivalves, gastropods, foraminifera, coral, and

the shells of birds. In addition, the gas chro-
matographic methods used for measurement are
relatively inexpensive, making the analysis of
large numbers of samples possible. Only about
2 mg of sample are needed.

Living organisms utilize amino acids only in a
left-handed (or L, for levo-) configuration of their
isomers. Upon the death of an organism, these
restrictive biological processes are terminated,
which frees the amino acids to racemize (flip) into
their right-handed (D, for dextro-) isomeric state.
The ratio of D/L configurations is, therefore, a
clock. The reaction behaves as a first-order revers-
ible chemical reaction in which an equilibrium is
established when the backward reaction (D to L)
balances the forward (L to D). As in all other
chemical reactions, the rate constants are
temperature-dependent, reflecting the Arrhenius
relation. Like the radioactive and cosmogenic
dating methods that entail both production and
decay, this method is commonly restricted to the
period of time over which the measurable ratio
(here the D/L ratio) is changing significantly. The
time to equilibrium varies from one amino acid to
another, and depends strongly upon the thermal

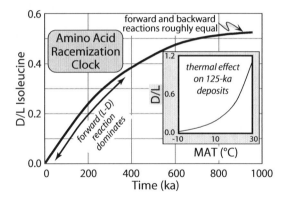

Fig. 3.15 Theoretical curve of amino acid racemization through time.
For early times, L–D transitions dominate, whereas back-reactions D–L become roughly equal at later times. Inset shows effects of higher mean annual temperatures (MAT) on reaction rates, and hence of D/L ratio expected at a given age (125 ka). Modified after Kaufman and Miller (1992) and Hearty and Miller (1987).

history to which a sample has been subjected subsequent to death. For instance, isoleucine attains its equilibrium value of about 1.3 in 100–300 kyr in equatorial sites, whereas in the Arctic it may take 10 Myr (Miller and Brigham-Grette, 1989) (Fig. 3.15). The thermal dependence is both a problem to be dealt with in the absolute dating of a sample, and a paleoclimatic tool in itself if the date is already known through other dating or correlation methods (inset in Fig. 3.15). Because the rate constant is apparently influenced by the type of animal (the taxonomic effect), researchers typically focus on a few common genera.

Typical application of racemization to absolute dating involves sampling a site of known age (say, a Holocene site that has been independently dated using ^{14}C) and a nearby site that is presumed to be Late Pleistocene. The same fossil genera are collected from each site. The D/L ratio of the Holocene sample yields an empirical rate constant that is both geographically and taxonomically specific. The interpretation of the Late Pleistocene D/L ratio takes into account the fact that the average temperatures since the Late Pleistocene are likely lower than the average since the Holocene sample was deposited. This temperature contrast indicates that the use of the Holocene rate constant should

yield an underestimate of the Late Pleistocene sample age. This tool has been recently exploited to tell the tale of the demise of giant birds in Australia upon the arrival of humans on that continent (Miller *et al.*, 1999, 2005).

Luminescence dating

The applicability of many of the dating techniques discussed thus far is often limited by the availability of suitable material for dating, such as carbonaceous debris, shells, or corals. In most terrestrial settings, shells or corals are absent, and even where carbonaceous material is preserved, the age of the deposit of interest may exceed the practical range of [14]C. Luminescence dating has the potential both to be broadly applicable to terrestrial deposits and to extend well beyond the age limitations of radiocarbon dating.

The basis of luminescence is the trapping and release of energy by electrons within crystals (Aitken, 1985, 1998; Duller 1996; Wintle, 1993). When energy is added to electrons through radiation, they tend to move from a lower energy level (the valence band) to a higher energy level (the conduction band). Some of these energized electrons can be trapped metastably by crystal defects between these two energy levels. Trapped electrons can be made to drop back to their valence band upon the addition of a small amount of energy, at which time they release photons of light equivalent to the change in their energy state. This released light is termed *luminescence*. The radiation that initially causes the energy state of electrons to increase within a crystal comes largely from the decay of nearby radioisotopes within a sedimentary deposit. That this rate of decay should be approximately constant is the basis of the luminescence clock. The electrons can be reset (zeroed) to their low-energy state by exposure to either intense heat or sunlight. When luminescence is caused by the addition of heat in the laboratory, it is termed *thermoluminescence* (TL), whereas when it results from the addition of light, it is termed *optically stimulated luminescence* (OSL).

In TL dating, luminescence is measured as the sample is incrementally heated (Fig. 3.16A). In OSL, the sample is exposed to light of a certain

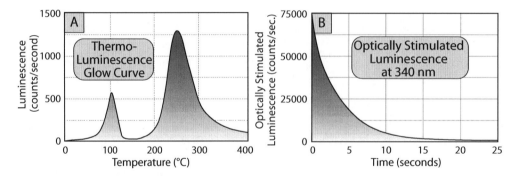

Fig. 3.16   Thermal and optically stimulated luminescence.
A. Example of a thermoluminescence "glow curve" that results from the progressive heating at a rate of a few degrees per second. In this quartz sample, most of the traps are emptied by 400°C. The peak at 100°C results from optical stimulation for 25 s at 420–560 nm (shown in B) and is not seen in the natural sample. B. Luminescence signal recorded at ~340 nm due to optical stimulation of the same sample as in A using a light source with wavelengths ranging from 420 to 560 nm. Modified after Duller (1996).

wavelength and the emitted luminescence is measured as a function of time (Fig. 3.16B). OSL has some important advantages over TL that are making it a generally preferable dating technique (Aitken, 1998). In particular, TL measures luminescence signals that are sensitive to light, heat, and any other types of energetic stimulation, whereas OSL measures only signals that are sensitive to light. In addition, TL destroys the signal of interest during measurement and, therefore, cannot be repeated, whereas OSL can be applied in short bursts that modify the total luminescence only slightly, such that multiple measurements can be made.

Both the number of electrons residing in "traps" in a crystal and the intensity of luminescence released in the laboratory due to exposure to heat or light are a function of the total dose of radiation received by the sample over time. If one can quantify that past radiation exposure, P, termed the "paleodose," and the rate (DR) at which the sample was irradiated, then an age, t, for the sample can be calculated:

$$t = P\,(\text{Gy})/DR\,(\text{Gy/kyr}) \qquad (3.3)$$

where Gy (grays) is the SI unit for radiation.

The dose rate is typically measured either *in situ*, by leaving a radiation detector for an extended time (one year, for example) in the sediment from which the sample was collected, or by laboratory measurements of the sample itself. Ideally, samples should be collected from the center of homogeneous beds that are at least 60 cm thick, because a 30-cm radius defines the approximate volume that will produce most of the radiation received by a sample. Collecting from such thick, homogeneous beds is especially important when the dose rate is to be determined in the laboratory.

The paleodose can be calculated in several ways (Duller, 1996). The additive dose method relies on measuring the luminescence resulting from different levels of irradiation in the laboratory. A curve drawn through these points and through the point defined by the sample's natural luminescence is extrapolated to zero to estimate the paleodose (Fig. 3.17A). In the regenerative method, several subsamples are measured for their natural luminescence and the remainder are zeroed through exposure to light, then exposed to known doses of radiation and remeasured for their luminescence. The curve developed from these measurements can be matched against the observed luminescence to estimate the paleodose (Fig. 3.17B). In the partial bleach method, some subsamples are subjected to the additive dose method, whereas others are exposed to a short burst of light (partial bleaching) in order to remove a proportion of their light-sensitive luminescence prior to measurement. The paleodose is defined by where these two extrapolated curves cross (Fig. 3.17C).

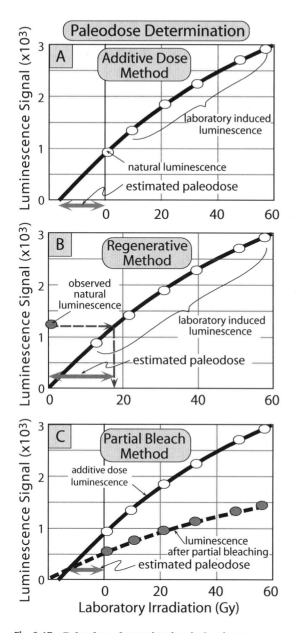

Fig. 3.17   Paleodose determination in luminescence dating using three methods (see text).
A. Additive dose method. B. Regenerative method.
C. Partial bleach method. Modified after Duller (1996).

When luminescence dating is applied to sediments, a key assumption is that the traps were emptied or "zeroed" prior to deposition, such that each grain began its life within the deposit with no record of previous radiation. For OSL, prolonged exposure to sunlight is sufficient to empty the traps, whereas for TL, sunlight will partially, but not completely, remove the thermally stimulated luminescence. In either case, one must assess the sedimentological setting in the field, asking whether it is likely that there was sufficient exposure to sunlight during transport to empty the traps. Loess and eolian sands, both of which blow around on the surface, are good bets to have been zeroed. Lacustrine silts and clays, which are transported as suspended sediments and should receive sunlight in the upper parts of the water column, also seem to yield reliable dates. On the other hand, sediment that traveled as bedload or was transported in mass flows would be generally poor candidates for luminescence dating.

The range over which luminescence dating can be successfully applied theoretically extends from less than 1 kyr to more than 1 Myr. Using OSL and very sensitive detectors, eolian sediments only a few hundred years old have been dated (Wolfe *et al.*, 1995). Unfortunately, for dating older strata, during prolonged exposure to radiation, the trapping sites may eventually become saturated; their luminescence signal will cease to increase linearly despite additional radiation. Sediments derived from highly radiogenic rocks, like granites, will experience higher dose rates and reach saturation long before sediments derived from rocks like carbonates that emit low levels of radiation. Practically speaking, high dose rates may limit luminescence dating in granite source areas to less than 60–200 kyr. Nonetheless, this technique has the potential to fill a critical gap beyond the range of radiocarbon dating. Moreover, because target minerals of quartz and feldspar are abundant minerals in many strata, these techniques can be applied to eolian and lacustrine beds where few other means of dating may be possible. Some studies have tried to use luminescence dating in alluvial fans and colluvial wedges associated with faults. In such settings, zeroing of the sediment by sunlight prior to deposition cannot safely be assumed. Use of an additional dating technique, such as radiocarbon, would have to be employed on at least some pairs of samples to validate the luminescence dates.

Cosmogenic radionuclide dating

Within the last three decades, the nuclear physics community has introduced the geological community to a new technology that allows dating of bare bedrock surfaces and of alluvial deposits that have been continuously exposed to cosmic radiation since formation. These techniques are evolving rapidly. For the first decade, much of the work was performed in geomorphic situations in which both the geomorphic and nuclear physics communities could benefit. Using surfaces previously dated using other methods, the production rate resulting from cosmic-ray bombardment of rock was calibrated, showing a strong dependence on both altitude and latitude. Since then, surfaces of unknown age have been dated in a wide variety of settings. Because the materials being dated are commonly available rocks and because the time scale over which the technique may be applied covers the entire Quaternary, this new technique is often the only method available. Here we briefly review the theory of the use of cosmogenic radionuclides (CRNs) in a variety of geomorphic settings. This treatment is by no means exhaustive, but is intended to serve as an introduction to this new and exciting field. More detailed reviews may be found in Lal (1991), Morris (1991), Bierman (1994, 2007), Cerling and Craig (1994), Granger and Muzikar (2001), and Gosse and Phillips (2001). We will see that the major problems facing the community lie in the interpretation of the CRN concentrations, that considerable care needs to be exercised in sampling appropriately, and that much effort must be put into developing a relevant geomorphic model of the site. Although much has been accomplished in this new field, the techniques continue to evolve, and the applications of the methods continue to broaden.

Background

We summarize here and in the cosmogenic primer (Box 3.1) the concepts that underpin the essential uses of cosmogenic radionuclides in geomorphological studies. Cosmic rays isotropically bombard the Solar System, meaning they come in essentially uniformly from all angles. Being charged, they are steered by the magnetic field of the Earth, generating a stronger beam of particles at high geomagnetic latitudes and reducing the downward flux at lower latitudes. These particles interact with atoms in the atmosphere and, thereby, create such familiar species as ^{14}C. These interactions reduce the number of energetic particles that penetrate to lower levels in the atmosphere: the production rate of radionuclides declines with a $1/e$ length scale of roughly 1.5 km within the lower atmosphere. Owing to this atmospheric attenuation, production rates of cosmogenic nuclides at altitudes of 3 km will be e^2 (about 7) times higher than those at sea level. These nuclides generated in the atmosphere have been dubbed "garden-variety" CRNs or meteoric radionuclides. Cosmic rays that survive to impact the surface of the Earth are capable of producing CRNs in near-surface materials: *in situ cosmogenic radionuclides*. The production rate is highest at the surface, and it decays with depth in bedrock with a $1/e$ scale of roughly 50–70 cm, the difference in attenuation length scales reflecting the relative density of rock and air. Minerals comprising atoms that are susceptible to the nuclear reactions that generate CRNs must be present in these near-surface materials to be useful. For instance, and quite fortunately, the very common mineral quartz (SiO_2) is a target mineral for both ^{10}Be and ^{26}Al: ^{10}Be being produced from ^{18}O, and ^{26}Al being produced from ^{32}Si. See the table within Box 3.1 for a compilation of the most commonly used species, their half-lives, and the sea-level, high-latitude production rates.

Note that the production rates for most cosmogenic nuclides are only a few atoms per gram of target mineral per year (see Box 3.1). These rates result in such low concentrations that conventional mass spectrometry and counting of decays are unreasonable methods for measuring concentrations. Analyses of the concentrations of those CRNs produced *in situ* requires separation of the target mineral from the rock, ridding the sample of garden-variety CRNs by a leach step, and subsequent chemical separation of the CRNs. The minuscule sample is then analyzed in an accelerator mass spectrometer (AMS), several of which have been converted from use in physics and defense industries for application in this field.

In situ CRNs

In situ CRNs have been used in two distinct settings: bare bedrock surfaces, where one is interested in either or both the exposure age of the surface and the erosion rate of it; and depositional surfaces. Each setting poses its own dilemmas, requiring differing sampling strategies. But, commonly, there's no other game in town.

Bedrock surfaces

Consider a bare bedrock surface exposed to the full cosmic-ray flux (no blocking by nearby outcrops or valley walls). The CRN concentration in a parcel of rock is dictated by the differential equation describing both production and decay of CRNs in the parcel:

$$dN/dt = P - \lambda N \qquad (3.4)$$

where N is the number of CRNs per unit volume of rock, t is time, P is the production rate, and λ is again the decay constant, which reflects the probability of decay of the nuclide in a unit of time and is related to the half-life by $t_{1/2} = \log(1/2)/\lambda$. Much of the complexity in the interpretation of the measured CRN concentrations resides in the history of the production rate, P, to which the parcel has been subjected. Let us address a few simple examples.

Consider a bare bedrock surface exposed at $t=0$ by a thick bedrock landslide. This approach to dating will equally well apply to a sample obtained from within the headscarp, or from a boulder on the surface of the landslide that we can safely assume to have been at great depth prior to the slide we would like to date. Assume that the rock involved is much older than the half-life of the CRN we wish to employ, assuring that all CRNs produced in a prior exposure at the surface have decayed. In other words, the "inheritance" is negligible. As it is bombarded with cosmic rays, CRNs will begin to accumulate within the rock, most rapidly at the surface, more slowly at depth. At early times, the concentration, N, is so low everywhere within the rock that the rate of growth is essentially linear at the local production rate, P, i.e., $N=Pt$. Because P falls off exponentially with depth (see Box 3.1), an exponential concentration profile develops, $N = P_0 t e^{-z/z^*}$, where P_0 is the production rate at the surface and z^* represents the depth in the rock at which the production rate is P_0/e. If a sample is collected from the surface ($z=0$), and we know the local production rate at the surface, then measurement of N allows solution for the exposure time, t. As the concentration builds up, however, the second term in Eqn 3.4, representing decay, begins to play a larger role. This decay results in a decline in the rate of increase in concentration, until ultimately a balance is achieved between new production and decay. This "secular equilibrium," represented by $dN/dt=0$, limits the concentration to a maximum of $N = P/\lambda$ (see Box 3.1). Note that, in these circumstances, when the sample has reached secular equilibrium, no information about the exposure time can be extracted from a measurement of N. Happily for the geomorphic community interested in dating surfaces in the Quaternary period (roughly 1.8 million years long), this secular equilibrium takes several half-lives to be achieved. Because [10]Be and [26]Al have half-lives on the order of one million years (see table within Box 3.1), they remain useful throughout the Quaternary and, in fact, well into the Pliocene. Dating of glacially polished surfaces (e.g., Nishiizumi *et al.*, 1989) in the Sierra Nevada is one example of dating surfaces that have seen negligible erosion since deglaciation – although recent work reveals that one must be cautious to find sites in which inheritance from past interglacials has been fully removed (Dühnforth *et al.*, 2010). Exposure dating of fluvial strath terraces along the Indus River, on which the fluvial polish is still intact and where the fluted and potholed nature of the fluvially carved bed has clearly not degenerated since abandonment of the strath, is another example of dating a surface considered to be pristine (Burbank *et al.*, 1996b).

More commonly, the rock surface being sampled is eroding at some rate that we would like to determine. This erosion can take place on

Box 3.1   An *in situ* cosmogenic radionuclide primer.

Cosmic ray particles are isotropically distributed as they enter the Earth's atmosphere. The atmosphere attenuates this production such that the particle flux is lower from higher zenith angles, θ (as shown by the curve in figure A). On a flat surface, they would produce cosmogenic radionuclides at a rate P_0. Any intervening rock also blocks production (called topographic shielding), reducing the effective surface production rate to below P_0 (see figure A). Measurement of this effect is accomplished in the field by documenting the shielding angle, φ, in four to eight directions. The sample site, at a depth $z=0$, may be eroding at a long-term rate of ε.

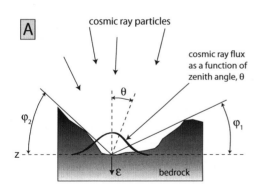

A. Topographic shielding of cosmic rays.

The case of steady erosion

As the depth of the sample decreases owing to erosion of the overlying rock, the production rate increases exponentially (see figure B). The concentration therefore increases exponentially. The concentration of a "parcel" of rock sampled at the surface is equivalent to the integral of the production rate history. The sample concentration is higher for slower erosion rates and for higher local production rates, P_0. Note that the sample

concentration is equivalent to that which the parcel of rock would have obtained had it sat on the surface for a time z^*/ε equal to the time that it takes for the rock to be exhumed from a depth of z^* at a rate ε (depicted by the gray box in figure B).

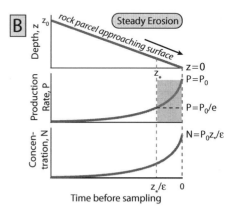

B. Sample histories, for the case of steady erosion.

Conservation of nuclides requires that the rate of change of concentration, N, equals the rate of production less the rate of decay:

$$\frac{dN}{dt} = P(t) - \lambda N$$

Production rate, P, decays exponentially with depth, z, below the instantaneous surface (see figure C) as

$$P = P_0 \, e^{-z/z^*}$$

where the surface production rate, P_0, is scaled by elevation and geomagnetic latitude. The rate of decay beneath the surface is scaled by the length scale z^* (see figure C), which is roughly 50–60 cm for most lithologies. The depth can vary with time, $z(t)$, owing to either erosion or deposition.

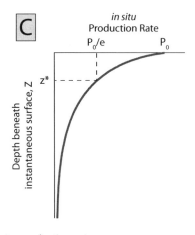

C. *In situ* production rate.

The case of no erosion: stable surfaces
When the nuclide is stable, the concentration simply increases linearly with time at a rate P_0. When the nuclide is radiogenic, the concentration approaches a steady value, or secular equilibrium, at which point the rate of decay equals the rate of production (see figure D). The achievement of secular equilibrium takes several mean lives, τ, where the mean life is set by the inverse of the decay constant ($\tau = 1/\lambda$).

The table shows some commonly used cosmogenic radionuclides, their *in situ* production rates (P_0) in atoms per gram of quartz per year at sea level in mid- to high latitudes, their decay constants (λ) and associated mean lives ($\tau = 1/\lambda$), and their half-lives ($t_{1/2}$) in years.

Properties of commonly used cosmogenic radionuclides.

Nuclide	Production rate, P_0 (atoms/ g_{quartz} yr)	Half-life, $t_{1/2}$ (yr)	Decay constant, λ (1/yr)	Mean life, τ (yr)
^{10}Be	4.6 ± 0.3 (quartz)	1.36×10^6	5.10×10^{-7}	1.96×10^6
^{14}C	16.5 ± 0.5 (quartz)	5.73×10^3	1.21×10^{-4}	8.27×10^3
^{26}Al	31.1 ± 1.9 (quartz)	7.05×10^5	1.42×10^{-7}	1.02×10^6
^{36}Cl	230 (Ca and K)	3.01×10^5	2.30×10^{-6}	4.34×10^5

[The following references have been used in preparing the contents of this box: Lal (1988, 1991), Bierman (1994), and Nishiizumi *et al.* (1989, 1991).]

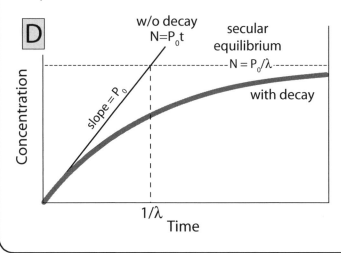

D. Evolution of concentration, for the case of no erosion.

a grain-by-grain basis, or through spalling of parcels of various finite depths representing joint spacing, for instance, or fire spall depths (e.g., Bierman and Gillespie, 1991). If the erosion is continuous and steady, the production rate in Eqn 3.4 can be rewritten

$$P = P_0 e^{-rt/z^*} \tag{3.5}$$

where r is the erosion rate of the surface, and t is time, $t=0$ being the time of sampling and t into the past being positive. Ignoring decay, for the moment, and solving Eqn 3.4 with this steady production-rate history yields

$$N = P_0(z^*/r) \tag{3.6}$$

which can be solved for the erosion rate, r. Here the term z^*/r represents the time it takes the sample to travel through the boundary layer in which the production rate is significant. Importantly, it is this time scale over which the measurement of erosion rate is being averaged in this application. The faster the parcel is exhumed, the lower the resulting concentration.

Including decay alters the equation to

$$N = P_0 z^*/[r + (z^*/\lambda)] \tag{3.7}$$

If, on the other hand, the erosion occurs in steps – due to fire spalls, for instance (Bierman and Gillespie, 1991) or to joint block removal (Small *et al.*, 1997) – then the calculation of a long-term erosion rate from employment of Eqn 3.6, and the CRN measurement of a surface sample, will vary depending on the time since the last spall event. The error in the estimate will depend on the thickness of the spall. In situations like those involving spalls, averaging of several sample sites is required to provide a more robust sense of the mean rate at which the surface is being lowered (Small *et al.*, 1997).

If one would still like to extract an exposure age from this site, and not just an erosion rate, Lal (1991) has shown that the use of two cosmogenic radionuclides with differing half-lives allows some additional constraint on the exposure age in certain scenarios. Working in the ^{36}Cl system, supplemented by ^{10}Be analyses, Phillips *et al.* (1997) have demonstrated the usefulness of this technique in constraining the age of large boulders whose surfaces have been slightly eroded since emplacement on the surface to be dated.

Depositional surfaces

Depositional surfaces, such as fluvial and marine terraces, present their own problems, even if the surfaces can be safely assumed not to be eroding. The principal problem lies in the likelihood that the clasts being used to assess the concentration of CRNs have experienced prior exposure elsewhere within the geomorphic system before being deposited on the surface to be dated. The nuclides accumulated during this exposure are called the "inheritance." Consider first a fluvial terrace. The inheritance derives from a combination of: (i) exhumation through the cosmogenic nuclide-production boundary layer as the hillslope surface is lowered; (ii) transport within the hillslope system; (iii) transport within the fluvial system, which will entail occasional burial in fluvial bars; and (iv) final deposition on the terrace to be dated.

Tactics must be employed in the sampling of the terrace materials that allow separation of the inheritance signal. In several fluvial systems studied to date, the inheritance can represent a significant (several tens of percent) portion of the total CRN concentration measured in clasts sampled from the terrace surface (e.g., Anderson *et al.*, 1996; Repka *et al.*, 1997) (Fig. 3.18). Because the inheritance differs markedly from one surface to another, one must employ some means of constraining this inheritance on a site-by-site basis. One effective strategy is to collect samples at varying depths in a terrace profile. Each sample should consist either of sand or of a contribution of an equal amount of rock from numerous clasts buried at the same level, for example, 10 g from each of 30 clasts. Assuming that the terrace aggraded quickly so that each sampled layer was rapidly buried to near its present depth, then the age of each sample represents a combination of the time since deposition and the inheritance at the time of deposition. When samples are drawn from deep enough in the terrace, say greater than ~2 m, their concentrations should reflect almost exclusively the inherited nuclides, because CRN production rates are negligible at this depth.

One other field situation bears mentioning. On many surfaces, the topmost few decimeters have been well stirred – turbated – by one or another mechanism: rodents, earthworms, tree roots, frost churning. This churning homogenizes all constituents within that layer, including the minerals

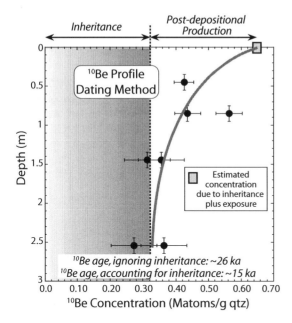

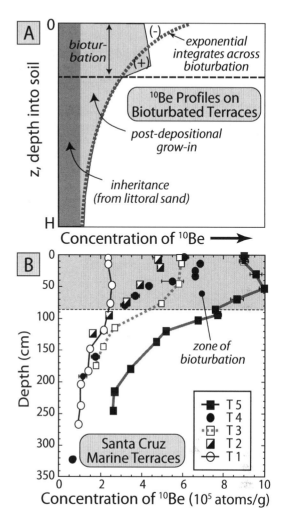

Fig. 3.18   Use of cosmogenic radionuclide concentration profile to deduce both inheritance and surface age.
Dated surface is the Stockton bar of Lake Bonneville, associated with the latest highstand of the lake, at roughly 14.5 ka. Grayed box represents the concentration due to inheritance of cosmogenic radionuclides by the quartzite clasts prior to deposition in the bar. Best-fit line is exponential shifted to account for inheritance. Age deduced from this method is 15 ka. If inheritance were not accounted for, and a single surface sample were used to deduce age, the estimated age would have been ~26 ka, too old by roughly 11 ka.

Fig. 3.19 ^{10}Be ages of bioturbated terraces.
A. Schematic profile of ^{10}Be developed in a marine terrace in the face of bioturbation of the topmost portion of the profile. The expected exponential curve of ^{10}Be concentration due to inheritance plus *in situ* production is blunted by this mixing. The integral of the post-depositional inventory of nuclides remains a faithful clock and can be used to date the terrace, because the total ^{10}Be inventory is unchanged, despite being redistributed by bioturbation. B. Examples of ^{10}Be profiles in five marine terraces near Santa Cruz, California. Lines connect measurement points on only the first, third, and fifth terraces. Each profile is blunted and shifted laterally by a different amount due to contrasts in the amount of inheritance, but the integral of the post-depositional portion of the inventory still increases monotonically with height of the terrace above sea level and, hence, with age. Modified after Perg *et al.* (2001).

carrying cosmogenic radionuclides. The resulting profile of ^{10}Be concentrations, for example, should therefore be blunted near the top. Whereas this mixing negates the possibility of fitting an exponential through the profile in order to deduce the component of post-depositional accumulation of nuclides, it has been shown that one can use instead the integral of the profile as a clock: as long as the inheritance can be constrained by deep samples, the age of the surface can still be obtained. This method was successfully employed on California's Santa Cruz marine terraces (Perg *et al.*, 2001), resulting in ages for each of the five major terraces that tell a story of roughly steady 1 mm/yr rock uplift of the coast over the last few hundred thousand years (Fig. 3.19).

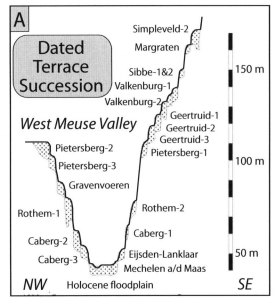

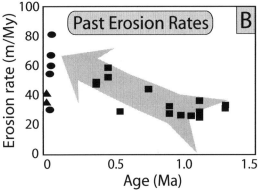

Fig. 3.20   Past erosion rates from ¹⁰Be inheritance.
A. Elevation cross-profile of the West Meuse valley, Netherlands, showing numerous fluvial terraces dated using ¹⁰Be produced *in situ*. B. Erosion rates deduced from the inherited component of the ¹⁰Be profile for each terrace, revealing a history of erosion rates in the catchment. Modified after Schaller *et al.* (2004).

One may also use the shift in the profile on a depositional surface – the inheritance – to deduce the erosion rate of the catchment at the time of deposition, using Eqn 3.5. The inheritance has been exploited (e.g., Schaller *et al.*, 2001, 2002, 2004) to derive paleoerosion rates through time from river terrace sequences on European rivers draining the Alps) (Fig. 3.20).

If for any reason a deposit containing an inventory of cosmogenic radionuclides is deeply buried, and taken out of the cosmogenic radionuclide production zone (say 4 m), another twist on the method may be employed. Here we take advantage of two radionuclides with different half-lives: ²⁶Al, with a half-life of 0.705 Myr, and ¹⁰Be, with its half-life of 1.387 Myr. We know their production rates at the surface in quartz differ by a factor of 6.75. Because the decay rates differ, this ratio will decline with time and becomes a clock once the sample is taken out of the production zone. The method based on this ratio clock is called the burial method (e.g., Granger and Smith, 2000; Granger and Muzikar, 2001). Burial dating allows us to date caves in which sediments exposed to nuclide production during hillslope and river transport in the headwaters are sequestered in the cave tens to thousands of meters below the ground surface (e.g., Granger *et al.*, 1997, 2001; Stock *et al.*, 2004, 2005) (Fig. 3.21). The burial method has also been used to date very old deposits that have been buried deeply since deposition, such as early tills in the North American glacial sequence (Balco *et al.*, 2005). Especially where recent excavations, such as deep road cuts or landslides, provide access to previously hard-to-date strata, burial ages provide a practical method to determine depositional ages (e.g., Harkins *et al.*, 2007). Because the age depends on a ratio of two different isotopic concentrations, and each concentration measurement has its own error, burial ages are not highly precise, but they allow a distinction to be drawn between strata that are, for example, 500, 700, and 900 ka (Fig. 3.21).

Landscape features may also be dated using garden-variety CRNs produced in the atmosphere and subsequently rained out in precipitation (see review in Morris, 1991). In particular, ¹⁰Be is found to be useful in that its chemistry is such that it is held tightly by clays in soils (Pavich *et al.*, 1984; see review in Willenbring and von Blankenburg, 2010). The soil, therefore, acts as a reservoir within which the ¹⁰Be slowly builds up with the age of the surface. If surfaces are well chosen to limit the role of soil erosion, i.e., are nearly flat, then the total ¹⁰Be inventory on grain surfaces

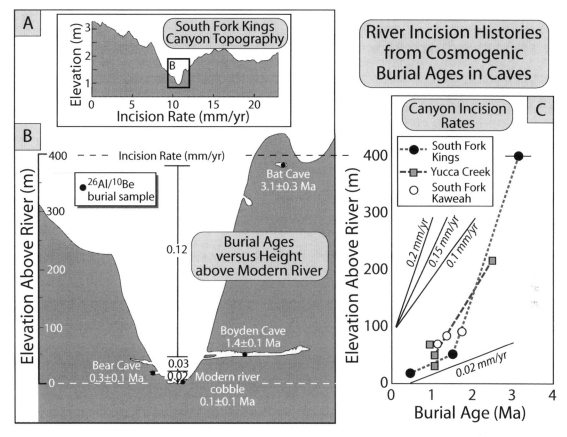

Fig. 3.21   Dating caves and estimating incision rates using the burial method.
A. Cross-section of the South Fork Kings River valley. B. Detail of canyon cross-section showing both locations of caves dated using the burial method and the incision rates derived from elevation–age data. C. History of river elevation deduced from the series of cave elevations and dates from nearby canyons in the southern Sierra Nevada. These data indicate rapid incision from 3.1 to 1.4 Ma, followed by an order-of-magnitude slower incision toward the present. Modified after Stock *et al.* (2004).

within the soil column can be used to constrain the age of the surface. Given that the map of rainout of meteoric [10]Be is now much better constrained than in the early 1990s, and that sample preparation of [10]Be targets from meteoric samples is much less time-consuming, we will likely see a resurgence of use of this method to establish timing in the landscape (Willenbring and von Blankenburg, 2010).

Using cosmogenic nuclide dates

The numbers one extracts from cosmogenic studies are, at best, only as good as the knowledge of the production rates over the age

of the surface at that site. These production-rate histories are difficult to know. Typical calibrations rest on independent dates from surfaces, such as glacially polished bedrock associated with moraines that can be dated with [14]C (e.g., Nishiizumi *et al.*, 1989). Two cautions are warranted: (i) these independent ages can be and are being reassessed with new techniques at new sites that may better constrain the true age of the glacially polished surface (e.g., Clark *et al.* (1995); and (ii) production rates estimated from these sites are, therefore, relevant to this altitude and latitude and are averaged production rates over the age of the sample. Use of these production rates for dating other surfaces must be done carefully, as the magnetic field intensity has been shown to vary significantly over the last 140 ka (e.g., Meynadier *et al.*, 1992). This record of field intensity variations will no doubt be extended by future research. The production-rate history at a sampling site should be calculated from a knowledge of the average production rate deduced from a sample of known age, the magnetic field history, and the relationship between field intensity and production rate (e.g., Clark *et al.*, 1995).

Knowledge of the production rate and its dependence on elevation (e.g., Stone, 2000), latitude, and time has greatly advanced over the last decade in part due to concerted and coordinated efforts in both Europe and North America. Among other activities, sites of independently constrained age are being used to calibrate the production-rate functions. One result is a calculator in which these functions are embedded that allows one to calculate the local production rate of a sample of given thickness and density (see Balco *et al.*, 2008). The half-life of ^{10}Be has also been reassessed (see table in Box 3.1), placing this method on yet more firm ground.

It remains the case, however, that the quality of the interpretation of the CRN concentration extracted from a sample, for either the exposure age or the erosion history at the site, is dependent on the degree to which the geomorphic processes active at that site over the history of the surface can be captured in a quantitative model that incorporates a reliable geomorphic history. Not only must the inheritance of a particular sample be assessed, but the post-depositional processes that might alter the production-rate history of the sample should be dealt with. On boulders presently at the surface, one must worry about

whether the boulder has always been at the surface, as well as the likelihood of erosion of the boulder once it was emplaced on the surface. Where range fires cause boulders to spall, samples must be collected to avoid or account for this process. One may also have to assess the role of burial by snow for some portion of the age of the surface. At present, this new set of dating techniques is generating much interest. Moreover, these new dating capabilities are opening a set of problems that provide an excellent opportunity for the marriage of new chronological tools with new modeling tools: a topic to which we turn later in the book.

Conclusions

In this chapter, we have sampled the dating methods available to provide the time scales for tectonic-geomorphic studies. The choice of method must be dictated by the availability of the proper materials, the details of the geomorphic setting, and, of course, the cost. Whereas relative dating methods are quick and dirty, they should not automatically be shunned. They do produce immediate results, whereas other techniques might take months to a year for processing of samples. In using any method, it is incumbent upon the researcher to document carefully both the geomorphic and depositional setting. Useful interpretation of the dates, sometimes painstakingly and expensively obtained, relies heavily on careful field observations: the field context and geomorphic history of a sample provide the critical framework for appropriate interpretations of any date or rate obtained from it.

4 Stress, faults, and folds

When rocks are subjected to stresses that exceed their strength, they rupture, fold, or flow. Different varieties of faults (strike-slip, normal, thrust) characterize contrasting tectonic settings and stress regimes in the upper crust. When faults break the Earth's surface either in a single earthquake or during many seismic events, they commonly create geomorphic features that can be associated with a particular type of fault. Sometimes earthquakes occur on faults that do not reach the Earth's surface, and, therefore, no ground ruptures are directly associated with the fault trace. Nonetheless, the Earth's surface will deform by folding in response to earthquakes along buried faults. The geometry of deformation and the evolving shape of a fold can reveal useful information about the nature of the subsurface faulting and the way in which the rocks adjacent to the fault respond to fault motions.

In order to take full advantage of the information to be gleaned from geomorphological surfaces, an understanding of the typical ways in which rocks deform due to both seismic and aseismic movements is helpful. A primary goal of this chapter is to examine the nature of faulting and folding and to discuss concepts related to scales of faulting and similarities between successive earthquakes on the same fault, the displacements of the ground surface that are expected for different types of faults, and the geomorphic imprint of faults of different types.

Stress, strain, and faults

Stress

All rocks in the crust are subjected to forces due to gravitational acceleration, plate tectonic motions, and the mass of rocks, water, and air around them. We typically identify two kinds of forces: *body forces* and *surface forces*. Body forces act equally on every element throughout a volume. The magnitude of a body force is proportional to the mass of the element on which it is acting, or its volume times its density. An example of a body force would be the weight of an object, which is the mass of the object times the acceleration due to gravity. Surface forces are forces that act across or along surfaces, and they include forces such as friction or pressure. Strictly speaking, these stresses are termed *tractions*, which are defined as the force per unit area acting on a surface.

A stress or traction acting on a plane with any orientation can be resolved into two components: a normal stress σ_n, acting perpendicular to the surface; and a shear stress σ_s, acting parallel to the surface. The unit of stress in the SI (meter–kilogram–second) system is the pascal. One pascal (1 Pa) represents the stress produced by one newton ($1\,N = 1\,kg\,m/s^2$) acting across or along a surface of one square meter ($1\,m^2$); thus $1\,Pa = 1\,N/m^2 = (1\,kg\,m/s^2)/m^2$. Pressure can also be expressed in bars, where

---

Tectonic Geomorphology, Second Edition. Douglas W. Burbank and Robert S. Anderson.
© 2012 Douglas W. Burbank and Robert S. Anderson. Published 2012 by Blackwell Publishing Ltd.

$1\,\text{bar} = 10^6\,\text{dyne/cm}^2 = 10^5\,\text{N/m}^2 = 10^5\,\text{Pa}$. Because one pascal represents a rather small force acting over a large area (one cubic meter of granite exerts a pressure of ~27 500 Pa on its base!), pressures are often expressed in megapascals ($1\,\text{MPa} = 10^6\,\text{Pa}$), which is also equal to 10 bars. A column of rock 1 km high would typically exert a pressure at its base of 25–30 MPa. The shear strength of many crustal rocks is in the range 10–100 MPa.

Each of the tractions acting on a rock surface and resulting from tectonic, lithostatic, buoyancy, or hydrostatic forces can be represented as a vector which can be summed with all other imposed tractions to define the total magnitude and orientation of the imposed stress on any specified plane. The total stress can be subdivided into three orthogonal components, which are typically labelled σ_1, σ_2, and σ_3, for the maximum, intermediate, and minimum principal stresses, respectively. For simplicity in discussing types and orientations of faults and associated structures, we will refer to three *orthogonal* stress vectors, in which one (σ_{zz}) is vertical and the other two (σ_{xx}, σ_{yy}) are contained in a horizontal plane (Fig. 4.1A). The vertical stress results from the rock overburden at some depth z, such that $\sigma_{zz} = \rho g z$, where ρ is the mean density of the overlying rock and g is the gravitational acceleration. This is termed the *lithostatic stress*. When the orthogonal horizontal stresses are equal to the vertical lithostatic stress, such that $\sigma_1 = \sigma_2 = \sigma_3$, this is termed a *lithostatic state of stress*. To the extent that stresses differ from this lithostatic condition, they are termed *deviatoric stresses*. In most of the situations involving faulting, the deviatoric stress ($\Delta\sigma$) results from a tectonic contribution.

Strain and faults

The presence of deviatoric stresses can cause a rock to undergo *strain*: any deformation that involves a change in size (dilation), shape (distortion), and/or orientation (rotation). Strain in rocks can be observed and quantified, whereas the deviatoric stresses that caused the strain commonly can only be inferred. Various components of strain, such as the amount of

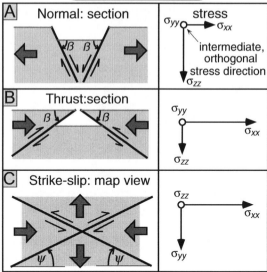

Fig. 4.1 Fault orientations with respect to principal stress orientations.
A. Normal faults: $\beta = 45° + \theta/2$, where θ equals the angle of internal friction. B. Thrust faults: $\beta = 45° - \theta/2$. C. Strike-slip faults. Modified after Turcotte and Schubert (1982).

rotation or the magnitude and orientation of extension, shortening, or stretching, can be defined through direct observations.

Not all strain in rocks is permanent. Below a certain stress threshold, termed the *elastic limit*, strain is recoverable. Thus, during the initial build-up of differential stresses, a rock will deform elastically, and, if the differential stress is eliminated during this stage, the rock will return to its original unstressed shape. Once a rock's elastic limit or *yield strength* is exceeded, it will either deform plastically or rupture; either case causes a permanent change in shape. When a rock deforms by rupture, discrete surfaces, or *faults*, are formed along which rocks are offset by movements parallel to the fault surfaces.

Consider the stresses that are responsible for normal faults. The vertical component of stress is the lithostatic pressure, $\sigma_{zz} = \rho g z$, where z is the thickness of overlying rock. For faulting to occur, an applied deviatoric stress $\Delta\sigma_{xx}$ that is tensile and, therefore, negative with respect to its lithostatic value, must exceed the yield strength of the rock. Therefore, $\Delta\sigma_{xx}<0$, the horizontal stress is less than the vertical stress, $\sigma_{xx} = \rho g z + \Delta\sigma_{xx}$, and $\sigma_{xx}<\sigma_{zz}$. If we assume that no strain (deformation) exists in the y direction, then the deviatoric stress in the y direction either is zero or is also tensile, but is some proportion (p) of the deviatoric stress in the x direction. Therefore, $\Delta\sigma_{yy} = p\Delta\sigma_{xx}$, where $p<1$. The total stress in the y direction, $\sigma_{yy} = \rho g z + p\Delta\sigma_{xx}$, is less than σ_{zz} (or equal to it, if $p=0$), but greater than σ_{xx}. Thus, for normal faulting, $\sigma_{zz}\geq\sigma_{yy}>\sigma_{xx}$. In theory, a fault plane should make an angle with the principal compressive stress of $45°-\theta/2$, where θ is equal to the angle of internal friction for the faulted material. Thus, given a typical θ value of $30°$, a normal fault formed in a relatively strong rock like granite should be inclined at about $30°$ to σ_{zz}, and the fault trend should be oriented perpendicular to σ_{xx} (Fig. 4.1A). Two different fault planes, each dipping in opposite directions at $60°$ from the horizontal, satisfy these conditions and thus represent *conjugate* fault planes. Given these stresses and this fault orientation, the *hanging wall* (the fault block that is located above the fault plane) moves down across the *footwall*, which is beneath the fault plane.

For thrust faults, a compressional deviatoric stress exists in the x direction, typically resulting from tectonic forces. Thus, $\Delta\sigma_{xx}>0$, and $\sigma_{xx}>\sigma_{zz}$. Once again, the deviatoric stress in the y direction can be considered intermediate, so that $\sigma_{xx}>\sigma_{yy}\geq\sigma_{zz}$. For a horizontally oriented maximum compressive stress, conjugate fault planes should be inclined at $45°-\theta/2$ from the horizontal and also at $45°-\theta/2$ from σ_{xx} (Fig. 4.1B). For thrusts, the hanging wall moves upward with respect to the underlying footwall.

Strike-slip faults are characterized by deviatoric stresses of opposite sign in the x and y directions. For example, if σ_{yy} is tensile, then $\Delta\sigma_{yy}<0$. Consequently, σ_{xx} must be compressive, such that $\Delta\sigma_{xx}>0$. Therefore, for strike-slip faults, either $\sigma_{yy}>\sigma_{zz}>\sigma_{xx}$, or $\sigma_{xx}>\sigma_{zz}>\sigma_{yy}$. For any stress regime in which the lithostatic stress is the intermediate stress, two conjugate vertically dipping fault planes oriented at $45°-\theta/2$ to the maximum compressive stress will accommodate strike-slip motion (Fig. 4.1C).

Clearly, faulting occurs in response to imposed stresses. But why should we care about the orientation and magnitude of local stresses? Perhaps the most important reason is that large gaps exist in our understanding of faulting within the truly heterogeneous rocks of the crust. Much of our current knowledge of the mechanics of faulting derives from laboratory studies, which commonly involve homogeneous rocks of hand-specimen size. In the real world, we need to know where and how slip occurs along a fault. How do real rocks actually deform when their yield strength is exceeded, and how are the resulting structures oriented with respect to prevailing stresses? Do stresses accumulate on a fault such that earthquakes tend to initiate in the same region and propagate in the same direction over multiple earthquake cycles? How does the stress across a fault surface change as a result of a faulting event? What is the role played by fluids? Until these questions are answered, our ability to quantify and predict deformation in earthquakes will be impeded.

The earthquake cycle

The classical model

Since at least the turn of the 20th century, geologists have attempted to understand the deformation that precedes, coincides with, and follows an earthquake. The time and deformation that encompasses an earthquake and all of the interval between successive earthquakes is termed the *earthquake cycle*. As originally described, the earthquake cycle had two parts: an interseismic interval and a coseismic one (Reid, 1910). Imagine two nearby pieces of the crust that are moving in opposite directions with respect to a fault that separates them. At some depth below the surface, the fault slips continuously and aseismically in a zone of ductile deformation, but in the brittle crust during the interseismic interval, the fault is "locked" such that no slip occurs along it (Fig. 4.2). At some distance from the fault, the rocks in the brittle zone are moving at the same rate as the crustal blocks. The amount of displacement decreases to zero at the fault, such that an originally

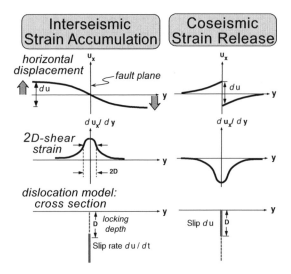

Fig. 4.2   Model of the earthquake cycle for a strike-slip fault.

The far-field strain (large arrows, top panel) of one block with respect to the other remains constant through time, as does aseismic slip (du/dt, bottom panel) in the ductile zone below the locking depth (D). Interseismic displacement is greatest farthest from the fault, but most of the shear strain (du_x/dy, middle panel) occurs within two locking depths ($2D$) of the fault. During coseismic displacement, the greatest displacement (du) occurs along and near the fault and compensates for the "slip deficit" developed during the interseismic interval. The coseismic shear strain is equal and opposite to the interseismic shear strain. Modified after Thatcher (1986b).

straight marker that was oriented perpendicular to the fault would be bent into a sigmoidal shape. The two-dimensional shear strain, which could be envisioned as the amount of bending of the formerly straight marker, is greatest near the fault (Fig. 4.2). This bending can be considered as "elastic strain" in that it is recoverable, rather than permanent. When the frictional strength of the fault is exceeded by the imposed stress, the fault ruptures during an earthquake. Coseismic displacement is greatest along and adjacent to the fault. In the context of the earthquake-cycle model, slip is just enough to balance the slip deficit and restore the marker to a linear trend, perpendicular to the fault, but now offset by the amount of relative motion of the two crustal blocks during the entire

interseismic interval. In this model, all of the elastic strain is recovered in each seismic event, so that no permanent strain remains within the blocks that slip past each other on opposite sides of the fault.

Data on pre- and coseismic deformation prior to and following the 1946 $M=8$ earthquake in the Nankai Trough of southwestern Japan provide support for aspects of this model (Fig. 4.3). Measured interseismic strain replicates the shape, but is opposite in vertical direction to the coseismic strain caused by that large subduction-zone earthquake. This behavior is clearly consistent with Reid's (1910) model. Considerable topography, however, exists across the deformed zone and suggests that coseismic subsidence does not fully compensate all of the interseismic uplift: some permanent strain is represented by this topography.

Alternative earthquake models

As a result of observations related to numerous earthquakes, Reid's (1910) simple model of the earthquake cycle has been significantly modified. First, the deformation observed along faults, such as folds and fault offsets, indicates that not all of the pre-faulting strain is recovered. In other words, as seen in the example from Japan (Fig. 4.3), some permanent, unrecoverable strain is commonly present. Second, the interseismic interval may sometimes be further subdivided into a post-seismic interval and an interseismic one. The post-seismic interval immediately follows an earthquake and is one during which strain accumulates more rapidly than in the subsequent interseismic interval (Stein and Ekstrom, 1992). In some recent earthquakes, the amount of post-seismic deformation and energy release has been shown to be equal to that of the coseismic event (Heki *et al.*, 1997). Newly recognized "slow" earthquakes (Beroza and Jordan, 1990; Kanamori and Kikuchi, 1993) in which deformation occurs over periods of hours to years can release vast amounts of energy and cause large-scale deformation without generating the catastrophic energy releases of typical earthquakes. Third, new

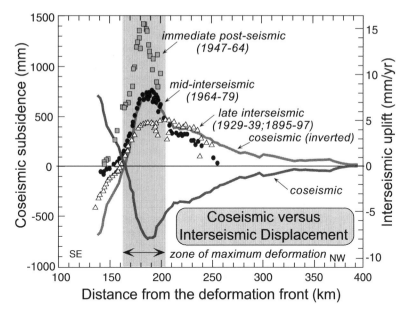

Fig. 4.3 Symmetry of interseismic, coseismic, and post-seismic deformation.
Three intervals of interseismic or post-seismic deformation are represented in the Nankaido region, Japan: immediate post-seismic (1947–64); mid-interseismic (1964–79); and late interseismic (1895–97, 1929–39). These data are compared with the inverted coseismic deformation profile. Note the spatial coincidence of the peak zone of deformation for each data set. The regional pattern of interseismic deformation closely mimics the inversion of the coseismic deformation. Modified after Hyndman and Wang (1995).

variations on Reid's (1910) periodic earthquake cycle have been postulated. In Reid's *periodic model*, the frictional strength of the fault, the stress drop (the difference in stress across the fault prior to and following the earthquake), and the slip associated with each earthquake are constant from event to event, such that both the time of each earthquake and its magnitude are predictable (Fig. 4.4A). In a *time-predictable model*, earthquakes always occur when a critical stress threshold is attained (Fig. 4.4B). The amount of stress drop and the magnitude of slip, however, varies from one earthquake to the next. Assuming that the accumulation rate of interseismic strain is constant and that the amount of slip in the previous earthquake is known, this model permits a prediction of the time until the next earthquake, because the stress needed to attain the failure stress is known. The displacement of that forthcoming earthquake, however, is unknown. Alternatively, a *slip-predictable*

model suggests that slip during an earthquake always terminates when the stress has dropped to a critical level (Fig. 4.4C). The stress level and strain accumulation at the time of rupture, however, vary between earthquakes, as does the time between successive ruptures. Thus, given a constant interseismic strain accumulation rate, knowledge of the time since the previous earthquake permits prediction of the amount of slip that would occur at any given time. But, it does not reveal when the next earthquake will occur, because no critical stress threshold is common to successive earthquakes. Finally, recent observations suggest that earthquakes may occur in temporal clusters, such that a cascade of earthquakes occurs over a relatively short interval followed by prolonged intervals of quiescence (Rockwell *et al.*, 2000; Wallace, 1987). The *clustered-slip model* (Fig. 4.4D), also called the "Wallace-type" model (Friedrich *et al.*, 2003), suggests that when a cluster will initiate is unpredictable,

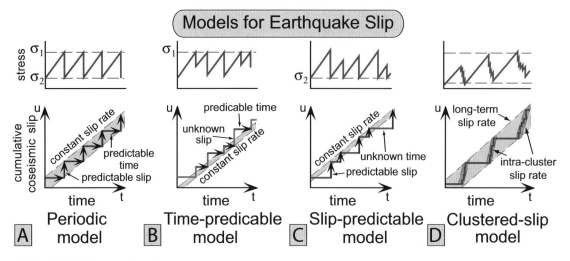

Fig. 4.4 Models for earthquake recurrence.
A. Periodic earthquake model in which stress levels at the time of rupture and after it are known. These thresholds yield a predictable time and slip for each earthquake. B. Time-predictable model based on a consistent stress level at which failure occurs. Stress drop and slip magnitude are unpredictable, but, given previous slip, time until the next earthquake (with unknown slip) is predictable. C. Slip-predictable model based on a consistent stress level at the end of an earthquake. Given time since the last rupture, magnitude of slip is predictable. D. Clustered-slip model has no consistent stress levels for the start or end of an earthquake, but predicts that earthquakes will tend to group together in time, and that short-term and long-term rates may vary significantly. Modified after Shimaki and Nakata (1980) and Friedrich *et al.* (2003).

but that, whenever an earthquake does occur, the likelihood of another earthquake occurring soon thereafter is high.

Asperities, barriers, and characteristic earthquakes

What controls ruptures?

In an effort to understand why and when earthquakes occur, many scientists have searched for patterns of ruptures that, if fully understood, could form a basis for predicting subsequent seismic events. Earthquakes clearly occur when the strength of a fault is exceeded. But, what controls that strength and does it change between earthquakes? In southern California, all of the recent large earthquakes ($M \geq 5$) have occurred on faults that were "loaded," that is, the stress on them was increased by a nearby earthquake in the previous 18 months (Harris *et al.*, 1995). Moreover, it seems

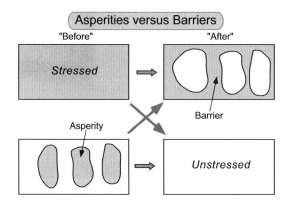

Fig. 4.5 Model of asperities and barriers along a fault plane.
Barriers (unruptured regions) can remain after faulting and may stop the propagating rupture. Asperities may exist before rupture and determine the strength of the fault and where ruptures initiate. Modified after Aki (1984).

probable that localized segments or patches of some fault planes control the strength of some faults, whereas other parts are at times relatively

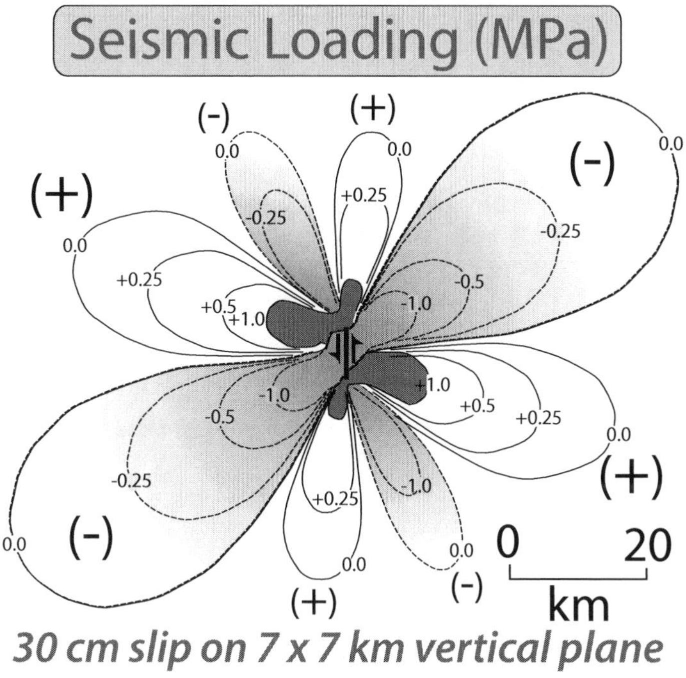

Seismic Loading (MPa)

30 cm slip on 7 x 7 km vertical plane

Fig. 4.6   Model of stress changes due to faulting.
Coulomb stress changes modeled for an $M=5.7$ sinistral strike-slip earthquake with maximum slip of ~0.3 m and a vertical, 7×7 km rupture patch. Note the symmetrical lobes of increased (+) and decreased (−) stresses that extend for tens of kilometers beyond the fault tips despite its small amount of slip. Depending on fault orientations within the area affected by stress changes, faults may be pushed closer to or farther from failure. Modified after Toda and Stein (2003).

stress-free (Fig. 4.5). These "sticky" patches of faults have been termed *asperities*. Such patches may be analogous to irregularities that jut out from a planar surface and which, if two planar surfaces were juxtaposed, would be the points where frictional forces would have to be overcome in order to permit sliding between the surfaces. Thus, asperities represent irregularities that can concentrate stresses along a fault, and their strength with respect to shear stresses exerted on them may control when faulting occurs. If the strength of asperities persists from one earthquake to the next, it might be expected that the recurrence pattern would follow a time-predictable model (Fig. 4.4B).

Alternatively, earthquakes may be thought of as causing ruptures along a heterogeneous fault plane in which some strong patches fail to break. These unruptured regions are termed *barriers* (Fig. 4.5). Aftershocks following the main earthquake will be concentrated around these barriers (Aki, 1984). If the fault plane is considered to be uniformly stressed prior to rupture, the presence of barriers after faulting

represents a "stress roughening" because the stress is less uniformly distributed after the earthquake. If the same strong barriers control successive earthquakes, such a fault might display recurrence patterns based on slip-predictable behavior (Fig. 4.4C).

From the simplest perspective, we envision stress changes on a given fault as being driven by far-field tectonic forces, such as those generated by plate motions. In such a case, we might expect steady rates of stress changes along a fault plane. Earthquakes on nearby faults, however, also change the state of stress in the crust. When a fault slips, it releases stress along most of its fault plane and reduces stress in some of the surrounding crust. But, the fact that slip goes to zero at the fault tip dictates that stress also increases in certain directions around the fault tip (Fig. 4.6). Following an earthquake, most aftershocks are confined to zones where the earthquake caused increases in stress. Although the stress changes are most pronounced in the region immediately surrounding the newly ruptured fault, an area perhaps 100 times larger than the rup ture patch itself can experience significant changes. Thus, slip on one fault can increase (or decrease) the stresses across other faults. Such loading can push other faults closer to failure, thereby causing *triggered slip* (Pollitz and Sacks, 2002). Alternatively, by lowering stress on some faults, the time until these faults rupture is expected to increase (Miller, 1996).

Characteristic earthquakes and fault models

It has been proposed that some faults are typified by *characteristic earthquakes*, in which a fault or a segment of a fault ruptures repeatedly and displays approximately the same amount and distribution of slip during each successive event (Schwartz and Coppersmith, 1984). If a fault did indeed display characteristic earthquakes, knowledge of a single faulting event would provide a remarkable understanding of both previous and future rupture patterns, because the strain build-up and release, stress drops during faulting, variations in displacements along the fault, and the length of the rupture would be approximately duplicated in successive earthquakes. Thus, tremendous predictive power may reside in faults that rupture via characteristic earthquakes. If ruptures along a fault were controlled by stable asperities and barriers (i.e., consistent failure stress and terminating stress) that persisted from one earthquake to the next, this stability could provide a mechanism for generating similar slip distributions along the fault in multiple events.

In order to test whether or not a fault is typified by characteristic earthquakes, either a suite of well-documented historical earthquakes or information from paleoseismic studies has to be used to reconstruct the temporal distribution of events and the spatial pattern of slip in any particular event in the past. The patterns of displacements of past ruptures need to be compared with respect to the length and position of the rupture and the distribution of displacement along the fault. Given a regional strain field, such as that controlled by relative plate motions, several different scenarios can be envisioned to accommodate the regional strain, only some of which would involve characteristic earthquakes (Schwartz and Coppersmith, 1984). In one scenario (Fig. 4.7A), displacements and rupture lengths are randomly distributed in such a fashion that, over time, all regional strain is accommodated, and an essentially uniform slip rate occurs along the length of the fault. In this *variable-slip* scenario, the displacement experienced at any given point, the size of the earthquake, and the position of the rupture segment each vary unpredictably between successive events. Consequently, no faults exhibit characteristic earthquakes in this scenario. A second scenario (Fig. 4.7B) suggests that, at any given point, the displacement is consistent from one earthquake to the next, although displacement can vary along strike. Over time in this *uniform-slip* model, a consistent slip rate also occurs all along the fault. In addition, the large earthquakes have a repetitive pattern in terms of rupture length and displacement variations along the rupture, and moderate-sized earthquakes occur more frequently. In the *characteristic earthquake* model (Fig. 4.7C), consistent displacement at a point also occurs from one event to the next. Over time, however, the slip rate varies along the fault because the cumulative displacement varies along the length of the fault. Each large earthquake represents a repetition of the rupture location, length, and displacement pattern of the previous large earthquake, and moderate-sized earthquakes are sufficiently infrequent that they only accommodate a fraction of the residual slip variation along the fault.

In viewing these models, it is easy to envision how characteristic earthquakes along normal or reverse faults could progressively build an irregular topography. If the site of greatest structural displacement along a fault were to be essentially fixed in the landscape through time, mountain peaks and basin depocenters would occur in predictable positions above the zone of maximum uplift or subsidence. Both structure contours and the landscape topography might be expected to be closely related to these repetitive cycles of displacement.

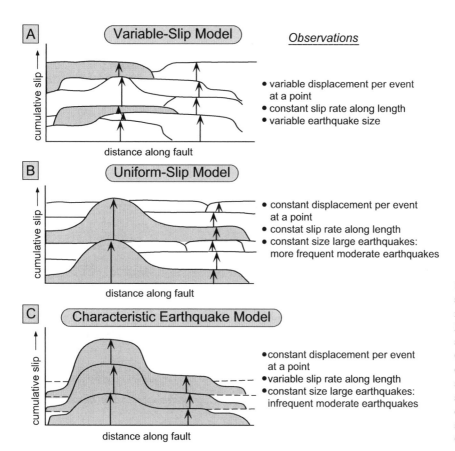

| A | Variable-Slip Model | *Observations* |

- variable displacement per event at a point
- constant slip rate along length
- variable earthquake size

| B | Uniform-Slip Model |

- constant displacement per event at a point
- constat slip rate along length
- constant size large earthquakes: more frequent moderate earthquakes

| C | Characteristic Earthquake Model |

- constant displacement per event at a point
- variable slip rate along length
- constant size large earthquakes: infrequent moderate earthquakes

Fig. 4.7 Models for slip accumulation along a regional suite of faults. Models for (A) variable slip, (B) uniform slip, and (C) characteristic earthquakes. Vertical arrows depict coseismic slip in individual earthquakes. Modified after Schwartz and Coppersmith (1984).

How could one test whether a fault displays characteristic earthquakes? In the vicinity of the "Parkfield asperity" on the San Andreas Fault, for example, five moderate thrust-fault earthquakes ($M \sim 6$) with rather similar rupture patterns have occurred during the 20th century (Bakun and McEvilly, 1984). Such repetition is suggestive of characteristic earthquake displacement patterns. Usually, however, the historical record is both too brief and too incomplete to permit reconstruction of the displacement patterns during several ruptures of an entire fault or a segment of it. Consequently, we often have to interpret the geological record of deformation and faulting to assess past rupture patterns. At least two different approaches can be used for such an assessment. In one, the variability of displacement along a fault is compiled for several past earthquakes using

measured displacements of strata, structures, and geomorphic features. We describe this approach in a subsequent chapter on paleoseismicity. In the second approach, the deformation observed to have resulted from a single, recent earthquake is compared with nearby deformed geological markers, such as the surface of a fold or deformed marine or fluvial terraces, that represent the cumulative deformation due to multiple earthquakes. If the spatial pattern of deformation shown by the long-lived structure could be created through repeated duplication of the deformation due to the individual faulting event (Stein *et al.*, 1988), then it is reasonable to suggest that the fault may have a characteristic behavior. An example (Fig. 4.8) of such a comparison can be made using (i) the strain resulting from the 1989 $M=7.1$ Loma Prieta earthquake south of San Francisco, and

Box 4.1   Magnitude–frequency relationships for earthquakes.

When extensive records of past earthquakes along a given fault zone exist, statistical analyses can reveal how earthquake magnitudes relate to their frequency. Along many fault systems, small- and intermediate-sized earthquakes show an inverse power-law fit known as the Gutenberg–Richter (G-R) magnitude–frequency relationship (see figure A).

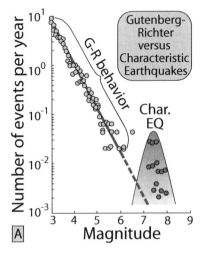

A. Earthquake magnitude–frequency distribution along the San Andreas Fault. Modified after Wesnousky (1994).

This relatio'nship provides a powerful predictive tool for assessing the frequency of an earthquake of any given magnitude. Along fault systems with extensive paleoseismic records, however, the frequency of large, apparently characteristic, earthquakes appears to violate the predictions derived from smaller earthquakes: large earthquakes occur more frequently than expected (Wesnousky, 1994). This apparently high frequency has sparked considerable debate about why the G-R scaling law seems to break down. Is the earthquake record too short or incomplete? Does the slope of the regression need to be modified? Do multiple, different modes of rupture exist? Because seismic hazard assessments are partly based on these statistical

relationships, the causes for this apparent bimodality are important to understand.

A combination of earthquake theory with some recent numerical models that simulate hundreds of ruptures of different sizes (Zielke and Arrowsmith, 2008) provides possible insight on why this bimodal distribution exists. First, recall some earthquake fundamentals. Fault slip begins when the shear stress (τ) along some portion of a fault exceeds the product of the coefficient of *static friction* (μ_s) and the effective normal stress (σ_n^{eff}) across the fault. Once slip initiates, it will persist as long as the shear stress along the fault adjacent to the failing region remains greater than the product of the coefficient of *dynamic friction* (μ_d) and the normal stress:

$$\tau > \mu_d \sigma_n^{eff}$$

The amount of shear stress released coseismically (commonly termed the *stress drop*, $\Delta\tau$) is proportional to the difference between the coefficients of friction: $\mu_s - \mu_d$. The larger the difference, the greater the stress drop and energy release, and the easier it is for more of the fault surface to fail progressively. In their model, Zielke and Arrowsmith (2008) propose that the value of $\mu_s - \mu_d$, and therefore the coseismic stress drop, changes with depth, mostly as a function of temperature. The maximum stress drop is predicted to occur at ~200°C, where laboratory friction experiments revealed maximum velocity-weakening behavior for granitic rocks (e.g., Blanpied *et al.*, 1991), equivalent to a depth z_p of 8–12 km (see figure B).

Three key results emerge from Zielke and Arrowsmith's (2008) numerical models. First, within any given rupture area, the greatest slip occurs where $\Delta\tau$ is greatest (see figure C). Second, small to moderate earthquakes are commonly confined to zones above or below the depth, z_p, where $\Delta\tau$ is a maximum. Coseismic growth of a rupture front toward z_p is impeded because higher rupture-induced

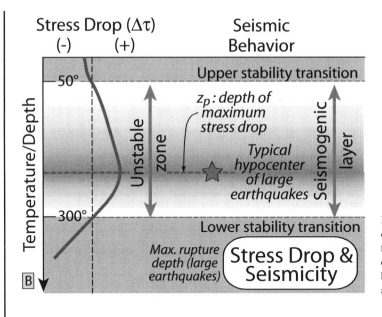

B. Difference between coefficients of dynamic and static friction as a function of depth and consequences for seismic behavior. Modified after Zielke and Arrowsmith (2008).

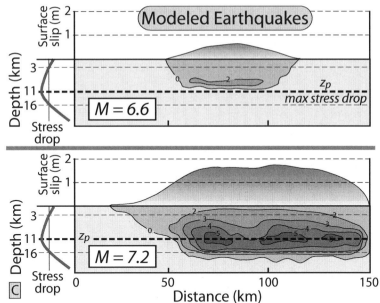

C. Representative models for slip during an M=6.6 and an M=7.2 earthquake. The smaller event does not propagate through z_p, but both events show maximum slip where $\Delta\tau$ is largest on the rupture surface.

stresses are required to rupture toward, rather than away from, the zone of maximum stress drop. These small to moderate ruptures follow a G-R distribution (see figure A). Third, when coseismic stresses are large enough to break through z_p, the rupture tends to cascade to the full width of the seismogenic zone. Such behavior has two outcomes consistent with observations: the frequency of large, similarly sized (characteristic) earthquakes rupturing the entire fault width and some distance along strike will be higher than predicted by a G-R distribution of smaller earthquakes; and relatively few earthquakes will fill the size gap between intermediate and large earthquakes (see figure A).

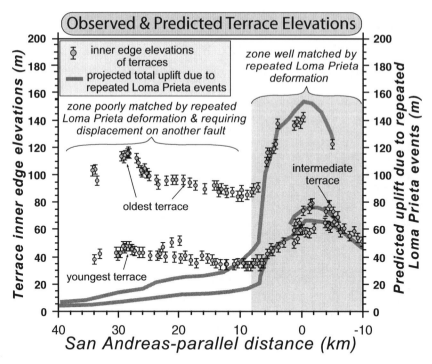

Fig. 4.8   Coseismic shoreline displacement compared with the warped marine terraces.
The variations in coseismic displacement due to the 1989 Loma Prieta M=7.1 earthquake and terrace altitudes
are projected orthogonally on to the trend of the nearby San Andreas Fault. The spatial pattern of coseismic
displacements (gray lines) in the southern half of the transect mimic the height variations of the warped marine
terraces. Such similarity suggests that warping of these terraces could result from multiple earthquakes with Loma
Prieta-like displacements. Modified after Anderson and Menking (1994).

(ii) deformed marine terraces (Anderson and
Menking, 1994). The Loma Prieta earthquake
caused deformation extending from the San
Andreas Fault to the California coast, where
shoreline displacement has been well documented by detailed surveying. A comparison of
the shoreline coseismic displacement with the
shape of the warped marine terraces clearly
indicates that the long-term terrace deformation
could result from repeated Loma Prieta-like
events. The match between the terraces and the
coseismic strain is very good for the southern
part of the terraces, but is unconvincing farther
north (Fig. 4.8). To explain that northern pattern of deformation, repetitive faulting on
another fault has been invoked (Anderson and
Menking, 1994). Whereas the excellent match
of the warped terraces with the pattern of
coseismic deformation does not prove that

the Loma Prieta rupture was an oft-repeated,
characteristic rupture, it certainly suggests that
this interpretation is reasonable.

Debate persists concerning whether or not
faults exhibiting characteristic earthquakes are
common or rare. Certain aspects of characteristic earthquakes appear incompatible with typical frequency–magnitude relationships for small
and intermediate earthquakes within a region
(Box 4.1). At present, with few possible exceptions, neither the historical nor the paleoseismic
studies provides a sufficiently robust database
to test the characteristic earthquake model.
Nonetheless, in comparison to the unpredictable
hazards posed by random, large earthquakes
on a fault system, the predictive value of
characteristic earthquakes certainly warrants
more focused study to ascertain if, where, and
why such earthquakes occur.

Displacement variations along a fault, fault growth, and fault segmentation

Fault length and displacement variations

The concept of characteristic earthquakes contrasts with studies suggesting that, as a fault accumulates a greater total displacement, it lengthens, such that, with each earthquake, the rupture is extended a bit more. But, does displacement vary systematically along a single fault, or is it unpredictable? Do spatial variations in total fault displacement provide insight into how the fault grew over time? In order to collect a robust data set on fault growth and displacement, at least three attributes are desirable: a widespread planar surface that can serve as a marker against which to calibrate fault displacements; a large array of faults offsetting the surface; and an absence of significant post-faulting erosion or deposition on the deformed marker. By exploiting these characteristics, a key study of fault growth (Dawers *et al.*, 1993) was undertaken on the Bishop Tuff (Fig. 4.9E), an extensive ashflow in eastern California that is ~770 ka (Bailey *et al.*, 1976; Crowley *et al.*, 2007). The normal faults that offset this tuff display similar displacement patterns for fault lengths spanning about two orders of magnitude (20–2000 m; Fig. 4.9A, B, and C). For all these faults, the most rapid increases in displacement occur nearer their tips than toward the central part of the rupture, and for most faults, the change in displacement with distance from a fault tip is initially quite linear. The region of the fault containing the maximum displacement may show either broad, slowly varying displacement, as is typical of the longer faults, or quite abrupt changes, as is typical of the shorter faults. The former display bow-shaped displacement profiles (Fig. 4.9C), whereas the latter show more triangular profiles (Fig. 4.9A).

The overall suite of faults cutting the Bishop Tuff has a broadly bow-shaped or semi-elliptical displacement profile (Fig. 4.9D), a shape that is considered to characterize many faults that have experienced numerous earthquakes. But, can such a displacement profile be developed by faults that lengthen with each earthquake and have a coseismic displacement gradient that is itself bow-shaped? In fact, the cumulative displacement profile is predicted to be triangular for coseismic displacement patterns that are either bow-shaped or box-like, whereby displacement is uniform along most of the fault in each seismic event. Studies of over 200 normal faults in the Afar Depression near the southern end of the Red Sea rift (Manighetti *et al.*, 2001) reveal both triangular and semi-elliptical displacement patterns, but these styles characterize only about 20% of all the faults (Fig. 4.10). Symmetrical triangular profiles develop when both tips of a fault progressively propagate outward without restriction. Much more commonly, however, barriers appear to impede propagation of either one or both tips of a fault. Whenever a fault continues to accumulate displacement without lengthening, the slip gradient becomes increasingly steep near the restricted fault tip (Fig. 4.10). Thus, if barriers remain effective in future earthquakes, the slip profile evolves from triangular to more and more semi-elliptical.

These models for fault profiles provide a basis for interpretations of where faults freely propagate, where barriers exist, and where former barriers have been breached. They also offer an attractive explanation for how the theoretically predicted triangular slip profiles evolve toward semi-elliptical ones. An important implication from this study in Afar is that, even if a fault is no longer lengthening, slip can continue to accumulate (Manighetti *et al.*, 2001). A key corollary is that, rather than steadily growing, faults may lengthen to their full lateral extent early in their history, as has been documented in studies in which the chronological history of fault growth is well constrained (Walsh *et al.*, 2002).

For the faults cutting the Bishop Tuff (Fig. 4.9), the maximum displacement (D) is about 1–2% of the fault length (L). A compilation of displacement data (Fig. 4.11) from a variety of settings and different types of faults (Davis *et al.*, 2005; Schlische *et al.*, 1996; Scholz, 2002) suggests that predictable scaling relationships exist between the length of a fault and the maximum displacement. For faults longer than a

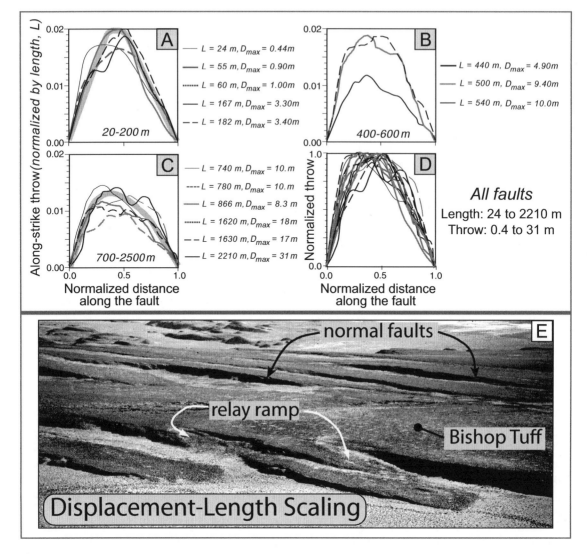

Fig. 4.9   Displacement–length scaling relationships.
Profiles of normalized fault length versus throw (normalized by fault length) for normal faults cutting the Bishop
Tuff, California. A. Faults 20–200 m long display roughly triangular profiles. Accumulated throw is ~2% of length.
B. Faults 400–600 m long. C. Faults 700–2500 m long display more bow-shaped profiles. Throw is ~1% of length.
D. All faults normalized by maximum throw. E. Aerial view of faults cutting the Bishop Tuff in the Volcanic
Tablelands of eastern California. Modified from Dawers *et al.* (1993).

few hundred meters, the maximum displacement is a power-law function ($n = 1.4$–2) of fault
length: the ratio of displacement to length for
longer faults is as much as 100 times greater
than for shorter faults. This increasing ratio of
slip to length for longer faults also implies that
long faults propagate less readily than shorter
faults; this conclusion is consistent with an early

phase of fault lengthening and a later phase of
slip accumulation with little lengthening
(Walsh *et al.*, 2002). If the tip propagation of
longer faults is restrained, the possibility
increases of characteristic earthquake ruptures
along them. Such behavior would have a useful
predictive value. On the other hand, the apparent evolution of cumulative slip distributions

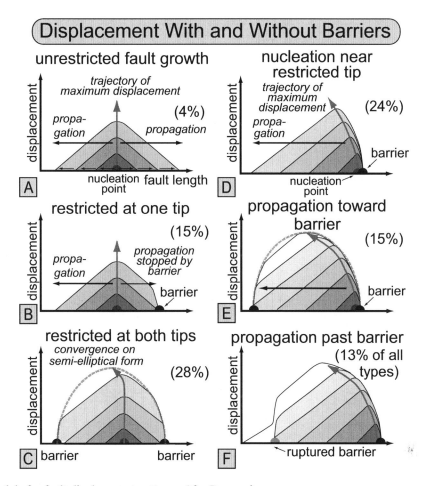

Fig. 4.10   Models for fault displacement patterns, Afar Depression.
Patterns of cumulative fault displacement resulting from multiple earthquakes. Darker to lighter shading represents displacement patterns from early to late in the fault history. The percentages in each box refer to the fraction of normal faults in Afar showing each class of profile. A. Triangular profiles result from unrestricted fault propagation with either bow-like or box-like displacement in individual seismic events. Note the nucleation point in the center of the fault and symmetrical propagation away from the nucleation point. Only 4% of 240 faults fall into this category. B. Symmetrical propagation becomes restricted at one fault tip by a barrier, causing a steepening of the displacement gradient. C. Barriers at each end of a symmetrically propagating fault create steep slip gradients on either end of the fault and continued slip causes the profile to evolve from triangular to semi-elliptical. D. When the nucleation point occurs near a barrier, propagation is primarily away from the barrier, creating an asymmetric triangular profile. E. If an asymmetrically propagating fault (as in D) encounters a barrier, it evolves toward a semi-elliptical slip profile. F. If a barrier is eventually breached at either or both tips of the fault, a tapered tail of slip can develop. Modified from Manighetti *et al.* (2001).

through time (Manighetti *et al.*, 2001) within fault systems (Fig. 4.10) suggests that, whereas a succession of earthquakes could rupture in a similar, characteristic fashion, overall the pattern of slip will change both with time and as a function of interactions with barriers or other faults.

Fault segment linkage

What happens when individual faults extend toward each other during the course of multiple faulting events? Are large faults the result of linkage between several smaller fault segments, or do they represent simple lateral propagation

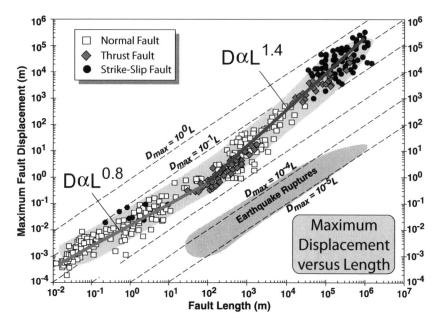

Fig. 4.11   Length–displacement ratios on faults.
Compilation of maximum displacement as a function of fault length for lengths spanning eight orders of magnitude.
Maximum displacement ranges from 0.3% to 30% of fault length. Regressions through these data indicate that, for
faults longer than 100 m, maximum displacements scale at approximately the 1.4 power of fault length.
Displacements on smaller faults may scale at approximately the 0.8 power of fault length. The dashed lines are
lines of equal strain (D/L). Note that the displacement in earthquake ruptures is 2–4 orders of magnitude less than
the accumulated slip on faults of the same length. Modified from Scholz (1990), Schlische *et al.* (1996), and Davis
et al. (2005).

of a single fault? How can we distinguish in the
modern record between different modes of fault
growth? If faults do not link up, how is regional
deformation accommodated among multiple
faults?

Simple models (Cartwright *et al.*, 1995; Walsh
et al., 2002) suggest several ways in which
elongate faults may develop. During fault
growth by radial propagation, an individual
fault simply lengthens through time and
accumulates more displacement, following the
idealized bow-shaped displacement gradient
(Fig. 4.12A). Displacement increases steadily as
the fault grows. Alternatively, fault segments
that ultimately link up may begin as smaller,
individual faults. But, at the moment their tips
link up, the total length of the fault suddenly
increases, whereas the total displacement does
not substantially change (Fig. 4.12B). This
linkage causes a departure from any previously

established displacement–length relationship.
Over time, the composite fault may smooth out
the slip deficit near the former segment boundaries, such that only small departures from
the expected displacement may mark these
boundaries. Eventually, the displacement–length
geometry for a fault resulting from segment
linkage may be indistinguishable from that of a
similarly long fault that grew from a single rupture. In such situations, patterns of subsidence
recorded by syntectonic sediments are probably
the most reliable discriminator among different
modes of fault growth (Anders and Schlische,
1994; Cowie *et al.*, 2000). In a third mode of
fault growth (Fig. 4.12C), the fault lengthens
rapidly early in its history, but then simply accumulates displacement with little to no fault
lengthening (Walsh *et al.*, 2002). This last model
supports the concept of strong, persistent barriers to fault propagation (Fig. 4.10C and E).

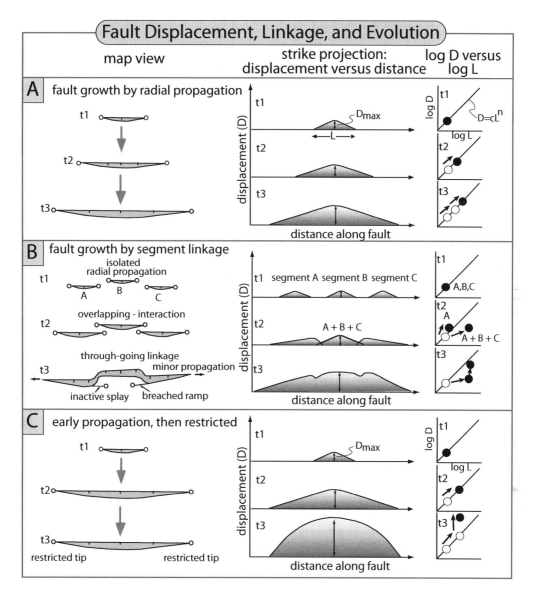

Fig. 4.12 Three models for fault growth.
A. During fault growth by radial propagation, an individual fault simply lengthens and accumulates more displacement through time. Plots of displacement versus fault length (right column) show a steady increase in displacement as the fault grows. The along-strike pattern of displacement versus length is approximately triangular and assumes uniform displacement along the rupture surface in each earthquake. B. In an alternative scheme for fault growth, small individual faults gradually link up to create one large, through-going fault. Whereas the accumulation of displacement follows a predictable path for the individual segments, when they link up, displacement becomes considerably less than that predicted for a fault of this length (see right-hand column). Through time, the slip deficiencies near the points of segment linkage are reduced. Ultimately, the only indication that the large fault resulted from linkage of smaller ones may be the presence of the perturbations to the smooth, bow-shaped displacement near the former zones of overlap and linkage. Modified after Cartwright *et al.* (1995). C. Fault growth with early propagation to full length followed by displacement without lengthening. Note that, for much of the fault's lifetime, the length is fixed, so the trajectory in length–displacement space (right column) is vertical. Once the length is fixed, the displacement profile evolves from triangular to bow-shaped.

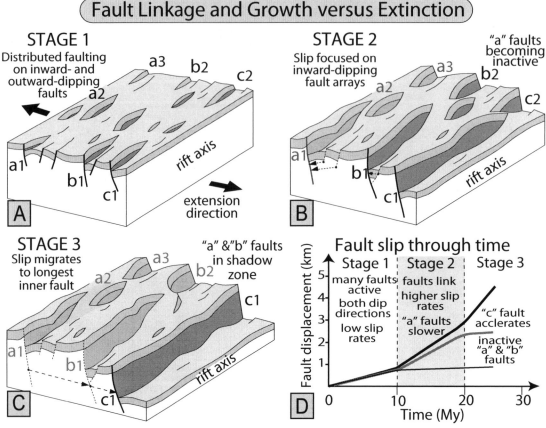

Fig. 4.13   Fault linkage, shadow zones, and slip rates during extension.
Synthesis of extensional fault development based on rifting in the Viking Graben, North Sea. A. Distributed faulting with dips in both directions and slow overall extension. B. Outward-facing faults die and "a" faults slow as "b" and "c" faults extend, link, and speed up. C. Through-going "c" fault absorbs almost all slip and accelerates; "a" and "b" faults in shadow zone become largely inactive. D. Changes in slip rates and behavior of interacting faults. Modified after Cowie *et al.* (2005).

Importantly, compared to the other models, this model predicts a vertical trajectory in length–displacement space through time (Fig. 4.11), a pattern at odds with common assumptions of how faults accumulate slip.

Another way to visualize linkage among faults is to imagine a competition among different fault segments in which the "big guy wins" (Fig. 4.13). Consider a homogeneous medium subjected to tensile stresses that cause numerous small defects or extensional faults randomly distributed across the surface. With continuing extension, these "defects" lengthen, and as they

grow laterally, some of them happen to encounter other lengthening faults. These segments then link up, creating a longer fault that accommodates increasingly more displacement. This linkage also creates a "shadow zone" where previously formed faults become inactive, because the larger fault is taking up all the strain in that sector of the deforming plate. Eventually, larger fault segments link together to form a through-going, master fault, and virtually all the other faults die. As slip gets concentrated on a single master fault, its slip rate may rapidly accelerate. Both numerical models and limited field

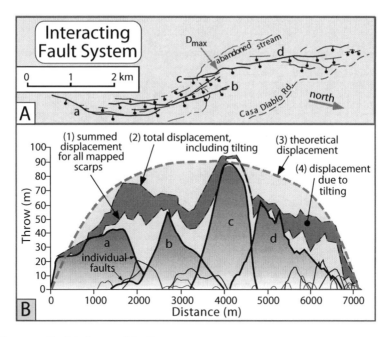

Fig. 4.14 Map of a zone of related normal faults.
A. The fault zone consists of four overlapping major faults (labeled "a"–"d") and more than 20 lesser faults. These faults cut the 0.77 Ma Bishop Tuff in east-central California. All of the faults are displaced down to the east, and some tilting of the original surface also occurs between the faults. B. Displacement versus distance along the fault zone. Thin lines show the displacement attributed to individual faults. The difference between the dashed lines equates to the contribution of tilting (up to 30%) to the total displacement. Note the broadly bow-shaped displacement along the total fault zone. A slip deficit occurs between faults "b" and "c", but none is apparent between faults "a" and "b" or between "c" and "d". Modified after Dawers and Anders (1995).

examples support this competition model (Gupta *et al.*, 1998; Cowie *et al.*, 2000).

Displacement compensation among multiple faults

Consider two parallel faults for which the ends of the faults are not linked, but they do overlap. Commonly, as the displacement decreases on one fault, the offset on an overlapping fault strand increases, such that the net slip across the overlapping region decreases only slightly (Willemse *et al.*, 1996). This compensating behavior can sometimes be more readily documented in a fault zone comprising multiple interacting faults and is clearly illustrated along a 7-km-long, linear zone of interrelated, overlapping normal faults that cut the Bishop Tuff (Fig. 4.14A). This fault zone

comprises four major, overlapping fault scarps, more than 20 subsidiary faults with lesser throw, and tilted blocks between faults (Dawers and Anders, 1995). When the displacement on individual faults and tilted blocks is summed along the length of the entire fault zone (Fig. 4.14B), a broadly bow-shaped displacement gradient emerges. This steady variation suggests that the whole array of faults is deforming as if it were a single fault and that overall displacement is being smoothly partitioned among multiple structures (Nicol *et al.*, 2006). In addition to the overall pattern of fault system displacement, three notable points emerge from this compilation of slip (Fig. 4.14B). First, tilting can account for up to 30% of the total slip at any one point, such that off-fault deformation should not be ignored. Second, slip deficits develop in the

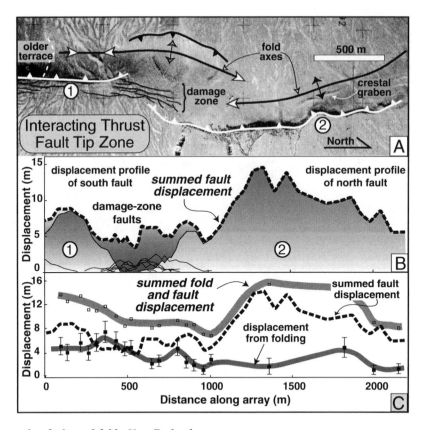

Fig. 4.15   Interacting faults and folds, New Zealand.
Compensating faults and folds in the Ostler Fault zone, South Island, New Zealand. A. Two, non-overlapping thrust
faults (1, 2) cut a 20 kyr old outwash terrace. Between their tips, numerous small thrusts define a "damage zone."
Active folding and a backthrust also accommodate shortening. B. Profiles of thrust-fault displacement for both large
faults, the backthrust, and damage-zone faults. Note that the slip deficit between faults 1 and 2 is partly filled by the
cumulative slip on the damage-zone faults and backthrust. C. When shortening due to fault-propagation folding is
added to the fault slip, the overall slip gradient becomes even smoother, indicating the importance of both folding
and faults in accommodating shortening.

overall profile near the overlapping tips of the
major faults, for example, between faults "b"
and "c". Finally, these slip deficits can be filled
by slip and tilting accumulated on arrays of
smaller faults spanning the larger faults, as
is seen between faults "a" and "b". One can
imagine that, with increasing displacement,
these faults may amalgamate into a single
irregular, but continuous, fault trace. At
present, however, based only on the summed
displacement profile alone, it would be difficult
to estimate how many faults contribute to
that profile.

Thus far we have largely used examples from
extensional settings to illustrate fault interactions. Similar compensatory trade-offs in displacement also occur in contractional settings.
For example, the Ostler Fault zone comprises
an elongate suite of active thrusts and fault-
related folds that extend for 70 km along the
eastern flank of the Southern Alps of New
Zealand (Read, 1984). Because these faults
deform young glacial outwash surfaces, a
remarkably clear record of recent deformation
is revealed by their displacement patterns
(Fig. 4.15). Here, the slip deficit between the

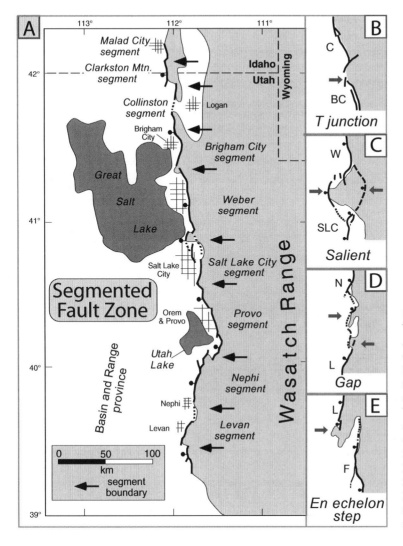

Fig. 4.16 Segmentation of the Wasatch Fault zone.
A. At least eight segments and four different types of segment boundaries have been identified along the 400 km long system of normal faults that delineate the Wasatch front. Arrows identify segment boundaries. B–E: Types of segment boundaries: B. T-junction; C. salient; D. gap; E. *en echelon*. Segment names – BC: Brigham City; C: Colliston; F: Fayette; L: Levan; N: Nephi; SLC: Salt Lake City; W: Weber. Modified from Machette *et al.* (1992a,b).

non-overlapping tips of larger thrust faults is filled via three mechanisms: an array of small forethrusts with displacements of less than 2 m define a "damage zone" between the fault tips; a single, parallel backthrust reaches its maximum displacement in the gap between the two larger faults; and the largest anticline also culminates in this gap. The overall pattern of faulting suggests that, during future shortening, the larger faults are likely to link together across the damage zone and simplify the fault trace. It is important to recognize that, until such linkage occurs, the entire array of faults and folds must be quantified in order to assess the displacement gradient along the fault zone.

Segmentation of range-front faults

Some large mountain ranges, such as the Sierra Nevada of California or the Wasatch Range of Utah, are bounded by a range-front fault system that is hundreds of kilometers long. The range-bounding fault never ruptures along its entire length in a single seismic event. Instead, smaller segments break in individual earthquakes, and multiple seismic events are required to displace the entire range front. Along some of these elongate range fronts, a recognizable *segmentation* of the bounding fault zone into smaller rupture lengths that have definable terminations can be discerned (Fig. 4.16). Along the Wasatch Fault system, four different types of segment boundaries (Machette *et al.*, 1992a) are recognized (Fig. 4.16). The most commonly observed boundary occurs where bedrock spurs or salients extend into the adjacent basin and major range-front faults terminate against them. Some of these spurs are bounded by Quaternary faults that are active, but have less displacement than the range-front faults. Other segment boundaries are delineated by an *en echelon* step in the range-bounding faults. Still others occur where the range-front faults are intersected by a cross-fault that trends at a high angle to the range front. Finally, some segment boundaries are simply defined by the absence of a rupture between two active range-front faults (Machette *et al.*, 1992a). Several of the segment boundaries display a combination of spurs, cross-faults, and *en echelon* offsets. If the positions of the terminations persist through time and if entire segments typically rupture in a single seismic event, each segment might be typified by characteristic earthquakes (Schwartz and Coppersmith, 1984). At present, however, neither the persistence of the terminations nor the tendency for an entire segment to rupture "characteristically" is well established. Moreover, recent studies suggest that no widely applicable rule yet exists to predict whether or not an earthquake will rupture across segment boundaries. A compilation of strike-slip earthquakes suggests that ruptures terminate when segments are separated by more than 3–5 km (Wesnousky, 2006), whereas a study of large earthquakes on continental thrust faults reveal that segments separated by up to 10 km can rupture in the same earthquake (Rubin, 1996), although this coeval rupture does not require a surface rupture to link the segments.

Geomorphic expression of faults

Faults in which the hanging wall moves only vertically, that is, directly up or down the fault plane, are termed *dip-slip* faults, whereas pure horizontal motion results in *strike-slip* faults. Few faults, however, are purely dip-slip or strike-slip. Most have some component of both horizontal and vertical motions.

Nonetheless, it is useful to describe the structures and geomorphic features that are expected for the primarily dip-slip and strike-slip end-members, because their contribution to a given natural fault setting can then be more readily recognized. Experimental results using homogeneous materials with known physical characteristics often depict an idealized array of structures that are associated with a particular stress field. Such structures are described in the following paragraphs. Natural heterogeneities in rocks, however, dictate that they will not deform uniformly, so that few natural situations exactly duplicate model predictions.

We typically envision a fault as an irregular, but singular, surface dipping into the crust. In fact, at the scales of less than 10 km, many faults consist of a tabular volume of typically unconnected or anastomosing smaller faults (Scholz, 1998). These component faults will span a broad spectrum of sizes, ranging from a few meters to several kilometers. Thus, whereas some faults appear to slip along a single plane at depth with a compact damage zone surrounding it (Fig. 4.17A), others display highly fractured zones hundreds of meters thick with multiple actively slipping fault surfaces (Fig. 4.17B) (Faulkner *et al.*, 2003). Even at the surface, however, the complexity of many fault systems is commonly difficult to delineate. Where fault traces on the ground are obscured by deposition in basins, erosion on hillslopes, or vegetation, remotely sensed data may serve to illuminate fault systems more completely. For example, aerial imagery can detect subtle changes in vegetation that are responses to variations in the water table or soil types due to faulting. In actively aggrading basins, high-resolution aeromagnetic surveys (Grauch, 2001) have been successfully used to expose intricate fault systems that were previously unknown (Fig. 4.17C and D). The complexity of surface deformation, therefore, is not solely attributable to the heterogeneous materials of the crust, but also to the fact that, during an earthquake, hundreds of small rupture surfaces actually accommodate the total displacement. Over

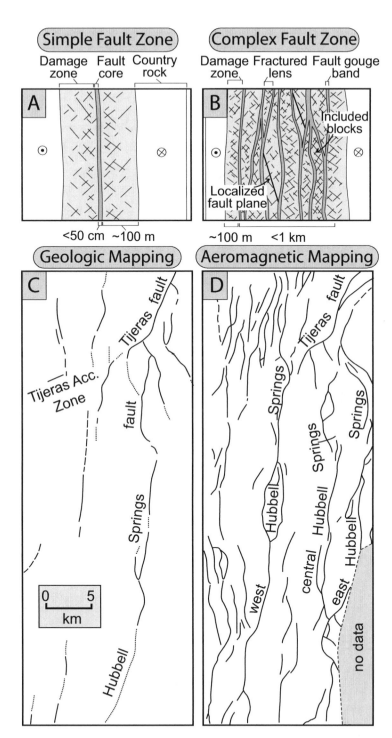

Fig. 4.17 **Fault zone complexity in models and maps.**
A. Traditional model of a strike-slip fault with a fault core of a few centimeters and a zone of damaged bedrock extending about 100 m on either side. B. Complexly fractured strike-slip fault zone with bands of fault gouge separating a fractured lens of commonly rotated and sheared bedrock within a broad (1 km wide) damage zone. A, B: Modified from Faulkner *et al.* (2003). C. Quaternary faults in the Rio Grande rift near Albuquerque, New Mexico, compiled from previous geological studies. D. Quaternary faults in the same area as interpreted from aeromagnetic anomalies that are attributed to thickness changes in Quaternary sediments across faults. C, D: Modified from Grauch (2001).

time, slip accumulates on numerous surfaces, and blocks between faults rotate and shear. Not surprisingly, the resulting deformation pattern can be complex, too!

Strike-slip faults

Some of the best-known faults in the world, such as the San Andreas, Altyn Tagh, and North

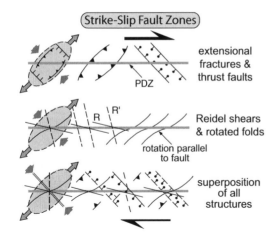

Fig. 4.18 Orientation of structural features formed in response to strike-slip shear couple.
Normal and thrust faults, folds (some rotated near the shear zone), Reidel shears and conjugate shears (R') form at predictable angles with the principal displacement zone (PDZ). Modified after Sylvester (1988).

Anatolian Faults, are strike-slip faults that have caused widespread destruction in 20th-century earthquakes. Strike-slip faults develop where the maximum compressive stress (σ_1) is horizontal and there is a horizontally oriented deviatoric tensile stress (Fig. 4.1C). Commonly, this stress orientation prevails where two crustal blocks are moving essentially horizontally and approximately parallel to the boundary between them, but in opposite directions. Thus, a shear couple is created across this boundary zone.

As observed in laboratory experiments and field studies, a predictable geometry of structures may form in shallow crustal rocks and alluvium in response to this stress field (Fig. 4.18). A principal displacement zone commonly forms parallel to the shear couple. This orientation is the one you would expect a strike-slip fault to have, if the crustal blocks on either side had a high rigidity with respect to a weak fault zone between them. Normal faults should form perpendicular to the direction of maximum elongation and maximum tensile stress. *Reidel* shears (R) form at an angle of $\theta/2$ (~15–20°) with the principal displacement zone, where θ is

equal to the angle of internal friction for the faulted material. These shears will have the same sense of strike-slip motion as the principal displacement zone, and they typically form a suite of *en echelon* fractures (Fig. 4.18). As shearing continues, the propagating tips of the Reidel shears tend to curve to become parallel with the normal faults and to display normal-fault displacements with little strike-slip motion in the tip zone. Conjugate Reidel shears (R') can form at $90° - \theta/2$ (60–75°) to the principal displacement zone. These conjugates have the opposite sense of strike-slip motion as the main fault. Finally, folds and thrust faults should form with their axes or traces, respectively, approximately perpendicular to the main compressive stress. The fold axes will initially be oriented at about 45° to the main displacement zone, but, as fault motion continues, they can be rotated into more complete parallelism with the main fault due to shearing adjacent to the principal displacement zone. In Nature, some of these structures may represent responses to different generations of motion, and all of them can be superimposed, making for a very complicated array of structures within a broad shear zone (Sylvester, 1988). Although the development of structures along a newly formed strike-slip fault does not follow an invariant pattern, a predictable succession of structures is common, whereby discrete folds and tension gashes form initially and eventually amalgamate into a through-going, anastomosing shear zone (principal displacement zone) (Sylvester, 1988; Anderson and Rymer, 1983).

Whereas all of these structures can be associated with a relatively straight segment of a strike-slip fault zone, many strike-slip faults have traces that bend such that slip between the adjacent blocks creates large compressive or tensile stresses in the curved fault segments. For example, in a restraining bend (Fig. 4.19), the fault trace curves into the path of the blocks on either side of the fault. The ensuing compression generates contractional structures that lead to development of thrust faults, folds, and ultimately mountains (Biddle and Christie-Blick, 1985). In the vicinity of a releasing bend, the fault curves away from the path of the blocks on

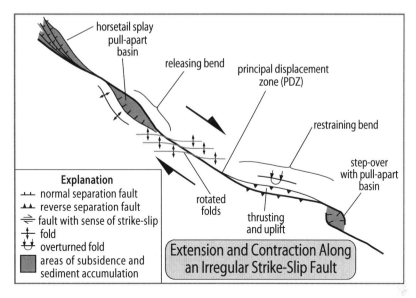

Fig. 4.19 Regional-scale structures along a strike-slip fault.
These structures can include restraining bends associated with thrusts and mountain building, releasing bends associated with basin development and rapid subsidence, and horsetail splays of either normal or reverse faults where deformation is spread over a broader zone. Right-stepping step-overs in a dextral shear zone (as shown here) create pull-apart basins as the fault tips curve toward the continuing fault trace and generate normal slip. Modified after Christie-Blick and Biddle (1985).

either side such that extensional structures, like normal faults and basins, form between them (Fig. 4.19). Alternatively, strike-slip faults may display major lateral step-overs, in which one fault trace ends and a second with the same sense of displacement commences. A combination of the sense of motion on the strike-slip fault and the geometry of the step-over determines whether this intervening zone will be one of compression or extension. With a compressional step-over, a zone of uplift forms between the fault tips, and thrust faults may form to accommodate shortening across the uplift. With an extensional step-over, normal faults form to accommodate subsidence within a *pull-apart* basin (Fig. 4.19).

The geomorphological expression of strike-slip faults can be viewed at the scale of individual seismic events or as a result of long-term strain accumulation. Because individual faulting events vary so widely in terms of the magnitude of offset, length of rupture, and material that they deform, a single set of geomorphic features that characterize them cannot be specified. At the scale of a few meters or less, *en echelon* faults commonly rupture the surface, small collapsed basins form at releasing step-overs, and small, thrusted uplifts appear at restraining step-overs. Apparent vertical displacements along the fault trace are also common. Some of these displacements result from simple horizontal translation of higher topography into an area of lower topography on the opposite side of the fault or vice versa. Many scarps, however, result from some component of vertical displacement along the fault that is a consequence of the fact that the fault is not a simple vertical plane or that the movement is not wholly strike-slip. Along a steeply dipping, but undulating, fault plane, one can visualize small-scale releasing and restraining bends that cause subsidence or uplift of one side of the fault with respect to the other. If such a configuration were to persist over long intervals, large basins and mountains would develop that would be restricted to one side of the fault. Along fresh ruptures of strike-slip faults, however, uplifted scarps commonly do not face in the same direction consistently,

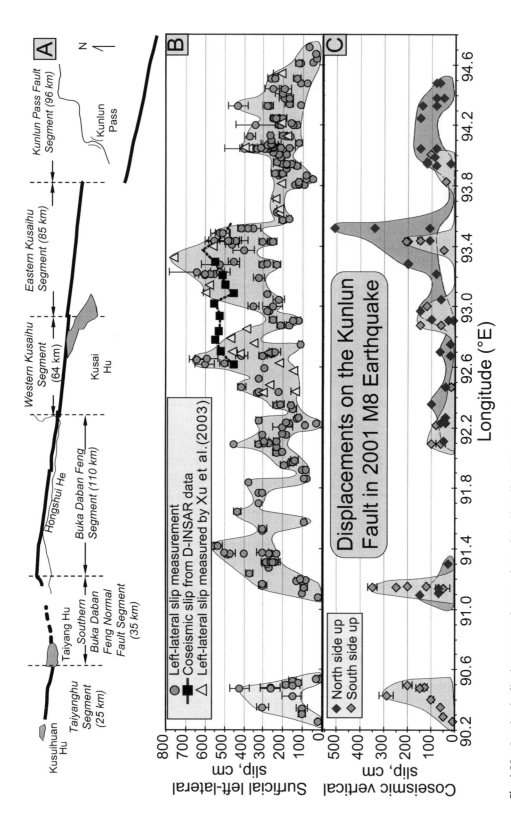

Fig. 4.20 Coseismic displacements due to a large (M = 8) strike-slip earthquake.

A. Simplified map of the surface ruptures of the 14 November 2001 M_s =8.1 earthquake on the East Kunlun Fault showing rupture segmentation. B. Left-lateral slip distribution as a function of longitude along the 400 km long rupture zone. C. Vertical slip shows a predominance of north-side up in the east and south-side up in the west. Modified after Chen et al. (2004).

but rather the upthrown side varies along the fault trace (Fig. 4.20). Whereas the vertical slip can sometimes equal the horizontal displacement, the ratio of vertical to horizontal slip is much more commonly less than 25% on large strike-slip faults (Beanland and Clark, 1994).

Over long time intervals, strike-slip faulting typically leads to some well-known geomorphological features (Fig. 4.21). A linear trough commonly forms along the principal displacement zone, because structural blocks are slipping past each other along this zone and because the fractured materials are more readily eroded along the fault zone. Recent numerical modelling suggests that simple deflection of streams across strike-slip faults can create linear valleys along the fault trace in the absence of any special "softening" due to brecciation (Braun and Sambridge, 1997). Irrespective of how the troughs are generated, within them, *sag ponds* may form in low-lying regions. Scarps can be preserved on either side of the fault. Linear features like streams and ridges become offset along the fault and can yield a clear sense of slip directions. Some care must be used, however, when inferring displacement directions from offset streams. Offsets determined from streams that are displaced in an uphill direction with respect to the local hillslope gradients are more reliable than downslope displacements because such offsets could result from stream capture. When a ridge that has been translated along the fault subsequently blocks a drainage, it is termed a *shutter ridge*. On the downslope side of strike-slip faults, *beheaded stream valleys* may be preserved (Fig. 4.21A and C). These are abandoned valleys that have been rafted laterally beyond the course of the stream that formerly flowed through them (Keller *et al.*, 1982). On both the upstream and downstream sides of a fault, river terraces may be systematically offset. Commonly, streams crossing a strike-slip fault will exit from a mountainous terrain into a gentler one. Upstream, their valleys will have been more confined, whereas downstream of the fault, they may build alluvial fans in the less confined topography. Horsts, grabens, small-scale pull-apart basins (Fig. 4.21B), and

various thrusts and folds can also have clear geomorphological expression and can often be understood in the context of the imposed shear couple and stress field (Fig. 4.18).

Normal faults

Normal faults form in settings where the maximum compressive stress (σ_1) is vertical and a deviatoric tensile stress in a horizontal orientation is present (Fig. 4.1A). Typically, normal faults cut the surface at high angles (~50–70°). In cross-section, coseismic displacements on normal faults are commonly asymmetric with respect to a horizontal datum, such that subsidence of the hanging wall is typically several times greater than uplift of the footwall. Although down-dropped keystone blocks (grabens) and uplifted horsts (blocks bounded by normal faults on both flanks) were once thought to typify regions of normal faulting, unpaired normal faults creating half-grabens or *fault-angle depressions* occur more commonly (Fig. 4.22). The down-dropped hanging wall of a normal fault commonly generates a basin in which sediments accumulate, whereas the footwall experiences uplift and, in terrestrial settings, becomes a site of erosion. Thus, many normal faults bound asymmetric ranges with one steep, fault-bounded flank and a gently sloping, largely unfaulted opposite flank (Fig. 4.22). Secondary *synthetic* (dipping in the same direction with similar sense of throw) or *antithetic* (dipping in the opposite direction with an opposite sense of throw) faults may develop within the hanging-wall block.

Rather than amalgamating into a single fault zone, displacement during extension can also be transferred between adjacent faults by a variety of structures (Gawthorpe and Hurst, 1993), including (Fig. 4.23): (i) *transfer faults*, which are oblique to the traces of the main faults, but which link them together; (ii) *relay ramps* (Fig. 4.9E), which may or may not be faulted and which bridge between two faults facing the same direction; (iii) *antithetic interference zones*, which are tilted (without or with faulting) ramps that develop between normal faults facing each other and dipping in opposite directions; and

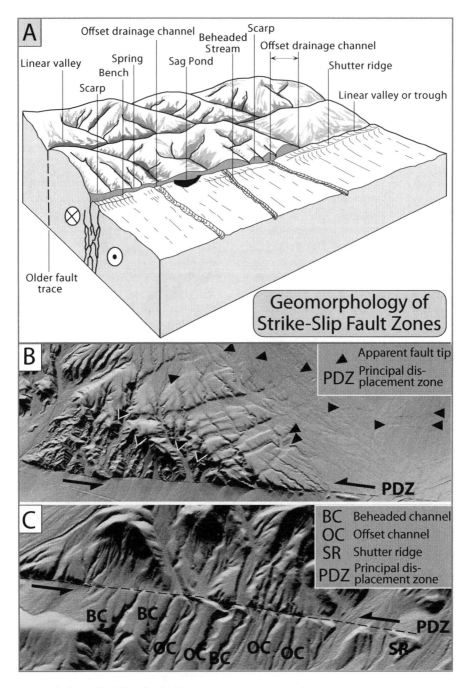

Fig. 4.21 Geomorphology of strike-slip fault zones.
A. A linear trough along-fault, sag ponds, shutter ridges, offset ridges and drainages, springs, scarps, and beheaded streams are typical geomorphic features indicative of strike-slip faulting. The older, abandoned fault trace displays analogous, but more erosionally degraded features. Modified after Wesson *et al.* (1975). B. Lidar-based shaded relief map of the sinistral Garlock Fault zone in southern California: 35°23′58″N, 117°48′1″W. Note complex faulting by normal faults oriented obliquely to the principal displacement zone (PDZ). C. Lidar image of offset channels (OC), beheaded channels (BC), and shutter ridges (SR) along the Garlock Fault: 35°26′22″N, 117°41′42″W. Images courtesy of the U.S. Geological Survey, Google Earth© and Earthscope Geon Open Topography Portal.

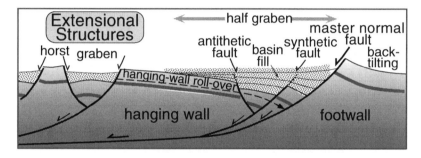

Fig. 4.22   Schematic of cross-section of normal faults in an extensional regime.
The master fault delineates footwall and hanging-wall blocks. A broad half-graben is cut by synthetic and antithetic faults. Note both the geometry of the syntectonic growth strata that fill the faulted basins and the curvature of the hanging wall toward the master fault (hanging-wall roll-over).

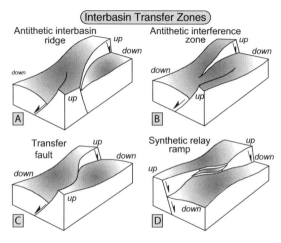

Fig. 4.23   Interbasin transfer zones.
Simplified geometries of structures that assist the transfer of deformation between two regional-scale, overlapping or non-overlapping normal faults. Modified after Gawthorpe and Hurst (1993).

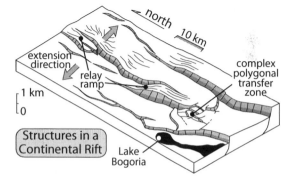

Fig. 4.24   Schematic fault geometry in the East African Rift system.
Relay ramps, transfer faults, and smaller-scale synthetic faults facilitate transfer of displacement between the major faults. Note that, for any cross-section perpendicular to the traces of the major faults, the net displacement across the region is nearly equal, despite large along-strike differences in displacement on individual faults. Modified after Morley (1989).

(iv) *antithetic interbasinal ridges*, which develop where normal faults change polarity and step into the footwall. Each of these structures provides a mechanism whereby the total regional displacement can vary smoothly along strike, despite the fact that individual faults die out, that displacement is variable along each fault, and that both fault lengths and amount of overlap change through time. Any relatively broad region that is undergoing extension will display some of these displacement-accommodation features. For example, during much of Cenozoic time, active extension has been occurring across eastern Africa. As a consequence, an elaborate network of mature normal faults has developed. Within many segments of the East African Rift, *en echelon* normal faults joined by relay ramps and complex transfer zones (Fig. 4.24) exemplify the spatially varied distribution of faults and strain that accommodates crustal extension. Where Quaternary strain within complexly faulted rifts has been particularly well quantified, such as in the Taupo Rift on the North Island of New Zealand (Nicol *et al.*, 2006), both the compensation that occurs between different subregions that slip at changing local rates through time and

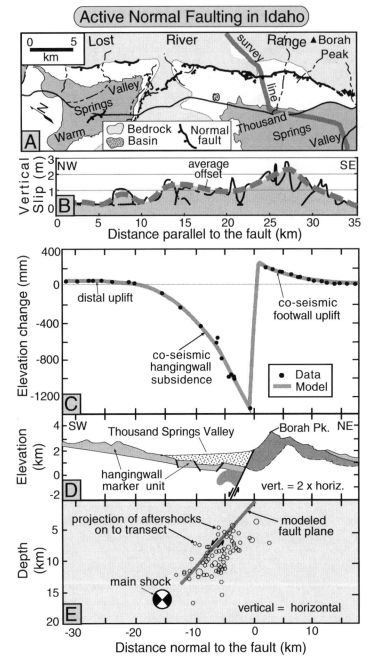

Fig. 4.25   Normal faulting in the Lost River Range, Idaho.
A. Map of the Lost River Range and the surface trace of the 1983 Borah Peak rupture. Location of leveling-line survey across the footwall and hanging wall is shown. B. Vertical surface displacements along the fault trace at the toe of the Lost River Range. C. Surveyed and modeled coseismic displacement resulting from the 1983 Borah Peak earthquake across a zone at least 35 km wide. Note the asymmetry of displacement across the fault, such that the hanging wall is down-dropped 4–5 times more than the footwall is uplifted. D. Schematic geological cross-section perpendicular to the fault across the Borah Peak footwall and the Thousand Springs Valley half-graben. Faulted 4 Ma volcanic rocks provide a marker in the footwall and hanging wall. E. Main 1983 shock and aftershocks define the rupture plane, which dips about 45° to the southwest. Modified after Stein *et al.* (1988).

the tendency for rates on individual faults to become steadier with time suggests that the entire rift acts as an integrated fault system.

Deformation caused by two recent earthquakes helps us visualize aspects of the actual vertical displacements and spatial variations in coseismic strain on active normal faults. The 1983 $M=7.0$ Borah Peak earthquake illustrates the typical asymmetry of footwall uplift and hanging-wall subsidence: the footwall rose a maximum of about 30 cm, whereas the hanging wall dropped about 130 cm (Fig. 4.25C). Following an earthquake such as this one, interseismic strain (Stein *et al.*, 1988) and

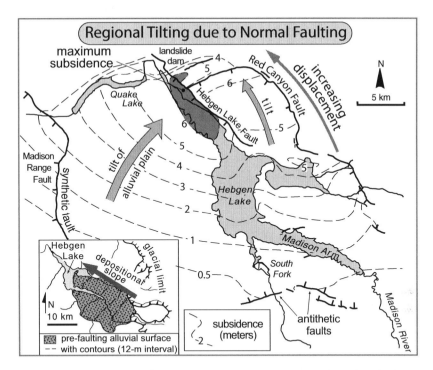

Fig. 4.26   Ground displacement resulting from the 1959 Hebgen Lake (*M* = 7.3) earthquake in west Yellowstone. Maximum offsets of more than 6 m occurred adjacent to the Hebgen Lake Fault. Note the complex three-dimensional geometry of displacement and the small antithetic faults in the south(east). Minor offsets are associated with the Madison Range Fault, whereas additional large offsets occurred along the Red Canyon Fault. The Madison River was dammed by landslides triggered by this earthquake. Given the sensitivity of rivers to small changes in gradient, fault-induced tilting should cause rivers to migrate toward the zone of maximum subsidence. Inset shows gradient of pre-earthquake, northwest-oriented depositional surface. Modified after Alexander *et al.* (1994).

isostatic uplift due to erosion of the footwall block often increase the total bedrock uplift within the footwall. Despite this footwall uplift, the load of sediments that accumulates in and depresses the hanging-wall basin guarantees that considerable asymmetry in net displacement usually persists. At Borah Peak, Thousand Springs Valley represents a half-graben that has accumulated the sediments eroded from the rising footwall (Fig. 4.25A and D). Crustal flexure due to the earthquake affected an area at least 35 km wide (Fig. 4.25C).

In three dimensions, the pattern of uplift and subsidence resulting from multiple surface ruptures during a single earthquake can be complex. As a result of the 1959 Hebgen Lake (*M* = 7.3) earthquake in western Yellowstone National Park, maximum subsidence (~7 m) occurred along the middle segment of

the Hebgen Lake Fault and decreased systematically toward either tip and away from the fault (Fig. 4.26). But, the nearby Red Canyon Fault has a highly curved trace, so that in places the contours of subsidence strike at high angles toward one margin of the fault and imply that this margin acted as a transverse structure or hinge along the side of a scoop-shaped zone of subsidence that strikes at right angles to the elongate zone of subsidence along the Hebgen Lake Fault (Fig. 4.26). In contrast to the observed displacement on the Hebgen Lake Fault and to fault models of decreasing displacement toward the fault tips, displacement on the Red Canyon Fault is highly asymmetric and suggests that the western termination of that fault acts as a "free boundary" where it abuts the Hebgen Lake Fault (Fig. 4.26).

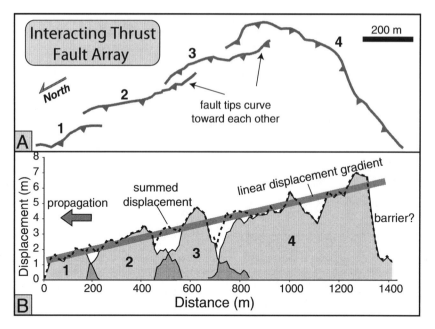

Fig. 4.27   Array of interacting thrust faults.
A. Fault traces of four small thrust faults that cut a late-glacial outwash surface in the Ostler Fault zone, South Island, New Zealand. The curvature of the fault tips is similar to elastic crack behavior observed on scales from millimeters in rock fracture experiments to scales of kilometers on mid-ocean ridge segments and inferred to result from out-of-plane crack interactions (Sempere and Macdonald, 1986). B. Along-strike displacement for individual faults. The summed displacement illustrates compensation between overlapping fault tips and shows a consistent taper toward the north. The abrupt termination to the south suggests the presence of a barrier to fault propagation, as in Fig. 4.10 D. Modified after Davis *et al.* (2005).

Thrust faults

Thrusts faults develop where the maximum compressive stress (σ_1) is horizontal and a vertically oriented deviatoric tensile stress exists (Fig. 4.1B). In theory, thrusts should cut a horizontal land surface at about 30° angles, but, in fact, thrusts and reverse faults can cut the surface at any angle and may occasionally be overturned at the surface. Owing to the commonly low-angle intersection of thrust faults with the Earth's surface, however, the traces of thrusts are more strongly affected by topography, such that they tend to be more highly sinuous than are normal faults, rendering it more difficult to measure offsets if the direction of fault motion is not well defined. Many thrust faults are "blind" and do not cut the surface, such that their displacement characteristics are more difficult to define than for many normal or

strike-slip faults. Nonetheless, where arrays of thrusts do cut the surface, they display the same types of compensatory, interacting patterns (Figs 4.15 and 4.27) as previously described for normal faults.

Geodetic measurements of coseismic deformation (Stein *et al.*, 1988) indicate that, during thrusting, hanging-wall uplift is typically considerably greater than footwall subsidence (Fig. 4.28). This pattern is the mirror image of the coseismic deformation associated with many normal faults (Fig. 4.25). The area affected by the coseismic deformation depends on the magnitude of displacement, the geometry of faulting, and the rigidity of the crust that is being deformed. In the 1952 Kern County earthquake in southern California, deformation extended for about 40 km on either side of the rupture (Fig. 4.28). Although the total vertical offset in this earthquake (~100 cm) was

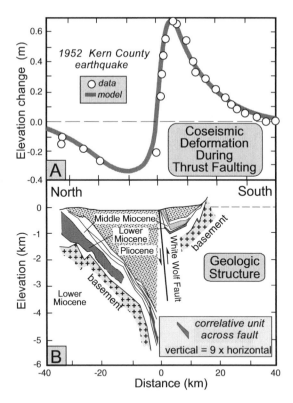

Fig. 4.28   Short- and long-term deformation due to thrust faulting.
A. Coseismic thrust displacement during the 1952 Kern County, California (M=7.3) earthquake. Note the asymmetry of hanging-wall uplift and footwall subsidence and the large wavelength of flexure. B. Geological cross-section of the footwall basin and hanging-wall block of the White Wolf Fault that ruptured during the Kern County earthquake. Miocene marker beds are offset about 5 km. Note that the Miocene units do not thicken appreciably as they approach the fault. This uniformity suggests that the fault did not become active until Pliocene time. Modified after Stein *et al.* (1988).

considerably less than in the Borah Peak rupture (~150 cm; Fig. 4.25), the width of the deformed zone was twice as wide, apparently due in large part to the higher crustal rigidity in the vicinity of the Kern County earthquake.

Convergent plate margins have produced most of the earthquakes of very large magnitude in the past 100 years. The interface between overriding and underthrusting plates commonly represents an irregular, earthquake-prone, regionally extensive surface. Since 1960, four megathrust earthquakes with moment magnitudes (M_w)

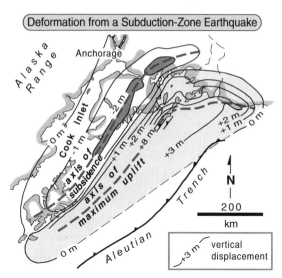

Fig. 4.29   Displacement due to the 1964 Alaska (M_w=9.2) earthquake in the Aleutian subduction zone.
Over 2.4×10^5 km^2 were deformed by this megathrust earthquake. Strong uplift (up to 10 m) occurred in an elongate zone in the proximal hanging wall, and lesser subsidence (up to 2.5 m) occurred in more distal parts of the hanging wall. Most of the deformation is determined from displaced shorelines. Footwall deformation (outboard of the trench) due to the earthquake remains unknown. Modified after Plafker (1972) and Lajoie (1986).

greater than 9 have ruptured subduction zones in the Pacific and Indian Oceans: the 1960 Great Chilean (M_w=9.5) earthquake; the 1964 Alaskan (M_w=9.2) earthquake; the 2004 Sumatra–Andaman (M_w=9.1) earthquake, and the 2011 Tohoku-Oki (M_w=9.0) earthquake (Satake and Atwater, 2007). Each of these large earthquakes had rupture lengths of 800–1600 km, coseismic displacements of 15–50 m, and deformed huge areas. For example, the 1960 Alaskan earthquake along the Aleutian subduction zone (Wyss and Brune, 1967) deformed a region over 350 km wide and about 800 km long (Fig. 4.29). The hanging wall experienced spectacular uplift (maximum of 8–10 m) across a 200-km-wide zone on the north side of the Aleutian Trench, whereas, still farther north, the hanging wall subsided up to 2.5 m within a 150-km-wide zone (Fig. 4.29). It may at first seem surprising that

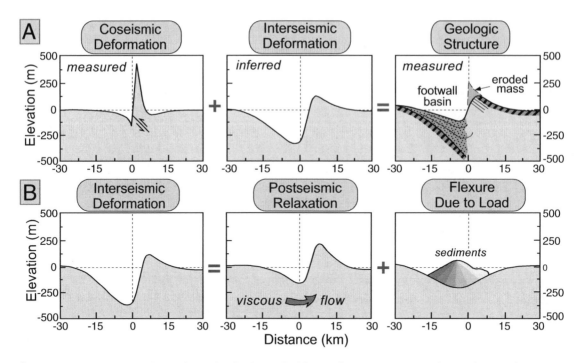

Fig. 4.30 Conceptual model of deformation in thrust-faulting regimes that produce observed geological structures.
A. The mismatch between the observed coseismic deformation and the geological structure (which integrates over many earthquake cycles) defines the pattern of interseismic deformation. B. The interseismic deformation is interpreted to result from viscous flow in the mantle and isostatic responses to sediment loading and erosion. Note carefully the horizontal offsets in the deflections from one panel to the next because these indicate how the distribution of deformation varies for each contributing variable. Modified after Stein *et al.* (1988).

the hanging wall, which is sitting above the thrust fault, would experience subsidence, as well as coseismic uplift. Two factors contribute to this pattern. First, thrusting of the hanging wall causes a redistribution of regional loads on the crust. Given an elastic crust with some specific wavelength of flexure, such a thrust load should produce a sinusoidal flexure that is dampened with increasing distance from the load (Turcotte and Schubert, 1982). Second, during the interseismic phase, part of the subduction interface is locked. As a consequence, part of the overlying crust bulges upward or downward as crustal plates on either side of the locked zone move inexorably toward each other. Much of this elastic strain is recovered by coseismic uplift or subsidence during the subsequent earthquake. In most earthquakes, the more distal, damped flexure is geomorphically undetected, but in a large earthquake (as in

Alaska), where an extensive, crenulated shoreline can serve as a marker of uplift, the variable flexure leaves a clear signature. Whereas the largest earthquakes appear to be limited to subduction zones, numerous destructive $M=8$ thrust earthquakes occur due to continent-to-continent collisions and within intracontinental deformation belts. Along with many smaller temblors, these earthquakes underpin the growth of topography in ranges like the Himalaya and Tien Shan and help produce the landscapes that tectonic geomorphologists like to study.

Despite the initial coseismic asymmetry during thrusting, which results in accentuated hanging-wall uplift compared to the footwall (Fig. 4.28A), erosion of this uplifted hanging wall, deposition in the depressed footwall, and isostatic adjustments to these changing loads and to the coseismic load lead to a geological structure in which subsidence of the footwall basin is commonly comparable to the bedrock uplift of the hanging wall (Fig. 4.28B). Thus, the preserved geological structure can be conceptualized as resulting from both coseismic and interseismic deformation (Fig. 4.30). The relative importance of the coseismic versus the interseismic deformation in creating the preserved geological structure depends on the size of the structure with respect to the strength of the crust and on the efficiency of surface processes in eroding and redistributing the load. If the load is small compared to the crustal rigidity, the isostatic response to the load will be less, and additional footwall subsidence due to sediment loading of the flexural depression will be limited. Conversely, if erosional processes are highly efficient, they will tear down the uplifted hanging wall and cause enhanced sediment loading (and subsidence) on the footwall.

Commonly, the amount of displacement on a thrust fault dies out toward the thrust tip, causing the hanging wall to fold adjacent to the thrust tip. The coseismic displacement during the Kern County earthquake displays this folding whereby the greatest vertical change occurs 8–10 km from the fault trace (Fig. 4.28A). Repetitions of this pattern of deformation through many successive earthquakes would generate a hanging-wall anticline.

In the arid terrain of northern Algeria where the 1980 El Asnam ($M=7.3$) earthquake occurred, patterns of hanging-wall folding and surface rupture are well illustrated by coseismic displacements (Philip and Meghraoui, 1983). Across the crest of hanging-wall anticlines, grabens due to bending-moment faulting (see description in next subsection) opened in several sites. The obliquity of the shortening vector with respect to the trace of the thrust can be judged by the orientation of these crestal grabens. Where the trend of the thrust trace and of the normal faults bounding the grabens are parallel, shortening occurred approximately perpendicular to the trace of the thrust (Fig. 4.31A and C; see also Fig. 4.15 for analogous graben orientation on the Ostler Fault). On the other hand, where the grabens trend obliquely to the anticlinal crest and to the thrust trace,

shortening is inferred to have been oblique to the thrust trace (Fig. 4.31B and D). This obliquity of the shortening vector is also consistent with the orientation of the conjugate shears that develop along the thrust front. In each case, the locally observed strain can be seen to result from the orientation of the local stress field (in the case of the grabens or conjugate shear zones) or regional stress field (in the case of the thrusts) with respect to the orientation of the fault surfaces (Fig. 4.32).

Flexural-slip and bending-moment faults

In reality, fault geometries are often far more varied than might be expected with simple models related to regional stress fields (Fig. 4.1). In part, this complexity is due to inhomogeneities in rocks: they have variable strengths, may be bedded, and may have had a diverse deformational history prior to the current deformation. The orientation of weaknesses within rocks that are subjected to stresses can exert a strong control on how they deform. Anyone who has observed books tilted on a shelf can recognize that tilting of rigid entities (like rock strata) requires slip between the rigid blocks. Thus, faulting along bedding planes is common whenever relatively rigid strata are tilted. Similarly, folding of a deck of cards results in relative slip between each of the cards. When strata are folded, if the initial length of each bed is preserved, then *flexural-slip faults* develop along bedding planes to accommodate the differential motion between adjacent beds.

When strata are folded, local stresses are created because the convex side of a folded bed is lengthened, whereas its concave side is shortened. The folding can be considered analogous to bending an elastic plate around a fold axis, such that equal and opposite moments are applied at the ends of the plate. The faults that result from the tensile stresses along the convex regions of the folded plate or from compressive stresses in the concave regions are called *bending-moment faults* (Fig. 4.33). Thus, normal faults are expected to form across the convex regions in order to accommodate length changes along these surfaces, whereas thrust faults will

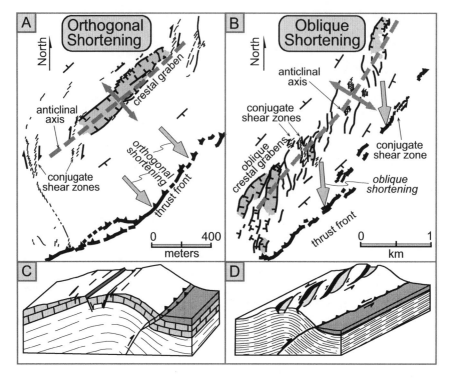

Fig. 4.31   El Asnam thrust front and crestal grabens.
A. Map of surface ruptures related to orthogonal shortening during the 1980 El Asnam earthquake. Parallelism of the crestal graben (bending-moment faulting in the hanging wall) and the frontal thrust and the orientation of the shear zones suggest that shortening was perpendicular to the thrust trace. Note that the orientation of the conjugate shear zones is consistent with the local bending-moment tensile stresses across the crest of the anticline, rather than with the regional stresses responsible for the thrusting. B. Surface ruptures due to oblique shortening, defining *en echelon* crestal grabens and shear zones with respect to the thrust trace. Note the differing orientations of the conjugate shear zones in the map area. In the crestal region, anticlinal flexure creates bending-moment tensile stresses, whereas shearing due to oblique shortening controls orientation of tensile stresses along the thrust front. C,D: Schematic block diagrams showing crestal graben orientation for orthogonal and oblique shortening. Modified after Philip and Meghraoui (1983).

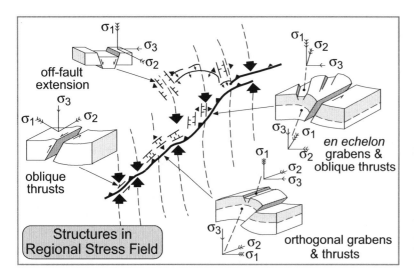

Fig. 4.32   Styles of deformation as related to stress patterns during thrust faulting.
Dashed lines represent the approximate orientation of the regional maximum compressive stress. Local stress variations due to flexing or shearing of hanging-wall anticlines result in crestal grabens. Modified after Philip and Meghraoui (1983).

form in the concave regions. Anticlinal grabens, such as those seen at El Asnam (Figs 4.31 and 4.32), and out-of-syncline thrusts are examples of bending-moment faulting.

Complex combinations of normal, thrust, bending-moment, and flexural-slip faults are exhibited by ruptures during the El Asnam earthquake (Fig. 4.34) in areas where a strong coupling existed between bedding orientations and fault geometries (Philip and Meghraoui, 1983). Folding of strata caused flexural-slip

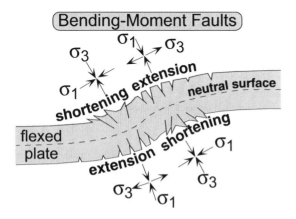

Fig. 4.33 Bending-moment faults resulting from flexure of an elastic plate.
Folding causes stretching along the outer, convex surface of the warped layer, whereas shortening is induced along the inner, concave surface.

faulting, which is expressed at the surface as a succession of high-angle reverse faults. Bending-moment stresses caused out-of-syncline thrusts to develop due to compression along the concave surface of folds. As the main thrust plane ramped toward the surface, it encountered steeply tilted strata and was diverted along weak bedding planes. In places as the fault followed bedding, it passed through a vertical orientation and became overturned. This overturning caused the displaced ground surface along the fault plane to appear analogous to a normal fault, rather than a thrust fault (Fig. 4.34). The diversity of structures associated with the El Asnam earthquake and the contrasts in their orientations clearly emphasize the importance of examining the full spatial array of rupture geometries and investigating the role of underlying bedding during faulting before using coseismic rupture patterns to specify the stress field at the time of the earthquake.

Folds

Relationships of folds to faults

Many faults do not rupture the Earth's surface. Typically, faults nucleate within the brittle crust at depths of several kilometers, and, as they

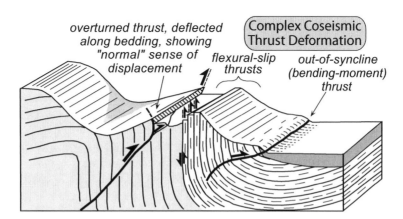

Fig. 4.34 Complex fault geometries in folded, stratified bedrock.
Multiple rupture patterns result from thrust faulting in folded strata, as exemplified by the 1980 El Asnam rupture. Coseismic slip along bedding planes creates flexural-slip faults and overturning of the thrust plane in places. Footwall folding leads to out-of-syncline thrusting. Modified after Philip and Meghraoui (1983).

accumulate displacement, they commonly propagate toward the surface. Until they actually break the Earth's surface, however, they are termed *blind faults*. In basins with a thick sedimentary fill, it is not uncommon for even large earthquakes to fail to rupture the surface. Instead, the highly localized strain that occurs along a fault plane at depth is accommodated by folding within the strata overlying the fault tip. Because sedimentary beds typically deform in predictable ways, patterns of folding can be intimately linked to the underlying fault geometry. Consequently, whenever a fold's geometry and its evolution through time can be documented, much can be learned about the geometry of faulting in the subsurface. This linkage is particularly important in some urban areas, such as Los Angeles and Seattle, where the seismic risk is high, but where many of the faults capable of causing destructive earthquakes are buried beneath thick Quaternary sediments. Learning to interpret the folds at the surface and to link them to coseismic displacements represents a significant, but worthwhile, challenge for tectonic geomorphologists.

The geometry of the faults themselves is commonly influenced by the mechanical properties of the rocks through which they rupture. As might be expected, thrust faults in the subsurface exploit zones of mechanical weakness, and, within sedimentary rocks, they often follow bedding planes. At irregular intervals, they ramp upward from an underlying bedding-plane decollement to an overlying one. In contrast to sedimentary rocks, igneous or metamorphic rocks commonly display more isotropic mechanical properties. Folding of formerly planar strata in metamorphic rocks distorts mechanically weak layers. As weak and strong layers become more spatially disorganized, they have no consistent orientation with respect to the regional stress directions, and overall the rock becomes more isotropic. Both planar and curving ruptures occur within isotropic bedrock. Kink-like changes in fault angle may occur where rock bodies with contrasting mechanical properties are juxtaposed.

The trajectory of a thrust fault within sedimentary rocks is often visualized as following a staircase-like pattern with long *flats* connected by shorter *ramps* along which the thrust steps upward through the stratigraphy. Thrusting of a hanging wall along a fault comprising ramps and flats causes uplift of the hanging wall above the ramps, whereas rocks are commonly translated laterally without uplift above the flats. The geometric consequence of this pattern is that folds will be created above each ramp. Folds also occur above faults that cut through more mechanically isotropic rocks. Deformation above the tip lines of faults and changes in the angle of the fault cause differential uplift at the surface and create folds.

In many cases, deposition occurs synchronously with folding. Newly deposited strata that are associated with an active fold are termed *growth strata*, and they commonly provide the best means of deciphering the history of fold growth (Fig. 4.35). The reason that growth strata are particularly useful is that we know their initial geometry quite reliably: their upper surfaces are essentially horizontal at the time of deposition. Consequently, even when we do not know the pre-folding geometry of the bedrock that cores the fold, growth strata provide robust markers that track deformation subsequent to their deposition (Suppe *et al.*, 1992). When aggradation is more rapid than the rate of uplift of the highest part of the fold, growth strata will cover the entire fold and will faithfully record the dips, limb lengths, and geometry of its upper surface (Burbank and Vergés, 1994; Vergés *et al.*, 1996). Under these conditions, the fold may have no topographic expression at the Earth's surface, but the subsurface record will clearly show changes in the fold shape through time. If the rate of aggradation is less than the crestal uplift rate, then the fold will become emergent, growth strata will offlap the fold, and its uneroded topographic shape may more clearly reflect the geometric controls exerted by displacement along an underlying ramp (Burbank *et al.*, 1996c).

Models of folding

From among the many types of folds that have been described, only a few of the popular models are described here. When rocks in the

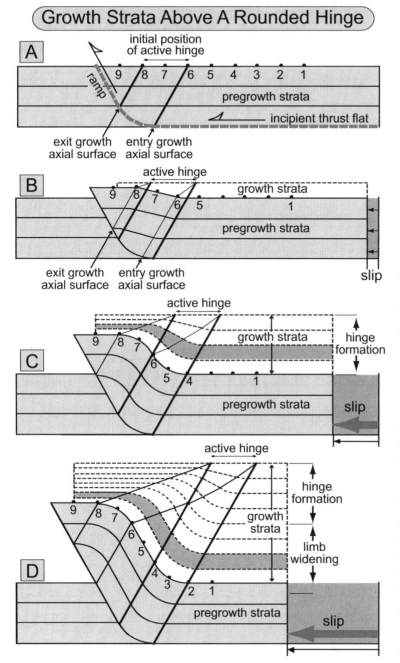

Growth Strata Above A Rounded Hinge

A
initial position of active hinge
9 8 7 6 5 4 3 2 1
ramp
pregrowth strata
incipient thrust flat
exit growth axial surface   entry growth axial surface

B
active hinge
9 8 7 6 5   growth strata   1
pregrowth strata
exit growth axial surface   entry growth axial surface   slip

C
active hinge
9 8 7 6 5 4   growth strata   1   hinge formation
pregrowth strata   slip

D
active hinge
9 8 7 6 5 4 3 2 1   hinge formation   growth strata   limb widening
pregrowth strata   slip

Fig. 4.35 Geometry of growth strata.
Model of accumulation and progressive folding of growth strata above a flat–ramp transition with a rounded hinge zone. Growth strata accumulate at a constant rate with initially horizontal top surfaces. Beds are deformed as they are carried through the hinge zone. A. Geometry as incipient fault with a rounded hinge forms. Numbers denote positions at the base of the growth strata for tracking deformation. Axial surfaces define the hinge zone through which folding occurs. B. First growth strata are deposited. Note that the bed thins over the uplifted backlimb of the fold. C. Sufficient shortening has occurred to completely rotate growth strata through the hinge zone. D. Older growth strata widen above the backlimb whereas younger growth strata are deformed through the hinge zone. Modified from Hubert-Ferrari *et al.* (2007) and Suppe *et al.* (1997).

hanging wall of a thrust fault are carried up and across a buried flat–ramp–flat transition, they are forced to deform into a *fault-bend fold* (Suppe, 1983) due to the nonplanar shape of the thrust surface. Both in front of and behind the ramp, apparently unfolded strata may be pre-sent in the hanging wall. But over the ramp itself, a fold will grow. For a given length and angle of footwall ramp and for a thickness of the hanging wall, the geometry of folding can be predicted, if it is assumed that bed length and thickness do not change during folding

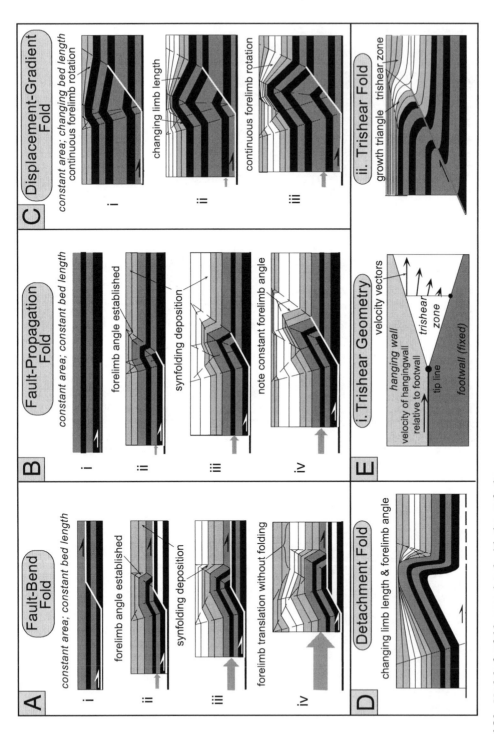

Fig. 4.36 Models for folds associated with thrust faults.

Each fold is depicted where uplift of the anticlinal crest is slower than the rate of sediment accumulation in the unfolded regions off the flanks of the folds. This difference in rates creates growth strata that record the development of the fold. Note that the final shape of the deformed bedrock can be very similar among these different models, but that the geometry of the growth strata differs markedly among them. Reliable reconstruction of step-by-step fold growth is commonly dependent on analyzing associated growth strata. A. Fault-bend fold. B. Fault-propagation fold. C. Displacement-gradient fold. D. Detachment fold. E. Trishear fold: (i) geometry of the deforming region of a trishear fold, showing rotation of slip vectors and change in magnitude from the top of the trishear zone to its base; (ii) model of trishear fold and growth strata above a thrust. Modified after Wickham (1995) and Allmendinger (1998).

(Fig. 4.36A). In the fault-bend fold model, *axial surfaces* are defined by kink bends in the deforming hanging-wall strata. Between any two adjacent axial surfaces, beds remain parallel that were parallel to each other prior to folding, and these parallel beds define a *dip domain*. Although fault-bend folding due to thrusting is emphasized here, folds obeying similar geometric "rules" have also been modeled for ramps along normal faults.

In a *fault-propagation fold*, a blind thrust creates a new ramp by progressively propagating upward toward the surface (Suppe and Medwedeff, 1990). The thrust never needs to reach the surface, but, through time, it accumulates more and more displacement. Folding occurs because there is a gradient in the amount of displacement along the ramp (Fig. 4.36B); at the tip of the thrust, no displacement occurs, whereas maximum displacement occurs at the base of the ramp. In both the fault-bend and fault-propagation fold models (Suppe, 1983; Suppe and Medwedeff, 1990), the dip of the forelimb is established during the first increment of folding, and it retains this dip throughout subsequent growth. Note that, for both models, the geometry of the growth strata changes with each increment of folding (Suppe *et al.*, 1992) and can become very complex (Fig. 4.36). If smaller-scale ramps and flats were incorporated into the overall ramp in either model, more axial surfaces and dip domains would be introduced and would create an increasingly complicated folding geometry.

Commonly, length and thickness of beds do not remain constant during folding. If the requirement for constant bed length is relaxed and only bed area is preserved, then *displacement-gradient folds* (Wickham, 1995) can be defined (Fig. 4.36C). As in fault-propagation folds, the amount of displacement varies systematically along the ramp. Because bed lengths change, however, the forelimb is allowed to rotate. Even if the thrust tip does not propagate, such folds can continue to amplify by accumulating more displacement along the ramp behind the thrust tip. In the absence of growth strata, the final geometries of fault-propagation and of displacement-gradient folds may be nearly

indistinguishable. Growth strata, however, can reveal whether the forelimb rotated or was fixed and will display distinctive differences both over the crest of the fold and in its forelimb (compare Fig. 4.36B and C).

A *detachment fold* forms by buckling above a fault that is subparallel or parallel to original layering (Fig. 4.36D). No ramp is needed to create such folds: they grow simply because displacement dies out toward a fault tip. At some stage, the fault may propagate upward through the overlying detachment fold, and the resulting final shape may be geometrically similar to a fault-propagation fold. Detachment folds are commonly associated with readily deformed strata, such as evaporites or shales, because such strata provide weak horizons along which the detachment can propagate, and these weak strata flow readily into the cores of growing folds. Limb rotation during folding is commonly observed in detachment folds, whereas dip domains and linear axial surfaces are typically difficult to define due to the more continuous curvature of the beds on the flanks of the fold (Hardy and Poblet, 1994; Poblet *et al.*, 1998; Vergés *et al.*, 1996).

A *trishear* fold (Fig. 4.36E) develops when a single fault at some depth expands outward to form a triangular zone of distributed shear (Allmendinger, 1998; Erslev, 1991; Hardy and Ford, 1997). This triangular zone is symmetrical with respect to the dip of the fault. Within this trishear zone, slip varies systematically in both orientation and magnitude. At its top, slip vectors match the slip of the hanging wall and are parallel and equal to that of the master fault. At the base of the trishear zone, the slip decreases to zero. In between these two boundaries of the trishear zone, the slip vectors systematically decrease in magnitude and rotate away from the hanging-wall slip direction toward parallelism with the lower (no-slip) boundary of the trishear zone. The shearing that results from differential slip causes both bed thickness and forelimb dips to change as the fold grows. The ratio between the rate at which the tip line of the fault propagates and the amount of slip on the fault itself controls the geometry of the forelimb. Low values of propagation-to-slip ratio cause the

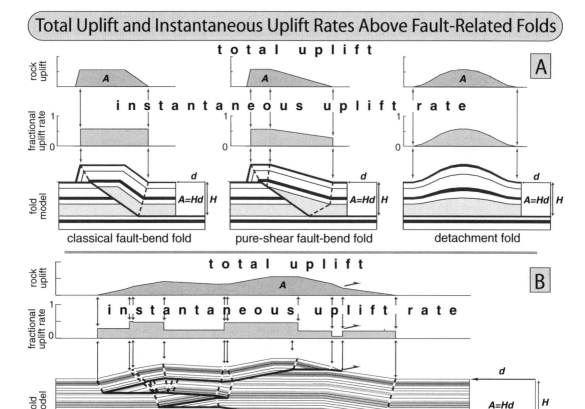

Fig. 4.37   Patterns of rock uplift above folds.
A. Depictions of rock uplift patterns associated with different types of fault-related folds. Instantaneous uplift rate represents the relative rate of vertical deformation predicted for an additional increment of folding at that instant of the fold's development. Abrupt rate changes across axial surfaces (dashed lines associated with bends in fault ramp angles) commonly indicate fold growth through kink-band migration, whereas gradual changes in uplift rates indicate limb rotation (compare the classical versus the pure-shear fault-bend fold models). Total uplift represents the sum of each increment of vertical uplift during the history of folding up to the stage depicted. Whereas total uplift and the instantaneous rate exactly mimic each other for the detachment fold, a mismatch exists for the other folds because fold shape changes through time. Note the equivalence between the area of total uplift (A) and the area of total shortening (displacement d times height H). B. Complex fault-bend fold model with growth and shortening accommodated by kink-band migration. Uplift rate equals sine of the underlying fault dip times the horizontal displacement as translated upward to the surface via axial surfaces. Pattern of total uplift is poorly correlated with instantaneous rates. Modified from Hubert-Ferrari *et al.* (2007).

forelimb to thicken and create tight folding in the trishear zone. Conversely, high values of propagation-to-slip ratio cause less thickening and more open folding (Allmendinger, 1998).

As originally described, both fault-bend and fault-propagation folding were characterized by axial surfaces that create kink bends in the hanging wall and that abruptly divide the

hanging wall into contrasting dip domains. In the real world, it is quite common to observe these dip domains or "panels" of folded strata that all display similar dips. The transition from one dip domain to the next, however, may not be abrupt (a kink bend), but may occur gradually through a change in curvature. In recognition of this curvature, fault-bend and fault-propagation

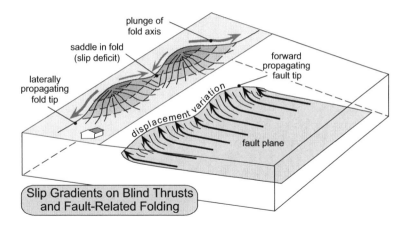

plunge of
fold axis

saddle in fold
(slip deficit)

forward
propagating
fault tip

laterally
propagating
fold tip

displacement variation

fault plane

Slip Gradients on Blind Thrusts
and Fault-Related Folding

**Fig. 4.38 Lateral propagation of
blind thrust faults and surface
folds.**
Displacement variations in the
subsurface fault are directly related
to the magnitude of rock uplift at
the surface. The noses of plunging
folds occur above fault tips, and
structural saddles mark zones of
fault linkage or slip deficits.

fold models have been modified to incorporate the possibility of rounded hinges (Suppe *et al.*, 1997), rather than relying exclusively on kink bands (Fig. 4.35). Bedding dips in the forelimbs of trishear folds tend to change continuously, rather than abruptly. Therefore, kink bands and dip domains do not commonly occur in such folds. These differences in the geometry of growth strata help to discriminate among different fold models.

Owing to differences in the ways that beds deform within folds of different types, the patterns of rock uplift that are associated with each fold model can vary considerably (Fig. 4.37) and can be diagnostic of key characteristics of both the underlying fault geometry and the way in which shortening is accommodated, such as by pure shear or simple shear. Depending on the fold model, spatial variations in uplift rate that result from an increment of shortening at any given moment in the fold's history may or may not mimic the pattern of total uplift over the life of the fold (Fig. 4.37). For fault-bend folds, the mismatch results from kink-band migration during fold growth, which causes a continual evolution of the spatial pattern of rates. When it is possible to document the changes in uplift rates through time, such as through surveys of multiple, deformed, and dated fluvial terraces, distinctions can be drawn between a fold that is undergoing limb rotation versus one growing by kink-band migration: the former causes gradual spatial changes in short-term uplift rates, whereas the latter creates abrupt rate changes (Fig. 4.37).

More complex fault geometries, such as might occur due to multiple wedge thrusts within layered rocks (Fig. 4.37B), or fault-bend folds with rounded hinges (Fig. 4.35), can create both diverse and intricate patterns of instantaneous uplift rates. The greater the complexity, the more difficult it becomes to confidently interpret subsurface fold geometries from patterns and rates of rock uplift.

Lateral fold growth of folds as three-dimensional features

Because faults grow by extending laterally, as well as forward, fault growth in the subsurface will cause lateral propagation of the tips of folds at the surface (Fig. 4.38). Consider the case of a single blind thrust fault. Folding at the surface will wane above either end of the buried rupture. The consequence of this decrease is to generate folds with doubly plunging terminations. Variations in displacement along the underlying thrust should also be reflected by the amplitude of folding at the surface. Thus, the structural crest of the fold should vary in height, and the fold axis should plunge toward zones where less displacement occurs on the underlying fault. As illustrated previously (Fig. 4.12), individual faults may simply lengthen, or they may link together with adjacent faults. In the latter case, if the linked structures are blind thrust faults, variations in displacement along them will create multiple plunging folds at the surface, with structural saddles marking the zones of linkage (Fig. 4.38).

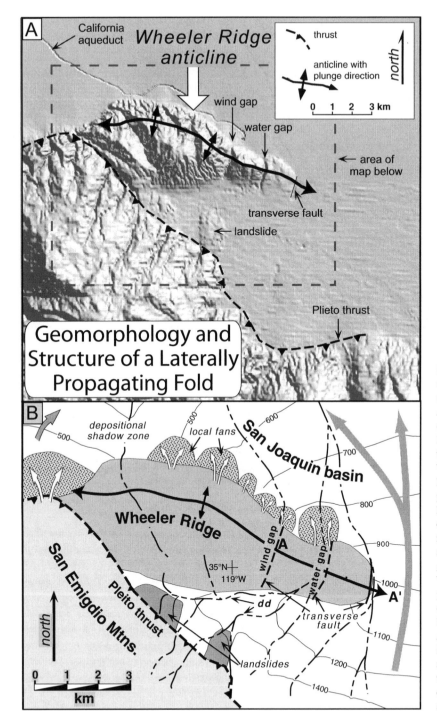

Fig. 4.39 Wheeler Ridge, a plunging fold in California.
A. Shaded relief 30-m DEM of Wheeler Ridge fold and Plieto thrust fault system in the southern San Joaquin Valley, California. Wheeler Ridge anticline plunges and propagates to the east. Dissection is more intense on the more steeply dipping forelimb of this asymmetric anticline. B. Map of geomorphology around Wheeler Ridge. Stream patterns indicate deflection of streams toward the east by the growing fold. Antecedent streams persist in water gaps, but these streams appear localized by transverse faults. dd: deflected drainage. Modified after Burbank *et al.* (1996c).

Why should one care about the shape of a growing fold or various folding models? Folds are a fundamental component of many orogenic settings. Folding is ubiquitous above blind thrusts, as well as almost any thrust fault that ruptures the surface. When trying to reconstruct the deformation that has occurred in a given region, most previous studies tended to focus almost exclusively on the displacement along faults. Although more subtle offsets may occur in folds, the deformation represented by them can constitute an important fraction of the total strain. Much of the landscape that is preserved today in active fold-and-thrust belts results from the initiation and amplification of growing folds (Fig. 4.39). The shape of those folds determines the uneroded geometry of the land surface. In such cases, the dip of the forelimb and backlimb determines the surface slope and the position of the drainage divide. The shape of the fold can, therefore, strongly influence both the nature and efficiency of surface processes that erode and redistribute mass. If we want to understand the modern landscape as a product of the long-lived interactions of surface processes with deforming structures, we need to know how those structures evolved through time.

Geomorphology provides one key to this understanding. In cross-section, most folds have a steeper forelimb than backlimb (Figs 4.36 and 4.37). Even when the uplifted surfaces have been dissected by erosion, contrasts in the dip of the limbs of the folds are commonly discernible in the modern topography and indicate the "facing direction" or orientation of the underlying blind thrust. For example, stream lengths are often asymmetrical across a fold. Shorter, steeper streams occur on the forelimb, causing the drainage divide to be initially displaced toward the forelimb. Information about how folds have grown laterally can also be revealed by stream patterns. As a plunging fold begins to grow laterally, rivers that flow across the axis of the fold must either incise into the uplifted area or be deflected out of their present courses and around the nose of the fold. Initially, rivers usually tend to maintain their course across a fold. To do so, they need sufficient stream power to erode enough of the newly uplifted material so that they can maintain a downstream gradient across the fold axis (Burbank *et al.*, 1996c; Amos and Burbank, 2007). Through the process of erosion, they create *water gaps* across the fold axis (Fig. 4.39). If at some time the rate of lowering of the stream bed by erosion is insufficient to keep pace with the rate of structural uplift of the fold, then the stream will be "defeated," a

wind gap will develop along the abandoned river course, and the river will flow subparallel to the limb of the fold until it finds a low point where it can traverse the fold (Fig. 4.39). The pattern of rivers adjacent to folds and the presence of wind and water gaps can be interpreted to indicate the direction of propagation of a fold (Jackson *et al.*, 1996). If the time of abandonment of the wind gaps or the age of the uplifted surfaces along the fold's flanks is known, then the rate of propagation can also be defined (Medwedeff, 1992).

It is becoming apparent in many areas that blind thrusts pose major seismic hazards for urban areas. Most of the recent destructive earthquakes in the Los Angeles basin, for example, occurred along blind thrusts. Even when enlightened city planners have prohibited building adjacent to known faults, the failure to recognize the potential for strong accelerations due to earthquakes on buried structures has led to inadequate building standards and huge loss of property and lives during recent earthquakes. To the extent that the geometry of folds can be used to infer the shape of underlying faults and to the extent that unravelling the history of folding provides insight into rates of deformation, analysis of folds can provide fundamental constraints on tectonic rates, patterns of past and probable future deformation, and seismic hazards.

Summary

A diverse array of faults and folds has formed at the Earth's surface in response to variations in stress fields, inhomogeneities in rocks, contrasts in crustal strength, and interactions among multiple structures. It seems clear that the build-up of stresses across and along rupture surfaces causes earthquakes. Unfortunately, direct measurements of stress are difficult to make, and understanding of changes in the distribution of stresses before, during, and after faulting is still incomplete. Nonetheless, earthquakes can be usefully thought of as part of a cycle in which strain that accumulated during interseismic intervals is fully or partly recovered during rupture events. Contrasting models for controls on stress release propose that: (i) faulting occurs whenever a certain stress threshold is attained, such that strong patches, or asperities, on the fault control the rupture pattern; or (ii) earthquakes release stress until a minimum threshold stress is achieved, in which case barriers, or patches of the fault plane that remain unruptured, may control the pattern of faulting. It is still not clear whether either model adequately describes the behavior of many faults.

It has been proposed that some faults are typified by characteristic earthquakes in that they display a similar rupture length, magnitude of displacement, and distribution of offset along the rupture in successive earthquakes. Rupture patterns along such faults could be controlled by long-lived asperities. If characteristic earthquakes exist, knowledge of their displacement history provides strong predictive capabilities. Given the societal importance of earthquake prediction, we need to improve our understanding of recurrence intervals, fault strength, characteristic behavior of faults, and controls on fault displacement and timing. Through field observations, patterns of displacements along faults and styles of fault growth are becoming better known.

In a given structural and geological setting, a predictable relationship commonly exists between the maximum amount of displacement on a fault and its length. Whereas a roughly triangular pattern of length versus displacement is predicted for faults that grow laterally without restriction, many faults display a more bow-shaped displacement gradient. Such a pattern can result from impediments to tip propagation that restrict fault lengthening, despite continued slip on the fault. In such cases, faults extend to their full length early in their history, rather than continually lengthening. Segmentation of faulted range fronts and the temporal persistence of segment boundaries provide support for the concept of

long-lived barriers to fault propagation. On the other hand, arrays of faults commonly interact such that slip deficits between major faults are filled by displacement on minor faults. Furthermore, compensation among multiple faults produces smooth variations in cumulative displacement across the array.

Our ability to refine and choose among the contrasting models for controls on earthquake cycles, accumulation of displacement, characteristic earthquakes, and fault-zone segmentation depends on developing more complete data for both past offsets and present rupture patterns. Part of these data will come from studies of the tectonic geomorphology of fault zones. Predictable patterns of fault rupture and geomorphic expressions of faulting are associated with strike-slip, normal, and thrust faults and can be interpreted in terms of imposed stresses. Nature, however, is complicated, and, in nearly every "compressional" or "extensional" setting, an overlapping array of growing folds and active normal, thrust, and strike-slip faults will be found. It is the job of the field geologist to sort out the relationships of these structures to local stress fields and to define how these stresses interact with inhomogeneities in the underlying rocks to produce the observed deformation.

Geomorphological studies can successfully document deformation following faulting when surface features have been offset. But what about subtle folding of the surface during faulting or far-field deformation? Such deformation is often undetectable in geomorphological studies. Similarly, the build-up of strain prior to faulting and the proportion of recovery of interseismic strain during faulting usually cannot be resolved through examination solely of coseismic displacements. Instead, precise surveying techniques are required to define both near-field and far-field deformation prior to, during, and after earthquakes. Such geodetic measurements are the subject of the following chapter.

5

Short-term deformation: geodesy

Owing to plate motions, radioactive heat production, and gravity, the surface of the Earth is in constant movement at a regional scale. Deformation of the Earth's surface as a consequence of faulting and folding results from differential motion of adjacent parts of the crust or from changes in thermal buoyancy. If contrasting velocity vectors can be resolved for two separate locations, then the amount of deformation that must be accommodated by the intervening region can be determined. Consequently, delineation of the regional pattern of crustal motion and the recognition of differences from one area to the next serve to pinpoint where and how much tectonic deformation should be occurring. A few decades ago, for example, transform plate boundaries were typically conceived of as narrow, elongate zones of focused deformation bounded by rigid plates. With this model, a major strike-slip fault like the San Andreas Fault was defined as the boundary between the North American and Pacific Plates. Today, precise surveys clearly indicate that this plate boundary is instead a diffuse zone of deformation more than 500 km wide (Minster and Jordan, 1987; Molnar and Gipson, 1994; Ward, 1990) and that the San Andreas Fault is only one among many faults that accommodate the relative motion between these plates.

Measurements of deformation can be made at both regional and local scales. Whereas regional patterns of relative and absolute crustal velocity indicate where strain should occur and may suggest the likely orientation of local stress fields, local measurements of deformation due to folding and faulting serve to characterize the style and magnitude of the displacements that accommodate differential crustal velocities. Such local information reveals how the land surface deforms due to individual seismic events, it provides data on the variability of displacement along the length of any surveyed structure, and it illustrates how the Earth's surface deforms between major earthquakes (interseismic deformation) or in the absence of significant seismicity (aseismic creep). Careful measurements of local deformation, therefore, provide insight into specific crustal responses, such as the wavelength, amplitude, and orientation of folding, to imposed stresses.

Geodesy, as used in geological studies, is the measurement of the exact position of geographic points. Typically, repeated surveys are made of the same locations, for example, before and after an earthquake, or of specific sites in the interior of two contiguous plates. In order to obtain the most reliable geodetic history of deformation, measurements in a long time series are preferable. The duration of such surveys is limited by the interval over which appropriate instrumentation has been available and the date when initial baseline measurements were made. By comparing the geodetically determined positions of various sites through time, both the magnitude of spatial changes and the rate at which they are occurring can be calculated.

Tectonic Geomorphology, Second Edition. Douglas W. Burbank and Robert S. Anderson.
© 2012 Douglas W. Burbank and Robert S. Anderson. Published 2012 by Blackwell Publishing Ltd.

Numerous unresolved questions motivate geodetic studies. The fundamental question of whether deformation rates that can be measured at decadal (geodetic) time scales are also persistent through time underpins many studies (Fig. 5.1). If rates were steady, this persistence would provide a potent basis for predictive deformation models based on presently measured rates. Plate-motion rates based on seafloor magnetic anomalies commonly indicate steady rates across spans of several million years (DeMets *et al.*, 1990, 1994). But, are they truly steady, and, if not, how much do they vary and at what time scales? When H.F. Reid (1910) propounded the *elastic rebound theory*, he postulated a gradient in crustal deformation across an active fault (Fig. 4.2). Then, as now, the steadiness of strain accumulation during the interseismic interval remains uncertain. Do rates accelerate or decelerate at predictable intervals within the seismic cycle? If so, could these changes be used to predict when an earthquake was going to occur? Many paleoseismic studies use observed displacements of stratigraphic or structural markers across faults to calculate coseismic slip. But, did all that slip accumulate during earthquakes, or were there two phases of slip: coseismic slip during the earthquake followed by afterslip in the weeks or months following the quake? If afterslip is important, how much of the total observed slip does it represent? Reconstructions of fault slip for many earthquakes suggest that coseismic slip at some depth in the crust is greater than at the surface (Delouis *et al.*, 2002; Simons *et al.*, 2002). If so, how does the surface catch up to the subsurface displacement? Does afterslip or creep occur on the shallow fraction of a fault or is strain distributed or absorbed in off-fault strain?

For phenomena such as afterslip that occur at time scales of months to decades, geodetic studies of similar or longer duration can potentially encompass the entire process (Johanson *et al.*, 2006) and, hence, provide durable insights on short-term crustal deformation. For problems spanning geological time scales that extend well beyond those of geodetic observation (Fig. 5.1), geodesy can never provide a

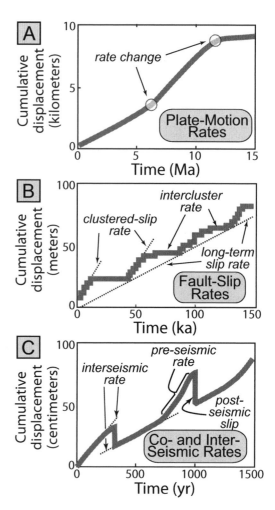

Fig. 5.1 Conceptual models of temporal variations in displacement rates at different time scales. Panels illustrate potential limitations to the extrapolation to longer time scales of geodetic rates measured at decadal time scales. A. To the extent that plate-motion rates are steady for millions of years, regionally measured geodetic rates may be reliably extrapolated. B. If earthquakes occur in clusters, geodetic measurements are likely to yield rates that are either faster or slower than the long-term slip rate. C. Geodetic sites that are primarily affected by individual faults can record a wide range of rates (including a reversal of direction) depending on the interval during the seismic cycle in which measurements are made. Modified after Friedrich *et al.* (2003).

complete answer. But, geodetic studies yield a critical current context of modern rates, as well as revealing their steadiness or variability in time and space.

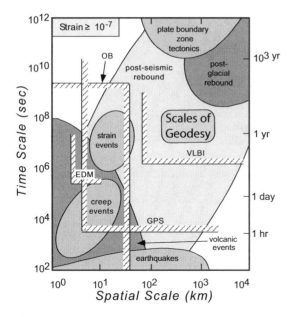

Fig. 5.2   Spatial and temporal scales of geological processes and various geodetic techniques.
Shaded ellipses depict typical ranges for classes of processes. Cross-hatched boundaries indicate the present range of various techniques for making geodetic measurements. EDM: electronic distance meter; GPS: Global Positioning System; OB: observation; VLBI: very-long-baseline interferometry. Modified after Minster *et al.* (1990).

The spatial scales (Fig. 5.2) across which geodetic measurements are collected vary from a few meters – for example, during calibrations of motion across a specific segment of an individual fault – to thousands of kilometers – such as when the relative motions of lithospheric plates are calculated (Dixon, 1991; Minster *et al.*, 1990). Clearly, different geodetic tools are appropriate for obtaining these highly varied scales of measurements (Fig. 5.2). The choice of tool and approach depends on both the scale of the problem and the precision and accuracy required. Based on the Global Positioning System (GPS) and very-long-baseline interferometry (VLBI), plate velocities can be calculated to ± 1–$2\,mm/yr$. This level of precision is unwarranted and is perhaps misleading when describing offsets of geomorphic features that have been displaced due to faulting. Because geomorphic surfaces typically have natural irregularities at scales of

greater than 1–2 cm, measuring and reporting geomorphic offsets with a resolution of 1–2 mm suggests an accuracy that simply does not exist.

In this chapter, various geodetic approaches are described, and examples of their applications are used to illustrate the types of positional information that can be obtained with these methods. *Near-field* techniques are those used to examine deformation on the scale of meters to a few tens of kilometers, whereas *far-field* techniques examine regional to global deformation fields. Simpler, traditional techniques of surveying are often appropriate for near-field observations and are described here first. Many far-field measurements require more complex, recently developed approaches and are described in the second half of the chapter. Several techniques, such as those using the Global Positioning System, are used in both near- and far-field applications.

Near-field techniques

Styles of strain

The geometry of displacement that occurs along a fault zone is strongly dependent on the strength of the crustal material both within the fault zone and on either side of it. Consider the situation in which the fault zone is very weak and is surrounded by strong rocks. Here, it is likely that nearly all of the displacement would be localized on the fault zone itself and that the adjacent regions would be relatively undeformed (Fig. 5.3A). Alternatively, if the fault zone is relatively strong, then strain will tend to be distributed across the terrain bounding the fault, and no strong change in the displacement gradient may be apparent as the fault is crossed (Fig. 5.3B). Simple models of displacement patterns (Fig. 5.4) clearly illustrate the influence of rock strength or rigidity. For example, when rocks bounding a fault have equivalent rigidities, a symmetrical pattern of strain is predicted on either side of the fault. As the rigidity of one side becomes increasingly high with respect to the other, less and less strain will occur within the more rigid block, and more will be focused

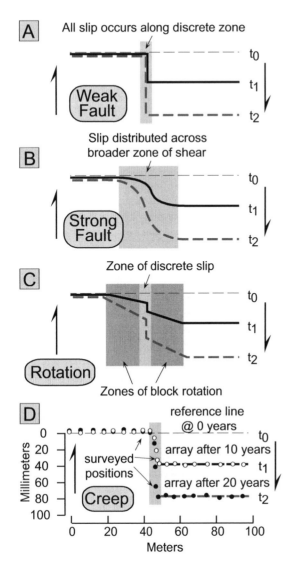

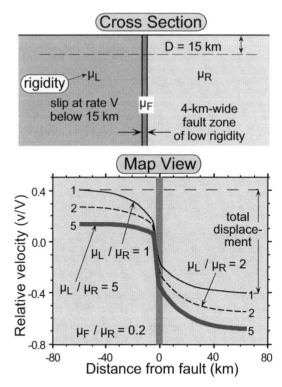

Fig. 5.4   Relative velocity as a function of contrasts in rigidity.

Patterns of strain across a fault. V is the rate of relative motion between the two blocks, and v/V is the fraction of the total motion (relative velocity) that is exhibited by any point. The rigidity of the fault (μ_F) is one-fifth that of the right-hand block (μ_R). The rigidity of the left-hand block (μ_L) is equal to, twice as great as, and five times as great as that of the right-hand block for the lines labeled 1, 2, and 5, respectively. In all cases, most of the displacement occurs along the weak fault zone. Note how the relative rigidity of the blocks affects the shape of the displacement within each block. For the highest rigidity (line 5, left-hand block), very little deformation occurs within the rigid block. Modified after Lisowski *et al.* (1991).

Fig. 5.3   Conceptual models for deformation associated with strike-slip faulting.

Displacement at three intervals, beginning with t_0, is illustrated. A. Fault zone is very weak and is creeping so that all motion occurs directly along fault. B. Fault zone is relatively strong and strain is distributed across broad zone on either side of the fault. C. Some creep occurs on the fault, and rigid block rotations occur on either side of the fault, taking up the additional motion. D. Schematic results from an alignment array along a creeping fault that is displaced about 80 mm in 20 years. Modified after Sylvester (1986).

along the fault zone and within the weaker block (Fig. 5.4). Thus, the asymmetry of displacement across a fault can be used to judge

the relative rigidities of the blocks on either side, whereas the relative proportion of the total displacement that is accommodated along the fault itself reflects its strength with respect to adjacent blocks. Faults that are continuously creeping will lead to lesser amounts of stored elastic energy in the adjacent blocks, whereas large earthquakes will be more likely to occur along faults that are "locked" and are relatively strong. Whereas the effect of differences in crustal rigidity on patterns of strain around faults is theoretically clear (Fig. 5.4), our knowledge of these differences is relatively limited. Because rigidity is only one of several factors that affect patterns of strain, it is challenging to tease out this single variable from any given data set. For example, viscoelastic responses to previous earthquakes can cause deformation for hundreds of years, such that patterns of strain can integrate post-seismic effects from earlier earthquakes on multiple faults (Dixon *et al.*, 2003). Most studies simply assume uniform rigidities for country rocks, or, where differences are defined, they are commonly based on inference, rather than observation. As more dense geodetic data sets become available, improved techniques to extract reliable viscoelastic properties should emerge.

Alignment arrays

Horizontal motions along strike-slip faults can sometimes be measured with alignment arrays. These arrays can consist of simple lines of nails that are hammered into the ground along a linear trend, typically less than 100 m long, that is oriented perpendicular to the trace of the fault. Because the initial orientation of the nails is well known, any subsequent displacement along the linear trend can be readily detected. Alignment arrays can reveal whether aseismic segments of a fault are creeping or whether they are locked. If they are creeping, alignment arrays can indicate whether a discrete offset occurs along the trace of the fault, and can document the breadth of the region within which significant shear is occurring (Fig. 5.3D). Sometimes man-made structures provide linear features that are serendipitously arranged across a fault. The offsets of fence lines, foundations, curbs, pipelines, railroad tracks, painted lines on highways, and even tire tracks, can provide useful indicators of both seismic and aseismic displacement.

Trilateration arrays

Traditional trilateration techniques utilized an array of triangulation monuments that were sited across a fault or region of interest and were typically spaced several kilometers apart, although they could be up to 50 km apart. A surveyed baseline would be determined between two benchmarks, such that their positions and separation distance were well known. Subsequently, measurements of the angles to other monuments from both benchmarks served to define the positions of these other points. These triangulation angles were measured with theodolites, which typically permit angular measurements with an accuracy of ≤ 1 arcsecond ($<1/3600$ of a degree). This uncertainty results in errors in the length calculations of about 3–6 in 10^6, equivalent to 3–6 mm/km. When considered in two dimensions, the length changes between each pair of benchmarks defines the distortion of the original geometry of the array and can be interpreted to result from differential tectonic motion within the array. More commonly today, distances within a fairly complex regional array are measured with some sort of laser-ranging electronic distance meter (EDM), which has a precision of about 1 ppm (1 mm/km). The laser beam is refracted as a function of the temperature of the air through which it passes. Consequently, measured distances also depend on air temperature, which, of course, varies during a survey. The resultant uncertainties dictate that, when several trilateration points are surveyed from any single site at any given time, the ratios of one distance to another are more accurate than the distances themselves.

Most trilateration surveys are used to investigate horizontal, rather than vertical, fault movements. Survey results can be used to assess both regional patterns of differential motion, as well as the magnitude and nature of slow, aseismic displacements on individual faults. As with alignment arrays, abrupt changes in displacements across an aseismic fault suggest that the fault is creeping (Fig. 5.3), whereas a smooth, undisrupted trend in the amount of displacement as a fault is approached or crossed suggests that considerable elastic energy is being stored in the volume of rock adjacent to the fault. In the context of large regional strains, identification of zones across which strain is most rapidly occurring and the differentiation between creeping and non-creeping faults using trilateration surveys can play an important role in defining those faults with the potential for destructive earthquakes.

Given a sufficiently dense array of surveyed benchmarks, distinctive contrasts in the style and magnitude of displacements at the regional scale can be delineated. With strike-slip faults, it is usually instructive to examine changes in the velocity field along transects oriented perpendicular to major faults and to the orientation of the regional strain field. In this situation, the data are most readily interpretable when the velocity data from many sites are projected on to a single plane perpendicular to the strain field. EDM surveys along the San Andreas Fault in California (Figs 5.5 and 5.6) clearly depict the contrast between locked and creeping segments of the San Andreas and adjacent faults (Lisowski *et al.*, 1991). Throughout this zone, the relative velocity of the North American Plate versus the Pacific Plate is about 50 mm/yr (DeMets and Dixon, 1999). A trilateration array east of Monterey Bay depicts sharp offsets in the velocity field across the San Andreas and Calaveras Faults, whereas there is little differential movement for 20–30 km on either side of these faults (Fig. 5.5A and B). Of the nearly 40 mm/yr of offset parallel to the San Andreas Fault that is recorded in this region, more than 75% of the offset is accommodated by creep along and immediately adjacent to the San Andreas and Calaveras Faults. In contrast, in the central Transverse Ranges of southern California (Fig. 5.6), <25 mm/yr of the differential crustal motion parallel to the San Andreas Fault is recorded along a 90-km-wide swath centered on the San Andreas Fault. Moreover, there is no evidence for creep (in terms of rapid spatial changes in the displacement gradient) as the San Andreas Fault is crossed. Given the absence of other major strike-slip faults in this transect and the considerable strain that is occurring across the fault, these data from the Transverse Ranges suggest that the potential exists for a major earthquake along this locked segment of the fault in the future. A weak fault zone is apparent in the Monterey array (Fig. 5.5), where most of the motion between the North American Plate and the Pacific Plate is accommodated in a

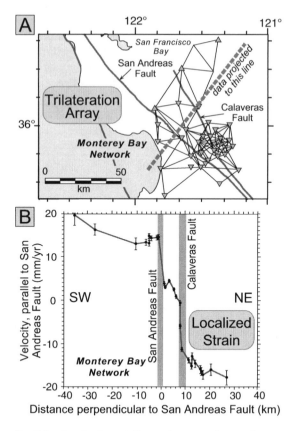

Fig. 5.5 San Andreas trilateration array in northern California.

A. Monterey geodetic network showing triangulation stations in the vicinity of the San Andreas and Calaveras Faults. B. Results of trilateration surveys between 1973 and 1989 across the San Andreas Fault zone near Monterey Bay. Data (projected on to a traverse oriented perpendicular to the trend of the San Andreas Fault) define the component of movement parallel to it. Long-term slip rates across this San Andreas region are about 35 mm/yr in this area, as indicated by the geodetic data. About 15 mm/yr of North America–Pacific relative plate motion accrues beyond the surveyed area. The San Andreas and the Calaveras Faults are clearly marked by abrupt changes in relative velocity. These discontinuities show that most of the relative motion across this 80-km-wide zone is accommodated by slip (creep) on these two faults. Little deformation occurs in the bounding blocks. Modified after Lisowski *et al.* (1991).

narrow swath that is 10–15 km wide. In fact, in that swath, 90% of the strain occurs on two highly localized fault zones (San Andreas and Calaveras Faults). In contrast, the broad regional strain pattern in the Transverse Ranges (Fig. 5.6)

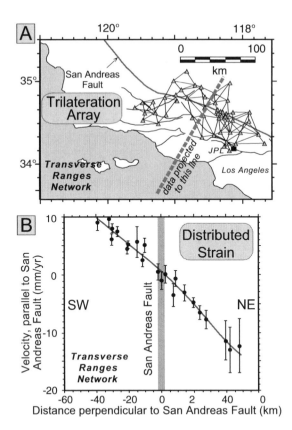

Fig. 5.6   San Andreas trilateration array in southern California.
A. Triangulation network in the vicinity of the Transverse Ranges of southern California. B. The Transverse Ranges data (1973–89) show no differential displacement across the San Andreas Fault, which can be interpreted as being "locked" in this region. Instead, strain is occurring across the entire surveyed zone. A suggestion of flattening at the ends of profiles could be interpreted as representing the expected sigmoidal shape of a deformation profile across a locked fault (see Fig. 5.3). Note that less than 25 mm/yr of relative plate motion is accommodated by displacements along the profile. This rate indicates that about half of the Pacific–North American relative plate motion occurs well beyond the San Andreas Fault zone. Modified after Lisowski *et al.* (1991).

emphasizes the potential for stored elastic energy within this array, whereas the fact that less than 50% of the relative plate motion occurs within the array itself indicates that considerable plate motion must occur along faults and folds beyond the surveyed area.

Precise leveling

Although trilateration arrays can successfully depict horizontal velocity fields, they have not been used extensively to examine vertical deformation at regional scales. Slow vertical movements can be defined using precise leveling surveys. The conceptual basis is similar to that for a trilateration survey. Established benchmarks and newly defined sites are resurveyed over the course of several years to decades in order to define the magnitude and rate of deformation in the intervening period. Because the absolute height of a benchmark can be poorly known with respect to the global reference frame, relative heights are used to calculate rates of vertical change in most leveling surveys. Naturally, the largest deformation signal will usually be recorded with the longest measurement interval. Thus, precise surveys of railway lines across the Alps in the first half of the 20th century provide a baseline against which more recent deformation can be calculated. On the other hand, the precision and accuracy of measurement techniques typically improve over time. Hence, a trade-off exists between the range of time spanned by a survey and the evolution of our instrumentation. Greater uncertainties commonly result from reliance on less accurate, older surveys. But, because the total signal is larger for longer intervals of measurement, these older surveys can sometimes provide excellent deformation histories, despite the imprecision of individual measurements.

As with any survey, the precision and accuracy of the measurements need to be evaluated. Slope-dependent errors can be assessed by plotting the tilt (the spatial derivative of the height changes) versus surface slope (the spatial derivative of the topography) (Jackson *et al.*, 1992). Because leveling lines typically comprise an extensive succession of surveyed sites, errors associated with each measurement accumulate and propagate through the entire survey (Fig. 5.7). With long survey lines, the accumulated uncertainty due to the imprecision of measurements can become considerably larger than the measured differential uplift. In such cases, it is sometimes possible to consider smaller subsets

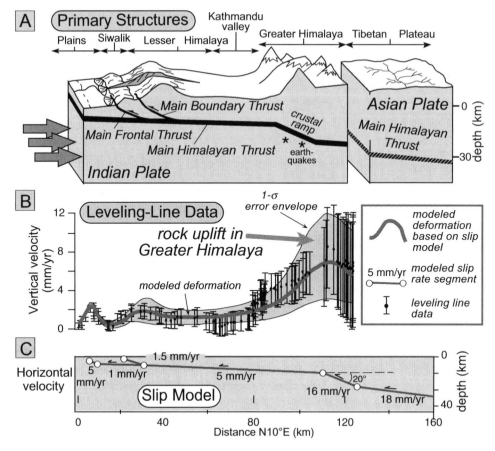

Fig. 5.7 Leveling-line data in an active orogen.
Comparison of (A) large-scale structure and topography with (B) observed and (C) modeled rock uplift rates. Relative rock uplift rates are calculated along a 250-km-long spirit-leveling line (B) oriented approximately perpendicular to the Himalayan Range in central Nepal. From the profile's origin at its southern end, errors become cumulatively larger to the north. The southernmost peak of uplift (~2 mm/yr) is interpreted as a response to a growing anticline above the Main Frontal Thrust in the foreland. No distinct topographic signature is associated with this deformation, probably due to ready erosion of the uplifting, but weak strata. Slow rock uplift occurs above the Main Boundary Thrust, whereas relative subsidence prevails in the intermontane Kathmandu region of the Lesser Himalaya. Relative uplift within the Greater Himalaya occurs at the highest rates (~6 mm/yr) and is associated with high topography, suggesting some permanent strain. C. Finite-element modeling of deformation of an elastic crust assigns variable slip to different fault segments. Note the striking rate change from 18 mm/yr in the north to 5 mm/yr to the south of the steeper crustal ramp lying beneath the Greater Himalaya (Bilham *et al.*, 1997). This abrupt gradient suggests that strain above and south of the crustal ramp could generate the observed pattern of uplift. In particular, the strong southward-sloping gradient in uplift within the Lesser Himalaya is consistent with a large component of elastic interseismic strain that should be released in future large earthquakes. Modified after Jackson and Bilham (1994a).

of the data by dividing the survey into segments. Within these segments, the uncertainty is smaller with respect to the signal, and differential uplift or subsidence can be more reliably determined.

Despite many researchers' preference for using new technologies, such as GPS, to collect geodetic data, leveling surveys can still provide some of the highest-quality, longest-duration measurements of vertical deformation, as well as compelling insights on patterns of crustal deformation. For example, comparisons of spatial variations in vertical uplift rates with mean topography or with large-scale geological structures along a survey transect reveal the extent to which they co-vary (Fig. 5.7). If the pattern of vertical geodetic uplift mimics the smoothed topography, such similarity could suggest that: (i) the present landscape results from the persistence of the present deformation field over a long interval of time; and/or (ii) short-term deformation includes some permanent, anelastic strain. On the other hand, a mismatch between vertical geodetic rates and topography may indicate that decadal rates are unrepresentative of the long-term pattern of deformation. Such a mismatch is not unexpected, because the interval between large earthquakes is commonly long when compared to the duration of geodetic surveys. Hence, these surveys typically span only some part of the interseismic period and do not include coseismic deformation (Fig. 5.1). During this interseismic interval, elastic (and sometimes permanent) strain slowly accumulates, whereas deformation during an earthquake may rapidly release the elastic energy stored in the crust and produce deformation in the opposite sense (Natawidjaja et al., 2004).

A mismatch between topography and uplift patterns along a leveling line may indicate that surface processes successfully reshape and modify any pristine topography. Under these circumstances, a differential vertical deformation field could be well recorded by an underlying geological structure, such as a fold, but the fold itself may be uplifting strata that are readily eroded as soon as they are raised above base level. This scenario appears to apply to the Himalayan foreland of Nepal (Jackson and Bilham, 1994b), where active folding is displayed by shallow structures and by the surveyed uplift rates associated with the Main Siwalik Thrust (MST: Fig. 5.7A), but the deformation is not well reflected by the topography (Lavé and Avouac, 2000). A topography–uplift mismatch is not very surprising here, because the fluvial strata of the Himalayan foreland commonly appear to be eroded nearly as rapidly as they are uplifted (Burbank and Beck, 1991). On the other hand, the highest rates of uplift along this same leveling-line transect coincide with the high

topography of the Greater Himalaya. In these bedrock ranges, rates of erosion and of bedrock uplift may be in a rough equilibrium (Burbank et al., 1996b; Lavé and Avouac, 2001), so that the average surface topography is changing very little despite the rapid bedrock uplift.

Patterns of vertical deformation can provide a data set against which models for ongoing deformation can be tested. The Himalayan survey, for example, traverses both known structures, such as the Main Boundary and Main Central Thrusts, and inferred ones, such as the crustal ramp beneath the Greater Himalaya (Fig. 5.7A). The survey reveals broad regions of more rapid uplift that are 20–50 km wide (Fig. 5.7B) and that appear to be related to deformation on these underlying structures. The overall pattern of surveyed deformation can be imitated using models that treat the crust as an elastic medium that will deform in response to differential slip along faults. The orientation of a fault, its slip rate, and the thickness of its hanging wall determine the deformation attributed to it in the model. One such model (Jackson and Bilham, 1994b) for this Himalayan transect suggests that the uplift pattern of the Greater Himalaya results from interseismic elastic strain due to a large decrease in slip rate between the crustal ramp and the main detachment farther south (Fig. 5.7B). Although the results of such models are not unique, they provide a possible explanation for why and where interseismic strain is occurring and can help guide thinking about deformation within orogenic belts.

Leveling-line surveys of coseismic deformation across structures underlain by thrust faults serve to examine the vertical deformation attributable to individual seismic events (Fig. 4.28A). Changes in displacement along serial traverses can be compared with theoretical models that predict variations in offset along the length of a fault from zero at fault tips to a maximum near the center. If one were interested in whether a particular earthquake represented a characteristic quake for a specific fault, a comparison of the pattern of coseismic uplift or subsidence along a traverse with the topography along the same traverse might provide useful insights. According to leveling-line data, the 1983 Coalinga earthquake

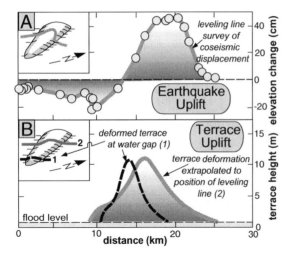

Fig. 5.8 Comparison of leveling-line data and deformed river terraces along a growing fold.
A. Coseismic deformation from the 1983 Coalinga earthquake ($M=6.5$) from leveling-line data in the central part of the anticline (line, inset). B. Deformed terrace profile surveyed near the nose of the same anticline (line 1, inset). For comparison with the coseismic deformation (A), the profile is scaled to the width of the fold in the vicinity of the leveling line (line 2, inset). Similar deformation patterns are displayed by both data sets (shaded areas in A and B), suggesting that the long-term pattern of terrace deformation could result from the accumulation of deformation similar to that caused by the 1983 Coalinga earthquake. Modified after King and Stein (1983).

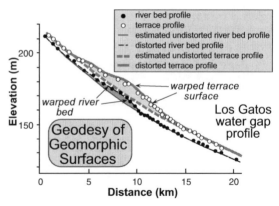

Fig. 5.9 Folding above a blind thrust fault.
Folding along the Los Gatos River water gap as shown by upward warping of both surveyed terraces and the river bed. The magnitude of warping (shown in Fig. 5.8B) is calculated by subtracting the estimated unperturbed river or terrace profile from their observed profiles. Modified after King and Stein (1983).

($M=6.5$) in central California (King and Stein, 1983; Thatcher, 1986a) caused a maximum of about 40 cm of uplift and 20 cm of subsidence along a 25-km-long profile (Fig. 5.8). This survey traversed the crest of a hanging-wall anticline that developed above a blind thrust fault. Nearer to the nose of this plunging fold, the Los Gatos river traverses the same fold in a water gap and has incised through a ^{14}C-dated 2550-year-old fluvial terrace. Whereas the terrace is about 1–3 m above the current bed of the river both above and below the water gap that cuts through the anticline, it lies about 10 m above the river in the center of the anticline (Fig. 5.9). Comparisons of the vertical coseismic deformation with the topography of the deformed terrace (Figs 5.8 and 5.9) suggest that they have rather similar shapes. This similarity could be interpreted as

evidence that this fault is typified by characteristic earthquakes (Fig. 4.7), which create comparable displacements in each seismic event. This similarity between displacement observed from a single earthquake and the shape of the folded river terrace also suggests that numerous repetitions of earthquakes similar to the 1983 quake could create the observed pattern of terrace deformation. In fact, knowing the height and age of the terrace (2550 years) and amount of coseismic uplift in 1983, one can calculate that about 25 similar earthquakes with a recurrence interval of about 100 years would be required to create the observed folding of the terrace.

Tide gauges

Sea level provides a reference surface against which height changes may be assessed. It might be imagined that sea level would change synchronously and in equal amounts all over the Earth's surface and, therefore, that sea level would provide a global reference frame. Although broadly correct, both the variable isostatic response of the crust to loading and the irregular configuration of the load on the Earth represented by the oceans dictates that a change in ocean volume will cause differences in isostatic compensation (and different subsidence or uplift) everywhere around the world. This poorly known isostatic response, therefore, makes it difficult to predict the site-specific effects of global sea-level change. Nonetheless, the use of local sea level as a reference surface can be exploited to assess vertical deformation.

Because tide gauges have been routinely used to record mean sea-level heights at numerous localities worldwide, a large database exists from which vertical changes can be extracted. Where tide-gauge records are available along an irregular, embayed coastline, a three-dimensional reconstruction of the regional uplift pattern can be determined. In order to use tide-gauge data to define vertical changes of the shoreline, two corrections are typically applied. First, changes attributed to *eustatic* sea level – that is, those changes that are due to changes in the volume of ocean water or in the shape of the oceans and result in changes in average sea level – are removed. At the time scale of years to decades, the volume of the oceans is affected by growth and decay of glaciers and by thermal expansion or contraction of the water column. At present, melting glaciers and warming temperatures appear to be increasing ocean volume at a rate that causes mean sea level to rise at about 2 mm/yr (Douglas, 1991). Second, local oceanographic effects resulting from changes in salinity, ocean temperature, atmospheric pressure, and ocean currents are removed. A remarkable inter-annual sea-level variability can result from these factors (Fig. 5.10), and they often cause local sea-level changes that are far greater than those attributable to tectonic effects. The oceanographic correction is usually determined through principal component analysis of a regional set of tide gauges (Savage and Thatcher, 1992). The objective of this analysis is to identify those components, termed *common mode signals*, of the observed sea-level variation that, after the eustatic effect is removed, are shared among all of the stations. Following scaling for each site, the oceanographic correction is removed from the locally measured tidal record. The resultant data are then interpreted to represent changes in relative sea level (Fig. 2.4) due to vertical rock movement.

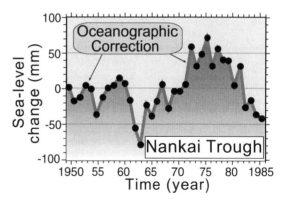

Fig. 5.10 Oceanographic correction from the southwest coast of Japan.
Note that both total variability (~150 mm) and year-to-year variability are large and unsteady compared to the rate of tectonic deformation. Modified after Savage and Thatcher (1992).

The use of tide-gauge data to reconstruct a regional pattern of vertical deformation is well illustrated by a study from the southwestern coast of Japan, where an excellent record spanning 1950–85 has been analyzed (Savage and Thatcher, 1992). In 1944 and 1946, two large (M_s=8) earthquakes occurred offshore along the Nankai Trough subduction zone. Given the nearly continuous tidal record since those earthquakes, questions concerning interseismic crustal deformation can be investigated. For example, does interseismic strain accumulate at a steady rate, or is there an interval of rapid, post-seismic deformation immediately after an earthquake that is followed by more steady deformation? Removal of eustatic and oceanographic signals (Fig. 5.10) from 27 Japanese tide-gauge records reveals clear site-to-site contrasts in the rate of vertical deformation (Fig. 5.11). Even without removal of the oceanographic effects, different long-term trends are visible in these data. Appropriate corrections, however, suggest that nearly constant rates of uplift or subsidence have been sustained at many sites during the past 30 years, but that the decade immediately following the earthquakes was characterized by more rapid rates of deformation (Fig. 5.11). These data suggest that an interval of more rapid post-seismic deformation that may indeed persist for years or decades following a major

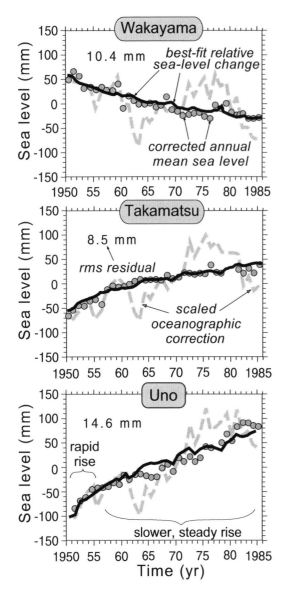

Fig. 5.11 Relative sea-level change and Japanese coastal deformation adjacent to the Nankai Trough. Subtraction of the oceanographic correction (which is scaled for each site) from the original measurements yields the corrected annual mean sea level. The line fitted to the corrected data defines the long-term tectonic rock uplift and subsidence. The root mean squared (rms) residual quantifies the deviation of the annual data points from the long-term fitted line. All sites show a steady trend in sea-level change between 1955 and 1985. Note that, at Takamatsu and Uno, an interval of rapid subsidence (seen as a rise in sea level) is subtly, but clearly, expressed prior to about 1955. Modified after Savage and Thatcher (1992).

earthquake (Fig. 5.1C). Moreover, they suggest that interseismic deformation in subduction-zone settings may be characterized by steady rates of deformation. A map view of the tide-gauge data (Fig. 5.12A) provides a coherent overview of the regional deformation. The maximum rates of uplift occur at 100–200 km from the trench. The trend of this zone of high rates is nearly parallel to the trench. Farther from the trench, rates diminish and even define a zone of subsidence west of Osaka (Fig. 5.12A).

This pattern of uplift and subsidence can be interpreted in several ways. One could assume that, following the post-seismic interval, elastic strain has been accumulating above a presently locked fault plane represented by the shallow part (<30 km) of the subduction zone interface. If this were the case, then elastic half-space models suggest that, above the trailing edge of the locked thrust, the overlying hanging wall should buckle upwards, whereas the leading edge of the hanging wall should be flexed downward toward the trench. If this elastic strain were released during a subsequent subduction-zone earthquake, strong subsidence in the southeastern coastal region and uplift west of Osaka would be predicted for these tide-gauge data. If the interseismic deformation is compared with the coseismic displacement resulting from the 1946 Nankaido M_s=8.2 earthquake (Fig. 5.12B), a striking spatial correspondence exists between the zones that have been uplifted most extensively during the interseismic interval and those that experienced the greatest coseismic subsidence. These data lend support to the concept that the pattern of interseismic strain accumulation is commonly mirrored by the coseismic deformation. An alternative explanation suggests that the deformation defined with the tide-gauge data represents a response to a down-dip migration of slip along the subduction zone following major earthquakes. If this were the case, then the observed deformation is more analogous to a migrating ripple of deformation (Pollitz *et al.*, 1998; Freymueller *et al.*, 2000) and would represent little storage of elastic strain. Comparison of the inter- and coseismic displacement maps lends some support to this idea, as well. Although the axis of maximum interseismic uplift coincides with the axis of maximum coseismic subsidence, coastal zones of strong coseismic uplift are still experiencing inter-seismic uplift (Fig. 5.12). Hence, there is no "mirror image" of subsidence and uplift along the outermost coastal areas. Finally, along accretionary prisms, both strata and marine terraces are commonly observed to dip toward the arc. This dip indicates that at least some component of non-recoverable deformation occurs within the overriding plate.

Despite the uncertainties in the interpretation of these tide-gauge data, the Japanese coastal data clearly demonstrate their usefulness in delineating vertical deformation patterns. On highly digitate coasts with numerous embayments and islands, a regional three-dimensional pattern of emergence and submergence can potentially be obtained. Because tide-gauge records are available from many harbors with a long history of use, these records provide a rich database for studies of temporal and spatial changes in vertical motions in coastal regions.

Tropical corals

In some of the world's tropical oceans, Nature has created its own geodetic recorder. Certain corals, such as *Porites*, produce annual growth bands (Fig. 5.13A) that are similar to tree rings, but develop in response to seasonal variations in the density of coral cells. Because these corals cannot survive continuous exposure to air even for relatively short durations, each year these corals grow up to the level of the annual lowest tide, which is also termed the "highest level of survival" (HLS). Thus, when corals become established in somewhat deeper water, they initially grow upward to the sea surface and then grow outward. The uppermost edge of their annual rings, therefore, serves as a natural sea-level gauge that can commonly record long-term upward or downward trends, depending on whether the coast is subsiding or rising, respectively (Fig. 5.13B and C). Because corals represent a dynamic natural system, they are imperfect recorders of subtle year-to-year tidal variations (oceanographic corrections;

Fig. 5.10), but they provide remarkable high-fidelity records at time scales ranging from decadal to millennial (Natawidjaja *et al.*, 2007; Sieh *et al.*, 2008; Zachariasen *et al.*, 2000).

Remarkable insights about earthquake behavior are emerging from coral studies above the Sumatra subduction zone. Rapid convergence (~50 mm/yr) along this plate boundary has produced the world's largest earthquake since 1964 (M_w=9.1 on 26 December 2004 (Chlieh *et al.*, 2007; Subarya *et al.*, 2006)) and several magnitude 7.5–8.6 temblors over the past century. A tropical archipelago sits above the equatorial segment of the Sumatran outer-arc high and hosts diverse coral communities that have been exploited for geodetic studies. For example, studies of coral heights along the fringes of Simeulue Island (Briggs *et al.*, 2006) reveal complementary patterns of uplift that occurred in two closely spaced, great earthquakes just a few months apart (Fig. 5.14A). When summed together, the uplift data reveal a saddle of low displacement (Fig. 5.14B), where no record of large displacements in past earthquakes has been uncovered. This saddle represents a deficit in coseismic displacement that has been interpreted to result from aseismic slip on a segment of the Sumatra megathrust (Briggs *et al.*, 2006). If true, little stress would be expected to accumulate on the megathrust in this specific region, and it might act as a barrier to propagation of large earthquakes.

In many tectonic settings, it is unknown whether cumulative deformation from repetitive coseismic displacements has resulted in the growth of the observed topography or geological structures. At Coalinga (Fig. 5.8), for example, the similarity of patterns of coseismic displacements and large-scale structures argues for earthquake-driven structures and topography. In striking contrast, coral data from Nias Island in Sumatra (Fig. 5.14C) support the opposite conclusion. There, the pattern of coseismic uplift bears no spatial relationship to the long-term uplift rate. Whereas the coseismic data reveal strong gradients in uplift, Holocene uplift rates deduced from uplifted corals are spatially homogeneous and clearly uncoupled from the coseismic patterns (Briggs *et al.*, 2008). This de-coupling argues that the coseismic deformation

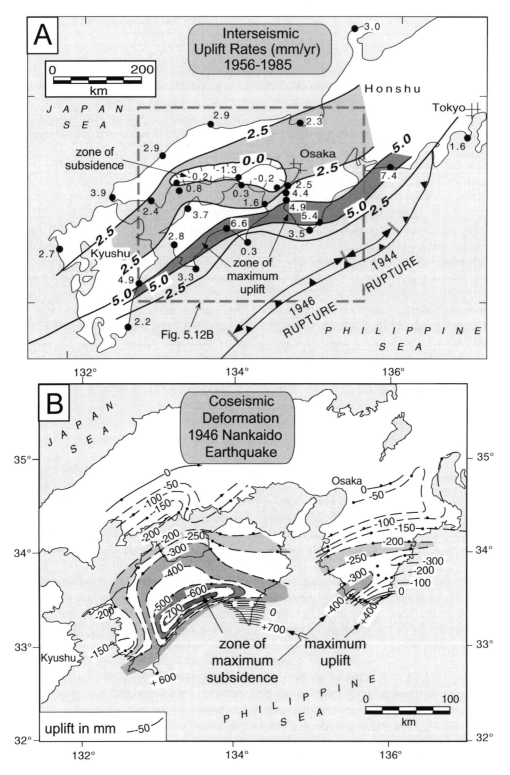

Fig. 5.12 Interseismic and coseismic deformation of southwest Japan.
A. Interseismic deformation of southwest Japan as determined from tide-gauge data. Rupture lengths of two
magnitude 8 earthquakes (1944, 1946) are shown along the Nankai Trough. Highly irregular coastline provides
excellent spatial distribution of tide-guage data. Note elongate zone of maximum uplift along the southeastern coast
and zone of subsidence to the northwest. Modified after Savage and Thatcher (1992). B. Coseismic deformation

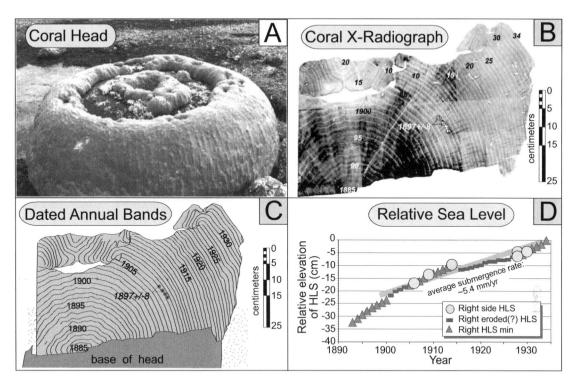

Fig. 5.13   Coral growth bands as tide gauges.
A. Coral exposed due to co-seismic uplift. Surrounding annulus and raised "micro-atoll" in center record interseismic submergence subsidence. Photo courtesy of D. Natawidjaja. B. X-radiograph of a slice through the right half of a tropical coral head above the Sumatran subduction zone shows clear annual banding. Note that the irregular upper surface is the actual surface of the living coral, which was limited in its upward growth by the level of the tide. C. Line drawing of annual growth bands in the same coral head. U–Pb ages yield a date of 1897±8AD midway through the coral head. Note the clear upper limits to annual rings after about 1915. D. Reconstruction of relative sea level based on the age and upper edge of annual rings. The quality and significance of each ring varies because some appear eroded, some are clear minima, and others appear to be robust indicators of the lowest annual tide or highest level of survival (HLS). After a 20-year initial interval of more rapid upward growth toward the surface, the coral grew obliquely outward and more slowly upward as the coast slowly subsided (5.5mm/yr) with respect to local sea level. Modified after Natawidjaja *et al.* (2004).

in this outer arc is a purely elastic process through which interseismic strain is recovered, a phenomenon consistent with the elastic rebound theory (Fig. 4.2). The ongoing coral-based geodesy in Sumatra promises continued insights on the nature and rates of tectonic processes because of three critical ingredients: broadly distributed and reliable geodetic markers (corals); excellent time control as derived from growth bands and high-precision U–Pb dates; and rapid tectonic rates driven by high convergence and numerous earthquakes. This last attribute dictates that the

Fig. 5.12   (*cont'd*) resulting from the 1946 Nankaido earthquake ($M=8.2$). Note maximum uplift along southeastern coast parallel to the Nankai Trough. Strong subsidence occurred just northwest of the zone of maximum uplift. When coseismic (B) and interseismic (A) deformation are compared, zones of maximum coseismic subsidence coincide with zones of maximum interseismic uplift and appear compatible with an elastic rebound model (Fig. 4.2). The southeasternmost sites in both data sets, however, show uplift both during (B) and after (A) the earthquake, and are not directly compatible with such a model. Modified after Fitch and Scholz (1971).

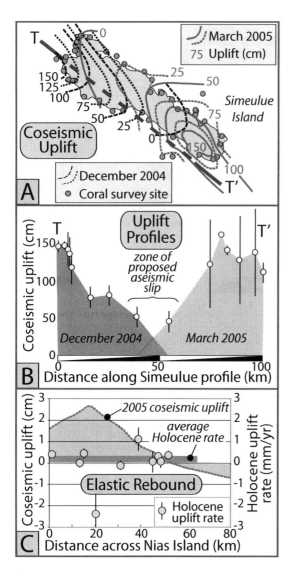

Fig. 5.14 Tectonic insights from coral-based geodesy.
A. Coseismic uplift patterns on Simeulue Island,
Sumatra, resulting from an M=9.1 (December 2004)
earthquake and an M=8.7 (March 2005) earthquake.
Note that peak deformation occurs at the opposite ends
of the island in each quake. B. Uplift profiles along the
length of the island (T–T′ in A) show strong
displacement gradients and reveal a saddle of low
displacement, where aseismic slip on the megathrust is
postulated. C. On Nias Island (~200 km southeast of
Simeulue), the pattern of coseismic uplift from the
March 2005 earthquake reveals no spatial correlation
with Holocene uplift rates. This decoupling between
coseismic and millennial deformation suggests that
almost all coseismic deformation represents elastic
rebound. Modified after Briggs *et al.* (2006, 2008).

tectonic signal will rise well above the geomorphic
noise and that the coral stratigraphy can yield
direct insights on seismic hazards.

Far-field techniques

Very-long-baseline interferometry (VLBI)

In order to define relative motion between
lithospheric plates, near-field measurement
techniques are both inadequate, inappropriate,
and time-consuming. To measure between plate
interiors that are assumed to be relatively
stable, a technique is required that permits
precise distances to be measured over distances
of hundreds to thousands of kilometers. VLBI
was developed in the late 1960s as a technique
to study compact extragalactic radio sources.
Since then, VLBI has evolved to become a
geophysical tool for study of real-time tectonics.
In the last three decades, the technology has
developed rapidly and is now in its third
generation. The discoveries using this new
technique have at times been surprising and
have helped change the way we think
about the integrity of the lithospheric plates. For
a summary of the development of VLBI
techniques and applications, see Ryan and Ma
(1998).

The VLBI system utilizes two or more widely
separated radio telescopes simultaneously
observing radiation from very distant (extragalactic) radio sources, typically quasars.
These sources are broadband in the sense that
they radiate in a wide range of frequencies.
Their extreme distance from Earth (billions of
light years) makes them very small, fixed
points in the sky, which therefore form good
reference points from which to measure
distances between sites on the Earth. The
complicated, random signal from the quasar
is recorded at very high rates and with very
accurate clocks at the two ends of each
baseline on the surface of the Earth.
The processing involves correlation of these
recordings in pairs, one pair for each baseline
in the network being observed. The signal
from one site is slowly shifted while the other

is held fixed until the correlation between the two signals reaches a maximum. The shift necessary to maximize the correlation is the time delay in the arrival of the signal from the radio source, which is directly proportional to the baseline distance. Today, approximately 40 VLBI antennas around the world systematically make such measurements. Horizontal velocity estimates with an uncertainty of about 1 mm/yr are available for about 60 sites, while uncertainties in the vertical dimension are approximately two to three times greater.

Prior to the advent of the Global Positioning System (GPS), VLBI was a critical tool for defining plate motions and resolving large-scale strain gradients (Herring *et al.*, 1986). Today, with much cheaper and more widespread use of GPS, VLBI still plays a critical role in defining a reliable reference frame within which GPS measurements can be interpreted.

Global Positioning System (GPS)

Two recent developments have revolutionized geodetic approaches for the accurate determination of positions at the scale of hundreds of meters to hundreds of kilometers. First, more than 30 US Department of Defense satellites have been launched that continuously transmit coded radio messages that specify the time of transmission and the satellite's position as a function of time. This constellation of satellites is referred to as the Global Positioning System (GPS). Second, highly sensitive receivers measure the transit time and phase of the radio signal in order to determine the distance to the transmitting satellite. Simultaneous solution of the distance from the receiver to several (usually four or more) satellites permits the location of the receiver to be specified with high precision. Given the orbital geometry and spacing of the GPS satellites, acquisition of positional data from four or more satellites is now usually feasible almost anywhere in the world. Thus, based on radio transmissions from satellites more than 20 000 km above the Earth's surface, this remarkable technology permits calculation of one's horizontal location to within less than

1 cm! Because of the satellite orbital geometry, the accuracy of GPS measurements in the horizontal dimension will typically be several times greater than their vertical accuracy. The number of global navigational satellite systems is growing. Russia, China, and the European Union are expanding, upgrading, or developing independent constellations of satellites that promise to densify global coverage and provide more geodetic capabilities over the coming years.

Two strategies are commonly employed for collecting geodetic data using GPS in order to define regional strain fields at scales of ten to several hundred kilometers. Permanent GPS receivers continuously record positional data, are commonly installed in broad arrays, and offer very precise positioning. These arrays can address problems, such as the character of deformation in the months before or after an earthquake, that require high geodetic precision and a dense time series of measurements. Permanent arrays have the drawback that each individual receiver (they are not inexpensive!) has to be wholly dedicated to a single measurement site. Since the mid-1990s, numerous permanent GPS arrays have been installed, including dense networks in Taiwan and Japan, and hundreds of more scattered permanent GPS stations around the world. More recently, the Plate Boundary Observatory (PBO) has installed new sites, taken over existing arrays in southern California (SCIGN) and the San Francisco Bay area (BARD), and now operates about 1000 sites in the western United States between the Rockies, the Pacific, and Alaska. All GPS data from the PBO are freely available on the web (www.pboweb unavco.org) to researchers worldwide. Unfortunately, such open-access policies have yet to be implemented by many other countries.

A second GPS survey strategy, known as *campaign mode*, involves periodic resurveys of an array of GPS sites that are placed at carefully chosen localities across a specific region. With this strategy, numerous sites can be visited with a small number of receivers, and the same receivers can be used for various projects in succession. Hence, the infrastructure costs for

Box 5.1 Episodic slow earthquakes.

One of the most striking recent seismological discoveries has been recognition of episodic events of relatively rapid slip along faults. Although hints of such phenomena had been recognized earlier, it was the advent of arrays of continuous GPS stations that revealed indisputable evidence for repeated episodes

of rapid slip. One of the earliest studies took place in the Cascadia subduction zone in Washington and British Columbia (Dragert *et al.*, 2001). Here, although the GPS record from a single station may show a steady, long-term slip rate, this long-term trend is periodically punctuated by brief intervals

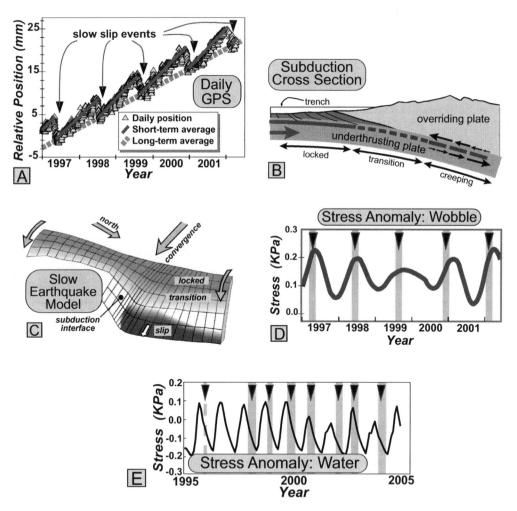

A. GPS data recording slow slip events on the Cascadia subduction zone. B. Cross-section of a subduction zone with locked to creeping sections. C. Model for the geometry of slow earthquakes. D. Calculated stress anomaly in Cascadia due to Chandler wobble. E. Stress anomaly on the Cocos plate due to water loading. Modified after Dragert *et al.* (2001), Shen *et al.* (2005), and Lowry (2006).

(see figure A) when the direction of slip is reversed! These slip transients span one to two weeks and occur about every 14 months.

Although the surface displacement is typically a few millimeters, modeling of these data suggests that each slow slip event represents 2–4 cm of slip along the subduction interface between the up-dip locked zone and the down-dip freely slipping zone (see figures B and C). Initially, these events were termed "silent earthquakes" because no seismogenic signature was recognized from them. Subsequent studies suggested that fluid flow and seismic tremors (low-frequency shaking) are caused by (or associated with) slow subduction zone earthquakes (Brown et al., 2005). The discovery of slow earthquakes on the Cascadia subduction zone prompted a search for similar behavior elsewhere using continuous GPS and seismic arrays, and, in less than a decade, either slow earthquakes or episodic tremor attributed to slow earthquakes have been discovered in subduction zones in Alaska, New Zealand, Chile, Japan, Mexico, and Costa Rica, as well as along the San Andreas Fault.

A big question now concerns the cause of the apparent periodic behavior of slow earthquakes. Is this an intrinsic property of slip on large fault interfaces, or are there external drivers? One might imagine that a process that periodically increases or decreases the load across the fault could modulate friction and fault slip. A potential source for such periodic loading is the *Chandler wobble*: the irregular wobbling of the Earth around its rotational axis due to the fluid nature of its core and oceans. In Cascadia, the periodicity of the Chandler wobble matches that of the slow earthquakes (see figure D). Stress anomalies that could be caused by the Chandler wobble were recently calculated (Shen et al., 2005) for the transition zone on the subduction interface where slow earthquakes occur (see figure B). Although little is known about the sensitivity of these faults to small stress changes, the magnitude of these anomalies may be sufficient to trigger episodes of slow slip.

A different study on the Cocos Plate subduction zone along the western coast of Mexico argued that climate may modulate slow earthquakes, which display a nearly annual periodicity there (Lowry, 2006). In this site, hydrologic loading (due to seasonal rainfall variations) is predicted to produce shear stress changes (see figure E) that are twice as great as those predicted for Chandler wobble. Neither of these models produces a perfect match to the slow earthquake record, but each tends to show stresses peaking before the slow-slip events, a pattern consistent with loading and slip on better-studied faults. A challenge in coming years as more examples of slow slip are uncovered will be to discover whether one or multiple causal mechanisms are responsible for this unexpected, but striking behavior.

campaign-mode GPS are much less than those using permanent GPS instrumentation for the same array. For rapid acquisition of geodetic data across a broad region, campaigns with mobile receivers are highly effective. Survey sites are usually marked by a monument fixed in an immobile substrate, such as bedrock, so that the GPS receiver can be positioned over precisely the same point on subsequent surveys. The precision of GPS measurements is in part a function of the duration of data collection. For any given occupation of a single GPS site, a longer interval of data collection will yield more precise positioning information. Good-quality data collected for 24 hours can have an uncertainty of greater than 4–5 mm. If data are collected for several days, the uncertainty can be reduced considerably more by smoothing the noise inherent in daily measurements. Continuously operating, permanent GPS receivers can reduce the uncertainties still further and can provide a temporal resolution that is impos-sible to achieve with occasional reoccupation of GPS sites.

The choice of whether to acquire GPS data using permanent or mobile GPS sites is some-times simply a matter of cost or practicality: campaign mode is cheaper, faster, and more portable between distant field sites. But the nature of the scientific problem being addressed should also dictate the choice. For example, if the goal is to delineate short-term crustal deformation before or after an earthquake, permanent GPS receivers are far superior to campaign-style acquisitions. Moreover, some of the most exciting discoveries in seismology over the last decade have been prompted by provocative data from continuous GPS monitoring (Box 5.1). On the other hand, if the goal is to define crustal deformation across a broad, rapidly deforming orogen, data acquired in campaign mode can commonly define multiple positions to within a few millimeters after two or three reoccupations of survey sites. For many GPS projects in active orogens nowadays, a compromise is reached: mobile receivers are used to define the broad deformation pattern, but a few permanent GPS receivers are embedded within the study area to provide a robust reference frame, as well as high temporal resolution.

Four factors contribute to the calculated GPS position. First, regional deformation is driven by large-scale plate tectonic movements and encompasses the relative velocity between interacting plates. Second, spatially localized deformation results from individual seismic events and aseismic deformation. Third, seasonal changes in groundwater levels can cause the Earth's surface to inflate or deflate. Fourth, measurement errors result from variations in atmospheric conditions, transmission and reception of the GPS data, and positioning of the GPS receiver. Obviously, considerable efforts are typically expended to reduce the measure-ment error to a minimum. Extraction of a plate-motion signal or a seismic signal from the data is then possible, if the strain related to one or the other is known. In fact, one check on calculated seismic displacements is to subtract independently measured or modeled ground offsets due to seismic phenomenon from the total displacement to see whether the expected rates of plate motion represent the remaining measured motion in the data set. Alternatively, by subtracting the expected plate motion, the resultant seismic displacements can be compared with measured offsets (Larsen and Reilinger, 1992).

Over the past decade, numerous GPS campaigns have been conducted across the Himalayan–Tibetan orogen. Although details are still lacking in many areas, a stunning pattern of crustal deformation has emerged from these studies (Fig. 5.15A). In the far field, the Indian subcontinent is converging with Siberia at ~40 mm/yr. As can be seen by comparing the length of the velocity vectors from north and south of the Himalaya, about half of this con-vergence is absorbed by the Himalaya. Both GPS and leveling-line data (Fig. 5.7) suggest that most (>70%) interseismic shortening is currently being absorbed in the Greater Himalaya. This Greater Himalayan deformation stands in contrast to the observation that, throughout the Holocene, shortening at some frontal faults within the Himalayan foreland has averaged ~20 mm/yr (Lavé and Avouac, 2000). This mismatch between long- and short-term deformation (deformation in the Greater Himalaya versus that in the frontal fault system in the foreland) suggests that most of the current deformation represents elastic strain that will be recovered in large earthquakes that rupture beneath the Lesser Himalaya and out to the foreland (Bettinelli et al., 2006; Cattin and Avouac, 2000).

Whereas the orthogonal convergence across the Himalaya is perhaps predictable, the GPS data depict extraordinary diversity north of the range (Fig. 5.15A). In the west, almost half of the convergence is absorbed in the Tien Shan (Abdrakhmatov et al., 1996), which, given its location some 1000 km from the collision front,

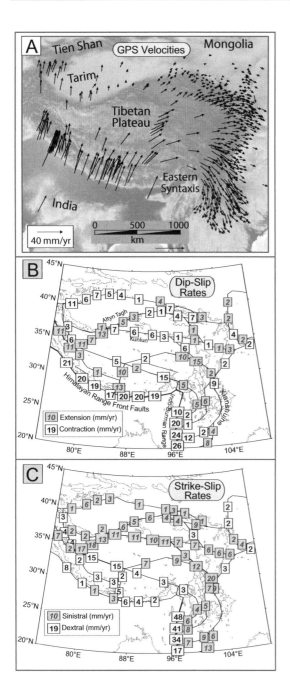

Fig. 5.15 Geodesy of the Indo-Asian collision and modeled fault slip rates.
A. GPS velocities in the Indo-Asian collision with respect to a fixed reference frame in north Asia. The diverse directions and rates of motion depict the striking complexity of crustal deformation in this orogen.
B. Best-fit slip rates on contractional and extensional faults derived from block models of GPS motions.
C. Modeled slip rates on strike-slip faults. This block model with all deformation concentrated on faults shows an overall consistency with the GPS velocity field. Modified after Meade (2007).

can be considered one of the world's greatest examples of intracontinental mountain building. Within the core of the 5-km-high Tibetan Plateau (Fielding *et al.*, 1994), the GPS data reveal east–west extension that is interpreted as ongoing gravitational collapse of the plateau. The amount of deformation north of the plateau decreases eastward, as major sinistral strike-slip faults play a larger role in the deformation. Whereas the GPS data indicate much slower velocities east of Tibet, the Eastern Syntaxis is characterized by striking corner flow, with rapid crustal flow toward the south! Almost anyone viewing these geodetic data will be struck by the dynamic and complex nature of deformation in the world's quintessential continental collision: a picture that has been so exquisitely captured by GPS.

An ongoing debate about the nature of deformation in the Himalaya has been heightened by these geodetic data. The generally smoothly varying velocity pattern has been likened to viscous flow that yields a continuum of deformation (Bendick and Flesch, 2007; England and McKenzie, 1982; Zhang *et al.*, 2004). Such crustal flow is considered compatible with a weak lower Tibetan crust, as inferred from seismic and magnetotelluric data that appear to indicate partial melting in Tibet's lower, thickened crust (Nelson *et al.*, 1996). On the other hand, the presence of major faults in Tibet and changes in velocity across them inspires some to argue that nearly all crustal motion is accommodated on faults that bound rather rigid crustal blocks (Meade, 2007; Thatcher, 2007). Modeling all deformation as occurring on faults (Fig. 5.15B and C) appears to explain the geodetic data as well as or better than a continuum model. Despite the apparent abundance of GPS data in Tibet, the large size of the orogen means that most active faults are not currently encompassed within dense geodetic arrays. Arrays targeted across such faults in the future will certainly help to clarify further the geodynamics of Tibet.

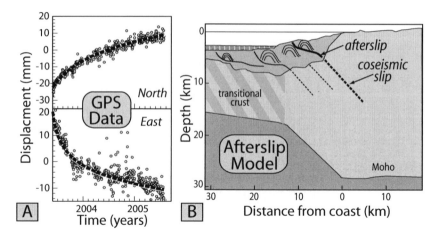

Fig. 5.16   Geodetic record of afterslip.
A. Continuous GPS data for two years following the 2003 $M=6.9$ Boumerdes earthquake in Algeria shows a time-dependent logarithmic decay in displacement. Notice that having continuous GPS data allows smoothing through the noise of day-to-day measurements that sometimes show large deviations (>15 mm) from their average position. B. Geological reconstruction of the geometry of afterslip. Although modeling of the GPS data indicates slip occurred up-dip of the coseismic rupture, no surface offset is observed. Instead, the afterslip deformation may be absorbed within the 5-km-thick sedimentary sequence overlying the rupture surface in the basement. Modified after Mahsas *et al.* (2008).

One persistent conundrum in tectonics emerged from geodetic observations showing that the convergence rates at some trenches were greater than could be accounted for by coseismic slip at these sites. Answers to this problem began to emerge in the late 1960s, when geodetic measurements after earthquakes (Smith and Wyss, 1968) revealed the occurrence of afterslip: accelerated, but slow, fault slip following an earthquake. Some subsequent studies have demonstrated that the amount of energy released by afterslip approaches the energy released coseismically (Heki *et al.*, 1997; Chlieh *et al.*, 2007). Whereas afterslip is now a commonly recognized phenomenon, where afterslip occurs in the crust and what controls its character are still debated. Does it occur along the very same part of the fault that ruptured coseismically, below the ruptured region, above it, or in the rocks enclosing the fault? Where afterslip occurs within the crust has important implications for paleoseismological studies. As described in the next chapter, paleoseismological reconstructions commonly rely on displaced stratigraphic, structural, or geomorphic features to calculate the amount of slip that occurred in past earthquakes. But, what if a significant fraction of the observed displacement occurred as afterslip? In such a case, coseismic slip and, hence, the magnitude of past earthquakes could be overestimated.

Geodetic studies, especially dense GPS arrays around recently ruptured faults, can provide insights on both the magnitude and location of afterslip. In general, the deeper in the crust that the afterslip occurs, the broader the anomaly at the surface. GPS studies following the 1999 $M=7.5$ Izmit earthquake on the North Anatolian Fault in Turkey concluded that afterslip in the subsequent three months amounted to ~10% of the coseismic slip, that almost all of it occurred below the coseismic rupture, and that the afterslip was indicative of the transition to stable fault creep at the base of the seismogenic zone (Reilinger *et al.*, 2000). In contrast, six continuous GPS stations installed following the 2003 $M=6.9$ Boumerdes earthquake in Algeria revealed a logarithmic decay in the rate of afterslip, with no apparent deep slip. Instead, the geodetic pattern could be best explained by shallow afterslip up-dip from the coseismic rupture (Fig. 5.16). More study is needed to understand what controls the location and magnitude of afterslip. Such studies will certainly emerge from future post-seismic geodetic studies and should help place useful uncertainties on paleoseismic slip reconstructions.

Radar interferometry

A new geodetic technique for measuring ground displacements over larger areas has been developed based on radar interferometry (Bürgmann *et al.*, 2000). Using *synthetic aperture radar* (SAR) carried aboard satellites at 785 km altitude, radar pulses are transmitted along west-pointing ray paths at an angle of 23° from the vertical. Based on the return signal, the distance from the satellite to the ground is calculated, and a phase shift due to the reflection from the ground is recorded. These measurements are made for each pixel (picture element; with dimensions commonly of 4×20 m). If the same area of the Earth's surface is contained within two different SAR images, if the positions of the satellites are well known, and if the ground and atmospheric moisture distributions are approximately the same for both images, then the differences in path lengths between the two images are attributable to some combination of: (i) the ground topography as seen stereoscopically from the satellite in slightly different orbital positions; and (ii) changes in the position of the ground in the time between acquisition of the two images (Massonnet *et al.*, 1993). If two images are used to construct the topography of a region, then a third image can be used to determine ground deformation with respect to that topography along the *look* direction of the radar. Alternatively, if a detailed digital topography already exists for the area, then the differences between two images can define the ground deformation. The resultant depiction of ground displacements is termed an *SAR interferogram*, and the overall approach is referred to as *InSAR*. The wavelength of the radar emitted by the satellite is 56 mm, and, using interferometry, displacements of 28 mm (half a

wavelength) or more can be readily defined. Under favorable conditions, even displacements of less than 10 mm with uncertainties of less than 5 mm can be delineated (Massonnet *et al.*, 1994). Radar interferograms for use in tectonic studies are probably best suited for arid settings where seasonal changes in vegetation and moisture are minimal. Because of the steep look angle of the radar pulse, InSAR's greatest sensitivity is to vertical change, rather than horizontal change – opposite sensitivities to those for GPS.

The unusual research attribute of radar interferometry lies not in its resolution (which is considerably less than that achieved through extensive GPS measurements), but rather in the tremendous spatial coverage it provides. One 60×60 km SAR image comprises more than 300 000 pixels measuring 100×100 m! The distance changes recorded by these many thousands of pixels can be displayed as a contour map of deflections, in which each successive *fringe* represents an additional 28 mm of displacement. In addition to seismic displacements, volcanic inflation or subsidence (Massonnet *et al.*, 1994), large landslides (Fruneau *et al.*, 1996), and the details of glacial flow (Dowdeswell *et al.*, 1999; Mohr *et al.*, 1998) can be observed and depicted through radar interferometry. Unlike most other geodetic techniques, radar interferometry can be done remotely, and it provides regionally extensive, high-resolution maps of interference patterns resulting from surface displacements. Although radar interferometry will have restricted applicability in highly vegetated regions, it can be extremely useful for quantitative analysis of deformation in remote, relatively arid areas.

An interferogram of the ground displacement associated with the Landers earthquake (Fig. 5.17) shows concentric, but asymmetric, fringes that extend at least 75 km in an east–west direction (Massonnet *et al.*, 1993). Note that the pattern appears to consist of several groups of distorted, concentric fringes. One explanation for the abrupt changes in displacement along the trace of the Landers rupture is that faulting occurred along a segmented fault. In fact, the rupture appears to have linked together individual fault segments that were not

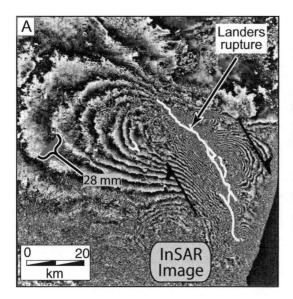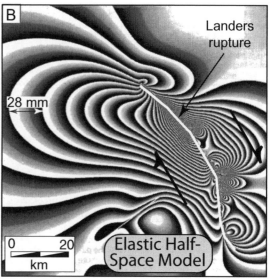

Fig. 5.17   Radar interferogram of the Landers earthquake.
A. SAR interferogram of ground displacement associated with the Landers M_w=7.3 earthquake of June 1992. Each fringe represents 28 mm of displacement, and at least 20 fringes are visible near the fault (equal to 560 mm of displacement). Coherence is lost as the ground rupture is approached, probably because the displacement gradient is greater than 28 mm/pixel. Note the broad, asymmetric deformation in an east–west direction covering >75 km and the abrupt termination of major deformation near the ends of the fault. The detailed deformation patterns seen here can be used to constrain models of surface displacement due to the Landers rupture. B. Modeled interferogram pattern based on eight fault segments rupturing along vertical planes in an elastic half-space. The excellent match between the observed and modeled results indicates that a simple half-space model can quite successfully mimic the observed deformation pattern. Modified after Massonnet *et al.* (1993).

previously known to be connected. Eight separate ruptures of varying sizes have been hypothesized along different segments of the fault. The complex suite of interferometry fringes, particularly along the eastern side of the fault, can be roughly interpreted to depict the effects and sphere of influence of these individual ruptures. The analysis also shows that the earthquake appears to have triggered slip on other faults as much as 100 km from the primary rupture. It is possible not only to look at the direct effects of an earthquake shortly after the event, but also to examine continuing ground deformation in the subsequent months and years. Continuing interferometric studies of the Landers region following the June 1992 earthquake have recorded deformation of up to 10 cm due to a magnitude 5 aftershock on a nearby fault and have shown that less than 28 mm of

post-seismic and interseismic slip has occurred in other areas (Massonnet *et al.*, 1994).

Recent InSAR studies have also delineated surprising seasonal changes in uplift and subsidence that are largely unrelated to tectonics. For example, wet winters and subsequent runoff fill aquifers in alluvial fans and inflate their surfaces. As the aquifers are drawn down, especially by groundwater pumping in summer months, the ground surface subsides. In some parts of the Los Angeles basin, for example, the height of the ground surface varies by more than 10 cm over the course of a year. In addition, the flexure induced by the groundwater loading and unloading causes nearby GPS sites to move horizontally as much as 15 mm. Such rates are much greater than any tectonic signal and have to be accounted for in order to discern the underlying

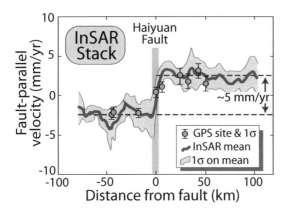

Fig. 5.18 Stacked InSAR data across an active fault. Average fault-parallel velocity profile across the Haiyuan Fault in north-central China derived by combining more than 20 InSAR images. The mean displacement was scaled in order to reconcile with the GPS data from the same area. This synthesis of stacked images suggests a relatively weak fault and rather block-like behavior for long distances (>50 km) on either side of it. Modified after Cavalié *et al.* (2008).

tectonic deformation (Bawden *et al.*, 2001). Some recent studies suggest that seasonal changes in water loading may modulate tectonism in surprising ways (Box 5.2).

InSAR is most successful for regions with rapid rates or large magnitudes of deformation and areas with little vegetation and relatively minor variations in surface moisture, such as deserts. In areas with abundant vegetation, widespread agriculture, or urbanization, seasonal or spatially abrupt changes in the surface can cause the image to decorrelate, such that coherence is lost between adjacent pixels. In such a case, a smoothly varying deformation pattern, such as seen with the Landers rupture (Fig. 5.17a), may be impossible to reconstruct. Even in desert sites with little vegetation or human interference, if rates of deformation are quite slow, say a few mm/yr, reliable detection of this deformation may be impracticable with traditional InSAR methods. Several new approaches have been developed to address such problems. In one approach, rather than using just two or three images, a temporal succession of images (perhaps several dozen) of the same region are stacked and smoothed

(Fig. 5.18) in order to detect reliable and sometimes small (<28 mm) amplitude or slow deformation (Peltzer *et al.*, 2001). In a second approach, the problem of decorrelation is circumvented by using *permanent scatterers*, which are robust features, either man-made or natural, that can be detected from one image to the next (Ferretti *et al.*, 2001). The relative movement of these features, whether they be outcrops, buildings, or utility poles, defines a deformation field that may lack the spatial detail of standard InSAR analysis in arid areas, but that instead can capture displacement in urbanized, vegetated, and mountainous areas (Bürgmann *et al.*, 2006) that are not commonly amenable to typical InSAR procedures.

Lidar imaging

Despite the attractive properties and scientific insights derived from InSAR studies, in many heavily forested regions such studies are impracticable, because the density and seasonal variation in the canopy simply does not provide a repeatable image of the ground. As mentioned in the introductory chapter, a new imaging technique has ushered in an era of very high-resolution topographic imaging that can define the ground surface beneath forest canopies (Fig. 1.5). Rather than relying on radar, lidar (light detection and ranging) uses concentrated pulses of light that can typically "see" the ground through even small openings in a forest canopy (Carter *et al.*, 2007). Both airborne and ground-based lidar are now being used to create "bare-Earth" DEMs. Their spatial resolution depends on the distance of the lidar instrument from its target, but commonly DEMs are created with 1-m spatial resolution from airborne instruments and 1-cm to 1-mm resolution with ground-based ones. As lidar has unmasked previously unknown faults and has imaged landslides, channels, and hillslopes beneath forest canopies, its popularity has grown. In the United States, two centers for lidar acquisition sponsored by the National Science Foundation (NSF) have been created: the National Center for Airborne Laser Mapping (NCALM) and

Box 5.2   Climate–tectonic linkages at seasonal time scales.

Although the largest load in a sedimentary basin derives from the mass of sediments within it, water that is either stored or passing through the basin creates an additional load. For basins where rainfall is highly seasonable, cyclic variations in that water load might be expected. Over the past decade, studies that combined remotely sensed altimetry with continuous GPS records have documented not only the variations in the height of water within a basin, but also coeval basement subsidence as the crust responds elastically to a seasonal water load (Bevis *et al.*, 2005). Remotely sensed gravity data now allow the changing mass of water in a basin to be quantified at daily to weekly time scales. As a consequence, we can now estimate the crustal flexure that would be expected to occur in response to changing seasonal water loads.

   Every summer the Indian monsoon dumps 1–4 m of rain on the Himalaya and the adjacent foreland basin. Rivers flood and the ground becomes saturated. In most of the foreland, about 70–80% of the annual rainfall falls during those summer months and causes the crust to subside under that temporary load. Surprisingly, this rainwater loading has a spectacular impact on crustal stresses, crustal motion, and seismicity within the nearby Himalaya (Bettinelli *et al.*, 2008). A comparison (see figure A) of the water load (or height) within the foreland, changes in the rate of north–south contraction in the Himalaya, and the frequency of earthquakes greater than magnitude 3 shows that they vary with similar periods, but that times of high water loading are anticorrelated with times of high seismicity and more rapid north–south shortening.

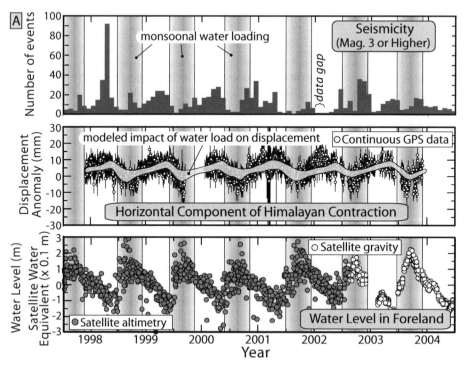

A. Continuous time series of seismicity and geodesy in the Himalaya and water levels in the adjacent foreland. Modified after Bettinelli *et al.* (2008).

Why is that? A simple flexural model (see figure B) predicts that downward flexure of the foreland due to seasonal water loading will cause extensional stresses at seismogenic depths (2–15 km) within the Himalaya. Such extension would partially counterbalance the overall contractional regime, thereby both slowing down the rate of contraction (as shown by the GPS) and inhibiting seismicity. With the end of the monsoon, water loading diminishes, the crust beneath the foreland rebounds elastically, contractional stresses are increased within the Himalaya, and seismicity accelerates. This stunning correlation suggests that faults are surprisingly sensitive to very small stress changes (2–4 kPa) and are most sensitive to changes in the rate of stress change. The success of this study highlights the insights that have emerged from nearly continuous and carefully coordinated measurements of geodesy, gravity, altimetry, and seismicity.

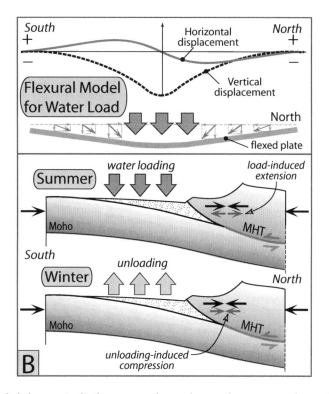

B. Modeled changes in displacement and crustal stress due to seasonal water loads.

GeoEarthscope, which, between them, have acquired thousands of square kilometers of lidar imagery over major faults in the western United States. Shortly after processing, most NSF-supported lidar acquisitions are available to the public (www.opentopography.org; www.calm.geo.berkeley.edu) and can be readily explored using *.kmz files in conjunction with Google Earth (see Fig. 4.21 for some examples).

Ground-based lidar is just beginning to be extensively used. This approach promises to provide tremendously detailed topographic imagery of geomorphic features that will enable detailed evaluation of many landscape attributes, such as terrace or hillslope geometries, or channel width, roughness, and gradient within diverse tectonic environments.

Following the 1999 Hector Mines $M=7.1$ earthquake, a helicopter-based lidar survey was conducted along the rupture (Hudnut *et al.*, 2002). These data provided a much more laterally extensive perspective on displacements than is commonly acquired by geologists on the ground. Not only can the three-dimensional topography and offset geomorphic features be imaged in significant detail (Fig. 5.19A), but topographic profiles that are parallel to the fault rupture but lie on either side of it can be used to calculate the average slip along any fault segment. To the extent that considerable geomorphic continuity existed across the fault prior to rupture, the amount of lateral and vertical translation of one profile that is required to match the other profile defines the magnitude of surface slip (Fig. 5.19B).

ASTER imagery for rapid or large-scale deformation

Although most geodetic studies are focused on tectonic deformation, geodetic analysis of rapidly moving geomorphological features, such as glaciers or earthflows, can delineate spatial variations in rates of movement. Such data can be used, for example, both to examine how flow rates vary as a function of slope, width, or climatic parameters, or to calculate volumetric fluxes or erosion rates as a function of flow rates. For geomorphological features that change at rates of meters per year, as well as for seismic offsets of several meters, much coarser-resolution satellite imagery can sometimes be used as a geodetic tool to delineate deformation rates or seismic offsets. For example, imagery produced by NASA's Advanced Spaceborne Thermal Emission and Reflection Radiometer (ASTER) or by the French Système Pour l'Observation de la Terre (SPOT) has a spatial resolution (pixel size) ranging from 15 to 5 m, respectively. Careful orthorectification based on topography derived from the Shuttle Radar Topography Mission (Farr *et al.*, 2007)

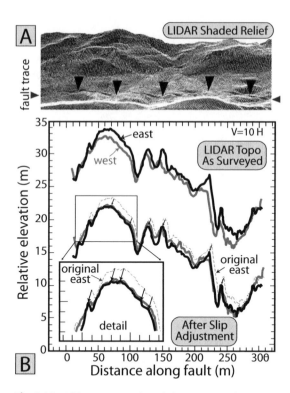

Fig. 5.19   Lidar topography of the Hector Mines rupture.
A. Shaded relief of high-resolution lidar topography of the terrain surrounding the surface trace of the 1999 Hector Mines rupture. The fault trace extends across the base of the image. Arrowheads point to geomorphic features that display dextral offsets across the fault.
B. Topographic profiles parallel to and on either side of the Hector Mines rupture. Upper profiles show topography projected perpendicular to the fault. The eastern profile has to be translated down and to the north in order to match the western profile. Inset magnifies one portion of the profiles after fault slip is removed and emphasizes both the displacement vector and the excellent topographic match following restoration. Note that the 10-fold vertical exaggeration distorts the slip vector: actual slip is 3:1 horizontal:vertical.

and co-registration of a succession of images permits cross-correlation among image pairs using a program such as COSI-Corr (Leprince *et al.*, 2007) in order to define sub-pixel displacements between the dates of the two images. Because many correlated pairs of individual pixels can be analyzed per image, a statistical average velocity or displacement can be obtained for a given longitudinal position on

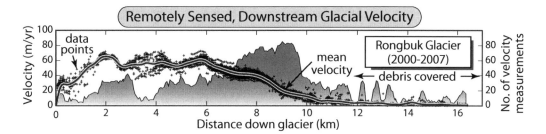

Fig. 5.20 Longitudinal glacier velocity from remote sensing analysis.
Velocity of the Rongbuk Glacier on the north flank of Everest as derived from analysis of ASTER and SPOT data.
The number of correlated data pairs varies along the glacier's length (gray shading). Commonly, many individual
measurements across the width of the glacier or over several time intervals provide a statistically robust measure of
the current velocity field (black crosses are individual measurements; central line is the mean). Note the concentration
of more rapid flow in the upper half of the glacier, whereas the debris-covered, lower half of the glacier is generally
stagnant. Modified after Scherler *et al.* (2011).

a geomorphic feature (e.g., Scherler *et al.*, 2011).
This approach can provide a remarkably
detailed view of spatial variability in glacial
velocity (Fig. 5.20), earthflow motion, or
coseismic displacement. Such patterns of spatial
variability on glaciers, for example, provide a
clear framework in which to examine how
velocity and flow vary as a function of ice
discharge, glacier width, ice temperature, or
surface slope, as well as enabling an improved
assessment of how glacial erosion relates to ice
dynamics.

Summary

Technological advances in measurement
techniques during the past decade are ushering
in a new era of high-precision geodesy. This
geodesy is providing a much clearer view of
partitioning of deformation at annual to decadal
time scales. Now, small differences in the
velocity vectors between nearby crustal blocks
and even within individual blocks can be detected.
These measurements help to pinpoint those
zones where these differential movements must
be accommodated. Through GPS and VLBI
observations, the regional-scale driving forces
represented by lithospheric plates moving with
respect to each other are becoming better defined.
Whereas continuously recording GPS networks
now can provide almost real-time records of

seismic deformation, radar interferometry is
providing an unprecedented, detailed view of
coseismic and interseismic deformation at a
regional scale. These data sets are generating new
perceptions of both the near- and far-field surface
effects of earthquakes, and they are guiding the
development of new models that more faithfully
mimic actual deformation. New techniques for
stacking radar images are permitting increasingly
subtle differences in deformation rates to be
illuminated. Over the past decade, GPS campaigns
have been conducted within many of the actively
deforming areas of the world, and we now have
a far better understanding of the regional
deformation field of the Earth. Numerous
permanent GPS networks have also been installed,
particularly in the vicinity of urban centers
confronted with a significant seismic hazard and
in many of the world's most seismically active
regions. Interferometric radar studies of lightly
vegetated, seismically active regions have become
more commonplace, whereas new automated
techniques for applying interferometry to urban,
vegetated, and mountainous areas are being
developed. Airborne and ground-based lidar
imaging is providing stunningly detailed topography of geomorphic surfaces in many actively
deforming terrains.

A place still remains for traditional geodetic
studies. Many important geodetic problems in
tectonic geomorphology can be addressed
without recourse to expensive, high-precision

instrumentation. Hand-held GPS units can very successfully define the deformation of most geomorphic features. Alignment arrays, trilateration surveys, leveling lines, and tide gauges all offer useful information on short-term deformation. Sometimes, better vertical resolution can be obtained through these approaches than through GPS measurements. Often the instigation for a more elaborate study will originate from simple observations of the deformation associated with faults and folds. Moreover, the database for these older measurements extends farther back in time and, thus, provides a longer temporal basis for calculating deformation rates.

The expanded geodetic knowledge that is rapidly emerging prompts a new array of important unanswered questions. We can now determine plate motions, rotations of crustal blocks, near- and far-field fault displacements, and folding rates with a greater precision and across a broader area than ever before. Armed with the knowledge of present-day rates of deformation, we would like to know the persistence of these rates in the past. Were the same faults and folds that are accommodating differential motion today also primarily responsible for accumulating strain in the past? What fraction of observed interseismic strain is retained as permanent crustal deformation and how much is simply elastic strain that is recovered during earthquakes? Does the shape of large-amplitude folds mimic geodetically measured, coseismic deformation? Were the displacement patterns and lengths of ruptures due to major earthquakes repeated consistently in the past, or were different patterns associated with each earthquake? Given a steady plate-tectonic forcing (or is it as steady as we assume?), what causes the formation and abandonment of faults? As the poles of plate rotation changed position in the past, how were the new stress orientations accommodated at plate margins? As discussed in the next chapter, some of the answers to these questions can be revealed through paleoseismological studies that seek to reconstruct deformational patterns associated with past earthquakes.

6 Paleoseismology: ruptures and slip rates

Geological maps of either Quaternary deposits or older bedrock often indicate the presence of faults in a given area. Through examination of displaced land features or of lithologic contacts in outcrops, we can tell that these faults have been active at some time in the past. But, how active have they been? Have they ruptured in large earthquakes that broke long sections of the fault, or are they associated with smaller displacements? Have the faults moved aseismically and continuously or in discrete events that generated earthquakes? When did the last rupture and the ones previous to that occur? Is there regularity to the timing of events? Is there a spatial pattern to the rupture that is repeated from event to event? Can the fault be divided into independent segments that rupture in characteristic events? Does an earthquake on one fault increase or decrease the likelihood of rupture on other nearby faults, and how are those changes dependent on the orientation of the fault? Do arrays of nearby faults exhibit systematic rupture patterns (Fig. 6.1)? Can we estimate the likelihood that a fault will produce an earthquake of a given magnitude within a specified period of time? What has the past slip rate on a fault been and how does that compare with geodetically determined rates? What can be done to estimate rupture histories when a fault is not exposed?

Answers to questions such as these constitute the focus of paleoseismological studies. Such investigations typically use stratigraphic, structural,

geomorphic, and biological evidence to reconstruct the sequence of displacements on a fault. When combined with dating of displaced features or of other indicators of faulting, the timing of past ruptures can also be determined. When multiple offsets can be dated on a single fault, it becomes possible to determine recurrence intervals and longer-term rates of displacement and to define the variation in the displacement that occurred during each of several earthquakes along the same fault.

Why do we want to know this? If we are interested in understanding how the brittle upper crust responds to imposed stresses, we need to know how and when it ruptures during earthquakes. In order to develop a basis for predicting the location and magnitude of future earthquakes, we need many details of past earthquakes. If we want to determine whether or not a fault exhibits "characteristic" or repeatable behavior, we have to know its rupture history and the variability of displacement along it in the past. We recognize that, in areas of active deformation, the interplay between tectonic movements and surface processes controls the geomorphology of these deforming landscapes. In order to quantify such interactions, we need to know both the magnitude and three-dimensional geometry of faulting or folding events, as well as the timing of these events.

A remarkably diverse array of approaches – ranging from determining the growth record of trees that grew along a fault trace to interpreting

Tectonic Geomorphology, Second Edition. Douglas W. Burbank and Robert S. Anderson.

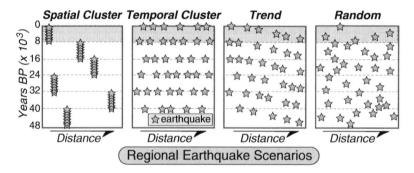

Fig. 6.1 Rupture scenarios for regional fault systems.
Contrasting spatial and temporal patterns for earthquakes on a regional system of faults. For spatial clusters, multiple earthquakes occur on a single fault over an interval of time and then that fault turns off as the rupture process switches to a different fault. For temporal clusters, multiple faults occur within a brief time window and then an interval of quiescence follows. Spatial–temporal trends in earthquakes suggest that, as one event releases stress on a given fault or section of a fault, it increases stress on nearby faults or segments, driving them to subsequent failure. In contrast to the other three models, random ruptures preclude predictability for any given fault. Modified after Burbank *et al.* (2002).

the stratigraphy of beds that have been affected by faulting and are now exposed in an artificial trench across a fault – have been employed in paleoseismological studies. The calculation of recurrence intervals and of rates of displacement depends both on the correct interpretation of the geological record of past offsets and on reliable dating of the timing of those offsets. Approaches to dating and some of the pitfalls and applications of various dating techniques were described in Chapter 3. Generally, these approaches are not discussed in more detail here. Instead, in this chapter, we focus on some of the many techniques that have been successfully used to reconstruct the record of seismicity along faults that have been active during late Quaternary times.

Seismic moment and moment magnitudes

Any time an earthquake occurs, it releases seismic energy, E_s. The energy released is proportional to the area A of the rupture plane, the average displacement d along it, and the stress drop $\Delta\sigma$ across the fault during the earthquake:

$$E_s = \frac{1}{2}\Delta\sigma dA \qquad (6.1)$$

The stress drop actually refers to the mean stress, σ_{mean}, acting across the fault during the

earthquake and is calculated as the average of the stresses before and after the earthquake:

$$\sigma_{mean} = (\sigma_{start} - \sigma_{finish})/2 \qquad (6.2)$$

If it is assumed (as is often done) that the fault is stress-free at the end of rupture ($\sigma_{finish}=0$), then the mean stress is equal to one-half of the stress at the start, $\sigma_{mean} = \frac{1}{2}\Delta\sigma = \frac{1}{2}\sigma_{start}$.

Although comparison of earthquakes on the basis of the energy released is now both possible and very useful, traditionally the size of an earthquake has been assessed on the basis of its magnitude. The first quantitative measure of size was the local magnitude (M_L) scale, which was based on Richter's (1935) observation that, with increasing distance from seismic sources in southern California, the maximum amplitude of ground motion decayed along a predictable curve. When the data for distance versus the logarithm (to base 10) of the amplitude of ground shaking were compared for several earthquakes, they followed parallel curves of decay with increasing distance. By measuring the amplitude, A, of shaking in a given earthquake and comparing it with the amplitude, A_0, of a "reference event," a local magnitude (M_L) can be defined:

$$M_L = \log A - \log A_0 \qquad (6.3)$$

For a reference earthquake with $M_L=0$, the amplitude of shaking at a distance, Δ, of 100 km from the source was 1 mm. A calibration of these curves using the reference event led to the following expression for local magnitude:

$$M_L = \log A - 2.48 + 2.76 \log \Delta \qquad (6.4)$$

The original Richter magnitude scale was developed based on measurements of southern Californian earthquakes during the 1930s using a certain type of seismometer (Wood–Anderson torsion instruments), and the equation above is only strictly applicable to these seismometers in this setting. Today, however, local magnitudes are calculated in many areas using a variety of seismometers. The coefficients in Eqn 6.4 have been modified to yield consistent estimates of the local magnitude, despite differences in instrumentation and regional contrasts in the transmission of seismic waves due to geological variability.

Several other magnitude estimates are commonly used. Surface-wave magnitudes, M_s, are typically calculated for events at distances exceeding 600 km by measuring amplitudes of surface waves with a period of approximately 20 s. Body-wave magnitudes, m_b, are based on the amplitude of direct compressional waves (P waves) measured from short-period seismograms, typically with periods of about 1 s. Empirical relationships have been established between these earthquake magnitudes and seismic energy, E_s, released:

$$\log E_s = 11.8 + 1.5 M_s \qquad (6.5)$$

and

$$\log E_s = 5.8 + 2.4 m_b \qquad (6.6)$$

A different measure of the energy released in an earthquake is the seismic moment (M_0, measured in dyne cm), which is equivalent to the product of the rupture area, a, average displacement, d, and rigidity, μ, or shear modulus of elasticity of the crustal material involved in the rupture:

$$M_0 = \mu d a \qquad (6.7)$$

where μ is commonly taken as 3×10^{11} dyne/cm^2 for the crust and 7×10^{11} dyne/cm^2 for the upper mantle. Thus, instead of measuring the amplitude of a deflection on a seismograph, observations of the rupture length, the probable total area of rupture, and the mean displacement are combined with estimates of rigidity to yield the seismic moment. Moreover, the magnitudes of many earthquakes today are reported as "moment magnitudes," M_w, and are based on the seismic energy released during an earthquake, that is, the seismic moment, M_0:

$$M_w = (2/3) \log M_0 - 10.73 \qquad (6.8)$$

In theory, the seismic moment is a single, definable quantity that reliably characterizes seismic energy release. As a consequence, moment magnitudes should theoretically provide a direct means for comparing different earthquakes. In reality, today, seismic moments are calculated both from the actual characteristics of the rupture and from empirically derived functions that relate body- or surface-wave forms to seismic energy and moment.

Because the amplitude of shaking in prehistoric earthquakes cannot be quantitatively reconstructed, the magnitudes of ancient earthquakes are difficult to estimate directly based on paleoseismological studies. What can be measured, however, are rupture lengths, mean displacements, and approximate rupture areas. These observations provide a means of determining the absolute size of past earthquakes for which no instrumental records exist. Such a quantification through paleoseismological studies represents a powerful basis for comparing ancient and modern earthquakes. Modern studies have defined relationships between the seismic moment and the amount of ground displacement and shaking. Building on these relationships, paleoseismological determinations of rupture lengths, fault geometry, and coseismic fault displacement provide key constraints on the assessment of modern seismic hazards along faults.

Direct observations of paleoseismic displacements

Two types of data can be brought to bear on the reconstruction of the history of past earthquakes. Direct observations of displaced or cross-cutting features provide unambiguous information about displacements. Such information can be stratigraphic, structural, or geomorphic in nature, and includes features such as faulted beds, offset stream channels, or uplifted shorelines (Plate 1B). Indirect indicators of earthquakes require a conceptual linkage between the observable data and the earthquake that caused it. In some instances, the evidence reflects the coseismic offset itself, whereas in others it reflects other parts of the seismic cycle. For example, drowned forests, stratigraphic evidence of tsunamis, or the chronology of rockfall deposits require an interpretation of their genesis in order to tie them into specific faulting events. We discuss the types of information that can be generated with direct observations first, and later we examine several kinds of indirect paleoseismological observations.

Surface rupturing versus buried faults

Faults that rupture the Earth's terrestrial surface are relatively easy to discern if they cut well-preserved geomorphic markers, are not significantly eroded, and are not masked by too much vegetation. Such faults provide ready targets for paleoseismological studies. On the other hand, buried faults tend to cause less obvious and commonly more diffuse deformation of the ground surface. Many thrust faults, for example, have both upward splays that rupture the ground surface, as well as buried ramps and faults that accommodate slip beyond the splay. Slip on buried ramps will tend to drive regional rock uplift that can be confidently identified only at its margins (where differential uplift occurs) or in the presence of robust markers with known positions with respect to sea level (Fig. 6.2 and Plate 2). Where such markers occur, they provide an opportunity to define a slip history for buried faults that lie far beneath the surface. More commonly, however, the presence of buried faults and the seismic hazard they represent is poorly known until these faults rupture in a unpredicted earthquake.

Trenching

In order to provide a detailed record of past displacements and their timing, excavated trenches across faults often provide key insights.

Overall, the basic practical objectives in trenches are: (i) to identify and date layers within a stratigraphic succession that either have been disrupted by faulting or overlie fault traces without disruption; and (ii) to document the amount of displacement in past faulting events. Sites for trenches have to be carefully chosen in order to maximize the useful information that they may generate. To the extent possible, trench sites should contain abundant datable material, and they should provide stratigraphic or structural markers that can be used to measure offsets. Typically, it is impossible to know what is likely to be found in the subsurface as a trench is excavated, so good judgment and good luck combine to create a data-rich excavation. Because radiocarbon dating is still the most frequently applied dating technique in trench analyses, trenches are commonly sited in swampy areas where fault displacements have dammed local streams or have ponded the groundwater table. In such circumstances, datable organic matter is more likely to be preserved within the young strata associated with the fault. Similarly, thinly bedded and/or channelized deposits are more likely to reveal discrete, measurable offsets than will massive deposits, such as debris flows, which are rather homogeneous in all directions. Thus, the likelihood of having a rich stratigraphic record can be enhanced by choosing sites of low-energy deposition and/or sites where linear stratigraphic markers with which to measure offsets are likely to be preserved. For example, if relict lake shorelines or small-scale channels are oriented approximately perpendicular to a fault, they can offer particularly good stratigraphic markers or piercing points across a fault.

In trenches, the analysis of the stratigraphic record of earthquakes is typically a time-consuming process. Commonly a week or more is spent analyzing a single trench. Thus, the choice of sites for trenches is not a trivial exercise: you want the maximum information for the time invested. Trade-offs have to be made between competing objectives: obtaining more detailed information on a single segment of a fault, comparing rupture histories in different areas, obtaining maximum information on the magnitude of offsets, and developing the most

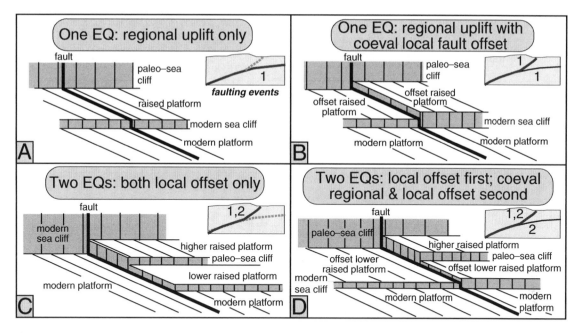

Fig. 6.2   Local surface uplift due to slip on surface-rupturing faults versus buried faults.
Displaced marine abrasion platforms (Fig. 2.1) provide a means to discern both local and regional uplift due to surface-rupturing faults and buried faults, respectively. A vertical, dip-slip, surface-rupturing fault (up on the right-hand side) occurs in the center of each scenario. The presence of a geomorphic marker (an abrasion platform) tied to sea level permits recognition of displacement due to buried faults. Inset on upper right of each figure depicts the activity and sequence of each fault: solid line for a ruptured fault plane; dashed line for an unruptured plane. A. A single earthquake on the buried fault causes regional uplift that yields a raised platform, a paleo-sea cliff, and modern sea cliff in a new position. No differential slip occurs on the fault at the surface. B. With coeval slip on both fault surfaces, the right-hand side is differentially uplifted, creating a higher modern sea cliff and a higher raised platform on that side. C. Two events that slip only on the splay create a second paleo-sea cliff and a second raised platform on the right-hand side only. D. Two events, the first on the splay only and then on both the splay and buried faults, create a complex pattern of features. Note that the alignment of the modern sea cliff all the way across the modern platform indicates that the buried fault slipped in the second event. Otherwise, the sea cliff on the left would align with the paleo-sea cliff on the right. Modified after Kelsey *et al.* (2008).

extensive chronology of past events. Two different strategies are typically used for orienting trenches with respect to faults. In most situations, a single trench is excavated perpendicular to the trend of the fault. The stratigraphy and structures revealed in the walls of the trench are meticulously surveyed and mapped, material for dating of various stratigraphic horizons is collected, and a paleo-earthquake history is interpreted based on these data. Alternatively, along strike-slip faults, two trenches parallel to the fault trace are sometimes excavated (Fig. 6.3A). The stratigraphy on the faces nearest to the fault trace is mapped, surveyed, and dated in each fault. Special attention is paid to linear features, such as channels, planar crossbeds, shoreline features, or unusual bedding configurations, that trend approximately perpendicular to the fault. After mapping the two trench walls parallel to the length of the fault, the intervening strata that are cut by the fault are "salami sliced" perpendicular to the fault trace (Fig. 6.3B), which is to say that they are incrementally cut back along vertical planes. Each stratigraphic feature that could act as a piercing point is traced to the fault and the magnitude of offset *vis-à-vis* the correlative feature on the opposite side of the fault is

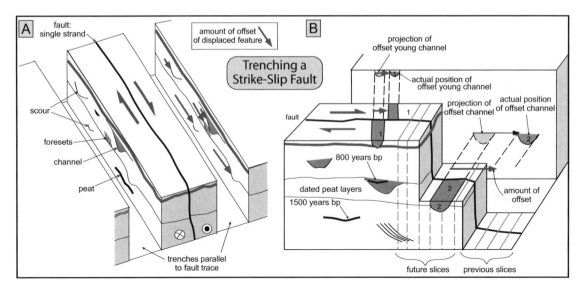

Fig. 6.3 Trenches on a strike-slip fault.
A. Orientation of trenches with respect to a strike-slip fault, when two trenches are employed to determine both the timing and magnitude of previous ruptures. On the walls closest to the fault trace, the stratigraphy is carefully surveyed and material, such as peats, for dating is collected. Hypothetical amounts of offsets of stratigraphic markers are depicted by the arrows on the right-hand trench wall. Note the general downward increase in displacement. Despite their difference in stratigraphic height, the top two offset markers, a channel and a scour, have identical offsets because they were both displaced in the most recent earthquake. B. "Salami slicing" across the fault trace between two surveyed trenches. Offsets of linear features like channels are clearly revealed, and displacements of the margins of these features are measured in order to determine the amount of offset in individual earthquakes. The uncertainty in the offset of features that are projected across the fault should also be estimated. Similar magnitudes of offset are expected for all of the features that were created between each successive pair of earthquakes. When the stratigraphy in the trench walls has been well dated, both the age and the displacement of past earthquakes can be quantified, such that both the slip per event and long-term slip rates can be reconstructed.

measured. Along a fault trace where multiple earthquakes are recorded and where the sense of offset has not changed between successive ruptures, the amount of the measured offset should increase with increasing stratigraphic age, that is, with depth in the trench. This increase, however, should not be steady; rather, it should be stepwise, such that, between two stratigraphic horizons that correlate with successive faulting events, all of the displaced features should display similar offsets (Fig. 6.4A). Each cluster of comparably offset features will be bounded by abrupt changes to smaller offsets above and larger offsets below, marking the accumulation of displacement in successive earthquakes. Ideally, the horizons that correlate with the abrupt changes will be well dated and, hence, will reveal directly the timing of previous earthquakes (Sieh, 1978).

The "salami slicing" approach can reveal detailed data on both the timing and magnitude of offsets in previous earthquakes along strike-slip faults. Even when the coseismic displacements in each rupture are more than several meters, a long trench can record multiple offsets. In fact, cumulative offsets of the oldest markers can be equal to the length of the trench. Many dip-slip faults that are of interest to paleoseismologists have displacements of 2–4m in individual earthquakes. In order to find the record of multiple ruptures in the past, trenches on dip-slip faults have to be deep, rather than long. A trench 10m deep might record only two ruptures, if the displacements were 4m in each event. The instability of trench walls (which must be shored up and braced) commonly imposes a practical limit on how deeply trenches can be dug, and, therefore, limits the number of previous rupture

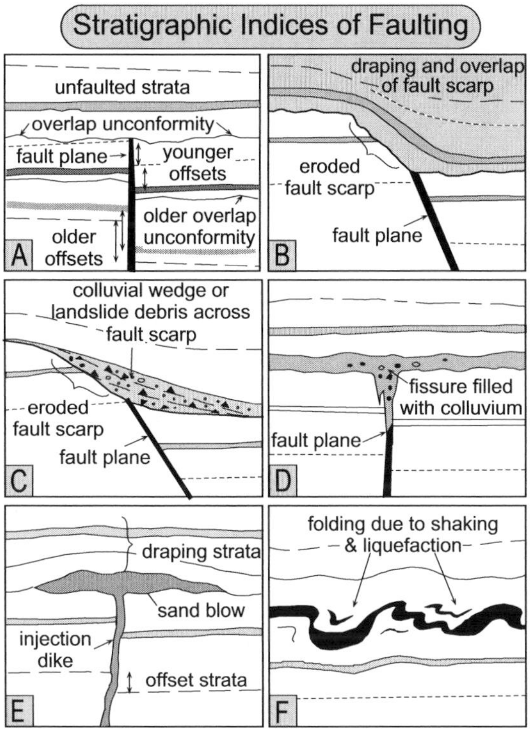

Fig. 6.4   Stratigraphic indicators of past faulting events.

A. Strata offset across a fault are beveled by an unconformity and overlain by undisturbed strata. The unconformity developed after the last earthquake. The amount of displacement increases downward. The two different amounts of offset of strata and the presence of an older offset unconformity provide evidence for a previous earthquake. B. If the topography across a faulted surface is incompletely beveled off, succeeding strata will drape across the eroded fault scarp. They will be unbroken, but may appear deformed due to the topography on which they were deposited. Commonly, they will show thickening above the downthrown fault block. C. Similar to B, but with a colluvial wedge derived from erosion of the upthrown block.
D. Fissures that open along a fault plane are typically filled with colluvial material shortly after faulting.
E. Injection dikes in the subsurface and sand blows or sand volcanoes on the surface provide evidence for past earthquakes. The age of the youngest strata cut by the dikes or underlying the sand blow provides a maximum age on the faulting. F. Liquefaction due to shaking can cause folding of weakly consolidated sediments near or at the surface. The age of the deformed beds provides a lower limit on the time of the earthquake that deformed them. Modified after Allen (1986).

events that can be examined. As a result, the paleoseismic records in trenches of large thrust or normal faults are commonly limited to one or two events (Rubin *et al.*, 1998; Lavé *et al.*, 2005).

Offset strata due to dip-slip faulting can present additional challenges that may not be present with strike-slip faults. Stratigraphy that is displaced by strike-slip faults is translated horizontally into new positions where the strata on both sides of the fault are commonly preserved. Thus, more sediments can accumulate above the displaced strata and piercing points recorded by the older strata will be preserved. On the other hand, when dip-slip faulting occurs, the strata on one side of a fault are upthrown with respect to the opposite side. In theory, the offset of the strata on opposite sides of the fault records the magnitude of displacement. Unfortunately for paleoseismologists, strata on the upthrown block are typically subjected to subaerial erosion. Consequently, the specific stratal layers that could define the offset precisely can be eroded away (Machette *et al.*, 1992a), such that reconstruction of past displacements becomes considerably more difficult.

In order to reconstruct the history of offset that is recorded by strata within a trench, stratigraphic and structural relationships revealed in the walls have to be carefully interpreted. One goal should be to reconstruct the direction of slip in previous earthquakes. Consequently, indicators of the slip direction, such as striae, slickensides on the fault plane, fold axes of deformed strata, or the displacement of piercing points, should be sought. As with traditional structural geological studies, cross-cutting relationships can provide unambiguous evidence of past deformation (Fig. 6.4). For example, if older strata are cut by a fault, but overlying younger strata are continuous across the trace of the fault, an earthquake is interpreted to have occurred between deposition of the youngest strata that are cut by the fault and the oldest strata that are not displaced by it (Fig. 6.4A). If these strata can be dated, they provide bracketing ages on the faulting event. Progressively larger structural offsets of stratigraphic horizons farther beneath an overlapping unconformity or the identification of older offset unconformities can indicate both the timing and magnitude of still earlier earthquakes (Fig. 6.3A).

Topographic scarps created by faulting events are rapidly "attacked" by erosional forces. If the scarp forms in an area where active deposition is occurring all around it, the scarp is likely to be draped with younger strata soon after the rupture (Fig. 6.4B and Plate 2E). If the scarp forms in a predominantly erosional environment, its upthrown side will erode and will commonly provide colluvial debris that accumulates along the degrading scarp as a *colluvial wedge* on the locally downthrown block (Fig. 6.4C). During strike-slip earthquakes, differential vertical motion is commonly small, but movement along the irregularly shaped fault plane will cause fissures to open in places (Plate 2C). These fissures are open to the surface and typically fill quite rapidly with colluvial, alluvial, or eolian debris. In repeated ruptures, fissures may reopen with each rupture (Sieh *et al.*, 1989). Consequently, a single fissure with multiple filling events may record several different earthquakes (Fig. 6.5 and Plate 2C). Seismic shaking can cause liquefaction of sediments having a high water content (Fig. 6.4F). Liquefied sand represents a slurry of sand and water that may be "erupted" on the surface to form sand blows or sand volcanoes (Fig. 6.4E and Plate 2A).

The interpretation of the stratigraphy and structures found in trenching sites is often not straightforward. In some fortunate circumstances, clear cross-cutting relationships and abundant material for dating are present (Fig. 6.6). At such sites, ruptures of older strata and draping or erosion across the upper termination of fault strands provide direct evidence for faulting. Even in these conditions, it is important to remember that faults do not always propagate to the surface and that new ruptures can form in successive earthquakes. Thus, a newly formed fault splay may branch out from the main rupture and terminate at some random level in the subsurface in an isolated outcrop or trench. In this case, the undisturbed layers overlying the fault termination will pre-date the faulting

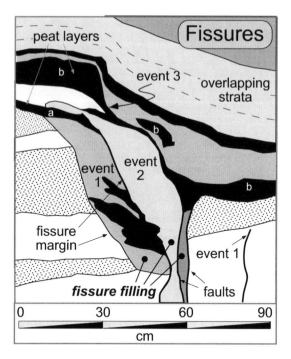

Fig. 6.5 Multiple fissure filling in successive earthquakes.
Event 1 opened a fissure. The fissure filling post-dates the event, contains some datable peat, and is overlain by peat "a." Event 2 reopens the fissure and cuts peat "a." Event 2 is overlain by peat "b," which draped the existing topography and is cut along a small fault along the fissure trend during event 3. Despite its complexity, this fissure reveals the detail that can be extracted from a good exposure and provides some insight into how overlapping, cross-cutting, and fracture-filling relationships can be exploited. Modified after Sieh *et al.* (1989).

event, which is the opposite of the case when undisturbed layers depositionally overlie a fault termination. Consequently, in evaluating fault terminations, it is important to seek evidence for: (i) surface erosion across the fault trace; (ii) bed thickness changes above the fault termination that indicate the filling of small-scale topography related to the fault; (iii) discrete groupings of the amounts of stratal offsets found in older layers cut by the fault (indicating that deposition was synchronous with faulting and that previous earthquakes had occurred on the fault); and (iv) significant displacement on the fault just below the undisturbed layers, rather than a progressive dying out of displacement

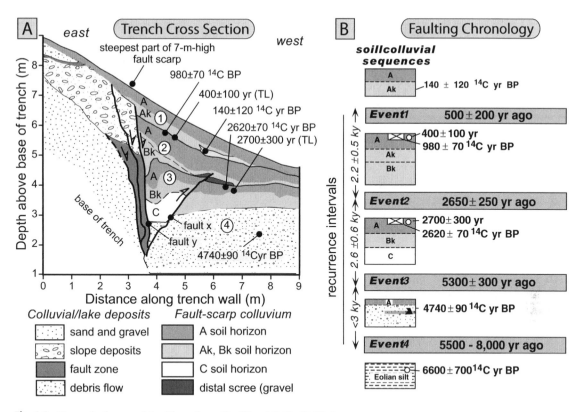

Fig. 6.6   Record of normal faulting along the Wasatch Fault, Utah.
Data from a trenched fault that bounds the Wasatch Range near American Fork Canyon, Utah. A. Evidence for three
faulting events is revealed in this trench. The excavation is >9 m deep, and yet, because markers typically show 1–3 m
of offset per earthquake, only three rupture events are recorded here. A repetitive soil zonation (A–Ak±Bk)
delineates the top of each of three colluvial wedges (labeled 1 through 3). The magnitude of offset for each wedge
can be estimated by restoring the top surface of the wedge back to the ground surface such that the "A" soil horizon
is continuous on to the upthrown block. Radiocarbon and thermoluminescence dates are used to constrain the timing
of faulting events. Cross-cutting relationships with faults and soils define rupture events. Fault "x" cuts soil sequence 3
and is overlain by sequence 2. Fault "y" cuts soil sequences 2 and 3, but is overlain by sequence 1. Note that, with the
exception of the lowest and topmost stratigraphic units on the downthrown side, none of the downthrown strata are
represented on the upthrown block. B. Interpretation of the rupture history based on the trench stratigraphy.
Evidence for the oldest faulting event is based on a nearby trench. Note that, on this trenched segment of the Wasatch
Fault, ruptures appear to be evenly spaced in time, but the magnitude of rupture varies, with the latest offset (~1 m)
being about half as large as the offsets during the previous two events. Modified after Machette *et al.* (1992a).

along the fault. If it can be shown that the same
stratigraphic level in several places overlies fault
traces, the likelihood increases that this rupture
represents a fault that did propagate to the
surface.

The stratigraphy of soils can provide useful
controls on the interpretation of trench stratigraphy. For example, colluvial wedges (Figs 6.4C
and 6.6) are often associated with dip-slip faults.
Each colluvial wedge is commonly interpreted

to result from a single rupture of the bounding
fault (Machette *et al.*, 1992b; Schwartz and
Coppersmith, 1984). Steepening of slopes,
localized uplift, or overthrusting of the land
surface along a rupture plane destabilizes
weakly consolidated surface strata. Quaternary
sediments that have been uplifted to a higher
topographic position along these steepened
slopes quickly erode and produce colluvial
wedges that accumulate along the toe of the

exposed scarp. If the faulted strata are well consolidated, they may maintain steep faces for many centuries. Nonetheless, it is commonly assumed that raveling of a steepened slope occurs rather quickly (about 100 years in many sediments), such that, not long after faulting, the surface of a colluvial wedge is at or below the angle of repose for unconsolidated debris (~30°) and the deposition rate is low enough for pedogenic processes to begin to develop soil zonations (A,B,K horizons). Recognition of such soil horizons in trench walls pinpoints the upper surface of colluvial wedges formed during earlier earthquakes and delineates some of the vertical extent of the wedges (Fig. 6.6A). Particularly in colluvium that is poorly stratified and contains few stratigraphic horizons with which to judge offsets, even subtle soil horizons provide markers for delineating displacements. In addition, the degree of soil development can be related to the interval between earthquakes, that is, the time from first stabilization of the colluvial wedge to its burial by the next youngest wedge (Fig. 6.6B).

Many fault zones actually have more than one fault strand at the surface. In order to account for all of the displacement along a fault zone, all of the strands or surface traces should be identified, and a paleoseismic record for each should be developed. It is not uncommon for strike-slip faults to splay into several branches as they approach the surface. Sometimes these splays all occur within a fairly compact zone, for which a single trench or small set of trenches will reveal the entire displacement record (Fig. 6.7A). However, when strands are separated by tens or hundreds of meters, each has to be examined separately. In compact zones with multiple splays (Fig. 6.7B), mapping the displacement of a distinctive marker bed across each of the fault strands can set useful limits on the total slip (Lindvall *et al.*, 1989).

Dating of fault-disrupted strata in trenches forms the basis for determining the timing of individual ruptures, which in turn is the basis for assessment of recurrence intervals and of slip rates, if displacement is known. A goal in many trenches is to generate as many reliable dates as possible for a given stratigraphy,

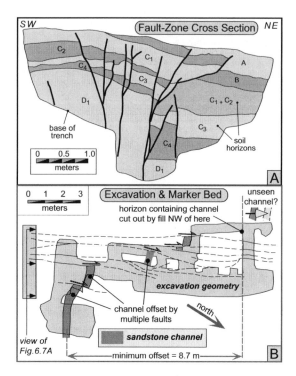

Fig. 6.7 Measuring slip on a fault with multiple splays.
A. Sketch of a trench wall oriented approximately perpendicular to a suite of splays along a strike-slip fault. This outward branching geometry or "flower" structure comprises 10 different splays across a zone ~3 m wide. Not all splays have been active in each earthquake. Note, for example, that two faults on the right cut the upper B_t soil horizon, whereas the other splays do not. Modified after Lindvall *et al.* (1989). B. Map view of the displacement of a marker horizon (a sandstone channel) across multiple splays of a strike-slip fault. A network of shallow trenches reveals the net offset of the marker channel. Across any strand, the offset is typically <2 m, but the cumulative offset is >8 m. Modified after Lindvall *et al.* (1989).

particularly if these dates will underpin the reconstruction of the seismic record. Radiocarbon dating of buried organics remains the most commonly used dating technique in trenches. Given analytical uncertainties in laboratory analyses, the possibility of contamination with young carbon, and the background variations in atmospheric carbon in the past (see Chapter 3), each radiocarbon date should be carefully interpreted. Combining different dating techniques, such as radiocarbon with optically stimulated luminescence, can provide a valuable check on the consistency of dates. In a trench along the normal fault along the Wasatch front (Fig. 6.6), for example, some radiocarbon dates vary by 400–700 years from their calendar ages. Although these temporal offsets can be assigned from the calibration curve for radiocarbon ages (see http://calib.qub.ac.uk/calib/), it is reassuring that thermoluminescence dates (Fig. 6.6A) reinforce the calibrated radiocarbon dates.

In most trenches, the strata that contain datable material do not consistently coincide with the strata that record rupture events. In such cases, dates on strata above and below the "event horizon" are used to bracket its age. To reduce the uncertainty in timing as much as possible, datable material should be sought as close as possible to the rupture horizon, and the highest laboratory precision available should be used to reduce the analytical uncertainty on the individual ages (Atwater *et al.*, 1991; Sieh *et al.*, 1989). Concerted efforts to provide the best possible time control are warranted, because reliable calculations of recurrence intervals are highly dependent on the quality of the ages assigned to rupture events.

Studies along the San Andreas Fault clearly illustrate the importance and utility of high-precision dating. Because the San Andreas Fault runs near or through several of the major population centers in California, and because it has generated "great earthquakes" (magnitude 8) in the past, there is great interest in knowing: (i) how often earthquakes occur along it; (ii) the length, displacements, and spatial patterns of past ruptures; and (iii) the recurrence interval between major earthquakes. During the past 30 years, much effort has gone into paleo-seismological studies along many parts of this major plate-bounding fault. Some of the earliest insights on recurrence intervals of San Andreas faulting came from trenches dug at Pallett Creek (55 km from Los Angeles) during Kerry Sieh's dissertation research (Sieh, 1978). Several subsequent studies have attempted to improve on the radiocarbon dating of the faulting events at this site, where 11 major earthquakes are recorded. A comparison of the faulting

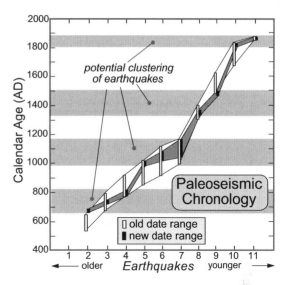

Fig. 6.8   Radiocarbon chronologies of earthquakes on the San Andreas Fault.
Two generations of radiocarbon dating along the San Andreas Fault at Pallett Creek, California. As opposed to the rather regular recurrence intervals suggested by the 1984 chronology, the much higher-precision dates published subsequently indicate a distinct clustering of earthquakes (horizontal shaded bands) into groups of two or three ruptures each that were separated from each other by a few decades, whereas the clusters themselves are separated from each other by 150–300 years. The higher-precision dates define a step-like event versus time curve (dark shading), as opposed to the smoothly changing curve (no shading) comprising the less precise dates. Modified after Sieh *et al.* (1989).

chronologies published in 1984 (Sieh, 1984) and in 1989 using higher-precision dates (Sieh *et al.*, 1989) is illuminating (Fig. 6.8). Based on the 1984 data, the recurrence intervals between major earthquakes fall into two groups: prior to about 1100 AD, the recurrence interval was about 100 years, whereas from 1100 AD to the present, the interval lengthened to 160–200 years. Within each grouping, the resolution of the radiocarbon dates suggests that the earthquakes may have been approximately evenly spaced in time. When radiocarbon dates with higher precision are used to date the earthquakes (black bars, Fig. 6.8), a rather different rupture history emerges (Sieh *et al.*, 1989). Rather than being evenly spaced in time, the earthquakes appear to cluster. Two or three earthquakes appear to have occurred during spans of 100–200 years, whereas each cluster is separated from the succeeding cluster by an earthquake-free interval of 150–300 years. The uneven temporal spacing of San Andreas earthquakes presents a challenge to seismologists trying to understand rupture physics and recurrence intervals. Moreover, the clustering of events suggests that seismic risks along this segment of the San Andreas Fault are very different than previously thought. Based on the 1984 data, it was deduced that the earthquakes were regularly spaced in time and had a recurrence interval of about 140 years. Given that the last large rupture in this area was in 1857, this scenario would suggest that the next big rupture could be imminent. In contrast, the 1989 data suggest that, because clustered earthquakes commonly occur within a few decades of each other and because no major rupture has occurred in the past 140 years, we are presently in one of the quiescent periods between clusters and that this quiet period might last for another 100 years!

Buried faults in terrestrial settings

Whereas faults that rupture the surface grab our attention, a larger seismic hazard is commonly posed by buried faults. Not only is their location poorly known such that seismic zoning for construction may ignore the hidden hazard, but when such faults do rupture, they tend to cause greater ground shaking (Somerville and Pitarka, 2006) because energy spreads outward above the buried fault tip rather than being guided and focused along the fault zone. Despite this hazard, quantifying slip on buried faults, particularly in the absence of clearly offset markers (a condition that is typical in most terrestrial realms), has posed a paleoseismological challenge. Some recent success has been achieved, however, using the shape of growth strata in the shallow subsurface to quantify the timing and magnitude of past slip events. The approach relies on the observation that, within the forelimb of fault-bend folds (Suppe, 1983), any increment of slip causes differential uplift of the hanging wall that lies above the ramp, whereas no uplift occurs

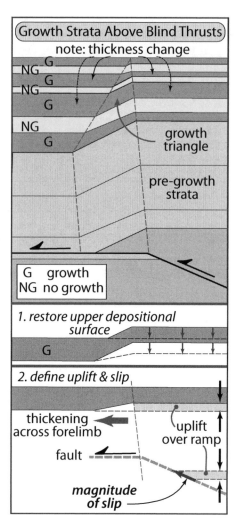

Fig. 6.9 Slip-controlled growth strata above a blind thrust fault.
The forelimb of a fault-bend fold is differentially uplifted and lengthened during each increment of slip, thereby creating a triangular accommodation space above the forelimb that can be subsequently filled with growth strata (G). Once deposition overtops the differentially folded forelimb, planar non-growth (NG) strata are deposited. A two-step process (lower panels) underpins calculation of the amount of slip related to one increment of growth strata. First, any post-depositional deformation of the top of the growth strata is removed in order to restore it to its depositional geometry (usually assumed horizontal). By maintaining the bed thickness, the base of the growth strata is also restored downward in this step to its position at the beginning of deposition. Second, the differential uplift of the base of the restored growth strata is measured and the amount of slip on the fault that is required to produce the observed uplift is calculated (lowest panel).

along a footwall flat (Fig. 6.9). Deposition that occurs following an increment of fault slip will fill the space above the newly tilted forelimb and will tend to restore a subhorizontal upper depositional surface (Dolan *et al.*, 2003). In the process, this deposition creates an irregular trapezoid of new growth strata that thins over the forelimb. Subsequent deposition in the absence of fault slip will cause simple vertical aggradation with no tapering of bed thickness across the forelimb. Thus, the magnitude of bed taper across the forelimb is indicative of the amount of relative uplift of the fold crest. If the dip of the underlying fault is known or can be deduced, then the amount of fault slip required to produce the observed differential uplift can be calculated. Dating of the base of each unit of growth strata will place bounds on the timing of the faulting events that caused each event of differential uplift. Because fold forelimbs are commonly hundreds of meters long, these geometries have commonly been investigated using boreholes, rather than trenches. This strategy, therefore, requires correlation of strata between boreholes, which introduces another source of uncertainty.

Buried faults in coastal settings

The largest earthquakes in the world in the past 60 years (Chile in 1960 and 2010, Alaska in 1964, Sumatra in 2004, and Japan in 2011) have occurred on subduction zones. Given the human population density in many coastal areas, and the fact that roughly half of the world's population lives along tectonically active plate margins, such earthquakes pose a major hazard. Most of these ruptures occur both at significant depths (commonly not breaking the surface) and far beneath the ocean surface. Hence, they are largely inaccessible. Given these facts, how can a record of past subduction-zone earthquakes be generated?

Displaced shoreline features typically provide the best coastal evidence of coseismic subduction-zone offsets. One appealing aspect of studying coastal features, in addition to the ocean vistas, is that detailed reconstructions of vertical deformation with respect to sea level

can be made. One obvious complication for such studies arises from changes in sea level. Whereas sea level has been quite steady since about 8 ka, sea level rose about 120 m as ice sheets melted during the preceding 10 kyr (Fleming *et al.*, 1998). Thus, a 20-ka terrace exposed at today's shoreline would actually represent 120 m of rock uplift since its formation. For most coastal records, data from a reliable sea-level curve must, therefore, be incorporated into any analysis attempting to determine the total rock uplift.

Along many uplifting coasts, wave-cut abrasion platforms (Fig. 2.1 and Plate 1A) provide the most abundant records of local rock uplift. Coseismic uplift causes a relative sea-level fall and exposes the newly emergent shoreline to wave attack (Figs. 6.2 and 6.10). The shape of the resulting platform depends on local rock strength, coastal orientation with respect to wave energy (Adams *et al.*, 2002, 2005), the time between successive uplift events, and the rate of eustatic sea-level rise. Broader platforms result from weaker rocks, higher wave energy, and longer interseismic intervals, because these factors prolong the exposure interval and increase the effectiveness of wave attack on coastal bedrock. If the platform-carving processes are sufficiently efficient, then steadily rising sea level also can generate wide abrasion platforms as previously formed platform is inundated and wave attack is focused progressively higher on the sea cliff. Except during intervals when sea level is constant, the change in sea level between seismic events must be added to the observed coseismic offset (usually measured with respect to sea level at the time of the earthquake) in order to determine the total rock uplift.

A staircase of raised abrasion platforms on the Peruvian coast (Bourgois *et al.*, 2007) reveals the importance of accounting for sea-level changes (Fig. 6.10B and C). Fifteen terraces span ~90 m of elevation and were created over the past 20 kyr during which sea level rose 120 m. The total rock uplift (>200 m) is, therefore, more than twice as large as indicated by the height (90 m) of the oldest terrace above current sea level. If ages and current elevations

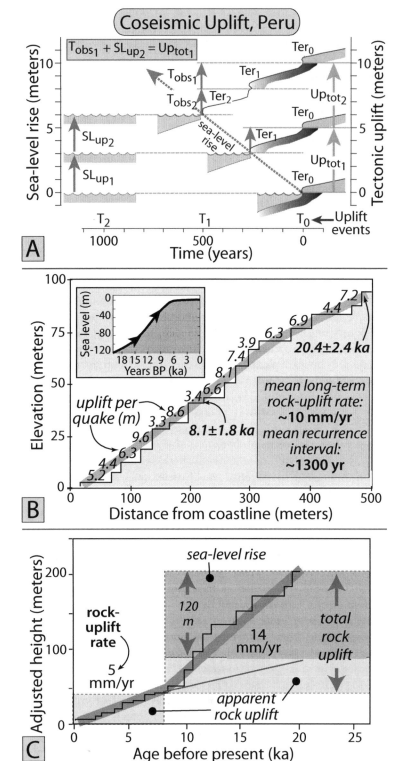

Fig. 6.10 Coseismically uplifted marine abrasion platforms.
A. Schematic pattern of abrasion platform formation and coseismic uplift during a rising sea level, yielding staircase of marine terraces (Ter). The total tectonic uplift (Up_{tot}) in each event represents the sum of the observed offset (T_{obs}) of the platform plus the sea-level rise (SL_{up}) that occurred since the preceding uplift event. In this example, 3 m of sea-level rise must be added to the 2 m of apparent coastal uplift shown by the vertical separation between successive terraces. B. Uplifted abrasion platforms on the Peruvian coast showing the vertical height difference between platforms, each of which is attributed to a large subduction-zone earthquake. Cosmogenic exposure ages are shown for two dated platforms. Modified after Bourgois *et al.* (2007). Inset shows late Quaternary sea-level rise. Modified after Fleming *et al.* (1999). C. Total rock uplift as the sum of the apparent uplift (observed terrace elevations) plus sea-level change. During the past 8 kyr when sea level was steady, the total and apparent rock uplifts are equal, but 120 m of sea-level rise between 20 and 8 ka more than doubles the total uplift during this earlier interval. Without accounting for sea-level change, the Pleistocene rate (dashed line) would appear slower than the Holocene rate.

of the terraces were used to estimate rock uplift rates, the Holocene rates would appear more rapid than the pre-8ka rate. Instead, when sea-level change is included, the Late Pleistocene rate is nearly three times faster than the Holocene rate.

In some tropical settings where reefs flourish, detailed reconstructions of both coseismic and interseismic deformation are sometimes possible (Natawidjaja *et al.*, 2007; Sieh *et al.*, 2008; Zachariasen *et al.*, 2000). Based on the annual growth banding of corals (Fig. 5.13), changes in relative sea level are recorded on an annual basis. Recall also that, if permanent deformation is excluded, vertical motions during inter-seismic intervals should be balanced by abrupt movement in the opposite direction during earthquakes. Whereas corals grow upward and outward during times of steady interseismic submergence (or downward and outward during emergence), earthquakes abruptly shift the position of the coral head with respect to local sea level (Fig. 6.11). If part of the coral head is lifted coseismically above sea level (Fig. 5.13A), those corals exposed to the air will die and slowly erode, whereas those that remain below the highest level of survival (HLS) will continue to grow outward. Such coseismic emergence will create a fringing annulus below the HLS and an unconformity above. After an earthquake, the height of the coral heads above the HLS provides an excellent measure of the vertical component of seismic deformation (Briggs *et al.*, 2006, 2008). If interseismic submergence then ensues, annual layers will grow upward and will unconformably overlap the eroded coral surface, thereby preserving the record of the minimum magnitude of coseismic emergence (Fig. 6.11). Moreover, with new high-precision protactinium-231 and thorium-230 dating techniques on carbonates (Cheng *et al.*, 2000; Edwards *et al.*, 1997), corals as young as a few decades old can now be dated with remarkably high precision. Such dates help to anchor chronologies based on counting annual bands. These dates permit synthesis of data from living coral heads with older, dead ones in order to produce paleoseismic records hundreds of years long (Sieh *et al.*, 2008).

The extensive, emergent, and coral-fringed forearc above the Sumatran subduction zone is a rich source of high-resolution data that permits studies that exploit dense two- and three-dimensional spatial coverage of seismological signals. For example, along transects oriented perpendicular to the trench, the detailed pattern of interseismic and coseismic deformation allows both testing of models for slip on the subduction interface (Fig. 6.12), as well as deducing lithospheric properties such as rigidity and temperature that affect flexural wave-lengths. The coral record yields well-defined spatial variations in the magnitude and rates of uplift and subsidence. In order to reproduce these data, an elastic dislocation model of the subduction zone (Fig. 6.12A and B) can be implemented in which slip occurs along both the upper and lower surfaces of the subducting slab (Sieh *et al.*, 1999). Given the duration of interseismic interval derived from the corals and their pattern of vertical motion, slip rates along the subduction interfaces can be varied until the best match is found to the observed interseismic deformation (Fig. 6.12C). When the stored elastic strain is abruptly released, the model should reproduce the observed, coral-based coseismic deformation (Fig. 6.12D). When successful, such data-to-model couplings can provide tremendous insight on the behavior of a subduction interface that is hidden many kilometers beneath the surface of the forearc.

In some tectonically active coastal settings, constructional beach ridges provide useful coseismic markers. For example, at Turakirae Head in New Zealand, beach ridges are built through the accumulation of clasts, shells, and organic debris during large storms. The modern beach ridge is about 2.3m above the average high-tide level. Nearly 6m above the modern ridge is a beach ridge that was uplifted in the 1855 earthquake along the nearby Wairarapa Fault (Fig. 6.13A and Plate 1B). Still higher above the modern shoreline are three addi-tional beach ridges that are interpreted to represent older coseismic offsets of 8, 6, and 2m and that probably extend back to about 6–8ka when global sea level stabilized. The correlated crests of these beach ridges indicate

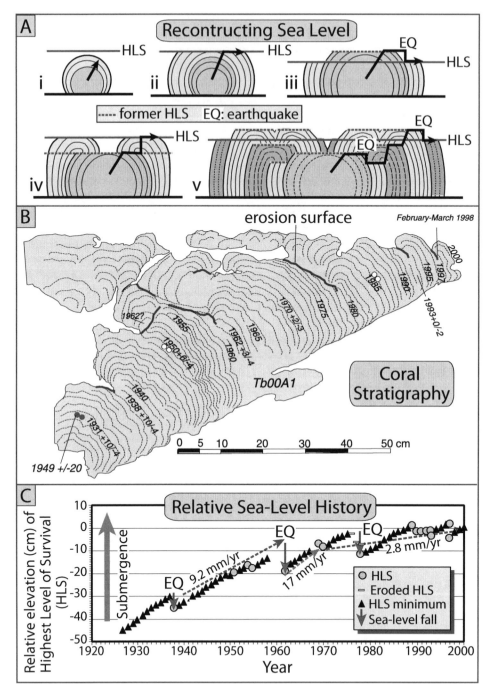

Fig. 6.11 Tropical corals as paleoseismic recorders.
A. Geometry of annual bands recording interseismic submergence and coseismic emergence of coral heads. The highest level of survival (HLS; tracked by black arrow) determines the height to which corals can grow and coincides with the height above which corals would die due to prolonged subaerial exposure. (i) Upward growth indicates the HLS is at or above the active layer. (ii) Outward growth in the absence of upward growth indicates the coral head is at the HLS. (iii) Abrupt drop in sea level due to coseismic emergence during an earthquake (EQ) causes death and erosion of the emerged part of the head, but outward growth of an annulus continues beneath the HLS.

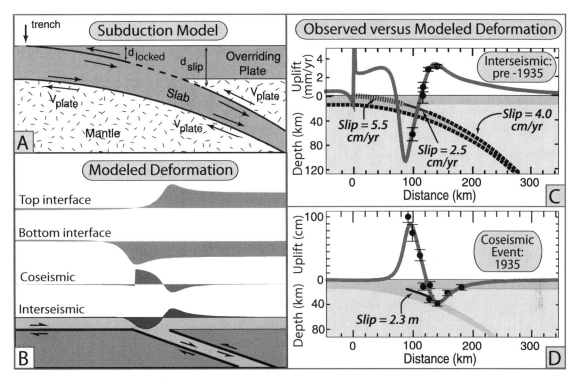

Fig. 6.12   Modeling observed deformation to deduce slip on the subduction interface.
A. Elastic dislocation model of a subduction zone with slip on both the upper and lower interfaces bounding the subducting slab. Modeled variables include the length and depth of any locked patch, as well as slip-rate variations along each interface. Modified after Sieh *et al.* (1999). B. Predictions of deformation in the forearc. Elastic strain occurs on both the upper and lower subduction interfaces. These strains are summed to produce the predicted interseismic and coseismic deformation patterns at the surface. C. Best match to the observed forearc deformation (black dots) prior to the 1935 earthquake in Sumatra predicts slow slip (2.5 cm/yr) on a patch that localizes the predicted elastic deformation. D. Coral data are well matched by 2.3 m of coseismic slip on the patch that experienced the slow interseismic slip. Modified after Natawidjaja *et al.* (2004).

that uplift rates increase to the northwest along the coast (Fig. 2.6).

Note, however, that one must exercise caution when deducing uplift from such storm berms, because the initial height of a storm berm above mean tidal position depends on the magnitude of the storm waves. This height must be known or deduced if we are to use such markers to reconstruct the amount of local coseismic uplift. An interesting lesson can be learned from the Turakirae succession (Plate 1B) in this regard (Grapes and Wellman, 1988, 1993). Until recently,

Fig. 6.11   (*cont'd*)   (iv) Intervals of relative sea-level rise or stability create a micro-atoll. (v) Complex relative sea-level history creates a unique coral stratigraphy. Modified after Natawidjaja *et al.* (2004). B. Sketch of annual banded coral from the Sumatran forearc. Heavy dark gray lines indicate erosion surfaces due to relative sea-level fall. Growth years of annual bands are indicated. A [320]Th date (1948±20) has been determined near the base of the head. C. Relative sea-level history constructed from the coral stratigraphy above. Note intervals of generally steady sea-level rise (submergence) followed by abrupt emergence during earthquakes (EQ) around 1940, 1960, and 1980. The 20th-century trend is one of overall submergence through several seismic cycles. Modified after Natawidjaja *et al.* (2004).

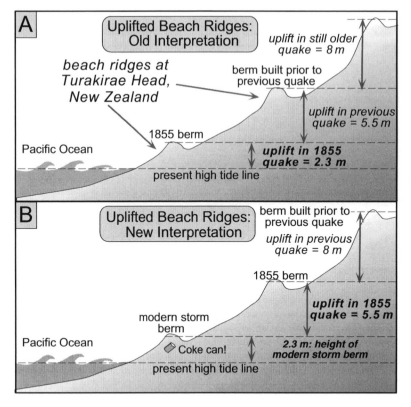

Fig. 6.13 Coseismically raised beach ridges at Turakirae Head, New Zealand.
A. In the earlier interpretation of these ridges, the beach ridge 2.3 m above the modern high-tide line was interpreted to represent the vertical coseismic uplift of the 1855 earthquake. B. At present, this ridge is recognized as the active storm ridge (+2.3 m), whereas the second ridge was raised ~6 m in the 1855 Wairarapa earthquake. Other, more inland, ridges resulted from additional earthquakes since about 8000 years ago when the Quaternary sea-level rise ended.

the beach ridge at 2.3 m above the modern shoreline was interpreted as representing uplift due to the 1855 earthquake (Fig. 6.13A). The underlying idea was that this ridge formed at the pre-1855 high-tide line, and presumably a new ridge was forming at the present-day high-tide line. The discovery of late 20th-century human litter in the lowest beach ridge indicates, however, that it formed during major storms long after the 1855 earthquake (Fig. 6.13B). Re-examination of the pre-1855 historical record confirms that there was indeed a beach ridge present 2–3 m above high-tide line prior to the earthquake and that this ridge was displaced nearly 6 m during the earthquake. The amount of coseismic uplift (2.3 m) calculated under the earlier interpretation is only half of the actual uplift (~6 m), as represented by the difference in altitude of the second ridge and the lowest (Fig. 6.13). Clearly, knowledge of the positions and geometries of potential geomorphological markers in modern conditions underpins reliable reconstructions of past earthquakes.

A fortuitous combination of repetitive creation of shoreline features, preservation of key aspects of their geometry through time, and unearthing of datable material commonly marks the quest to discover useful paleoseismic records in typically high-energy shoreline environments. For beach ridges, abrupt uplift by seismic events is a key to their preservation. If they slowly emerged, a broad shingled bench of composite age would be created, rather than a discretely crested berm. A surprisingly rich paleoseismic record has been recently described (Bookhagen *et al.*, 2006) from a suite of 20 uplifted beach ridges from Santa María Island along the coast of Chile (Fig. 6.14). At the time of their formation, these ridges were only 0.5–1 m high, and, today, the entire succession lies less than about 10 m above sea level. Their preservation depends on their orientation: they face eastward, away from the open Pacific Ocean, and they have been gradually uplifted above an east-vergent thrust fault. Dates from optically stimulated luminescence on the buried sands within the berms define a late Holocene

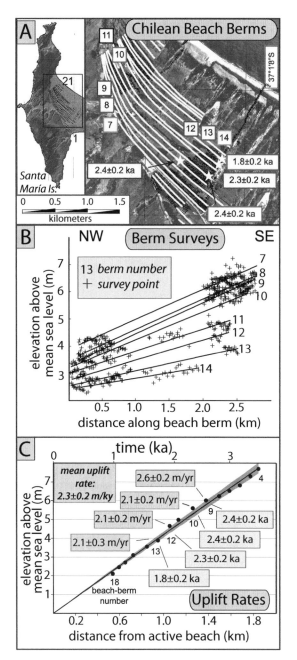

Fig. 6.14 Raised Chilean berms recording 20 late Holocene earthquakes.
A. At least 20 late Holocene beach berms are preserved on Santa María Island along Chile's west coast. These rather low ridges (up to 1 m high) are protected in a northeast-facing embayment lying within a few meters of modern sea level (see inset). White lines depict berm crests. Stars show the locations of OSL dates ranging to 2.4 ka. B. Surveys along berm crests reveal both their

chronology of earthquakes (Fig. 6.14), and their regular spacing and apparently steady uplift rate suggests a rather regular (characteristic?) rupture behavior on the underlying fault.

Stratigraphic evidence

A key question in many paleoseismic studies of subduction zones concerns the lateral extent of previous ruptures. The largest subduction-zone quakes have rupture lengths of 1000 km or more. The annual detail of some coral records in tropical settings permits earthquake-by-earthquake correlation of multiple sites along a subduction zone, thereby enabling reconstruction of rupture lengths and magnitudes (Sieh et al., 2008). But, even this record only goes back a few hundred years.

In efforts to reconstruct the magnitude and lateral extent of subduction-zone earthquakes over Holocene time scales, the stratigraphic record has been used with considerable success outside of the tropics. As shown so clearly by the coral records (Fig. 6.12), a predictable sequence of submergence and emergence characterizes the seismic cycle. Whether interseismic emergence or submergence prevails at a given site depends on its position with respect to the locked patch of the subduction zone. Sites farther from the trench (or closer to the down-dip limit of the locked patch) tend to slowly rise and then abruptly sink during interseismic and seismic intervals, respectively. At sensitive coastal sites, we might expect that these ups and downs with respect to sea level would be reflected in the stratigraphy.

Indeed, this sort of record has been discovered along the Cascadia subduction zone of Washington and Oregon, where the record is being used to

continuity and their subtle tilt toward the northwest that is interpreted to result from differential uplift above a westward-dipping thrust fault. C. OSL ages indicate the middle suite of berms range from about 1.8 to 2.4 ka. Given the number of berms, a recurrence interval of ~180 years with ~0.5 m of uplift per earthquake can be defined. In the context of the OSL ages, berm crest heights above sea level define an average uplift rate of 2.3±0.2 m/kyr.

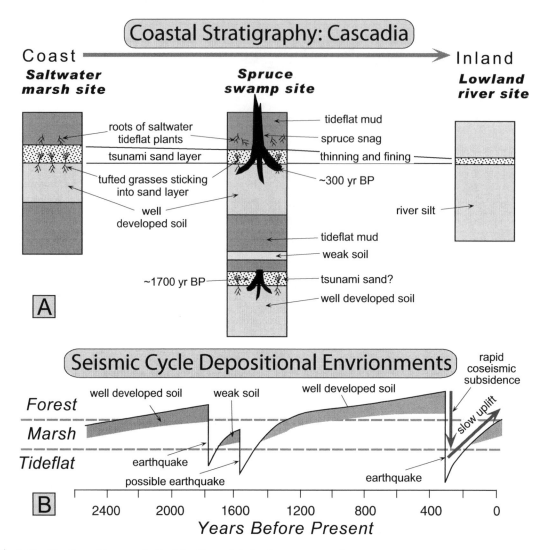

Fig. 6.15   Stratigraphic record of subduction-related submergence and emergence.
A. Schematic stratigraphy of buried soils and associated plants in saltwater tideflats, spruce swamps, and riverine settings
along the Washington coast above the Cascadian subduction zone. Abrupt drowning of trees and burial of grasses in
growth position result from coseismic submergence and an accompanying tsunami. Buried soils can indicate coseismic
submergence, but also could result from slow, interseismic subsidence. B. Schematic chronology of depositional
environments created by subduction-zone deformation across several seismic cycles. Modified after Atwater (1992).

address what had been a major unsolved question: Is there a major seismic hazard associated
with this rather slow subduction zone? In the
mid-1980s, Thatcher (1986a) wrote: "No great
earthquake has been recorded off the Washington
coast and the hypothesis that subduction occurs
seismically will not be proven until one does."
Today, as a result of stratigraphic studies by

Atwater and co-workers (Atwater, 1992; Atwater
et al., 1991), we now know that Cascadia produced
very large subduction-zone earthquakes as recently
as 300 years ago.

   In estuaries along the coastal areas of
Washington and Oregon, several superposed,
buried lowland soils crop out at low tide
(Fig. 6.15). The estuaries are typically within

0.2–1.0 km of the coastline. The uppermost soil contains rooted snags of dead, but still-standing, spruce trees (ghost forests), even though the soil is presently 1 m below the modern tide-marsh surface. Above the soil, a layer of fine-grained sand can be traced inland, where it thins and becomes finer-grained along the banks of a coastal river. Lying directly above the sand, mud contains roots of pioneer species from saltwater tideflats. Soils of the buried spruce swamps can be traced seaward into buried saltwater marsh soils, some of which contain rooted stems and leaves of a tufted grass. Tufts project up out of the soil and are encased in a sandy layer that is analogous to that found above the spruce swamp soil. Preservation of the tufted grasses indicates that they were buried in less time than that needed for subaerial decomposition (<2 years). Similarly, study of tree rings from the snags suggests that they died suddenly, rather than being gradually submerged and killed (Atwater and Yamaguchi, 1991; Yamaguchi et al., 1997). Given the elevation of modern coastal spruce forests above sea level, the presence of the tideflat species above the soil that bears the rooted trees dictates that these forests were submerged by at least 0.5 m. Detailed investigation of ostracod assemblages, which are zoned vertically within the tidal marsh, corroborate this estimate of vertical displacement (Hemphill-Haley, 1995). The overlying sand layer, which becomes finer-grained and thinner inland and which entombs grasses in growth position, is interpreted to result from a tsunami that swept across the coastal region. (Note that, along rivers, the normal trend is for downstream fining (Paola et al., 1992), whereas tsunami deposits become coarser toward the coast.) The two underlying buried soils are similarly associated with roots, stems, overlying sands, and tideflat muds. All of these observations are consistent with repeated subduction-zone earthquakes that caused instantaneous coastal submergence followed by gradual emergence (Atwater et al., 1995).

As a result of these stratigraphic studies, which documented the presence of large Cascadian earthquakes, the key remaining questions changed from whether they occurred in the past, to "When did they occur?" and "Did the entire length of the subduction zone rupture (a single night of horror) or were there several individual earthquakes (a decade of terror)?" High-precision radiocarbon dates (Atwater et al., 1991) and studies of the tree rings of the spruce snags (Yamaguchi et al., 1997) suggested that the most recent earthquake occurred about 300 years ago, probably in the winter according to the tree rings and local oral histories. But the resolution of the dating was insufficient to tell whether the earthquakes recorded from southern Oregon to Washington represented one or several closely spaced events. A solution to this uncertainty came from across the Pacific, where historical records of tsunamis in Japan indicate that a large tsunami with no known local source inundated the east coast in January 1700 (Satake et al., 1996). The timing of the tsunami and the pattern of coastal inundation in Japan suggest the Cascadian subduction zone as a likely source. In addition, numerical models have been developed to describe how tsunami waves dissipate as a function of original size and distance while traversing the Pacific. By comparing the model results for different lengths of Cascadian ruptures with different sizes of waves and run-up on the Japanese coast, it appears quite possible that nearly the entire locked portion of the Cascadian subduction zone ruptured 300 years ago. A similar rupture today could produce a magnitude 9 earthquake perhaps analogous to the 2004 Sumatra earthquake.

Although the coastal stratigraphic record clearly records as many as three paleoearthquakes, a longer time series is needed to examine recurrence intervals for the Cascadian subduction zone. The violent shaking from large earthquakes also triggers large-scale failures on the edge of the continental shelf. Such failures generate turbidity flows that rush down submarine canyons and produce recognizable turbidite beds as they come to rest (Adams, 1990). Over the past 15 years, extensive coring of numerous channels that drain the continental shelf along the Cascadia margin has revealed an extended sequence of Holocene turbidites (Goldfinger et al., 2003). Radiocarbon dating, identification of well-dated volcanic ashes, and stratigraphic superposition has permitted correlation among these turbidites, but such correlation cannot readily distinguish between one large event and several events spaced very closely in time. One clever way to address this issue is to determine the turbidite record both above and below confluences of submarine channels. If the upstream channels host turbidites that were not synchronous with each other, the turbidite record below their confluence should reveal the sum of the two upstream records. Instead, for almost all channels tested, the records show the same number of turbidites within equivalent stratigraphic sections, thereby implying that singular earthquakes affected both tributary channels. This test of synchrony lends strength to the correlation of turbidites along the Cascadia margin that has yielded a time series extending from the 1700 AD event to the base of the Holocene and contains 18 events (Goldfinger et al., 2003). A similar methodology has been applied to the turbidite record offshore from the northern San Andreas Fault (Goldfinger et al., 2007). There, 15 events over the past 2800 years define an average recurrence interval of about 200 years (similar to the 230-year recurrence interval estimated from onshore data for the northern San Andreas Fault). The distribution of turbidites argues that at least eight out of the last 10 earthquakes ruptured over 300 km of the northern San Andreas Fault. Although acquiring sea-floor samples is not inexpensive, the data synthesized along many hundreds of kilometers of coast provide a potent perspective on high-magnitude earthquakes.

Displaced geomorphological features

Although detailed stratigraphic records in trenches can provide constraints on multiple earthquakes in the past, each trench requires a large investment of time and is most relevant only to those segments of the fault directly adjacent to it. Moreover, some faults have very large offsets in individual earthquakes, e.g., up to 18 m on some strike-slip faults (Rodgers and Little, 2006), such that extraordinarily long trenches would be required to document multiple earthquakes. Offset geomorphic features, on the other hand, can be readily observed and surveyed, and they are often distributed along the length of a fault. Geomorphic studies following recent earthquakes have clearly depicted variations in coseismic slip along a fault's length (Fig. 4.20). The same methodology provides a means to document the spatial variations in cumulative deformation during past ruptures and, hence, underpins reconstructions of longer-term slip rates. In alluvial environments, typical displaced geomorphic features include stream channels, terrace risers, channel walls, debris flows and their raised levees, small alluvial fans, ridges, and gullies. In coastal environments, displaced features could include beach ridges, coral platforms, solitary coral heads, delta plains, and wave-cut notches (Figs 6.2, 6.10, 6.13, and 6.14).

Although a rough estimate of the amount of displacement of an offset feature can often be made with a tape measure, detailed topographic maps of displaced features and the area surrounding them is usually preferable, because such maps permit a more rigorous geometric reconstruction (Fig. 6.16A). Typically, such maps are constructed using a theodolite with a built-in electronic distance measuring device (an EDM or "total station"), with differential GPS, or with terrestrial laser scanners. In many field situations, displaced features no longer directly intersect the fault plane because slope processes have either eroded or buried part of them. In such cases, the trend of a feature – as represented by planar surfaces, such as terrace risers, or by linear features, such as debris-flow levees or the intersection between a terrace tread and riser – has to be projected on to the fault plane (Fig. 6.16A). To the extent possible, a structure contour map of the fault plane should be constructed, especially with dipping faults. In order to measure a horizontal offset, once the fault plane is specified, the distance is measured between the projections of linear features on to the fault plane after any vertical component of offset is subtracted. With a detailed topographic map, often the uncertainties in such a projection can be estimated and incorporated into the displacement estimate (Fig. 6.16A).

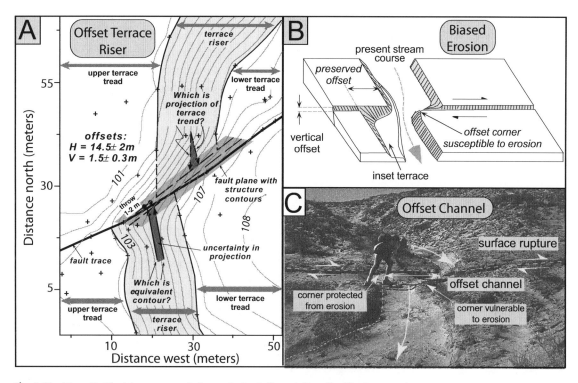

Fig. 6.16 Use of offset terraces and channels to define strike-slip displacements.
A. Gridded and contoured topographic map of late Holocene terrace riser offset by Awatere Fault at Grey River,
Awatere Valley, New Zealand. Map derived from total-station measurements. Dashed lines subparallel to the fault are
two structure contours on the fault plane, which is dipping to the south-southeast (contour interval 1.5 m). Top
structure contour on the fault plane is at the same elevation as the mid-riser contour on the south side of the fault
that is used to project the position of the terrace riser on to the fault plane. Lower structure contour on the fault is
1.5 m lower down (mean throw on both the lower and upper terrace surfaces). Shaded region on south side of fault is
uncertainty in projection of riser contour on to eroded fault surface on upthrown (southeast) side of fault. Shaded
uncertainty belt on north side of fault represents uncertainty in knowledge of which contour on north side of fault
originally corresponded to the south-side riser contour prior to displacement. This uncertainty is the same as that in
throw for the lower terrace surface (±0.3 m). Modified after Little *et al.* (1998). B. Sketch of the geometry of the
displaced channel wall and terrace and the optimal position in which to measure offset along the fault. The offset
"corner" to the right of the stream is likely to be modified by erosion as the stream impinges on it. Therefore, it is far
better to measure the displacement based on the protected channel wall on the opposite side of the channel. Modified
after McGill and Sieh (1991). C. Offset channel along the Superstition Hills Fault.

Along strike-slip faults with offset fluvial features, erosion along an active channel can continue to modify offset markers (Cowgill, 2007).
Consider, for example, terrace risers that are
coseismically offset on each bank of a river. The
riser on the downstream side of the fault will be
moved toward the thalweg of the active river,
whereas the other riser will be moved away
from the thalweg (Fig. 6.16C). Any subsequent
erosion of the marker moved toward the channel will tend to reduce its apparent amount of

offset (Fig. 6.16B), whereas the opposite riser
should be protected from erosion and is more
likely to preserve the initial offset. As with nearly
all efforts to measure fault slip with surface offsets, careful reconstruction of the history of the
landscape provides the critical underpinning for
reliable assessments of fault offsets (Box 6.1).

Clearly, vertical displacements can also be
measured using a similar methodology, but one
that relies on projections of subhorizontal
features, such as terrace treads or channel

Box 6.1   Slip rates from fluvial terraces.

Conceptually, calculations of slip rates on fluvial terraces depend simply on two attributes: the age of an offset geomorphic feature, and its magnitude of offset. Displaced terrace risers are commonly used to determine slip magnitude, but along strike-slip faults, a choice arises as to whether the age of the upper tread (at the top of the riser) or the lower tread (at its base) should be used to calculate a rate. Because the lower tread is always younger, its age will yield a faster rate for any given offset. But, is that rate accurate?

Many studies have systematically used lower tread ages and have typically justified this choice by arguing that, as long as the active channel is flowing on the lower tread (the channel has not incised to a still lower level and abandoned the terrace), the channel can bevel laterally into any offset riser and modify its geometry. In the extreme case, such beveling can remove any accumulated offset. In such a case, the riser begins to record slip only after the lower tread is abandoned, such that the age of lower-tread abandonment would be the appropriate age to use in slip-rate calculations. On the other hand, even when the lower tread still has an active channel somewhere on it, if an offset riser on the downstream side of a strike-slip fault is moved laterally away from the zone of erosion by the active channel, the total magnitude of offset may be well preserved, such that the age of the upper tread would be the appropriate one to use in a rate calculation.

Hence, any field geologist has to ask whether specific criteria can be used to choose between using an age for the upper tread or for the lower tread. Several studies have focused on how

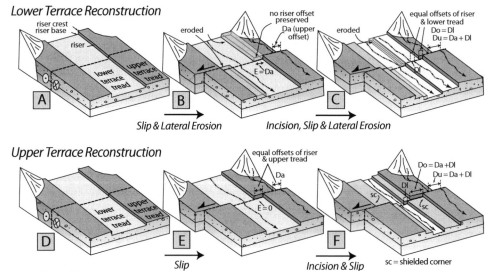

Do = total observed riser displacement
Du = total displacement of the upper tread after its abandonment
Dl = total displacement of the lower tread after its abandonment
Da = displacement of the upper tread after its abandonment but before incision of the lower tread
E = lateral erosion of the displaced riser after abandonment of the upper tread but prior to incision of the lower tread

Conditions that favor either using the age of a lower terrace tread (A–C) or an upper tread (D–F) to calculate slip rates along a sinistral strike-slip fault. Modified after Cowgill (2007).

to make this choice (Cowgill, 2007; Harkins and Kirby, 2008; Lensen, 1964). For example, if preserved riser offsets are equal to observed offsets of features that act as "passive markers" (such as small channels) on the upper terraces (see figure E), then the age of the upper tread should be used. Similarly, if an age gradient can be demonstrated from (i) older ages in sheltered positions adjacent to the offset riser to (ii) younger ages farther from the riser, the age progression argues for riser preservation despite continued occupation of part of the lower tread (see figure F). In contrast, if offsets of passive markers on an upper tread are larger than riser offsets (see figure B), or if the riser offset is equivalent to that of a passive marker on the lower tread (see figure C), then the riser has been modified during occupation of the lower tread and the age of abandonment of the lower tread should be used (Cowgill, 2007).

Once risers are formed due to channel incision to a lower level, risers begin to degrade. If left undisturbed, diffusion is likely to dominate the degradation process, such that a predictable topographic profile will develop (Hanks *et al.*, 1984). Subsequent erosion of the toe of the riser will cause it to steepen and depart from the predictable diffusion profile, such that the signature of episodic erosion by a channel on the lower tread could be embedded in the topographic profile. In the context of regional studies, comparisons of riser profiles among dated and undated terrace treads can serve as a guide to whether the treads are the same age (Harkins and Kirby, 2008).

No single strategy for choosing between lower and upper terrace treads is likely to work, because so much depends on what data can be extracted from the local geomorphology. The recommended approach combines careful attention to detail and multiple criteria that are supportive of one interpretation or the other. See Cowgill (2007) for a more detailed discussion.

bottoms, on to the fault plane. Each such projection is commonly associated with some uncertainty. Consider, for example, trying to calculate the slip on a thrust fault based on a survey of a displaced terrace surface that has been projected on to a cross-section perpendicular to the fault trace (Fig. 6.17A and B). The slip calculation should include uncertainties related to the slopes of the upper and lower offset surfaces, the dip of the fault, and the position of the fault tip beneath the scarp. One of the easiest ways to incorporate such uncertainties is via Monte Carlo simulations in which a probability distribution is assigned to each variable in the calculation (Fig. 6.17). Many of these probabilities may have Gaussian distributions, such as would be expected from a regression on a surveyed topographic slope. Some other probabilities may be harder to quantify, such as the position of the buried fault tip beneath the scarp. If its position seems likely to be random along the scarp, a box-like probability distribution could be assigned. Or, if experience suggests that, 95% of the time, the tip lies beneath the lower third of the scarp, then a corresponding probability distribution can be defined. In a Monte Carlo simulation for fault slip, a value for each variable is randomly drawn from each variable's associated probability distribution and then is used to define a specific geometry from which the fault slip is calculated. This computerized process is repeated thousands of times and rapidly produces a probability distribution for the fault slip (Fig. 6.17C).

In theory, every large earthquake on a fault could be represented by a suite of displaced geomorphic features. In order for this to be true, some new geomorphic markers would have to form in the interseismic interval between faulting events, and older markers that were displaced previously would have to be preserved until the time of measurement. Such conditions are not uncommon, because each seismic displacement commonly forces some aspects of the geomorphic system to adjust, thereby producing new markers. In many landscapes,

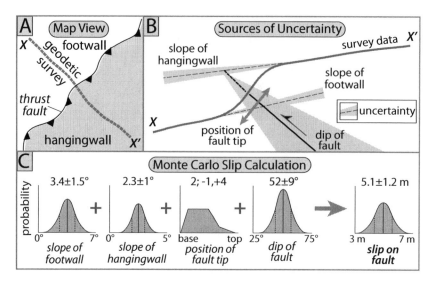

Fig. 6.17 Fault-slip calculations using Monte Carlo simulations.
A. Map of geodetic survey oriented perpendicular to the trace of a thrust fault. B. Projection of survey data on to a vertical cross-section and associated types of uncertainty (gray zones). C. Some features, such as slopes of geomorphic features, may have formal Gaussian uncertainties deriving from analysis of the surveyed data. For others, such as the position of the fault tip, a probability distribution may have to be assigned based on observations of other faults. Randomly chosen values from each distribution define a geometry from which a single slip calculation can be made. Based on the outcome of thousands of simulations, the probability distribution of fault slip can be defined. Modified after Davis *et al.* (2005).

therefore, one might expect to find numerous markers that had been displaced in the most recent earthquake, a lesser number displaced during the penultimate rupture, and so forth. Such landscapes could be described as *palimpsest* landscapes in which the imprint of older features is only partially overprinted or obscured by younger features. Quite commonly, the size of an offset may not be large enough relative to the scale of the geomorphic feature to have altered it significantly, or to have caused its abandonment. This ratio of scales is, therefore, important in determining if a particular landscape will record individual coseismic offsets, or if instead it will record only the long-term average slip on the fault, because it smears out the effects of individual events.

Clearly, the record of offset features represents an interaction between the processes that create geomorphic markers and the tectonic events that displace them. The variability of climate suggests that some intervals will be more conducive than others to the formation of clear geomorphic markers (Fuller *et al.*, 2009; Harkins and Kirby, 2008; Pan *et al.*, 2003). For example, during times of fluvial incision, many steep-sided geomorphic features, such as channel walls, terrace risers, and gullies, are etched into the land surface and gently dipping features like terrace treads are left behind, whereas during aggradational phases, broad alluviated surfaces provide fewer distinctive features that could serve as geomorphic markers. Moreover, aggradation can bury offset features and thus obscure the record of older earthquakes. As a consequence of these climatically controlled contrasts, earthquakes that occurred after intervals of incision are likely to be more clearly represented in the geomorphic record than are those that occurred during times of aggradation. Paleoseismologists exploit these contrasting regimes of incision and deposition. The offset geomorphic features of incisional areas allow ready identification of faults, whereas places where faults cut areas of active deposition are ones where the fault trace may be obscure,

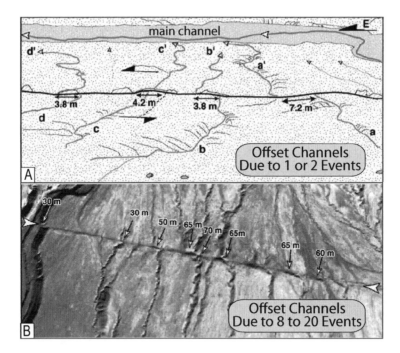

Fig. 6.18   Channel offsets along the Kunlun Fault, Tibetan Plateau.
Offset stream channels on the Kunlun Fault in northern Tibet provide evidence for slip in one to more than a dozen earthquakes. A. Sketch of channel offsets that record one or two earthquakes. Displacement in a single earthquake is ~4 m, as shown by the three left-hand channels, whereas the channel on the right records two earthquakes. B. IKONOS image of channels with offsets of 30–70 m. If the average coseismic slip is 4 m (as in top panel), these offsets appear to record up to 20 earthquakes. The mix of channel offsets (30, 50, 60, 65, 70 m) suggests that (i) this region is characterized by punctuated channel-forming processes, and (ii) the most recent resetting occurred when the channels with 30 m offsets were initially formed as straight channels. Modified after Fu *et al.* (2005).

but the past record of earthquakes can be discerned through trenching.

Climatically induced unsteadiness in the production of geomorphic markers may create landscapes with different aged features that show differential offset along a single fault. For example, along the Kunlun Fault in the northern margin of the Tibetan Plateau (Chen *et al.*, 2004; Fu *et al.*, 2005), certain segments of the fault displace channels that record only one or two earthquakes, whereas other segments preserve channel offsets that result from up to 20 faulting events (Fig. 6.18). The offsets due to multiple earthquakes provide an opportunity to determine long-term slip rates, but they require reliable dating of offset features.

How is a catalog of displaced features along a fault transformed into a useful estimate of paleoseismicity? First, the magnitudes of

individual displacements can be plotted as a function of distance along the fault (Fig. 6.19). Such a plot should indicate whether consistent patterns of offset exist along the fault, whether the displacements fall into discrete groups, and whether evidence exists for decreases in displacement toward the terminus of the rupture (Figs 4.9 and 4.29). On strike-slip faults, displacements can be measured directly in the field or from remotely sensed data. For example, both lidar topographic data and Quickbird imagery (each with ~1 m spatial resolution) have been used to assess strike-slip displacements (Frankel *et al.*, 2007; Zielke *et al.*, 2010; Klinger *et al.*, 2011). The power of these high-resolution data is that offset features can be sequentially restored such that multiple slip events can be extracted from an image even if the ages of each event remain unknown.

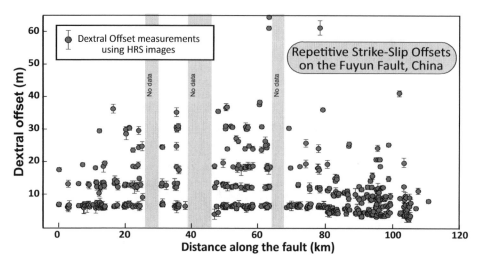

Fig. 6.19   Displacement of geomorphic markers along a strike-slip fault.
Plot of displaced markers versus distance along the Fuyun Fault in northern China. Along more than 100 km of fault length, Quickbird imagery was used to identify and sequentially restore offset features in order to define multiple slip events. Data gaps tend to occur near segment boundaries and changes in fault orientation that cause contractional features near bends. Modified after Klinger *et al.* (2011).

Second, histograms of the magnitude of displacement are constructed for all of the displaced features along a fault. When the reliability of each measurement can be assessed, this somewhat subjective variable should be cataloged as well. If a segment of a fault were typified by repetitive ruptures of similar magnitudes, then histograms of displacements along that segment would tend to display groupings of offsets that were multiples of each other. For example, repetitive offsets of comparable amounts occur along much of the Fuyun Fault in northern China (Figs 6.19 and 6.20) or along individual segments of the Garlock Fault in California (McGill and Sieh, 1991). Along much of the Fuyun Fault, similar displacements occur in as many as five slip events (Klinger *et al.*, 2011), thereby suggesting a repetitive rupture pattern that would be typical of faults dominated by characteristic earthquakes (Schwartz and Coppersmith, 1984). In contrast, along the Garlock Fault, displacements within individual segments commonly appear repetitive for a few events, but neighboring segments display differing slip-per-event histories. Such contrasts have been interpreted to suggest that rupture patterns may vary significantly from one

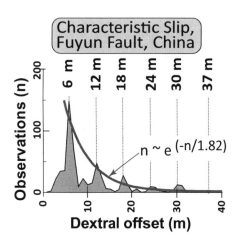

Fig. 6.20   Histogram of strike-slip displacements along the Fuyun Fault.
Compilation of displaced features along >100 km of the Fuyun Fault (data in Fig. 6.19). The most recent rupture averaged 6 m of displacement and corresponds to the 1931 M_s=7.9 earthquake. Previous ruptures cluster near multiples of 6 m, and are interpreted to indicate that the Fuyun Fault hosts characteristic earthquakes. Geomorphic degradation appears to cause an exponential decrease in the preservation of offsets as a function of time. Modified after Klinger *et al.* (2011).

earthquake to the next on the Garlock Fault (McGill and Sieh, 1991). Obviously, if the displacement histograms display more than one peak and are interpreted to record more than one earthquake, some geomorphic evidence should be sought to indicate that those features showing greater displacements relate to older earthquakes in comparison to features along the same fault segment with smaller displacements.

Third, given the length of each ruptured fault segment and the mean displacement along it, a seismic moment can be estimated, if the depth of the fault and a crustal rigidity are assumed or known (Eqn 6.3). Often, modern seismicity defines the local depth of the seismogenic layer rather well, and, because most large earthquakes will rupture through the entire seismogenic layer, reasonable estimates of the seismic moment can be made. Various rupture scenarios can be evaluated along a segmented fault, ranging from rupture of its entire length to rupture of the shortest segment (McGill and Sieh, 1991). In each case, the mean displacements for the relevant fault segment are fed into the calculation, and the moments for each segment are summed to yield a total seismic moment and moment magnitude for the rupture scenario. Such an approach clearly provides a very useful data set for assessing seismic hazards along such a fault. No consensus exists, however, on when earthquakes tend to rupture beyond segment boundaries. No historical strike-slip faults have been observed to rupture across steps of more than 3–4 km between segment terminations (Wesnousky, 2006), whereas quite a few historical thrust faults have ruptured multiple fault segments lying as much as 15 km apart (Dong et al., 2008; Rubin, 1996). Hence, assessments of simultaneous ruptures of multiple segments during prehistoric earthquakes and their implications for future temblors will require exceptionally tight dating control to demonstrate rupture synchrony in the past.

One potentially controversial aspect of using multiple offset features to characterize fault displacements in several different earthquakes is the fact that the age of most, if not all, of the features is commonly unknown. At some sites, it is possible to show that two or more sets of features are displaced by distinctly different

amounts (Fu et al., 2005; Liu et al., 2004; Sieh and Jahns, 1984) and can, therefore, confidently be interpreted to represent two or more earthquakes. More commonly, individual displacements all along the fault are amalgamated and compared, with little knowledge of their relative age. If they seem to fall into clusters, especially ones with displacements that are multiples of each other (Fig. 6.20), a natural tendency is to interpret each cluster as representing the coseismic slip that was added to the penultimate earthquake's cumulative slip. Whereas this strategy seems reasonable, data from the Landers earthquake (June 1992, $M_w = 7.3$) in southern California suggest some pitfalls with this approach.

Although the total Landers rupture was approximately 85 km in length, the map pattern clearly indicates that the rupture comprises an elongate zone of several known faults that were linked together either by previously unrecognized or by newly formed faults. Mapping of offsets along one of the previously known faults (central Emerson Fault; Fig. 6.21A) showed abrupt changes in the amount of offset along its 5-km-long trace (McGill and Rubin, 1999). Numerous ephemeral features, such as offset tire tracks, indicate that the displacements all occurred during the 1992 earthquake. When the displacements were synthesized in a histogram, however, they showed two discrete peaks (Fig. 6.21B) at about 2 m and 4 m. In a paleoseismic study, such peaks would probably be interpreted as being due to two different earthquakes, rather than resulting from variations along a short rupture during a single event. One approach to try to determine if more than one event is encompassed within the suite of offset features is to subtract a running mean from each data point (McGill and Rubin, 1999). For many data sets, the remaining anomalies will define a single-peaked probability distribution if the data derived from a single event, whereas they may display multiple peaks for more than one event. Nonetheless, when the data are particularly noisy and vary abruptly in space, this subtraction procedure can still produce two peaks for a single earthquake (as is the case with the Emerson Fault data), Clearly, some caution is warranted when interpreting histograms of offset features.

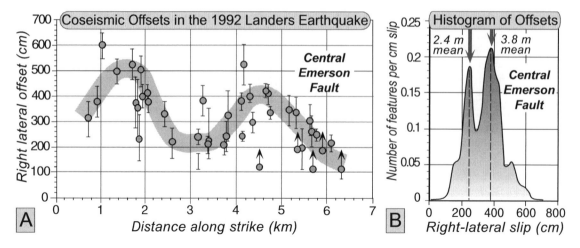

Fig. 6.21   Coseismic offsets in the 1992 Landers earthquake.
Measured displacements resulting from the 1992 Landers M_w=7.3 earthquake along a 6 km segment of the central Emerson Fault. A. These data were collected shortly after the rupture and include ephemeral linear features, such as tire tracks, that were offset by the earthquake. The 85 km long Landers rupture involved several nearby faults. The 6 km long segment of the Emerson Fault shown here represents only 10% of the total length of the Emerson Fault that ruptured in 1992. Surprisingly, this short segment appears to display a sinusoidal pattern of displacements that is reminiscent of merged bow-shaped displacement gradients (Figs 4.9 and 4.14) that would result from slip on separate rupture patches. B. Histogram of displacements (represented as Gaussian probability density functions) along the central Emerson Fault. Note the bimodal nature of the histogram, which would typically be interpreted to have resulted from two earthquakes, rather than one. Modified after McGill and Rubin (1999).

Whereas the data from trenches often allow determination of when past ruptures occurred and sometimes permit the magnitude of offset to be determined, geomorphic studies of offset features can clearly define displacements, but often say little about timing, recurrence intervals, or slip rates. At least two approaches can address the issue of timing. The most obvious approach is via direct dating of the offset features (Fig. 6.22). For older dated surfaces, the accumulated displacement can be used to estimate a long-term slip rate. With several ages on offset, but pristine and, therefore, young, geomorphic features, such as on small channels, rills, or debris-flow levees, the intervals between past earthquakes can sometimes be defined and constraints can be placed on average recurrence intervals. Alternatively, if mean slip rates across a fault are known, the magnitude of measured displacements can be divided by the long-term slip rate to obtain an estimate of the recurrence interval. A mean slip rate could be derived either from geodetic measurements at the decadal scale or from longer-term rates based on displaced

and dated Quaternary features, such as strand lines, terraces, or alluvial fans (Fig. 6.22).

Whereas the concept of dating offset features is straightforward, such dating almost always requires diligence, luck, and interpretation. Two recent studies of the same offset fan surface that is cut by strands of the southern San Andreas Fault reached rather different conclusions on the age of the fan surface and the amount of fault offset. These studies provide an illustration of two challenges that paleoseismologists must address: deriving reliable ages for an offset feature; and determining the actual amount of fault slip. A study by van der Woerd et al. (2006) used cosmogenic exposure ages on cobble-sized clasts, determined a surface age of 35.5±2.5 ka (Fig. 6.22C), assessed the fault displacement as 565 m (Fig. 6.22B), and calculated a mean slip rate of ~16 mm/yr. A second study by Behr et al. (2010) also used cosmogenic dating, but found that, for boulders less than ~0.5 m high, exposure ages varied from 30 to 45 ka as a function of clast size (Fig. 6.22D). Because taller boulders gave more coherent ages around 50 ka, Behr

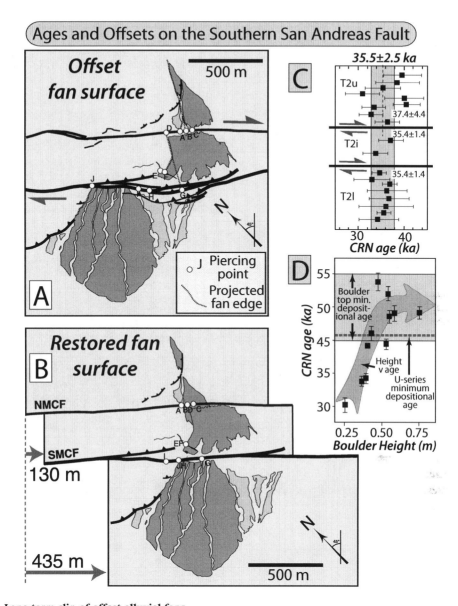

Fig. 6.22   Long-term slip of offset alluvial fans.
Dating and slip rates for a faulted alluvial fan along the southern San Andreas Fault, California. Although this offset fan was recognized in the early 1980s (Keller *et al.*, 1982), no slip rate was determined for it because the fan surface was undated. Two recent studies based on cosmogenic exposure ages reached different conclusions on both ages and the magnitude of slip. A. Offset fan surface displaced by two fault strands. B. One restoration of the fan to a pre-deformed state suggests 565 m of total offset (van der Woerd *et al.*, 2006). C. Cosmogenic exposure age dating of 19 cobble-sized clasts on the fan surface yielded an age of 35.5±2.5 ka and an estimated Late Pleistocene slip rate of ~16 mm/yr. D. Another cosmogenic study (Behr *et al.*, 2010) suggests exposure ages that scale with boulder size for clasts less than ~50 cm high, a mean age of ~50 ka for the fan surface, a cumulative fan offset that ranges from 680 to 980 m, and a preferred slip rate of 14–17 mm/yr. Modified after van der Woerd *et al.* (2006) and Behr *et al.* (2010).

et al. (2010) surmise that the shorter boulders – such as those used in the van der Woerd (2006) study – have been exhumed since deposition and, consequently, yield minimum exposure ages. Because the margins of fans are commonly indistinct and digitate, rather than linear, and because channels within the fans may also be somewhat sinuous, determinations of the magnitude of fan offset typically have inherent uncertainty. Whereas van der Woerd *et al.*'s (2006) study estimated slip of about 565 m, Behr *et al.*'s (2010) reconstructed slip ranged from 680 to 980 m. Ironically, when the differences in slip are combined with the differences in exposure ages for the two studies, they both happen to yield similar long-term slip rates – but only one of them (at most!) is correct in both its estimated age and total slip.

Despite the uncertainies, these slip-rate estimates have important implications. For example, even though the San Andreas Fault has traditionally be viewed as the most dangerous and seismogenic fault in southern California, this long-term slip rate is less than half of the Pacific–North American plate rate. Recent paleoseismic and geodetic studies (Bennett *et al.*, 2004; Dorsey, 2002) suggest that the nearby San Jacinto Fault may be absorbing much of the plate-boundary slip, and that rates along these two faults have varied synchronously and in opposite senses for the past 100 kyr or more.

Indirect observations of faulting

Vigorous seismic shaking can generate clusters of landslides, damage the roots of trees, create tsunamis, and cause weakly consolidated sedimentary beds to deform. Each of these seismically induced consequences produces a preservable, geological or biological record that could be interpreted for paleoseismic purposes. Because non-seismic events, such as rainfall-driven landslides, high winds, or abrupt sediment loading, could produce similar outcomes, however, their interpretation as paleoseismic recorders commonly becomes more ambiguous.

Earthquake-related stratigraphy

Some of the clearest indirect paleoseismic records come from settings where a stratigraphic signal can be discerned that appears clearly linked to seismic deformation, rather than to variability in climate or sediment supply. For example, the 1980 ($M=7.3$) El Asnam earthquake in Algeria caused hanging-wall uplift and temporarily dammed the Cheliff River (Meghraoui *et al.*, 1988a,b; Philip and Meghraoui, 1983), where it traversed the folded hanging wall through a water gap (Fig. 6.23A). Eventually, the river eroded through the dam represented by the upthrown fault block, and, at that time, the lake drained. Until that time, lacustrine sediments were deposited in the earthquake-dammed lake. This stratigraphic response to a thrusting event provides a conceptual model for interpreting the stratigraphy in trenches located on the footwall at some distance from the main rupture. Seismic shaking may also induce local faulting within the lacustrine sediments and cause sand blows or other features indicative of faulting, which can be exposed in trenches. Most importantly, however, lacustrine deposits will mantle the flooded topography and form identifiable seismo-stratigraphic markers. With such a model in mind, it is fairly straightforward to interpret the Holocene stratigraphic record (Fig. 6.23B). At El Asnam, a trench reveals flood deposits alternating with paleosols (Meghraoui *et al.*, 1988b). For six of the lacustrine layers, some associated indicators of faulting or seismic shaking are present. Therefore, six stratigraphically defined earthquakes have been identified as occurring between 6000 yr BP and 1900 yr BP, after which tilting apparently shifted the position of the earthquake-dammed lake away from the trench site. This interpretation is, however, not without uncertainties. Because the Cheliff River passes through a water gap, non-seismically induced landslides could also dam the river and trigger lacustrine sedimentation. Within these sediments, a careful search for features indicative of seismic shaking helps to rule out non-seismic causes.

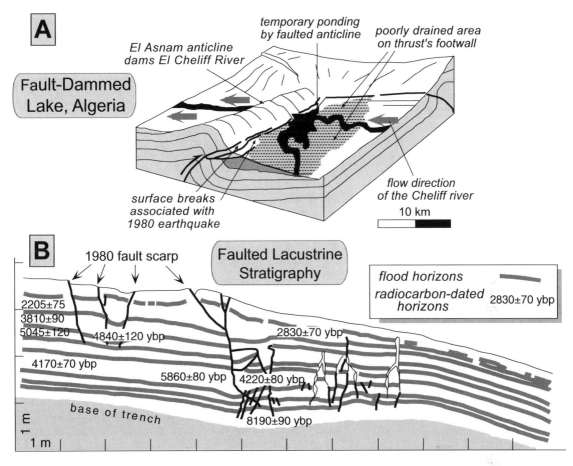

A

El Asnam anticline
dams El Cheliff River

*temporary ponding
by faulted anticline*

*poorly drained area
on thrust's footwall*

Fault-Dammed
Lake, Algeria

*surface breaks
associated with
1980 earthquake*

*flow direction
of the Cheliff river*

10 km

B

1980 fault scarp

Faulted Lacustrine
Stratigraphy

flood horizons

radiocarbon-dated
horizons   2830±70 ybp

2205±75
3810±90
5045±120   4840±120 ybp

2830±70 ybp

4170±70 ybp

5860±80 ybp   4220±80 ybp

base of trench

8190±90 ybp

1 m

Fig. 6.23   Earthquake-dammed lakes along the Cheliff River, Algeria.
A. Oblique view of the damming of the Cheliff River by the upthrown hanging wall during the 1980 El Asnam
earthquake. The river flows through a water gap, where uplift due to thrusting temporarily dammed the river as it
crossed the upthrown hanging wall. B. Stratigraphy of a trench in the footwall of the El Asnam rupture. Complexly
faulted, but formerly continuous, lacustrine beds are interpreted to result from previous Holocene earthquakes. The
presence of faults cutting the layers and sand blows is used to distinguish those layers that are clearly associated with
major thrust faulting events from those that might be due to natural flooding or landsliding. Note that, although the
master fault is a thrust fault, it occurs at some distance from this site, and all the faults here are small normal faults
developed in the footwall of the thrust. Modified after Meghraoui *et al.* (1988a,b).

Tree rings

As discussed earlier in this chapter, the length of
surface ruptures in past earthquakes needs to be
known in order to calculate the seismic moment
that a fault has produced. Hence, reconstruction
of rupture length represents a key objective in
many paleoseismological studies. The geomorphic evidence for ground ruptures is ephemeral in many settings. If the scarps were
formed in weakly consolidated sediments, if

surface fissures have been filled with colluvium,
or if abundant vegetation surrounds the surface
trace, then the evidence of a surface rupture
may be removed within a few decades of an
earthquake. Clearly, trenching and radiocarbon
dating of offset layers is one approach to
determine the length of surface rupture. But
trenching is a time-consuming and commonly
expensive undertaking, and several trenches
will probably be needed in order to document
where the rupture terminates.

A rather different technique has been applied in some areas where faults pass through forested landscapes. During the sudden displacements that occur in an earthquake, the roots of trees growing along the fault line are often disturbed and can be sheared off or the trunk itself may be cracked (Plate 2D). If ground shaking is sufficiently vigorous, some trees can literally have their crowns shaken off. The branching pattern of such trees can indicate that they lost their crowns at some time in the past: the main trunk abruptly ends and forks into lesser trunks, so that it looks like a tuning fork. Clearly, any of these events (breaking off the roots, splitting the trunk, or lopping off the top) traumatize a tree. In response to such a disturbance, the rate at which a tree grows can plummet (Jacoby *et al.*, 1988; Sheppard and White, 1995; Van Arsdale *et al.*, 1998). One could imagine, therefore, that by simply examining the temporal pattern of annual ring widths produced by trees growing adjacent (±20m) to a major fault, one might be able to discern past earthquakes as represented by times of decreased ring widths.

Unfortunately, analyzing ring widths is somewhat complicated (see Chapter 3), because ring widths are also a sensitive function of climatic conditions. In fact, most studies of tree rings have been undertaken in an effort to reconstruct past climate. Many factors affect ring widths, including moisture availability, air temperature, and seasonal variability. In general, when trees are stressed, they produce narrow rings (little growth). So, how does one separate narrow rings following earthquakes from those formed in response to climate variations? The solution requires a two-step procedure. First, a master tree-ring chronology has to be developed for an area (Fig. 6.24). This time series would represent the average ring-width variation for trees not located on the fault and would be interpreted as the climatically induced signal. Second, the ring-width chronologies from trees growing along the fault are compared with the master chronology. Parts of the record that display distinct mismatches and, in particular, prolonged intervals of reduced ring width with an abrupt beginning, can be interpreted to have resulted from past earthquakes (Fig. 6.24). Whereas

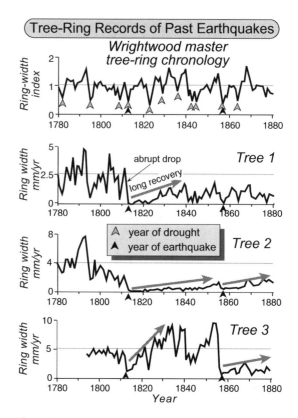

Fig. 6.24 Tree rings as seismic recorders. Tree-ring time series near Wrightwood, California, along the San Andreas Fault. The master tree-ring chronology portrays the climatically induced variations in ring widths as determined from a regional synthesis of many trees unaffected by faulting. A low index corresponds with narrow rings (low growth) and correlates with years of drought (marked by open arrowheads). Ring-width chronologies for trees 1 to 3 display an abrupt decline in ring width (to lower than any previously formed ring) in 1812–13. Although this year also corresponds to a year of drought, the accentuated decrease in ring width and the long period of recovery after 1813 indicate that an earthquake, rather than climate, caused the tree to be highly stressed. These data provide the first clear evidence that the earthquake of 1812 (known primarily from coastal records) actually occurred along the San Andreas Fault. The better known San Andreas earthquake of 1857 (Ft.Tejon earthquake) is also well represented in trees 2 and 3, thereby suggesting that the 1857 rupture propagated at least this far south. Modified after Jacoby *et al.* (1988).

climatic conditions change on a year-to-year basis and ring widths responding to climate show similar year-to-year variability, recovery from coseismic shearing of the root system or trunk can take many decades.

Ring-width studies along the San Andreas Fault provide clear evidence for an earthquake in 1812–13 and a less well-expressed earthquake in 1857 (Fig. 6.24). Coincidentally, both of these years also coincide with drought years in the master chronology. In the regional master chronology, however, ring widths bounce back to average within a few years. In contrast, all of the trees damaged by earthquakes show a slow recovery over several decades; some never recover fully to the previous rates of growth. Such studies are particularly useful because they serve to clarify the paleoseismic record. In 1812, California was sparsely populated, and, due to seismic damage in some coastal communities, it was thought that the earthquake occurred on some fault near the coast. The tree-ring data show that it occurred instead along the San Andreas Fault, and, when combined with trench data from distant sites, these data help define the extent of ground rupture and permit an estimate of the magnitude of these 19th-century earthquakes (Jacoby et al., 1988).

Rather than relying on trees that were directly disturbed by earthquakes, innovative tree-ring studies along the Alpine Fault in New Zealand have relied on a process–response rationale to date prehistoric earthquakes (Wells and Goff, 2006). As witnessed with many recent earthquakes, strong seismic shaking can induce thousands of landslides that lead to large increases in downstream sediment fluxes (Dadson et al., 2004; Harp and Jibson, 1996; Meunier et al., 2008). Following each major earthquake on the Alpine Fault, the enhanced sediment flux to nearby river mouths has been reworked by powerful Pacific storms into a new dune ridge that accretes successively seaward. Once stabilized out of the surf zone, the dune ridge has soon been colonized by rainforest, such that a cohort of similarly aged old trees is found on each ridge (Wells and Goff, 2006). In general, the oldest tree on each ridge post-dates by 20–50 years the earthquake that is interpreted to have precipitated dune formation. Overall, the tree-ring record extends back 500 years and records four large earthquakes with a precision similar

to (or better than) that derived from individual radiocarbon ages for each earthquake.

Rockfalls

Earthquakes near rugged mountains trigger rockfalls. Clouds of dust due to tumbling rocks enveloped hillslopes in southern California even during the larger aftershocks (typically $M_L \sim 4$–5) of the 1994 Northridge earthquake (Harp and Jibson, 1996). The 1999 $M_w = 7.6$ Chi-Chi earthquake in Taiwan triggered over 20 000 landslides (Dadson et al., 2004). The effects of seismic shaking are not confined to areas directly adjacent to a rupture. John Muir reported rockfalls in the Sierra Nevada that were associated with the 1872 Owens Valley earthquake, which occurred over 100 km away. In New Zealand, rockfalls have occurred greater than 300 km from the epicenter of large earthquakes (Bull et al., 1994). Consider for a moment a talus cone or other accumulation of fallen blocks. Over time, a cone accretes blocks that have tumbled as a result of avalanches, freeze–thaw cycles, or other biological or meteorological events that loosen and dislodge blocks. These additions to the talus are considered to be randomly distributed with respect to time. Against this background of intermittent, randomly spaced additions of blocks, rockfalls triggered by earthquakes will add a pulsed signal in which many tumbling blocks are added simultaneously to the talus deposits. Presumably, if we could date the time of addition of each block exposed on the surface of a talus cone, we would expect to discern a "spiky" record of seismically triggered rockfalls poking through a background of steadier talus growth unrelated to earthquakes (Fig. 6.25). This interpretation, of course, assumes that earthquakes, as opposed to very large storms, avalanches, or spontaneous rockfalls, are responsible for episodic addition of large numbers of blocks. Avoidance of rockfall deposits fed by avalanche chutes and selection of sites situated beneath short, steep slopes of weakly consolidated bouldery material, such as young glacial moraines, can enhance the likelihood that earthquakes will have caused most of the major additions of new blocks (Bull and Brandon, 1998).

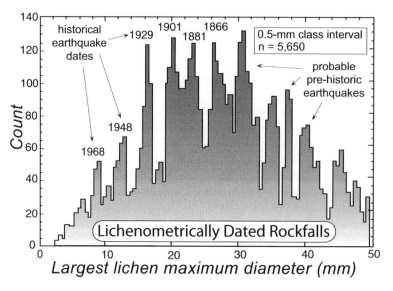

Fig. 6.25   Histogram of maximum lichen diameters, New Zealand.
Data were collected from more than 5000 rockfall blocks distributed across 20 000 km² on the South Island,
New Zealand. Note the clustering of maximum diameters that rise above the background level. This clustering
suggests that discrete events caused episodic input of rockfall blocks. Despite the spatially broad collection area, the
presence of peaks that are separated by less than 3–4 mm suggests that lichen growth rates cannot vary significantly
between or within the study sites. The fact that several historical South Island earthquakes can be correlated to
discrete peaks in the histogram suggests that the lichenometric dating is robust. Modified after Bull *et al.* (1994).

Box 6.2   Ground acceleration during earthquakes.

Although the vertical acceleration in past earthquakes is of great interest to construction engineers and urban planners, measurements of vertical accelerations in prehistoric earthquakes have been rare. Neither the stratigraphic nor the geomorphic record commonly yield data that reveal peak vertical accelerations, because these data sets record net ground displacement, rather than the rate of slip, or its derivative, the acceleration. Two geomorphic approaches using boulders can sometimes provide useful limits on vertical acceleration. If seismic accelerations exceed gravitational acceleration, unattached objects will be thrown in the air. In the 1989 oblique-slip Loma Prieta earthquake in California, there were numerous reports of massive objects (cars, stoves, fireplaces) that were thrown into the air, indicating that at least

locally the ground accelerations exceeded 1g. Boulders that are lying on the surface can actually be moved some distance laterally if the accelerations are oblique and approach or exceed 1g in the vertical direction. During the 1992 (M_s=7.4) Suusamyr earthquake in Kyrgyzstan, boulders weighing up to 1.5 tons were displaced 2 m horizontally and were overturned on a nearly horizontal surface (Ghose *et al.*, 1997). The overturning of a boulder suggests that it partially rolled into its new position, but first it had to move upward out of the depression (~30 cm deep) in which it had been situated.

A second approach to estimating vertical and horizontal acceleration in past earthquakes uses "perched" boulders to place limits on the amount of shaking in the past (see figure). In arid regions, jointed bedrock can weather to produce fields of perched boulders. Initially, the bedrock weathers along the joint surfaces. Erosional removal of the disaggregated debris creates core stones that rest and sometimes balance on underlying bedrock. When such boulders have a long vertical dimension, they appear to be precariously perched and should be highly susceptible to shaking (Brune, 1996). In fact, some studies show that, if such boulders are less than 0.3 m high, they tend to be blown over by strong desert winds (Shi *et al.*, 1996). Boulders larger than this appear to require seismic accelerations to topple them.

Precariously balanced boulders act as a paleo-accelerometer.

Field tests and modeling of the forces required to topple the boulders indicate that accelerations of greater than 0.2*g* would knock over the more precarious boulders, whereas accelerations of 0.3*g*–0.4*g* would be required for the "semi-precarious" boulders (Brune, 2002). Such numbers provide very useful paleoseismological limits on the magnitude of past shaking. If these boulders are still standing, then they have not been subjected to sufficiently strong ground motions to knock them down since the time they became precariously perched. A key to the analysis thus becomes the determination of the time when the boulders became susceptible to shaking and toppling. Presumably, a precarious state was achieved when the boulder was significantly undercut to create a "pedestal" at its base. Radiocarbon dating of the desert varnish around boulder pedestals has provided time control in several areas of the southwest United States. Surprisingly, these data (Brune *et al.*, 2006) suggest that the amount of ground shaking during the past 10–30 kyr in the areas of perched boulders has been much less than that predicted by the Working Group on California Earthquake Probabilities (1995), which has tried to assemble all of the relevant seismic, geodetic, and paleoseismological data into quantitative models of earthquake hazards. Not all areas have perched boulders, but where they are present (or where they have been toppled over), they represent a useful paleoseismic tool.

But how do we date the time of addition of individual blocks? One recently developed method relies on the concept that lichens colonize freshly exposed, stable rock surfaces and that, after an initial period of more rapid growth, lichens grow at a steady and predictable rate (Bull *et al.*, 1994; Locke *et al.*, 1979) (see Chapter 3). A lichen growth curve can be developed based on maximum lichen sizes of a particular species found on rock surfaces that were exposed in a known year. Such surfaces can

include both geomorphic features and cultural ones, such as walls, buildings, or gravestones. Careful choice of the rockfall site combined with knowledge of historical earthquakes can help establish growth rates, if not previously known. For example, on rather stable slopes, only intense shaking due to rupture on a nearby fault is likely to dislodge boulders (Box 6.2). Commonly rockfall deposits in such sites may only include blocks that were added during earthquakes on that fault. Peaks in histograms of lichen sizes corresponding to those events might provide calibration points for a growth curve that can be used to date other sites.

Recent studies in the South Island of New Zealand have demonstrated the utility of lichenometric dating of rockfall deposits (Bull and Brandon, 1998; Bull *et al.*, 1994). For each rockfall deposit studied and on each block that meets a certain size criterion, the largest lichen diameter was measured, such that hundreds of measurements were assembled at a site. The data collection was aided greatly by the use of an automatically recording digital caliper. When several thousand lichen measurements from a large region are synthesized into a histogram (Fig. 6.19), a very irregular and spiky distribution of lichen diameters is apparent. If lichen growth rates were highly variable from place to place or if rockfall events were randomly distributed through time and space, then the histogram should not exhibit discrete peaks in lichen diameter. In fact, given the density of these data, it is clearly possible to differentiate peaks separated by only 2–3 mm. When converted to ages, these size increments represent age differences of approximately 30–40 years. In the New Zealand study, six of the peaks can be assigned to specific historical earthquakes; it is probable that five other earthquakes pre-dating 1866 are also represented (Bull and Brandon, 1998). When interpreting such data, one caveat should be kept in mind: non-seismic triggers, such as exceptional storms or avalanches, also have the potential to produce bulk additions to a rockfall deposit.

Beyond delineating past earthquakes, dated rockfall deposits can be used to estimate the amount of shaking that occurred at a site (Bull *et al.*, 1994). All other variables being equal, the number of rockfall blocks resulting from a given earthquake might be expected to be a function of both the size and the proximity of an earthquake. Larger earthquakes produce more shaking than small earthquakes on the same fault, and the greater the distance from the epicenter, the lower the shaking at a site. Based on the dates of historical earthquakes and the lichen diameters associated with those

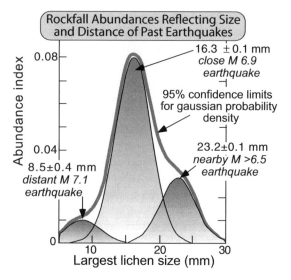

Fig. 6.26 Gaussian distributions of lichens representing three earthquakes.
Histogram of lichen sizes at a local site after decomposition into Gaussian curves, each associated with a separate earthquake. The curves sum to yield the approximate original distribution of sizes. The amount of locally felt seismic shaking determines the height of each Gaussian curve and is interpreted as a function of the distance to the epicenter and the size of the causative earthquake. Note that the largest earthquake ($M = 7.1$) produced the smallest peak because of its distant epicenter. Modified after Bull *et al.* (1994).

events, the lichen data for a given site can be decomposed into a suite of Gaussian curves, each of which is centered on the mean diameter for the age of a specific earthquake and is assigned a height and width, such that the suite of curves will sum together to yield the overall distribution of measured lichen sizes (Fig. 6.26). The relative height of the individual Gaussian curves is used to infer the amount of shaking at the site during each earthquake. By analyzing the local lichen records from many areas, a seismic shaking index may be determined for each site for each earthquake. Moreover, for prehistoric or poorly known earthquakes, these shaking indices can be contoured for individual earthquakes (Fig. 6.27) in order to define the subregion where the epicenter was located (Bull and Brandon, 1998).

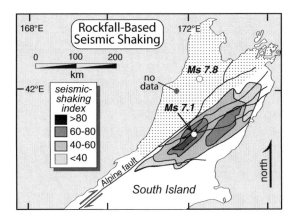

Fig. 6.27   Seismic shaking index map for the effects of two earthquakes.
Two large earthquakes occurred in 1929 in the northern South Island, New Zealand. Contoured map shows shaking indices for the 16mm (1929) diameter lichen peak for 53 sites. The epicenter and magnitude of the earthquakes are shown. The M_s=7.1 earthquake (depth ~15km) caused strong shaking in an area concentric to the epicenter, whereas the larger M_s=7.8 earthquake affected a broader region. Note that no data are available to the northwest of the shaded areas. Modified after Bull *et al.* (1994).

This remarkable result has significant social applications. When environmental assessments are made for construction of buildings intended for "high-risk" usage (nuclear power plants or chemical manufacturers), the question is often posed: "Are there any faults within a certain distance that have been active during Holocene times?" When faults do not break the surface, this question can be particularly difficult to answer. If lichenometric dating of rockfalls resulting from seismic shaking can quantify prehistoric events, and even specify previous epicentral areas, this analysis could provide valuable insights for paleoseismic studies.

Paleoseismic insights on fault behavior

Given the wealth of new paleoseismological data accruing from studies around the world, one would hope that clear answers would emerge to some of the overarching questions about fault behavior: Are faults slip-predictable or time-predictable (Fig. 4.4)? Do seismic events

tend to cluster in time or space (Fig. 6.1)? How variable are fault slip rates and how do they compare with geodetic rates or million-year geological rates (Fig. 5.1)? Some of the most potent insights on the seismological behavior of the shallow crust seem to derive from two regions: the region lying above the extensive Sumatran subduction zone, and the region encompassing the North America–Pacific transform plate boundary in southern California. The former region is remarkable for the widespread, detailed chronologies of seismically driven emergence and submergence as shown in the coral records, whereas the latter region has benefited from more than 30 years of extensive study, much of it under the aegis of the Southern California Earthquake Center. In this latter region, not only has the longest age-constrained, earthquake-by-earthquake record of slip been developed (Box 6.3), but many dozens of paleoseismological studies on the active faults of southern California have rendered this the most broadly studied and best-dated region in the world.

Time-predictable or slip-predictable earthquakes

Very few records have sufficient length, reliable limits on coseismic slip, and precise enough dating to provide robust tests of time or slip predictability. The two best data sets seem to provide different answers. The best-dated earthquake sequence of the long Wrightwood time series suggests that faulting on the southern San Andreas Fault is neither slip- nor time-predictable (Box 6.3). On the other hand, the coral-based data from Sumatra (Sieh *et al.*, 2008) for the past seven centuries suggest an overall time-predictable behavior (Fig. 6.28). Gathered from sites spanning more than 100 km along strike above the subduction zone, the Sumatran data argue for megasequences of great earthquakes spaced about every 200 years. Sometimes the largest earthquake has been preceded over several decades by one or several smaller ones. The overview of the entire sequence from all three sites, however, indicates

Box 6.3   Lessons from the San Andreas Fault.

In the 35 years since the pioneering paleoseismic studies of Kerry Sieh on the San Andreas Fault as a doctoral student in the 1970s, this fault has provided a treasure trove of insights on how large strike-slip faults behave. Dozens of trenches have been excavated along the length of the fault to determine when earthquakes occurred during the past millennium and to try to deduce the patterns of past ruptures. Despite these many studies, the imprecision of radiocarbon dating introduces sufficient temporal ambiguity that correlation of individual rupture events from one site to another remains somewhat uncertain. Nonetheless, a consensus has gradually emerged about the lateral extent of many of the earthquake ruptures of the past millennium. Surprisingly, despite numerous paleoseismic records from those many trenches, until recently relatively little was known either about how much slip occurred at a given site in each rupture or about overall slip rates at a site. (The lack of information on slip per rupture is a typical consequence of trenching perpendicular to a strike-slip fault and not having complementary trenches parallel (Sieh, 1984) to it for salami-slicing (see Fig. 6.3) in order to reconstruct the slip record.)

Dating of offset geomorphic features is now providing a more detailed record of slip variations (e.g., Weldon and Sieh, 1985; Liu *et al.*, 2004; van der Woerd *et al.*, 2006). Several of the longer paleoseismic records (see Fig. 6.8) on the San Andreas Fault contained as many as 10 events (Sieh *et al.*, 1989) and allowed geologists to calculate recurrence intervals and to consider whether San Andreas earthquakes occurred in clusters or not. But time series of 10 or fewer earthquakes (most trenches had fewer) are too short to make a statistically reliable assessment of whether slip rates on the San Andreas Fault are steady, whether the fault behaves in a time-predictable or a slip-predictable manner, or whether a predictable pattern for a sequence of San Andreas earthquakes exists at any given site.

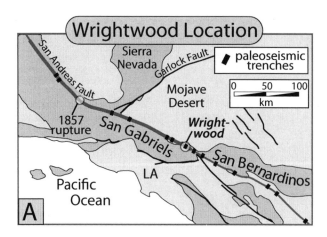

A. Location of Wrightwood, CA, studies.

All these uncertainties are beginning to be addressed through studies by Ray Weldon and his associates along the San Andreas Fault near Wrightwood (see figure A) at the eastern end of the San Gabriel Mountains (Weldon *et al.*, 2004). Over an 18-year period, they have excavated more than 40 trenches and radiocarbon-dated over 50 peat layers in order to place tight chronological constraints on at least 30 earthquakes that were recognized in the stratigraphic record. The record stretches back for some 5000 years (see figure B), and, although the middle interval from ~1500 BC to 500 AD is not yet fully explored, this reconstruction is currently the world's highest-quality, long-term earthquake reconstruction!

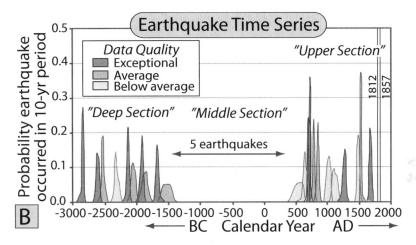

B. Earthquake time series, Wrightwood.

We can use these data to test whether the earthquake behavior here is either time or slip predictable (see Fig. 4.4). For example, for slip-predictable behavior, a one-to-one correlation should exist between the duration of the previous interseismic interval and the magnitude of slip (see figure C), whereas for time-predictable behavior, the magnitude of slip should linearly correlate with the time until the next earthquake (see figure D). Neither of these models does a good job of describing the behavior of the San Andreas earthquakes at this site.

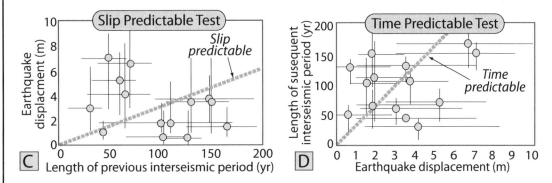

C. Wrightwood data on slip predictability. D. Wrightwood data on time predictability.

With large numbers of events for which the date and amount of slip are known, we can examine how the slip magnitude varies through time and whether any consistent pattern of rupture behavior is apparent (see figure E). The short answer is that no clear pattern exists. Large earthquakes can follow each other, sometimes in close succession, or may have multiple small earthquakes separating them. Sometimes, interseismic strain accumulation and coseismic strain release are about equal (from 1500 to 1800AD), whereas for long intervals between 600 and 1500AD, strain accumulation and release are persistently out of balance. During one such unbalanced interval (between 690 and 850AD), four large earthquakes occurred and represent an apparent cluster.

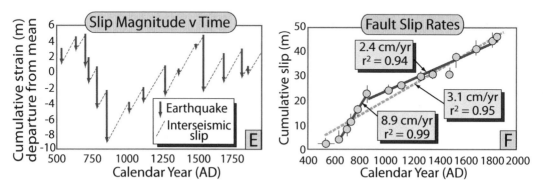

E. Slip per event versus time at Wrightwood. F. San Andreas slip rates through time at Wrightwood.

When slip rates are reconstructed in this well-dated site, the data strongly suggest that slip rates have changed over time (see figure F), such that, for some intervals, slip was more than three times greater than during others! Such changes in slip rate along a major transform fault at sub-millennial times are unexpected and not yet well explained. Perhaps most surprising, the apparent slip rate of 8.9cm/yr between 690 and 850AD is about twice as fast as the accepted relative rate between the Pacific Plate and the North American Plate! Because we have no indications from elsewhere that the relative Pacific–North America rate accelerated during this interval, let us assume that no such acceleration occurred. We then must consider that, over multiple earthquake cycles prior to the cluster of large earthquakes, large amounts of uncompensated elastic strain accumulated in the crust, i.e., each successive earthquake only partially released the elastic strain that had been built up since the previous earthquake. Given that the rate of slip (2.4mm/yr) in the last millennium has been less than the long-term slip rate of 3.1mm/yr, "excess" elastic strain may again be present in the crust near Wrightwood, such that we may be poised for one or several large earthquakes in the near future.

persistent time predictability. These data clearly imply that major earthquakes should be expected over the next few decades because the threshold has been reached that preceded all of the previous large earthquakes in the past 700 years.

Temporal or spatial clustering of earthquakes

Within the drumbeat of regular supercycles along at least part of the Sumatra subduction zone (Fig. 6.28), temporal clusters of smaller

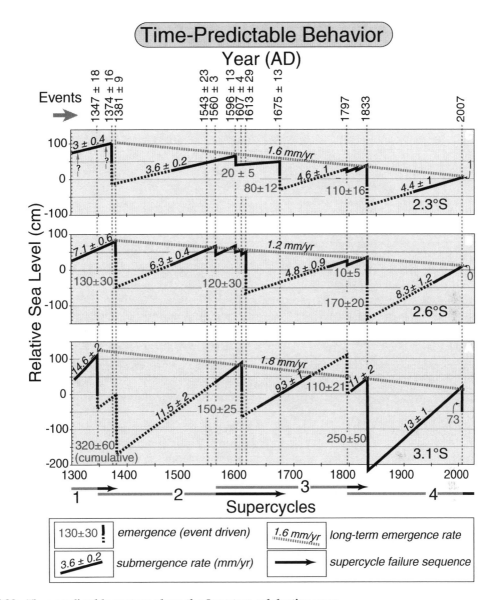

Fig. 6.28   Time-predictable rupture along the Sumatran subduction zone.
The coral-based chronology of subduction-zone earthquakes from the Mentawai Islands, Sumatra. A background emergence rate of 1–2 mm/yr at each site indicates permanent strain in the forearc. Megathrust earthquakes occur each time this long-term trend of local sea-level fall (dotted gray line) is reached. Four supercycles of megathrust earthquakes with recurrence intervals of about 200 years are clearly delineated at each site. The sites at 2.3°S and 2.6°S commonly show a sequence of smaller offsets in the decades preceding a very large earthquake. At 3.1°S, where 2.5m of subsidence occurred in the 1833 earthquake, the 2007 M=8.4 earthquake appears likely to be the first in a series of major earthquakes that would be expected to affect all three sites in the coming decades. Modified after Sieh *et al.* (2008).

earthquakes tend to occur in the last few decades of a cycle prior to a much larger event. If earthquakes cluster in time, not only does this behavior provide some predictive power, but it also provides some insight on seismic behavior. Such clustering signifies a pulse of regional strain release. Clustering might result from interactions, such as stress transfer (Stein, 1999), among

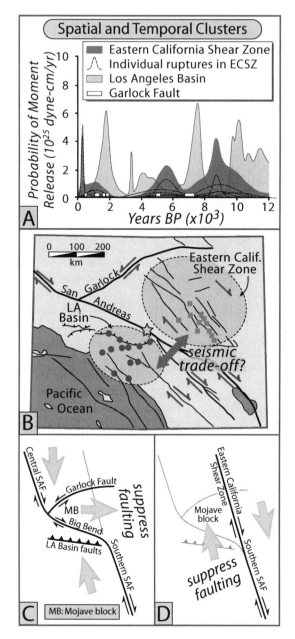

relatively closely spaced faults that lead to a spate of earthquakes over a restricted time interval. Recent studies in the western part of the Eastern California Shear Zone (ECSZ), which lies just east of the San Andreas Fault (Fig. 6.29), indicate that clusters of Holocene earthquakes occurred at about 9, 5.5, and 1 ka (Rockwell *et al.*, 2000). These data suggest that, about every 4000 years, multiple faults rupture in the ECSZ over a fairly brief interval (~1000 years). When viewed in the broader context of southern California (Dolan *et al.*, 2007), these Holocene events in the ECSZ do not simply represent temporal clusters, but they appear to be part of a spatial cluster that alternates through time, with clustered earthquakes in the broader Los Angeles basin to the west of the San Andreas Fault (Fig. 6.29). When viewed together, the integrated seismic moment release during the Holocene is much smoother than would be perceived if the rich data sets from either side of the San Andreas Fault were viewed in isolation. This holistic overview of fault clustering suggests that kinematic linkages exist across this wide (400 km) zone (Dolan *et al.*, 2007). One possibility is that, when slip is focused along the Big Bend and central San Andreas Fault, it both drives compression in the Los Angeles basin and promotes eastward extrusion of the Mojave block, which, in turn, suppresses fault slip in the ECSZ (Fig. 6.29C). This mode of slip may alternate with one in which more of the southern San Andreas slip is fed into the ECSZ, thereby decreasing both slip in the Big Bend and compression across the Los Angeles basin, while increasing faulting in the ECSZ (Fig. 6.29D).

This regional overview of southern California paleoseismicity could only emerge after dozens of faults had been previously trenched. Despite these studies, the picture remains incomplete. Few of the buried thrusts in the Los Angeles basin are well documented, and many faults

Fig. 6.29   Earthquake clustering in Southern California.
A. Chronologies of Holocene faulting and estimates of seismic moment release reveal clustered seismicity based on data from 10 or more faults on either side of the San Andreas Fault. B. Map depicting location of studied faults in the Eastern California Shear Zone (squares) and the Los Angeles basin (circles). Star indicates location of Wrightwood (Box 6.3). C. Scenario for suppressed faulting in the ECSZ when slip is focused through the Big Bend and central San Andreas Fault (SAF).

D. Scenario for slip being fed into the ECSZ causing decreased slip rates through the Big Bend and suppressed Los Angeles basin faulting. Modified after Dolan *et al.* (2007).

within the ECSZ have received little detailed study. As a consequence, the possibility exists that seismic moment release is temporally smoother than it now appears in either the ECSZ or the Los Angeles basin. Paleoseismologists will be happy to know that the rupture history of many more faults needs to be known before the regional dynamics are fully understood.

Fault slip rates

When a paleoseismic history delineates the slip for multiple earthquakes on a fault, cumulative slip versus time plots serve to delineate the steadiness of deformation. If additional constraints on slip rates can be incorporated at longer time scales – such as may be derived from offset geomorphic markers, larger-scale structural reconstructions, or even bedrock cooling and erosion histories – then slip rates across many time windows can be examined. Several such compilations that span Late Pleistocene to Holocene time suggest significant slip-rate changes have occurred. Across the normal fault system spanning the easternmost Basin and Range (extending ~80 km westward from the Wasatch Range), data over the past 15 kyr (Fig. 6.30) suggest that Holocene slip rates were up to 10 times faster than Late Pleistocene rates (Friedrich et al., 2003). This acceleration may represent a predictable, but surprising, response to climate change. During the Late Pleistocene, the hanging wall lying to the west of the Wasatch Fault was covered by Lake Bonneville (Gilbert, 1890), which was about 300 m deep and covered some 50 000 km^2. When the lake abruptly drained ~15 ka, the removal of the water load should have caused a decrease in normal stresses on the Wasatch Fault (Hetzel and Hampel, 2005) and could have initiated the subsequent interval of increased slip.

Perhaps an even more surprising slip-rate result comes from the long time series on the San Andreas Fault at Wrightwood. Across 15 earthquakes spanning 1300 years, not only were rates unsteady, but for one 200-year-long interval when four large earthquakes occurred, the average slip rate was twice as fast as the North America–Pacific plate rate (Box 6.3). This

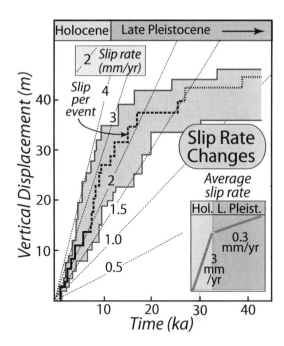

Fig. 6.30 Slip-rate changes on the Wasatch Fault system.
Cumulative vertical displacement comprises the summed offsets from four normal fault systems (Wasatch, West Valley, Oquirrh, and Stansbury) in the eastern Basin and Range of Utah. Inset in lower right depicts average Holocene versus Late Pleistocene slip rates. Modified after Friedrich et al. (2003).

unexpected result suggests that significant elastic strain can be stored in the crust and then released over multiple seismic cycles.

Continental collisions and megathrust earthquakes

Many continent-to-continent collisions are characterized by fold-and-thrust belts underlain by a gently dipping sole thrust that extends from the foreland to far beneath the hinterland, where it steepens. This sole thrust roughly coincides with the modern plate interface. In some ranges, such as the Himalaya, contemporary GPS, leveling-line measurements (Fig. 5.7), and ongoing moderate seismicity indicate that current deformation is focused in the hinterland. Some studies of deformed Himalayan Holocene terraces in the foreland (Lavé and Avouac, 2000),

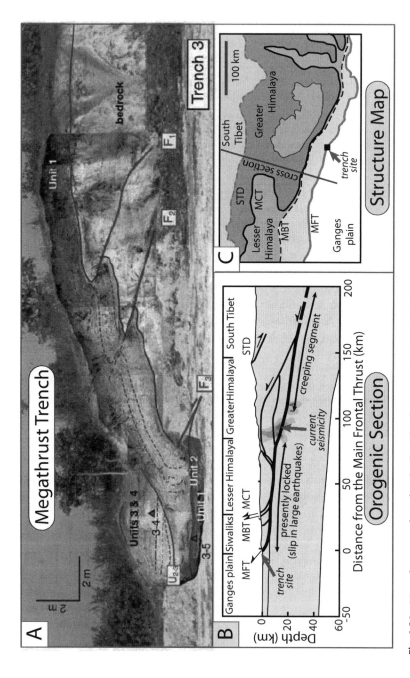

Fig. 6.31 Megathrust earthquakes in the Himalayan continental collision.

A. Excavation across the scarp of the Himalayan megathrust in eastern Nepal. A cumulative slip of ~17(+5/−3)m deforms the overlying units that date from ~1100 BP. Note the bedrock uplift in the hanging wall. B. Simplified cross-section of the Himalaya showing the creeping zone along the sole thrust in the north that generates current seismicity beneath the Greater Himalaya, and the presently locked sole thrust underlying the Lesser Himalaya and breaking the surface at the Main Frontal Thrust (MFT). Elastic strain currently accumulating in the Greater Himalaya is proposed to be transferred to the MFT during megathrust earthquakes. MCT: Main Central Thrust; MBT: Main Boundary Thrust; STD: South Tibetan Detachment. C. Location map of the cross-section (B) and the trench site (A) with respect to major Himalayan structures. Modified after Lavé *et al.* (2005).

however, have concluded that essentially all of the geodetic convergence is ultimately passed along the sole thrust to the leading edge of the plate interface during megathrust earthquakes (Avouac, 2003). Although the rupture patches can apparently span thousands of square kilometers during $M \geq 8$ earthquakes and thereby pose a major hazard (Bilham *et al.*, 2001), no surface ruptures have been previously described during three megathrust earthquakes ($M_w = 7.8–8.5$) in the 20th century. Until recently, neither the amount of slip during these megathrust events nor the earthquake magnitude of pre-20th-century earthquakes have been known. New paleoseismological studies along the Himalayan Main Frontal Thrust have helped to resolve these questions. Trenches across this fault (Fig. 6.31) have revealed a Medieval earthquake (~1100 AD) with dip slip of ~17 m, a lateral extent probably exceeding 240 km, and an estimated moment magnitude of 8.8 (Lavé *et al.*, 2005). Although not as large as the 2004 Sumatra earthquake, a repeat of this Medieval rupture would wreak havoc with the lives of several million people who currently live directly above the sole thrust.

Remaining problems

Among the paleoseismic approaches described in this chapter, many are generally suitable for assessing past activity along faults that break the surface or for defining deformation that occurred where a good spatial reference frame against which to measure displacement is available. When successful, such studies reveal when earthquakes occurred in the past, the magnitude of offset, and the extent of rupture associated with each earthquake. Numerous paleoseismological problems remain to be solved, however, and some will require innovative approaches that are presently untested.

One major unresolved problem is to define the seismic record of faults in terrestrial settings that do not rupture the surface. Unlike coastal records, terrestrial regions commonly lack a reliable reference frame against which to

delineate deformation. Although some innovative techniques for defining displacement on blind thrusts have recently been described (Dolan *et al.*, 2003), relatively few buried faults have been evaluated and, for those that have, the interpretation is highly dependent on a geometric model for growth strata *vis-à-vis* coseismic rock uplift. Such models may need to be constructed independently for each fault that is studied.

We have seen that trenching along strike-slip faults can yield a history of multiple earthquakes, even if each earthquake causes displacements of several meters (Fig. 6.8). What practical means can be used to develop similar histories for dip-slip faults with large offsets during each earthquake? Deeper trenches could help, but such trenches are often dangerous places in which to work. Can aspects of the stratigraphic record be tied unambiguously to recurrent fault movements, as has been attempted at El Asnam in Algeria (Fig. 6.23)? Can high-resolution lidar topography reveal event-driven hanging-wall deformation that could be tied to individual earthquakes?

Determining precise timing for deformational events is a primary concern and an almost never-ending problem for paleoseismologists (and for many geomorphologists!). In this chapter, applications of high-precision radiocarbon dating, ^{230}Th dating, cosmogenic exposure ages, lichen diameters, and annual growth rings in corals and trees have been discussed. A perennial search is ongoing for ways to improve existing techniques and for new ways to date the record of deformation. The limitations, assumptions, applicable age range, accuracy, and precision should be evaluated prior to applying any technique to a particular dating problem. As researchers confront the chronological problems and possibilities of trenches and outcrops, new ways to use existing techniques, as well as new techniques themselves, should be forthcoming. Similarly, more ways to assess vertical and horizontal accelerations and the distribution of seismic shaking in prehistoric earthquakes need to be developed.

Despite the many challenges and problems confronting paleoseismologists, tremendous progress has been made in the past few decades in understanding the past history of faulting on many of the major faults near population centers. We now know much more about the timing, displacement, and rupture length of many prehistoric earthquakes. As more faults that rupture the surface are studied, as new techniques are developed for investigating buried faults, and as interdisciplinary studies integrate increasingly diverse data, our knowledge will continue to expand, providing constraints for engineers and city planners and for the broader geological community that is just trying to understand how the Earth works.

7

Rates of erosion and uplift

We recognize that the landscape in actively deforming areas results from interactions among the processes of rock uplift, subsidence, and surface processes that can lead to local erosion and deposition. It is useful, therefore, to try to calibrate the rates and relative contributions of each process at a particular time or over some span of time. The goals of this chapter are to describe approaches that can be employed to define geomorphic and tectonic rates at diverse spatial and temporal scales. In particular, the focus is on: (i) determining rates of erosion; (ii) calculating mass balances and material fluxes; (iii) determining rates of rock and surface uplift; and (iv) reconstructing the past topography of tectonically active landscapes.

Studies that define rates of erosion and uplift lie at the heart of tectonic geomorphology and represent some of the most interdisciplinary research that occurs today. Successful studies typically include geochronology, structure, stratigraphy, geomorphology, and numerical modeling. Those studies that address the cause of changes in rates also encompass paleoclimatology, tectonics, and geochemistry. Much current debate revolves around whether climate or tectonics is responsible for changes in either erosion rates or the shape of the landscape (Masek *et al.*, 1994b; Molnar and England, 1990; Whipple *et al.*, 1999; Whipple, 2009). Potential feedbacks between climate and tectonics have been hypothesized, but remain poorly docu-

mented. For example, the following questions remain. Did surface uplift of mountain ranges globally accelerate in late Cenozoic times and drive climate change, or did climate change cause accelerated erosion that drove isostatic uplift (Zhang *et al.*, 2001)? Has glacial erosion sufficiently changed the depth and shape of valleys (the cross-sectional difference between a V-shaped and a U-shaped valley) that peaks have risen isostatically in response to the eroded mass (Stern *et al.*, 2005)? Are the aprons of Quaternary conglomerates that bound many mountain ranges indicators of tectonic or climatic events? Did uplift of the Tibetan Plateau cause strengthening of the Asian monsoon, changes in rates of chemical weathering, and changes in atmospheric CO_2, thus precipitating major Northern Hemispheric glaciation (Edmond, 1992; Raymo and Ruddiman, 1992)?

Calibration of rates of erosion and uplift permits a quantification of the interplay of the surface processes that shape landscapes. Because rivers set the base level for adjacent hillslopes, we hypothesize that they are coupled (Fig. 7.1). In the absence of well-defined rates, however, the nature of this coupling remains uncertain. How quickly do knickpoints migrate through a river system? How rapidly do adjacent hillslopes respond to changes in river incision rates? Do fundamental differences in the nature and rate of response occur when soil-mantled hillslopes are compared with bedrock-dominated

Tectonic Geomorphology, Second Edition. Douglas W. Burbank and Robert S. Anderson.
© 2012 Douglas W. Burbank and Robert S. Anderson. Published 2012 by Blackwell Publishing Ltd.

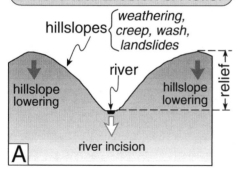

Differential Erosion & Relief

hillslopes { weathering, creep, wash, landslides

A

river incision

hillslope lowering > river incision

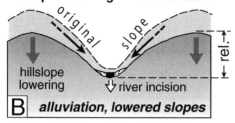

B *alluviation, lowered slopes*

hillslope lowering < river incision

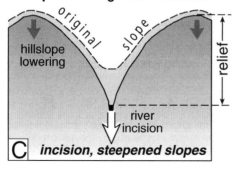

C *incision, steepened slopes*

Fig. 7.1 Hillslope–river coupling.
River-incision rates define changes in base level for adjacent hillslopes. Differences in the relative rate of river incision versus hillslope lowering control changes in topographic relief. A. In steady-state topography, rates of river incision and of hillslope lowering are balanced. B. If hillslope lowering exceeds river incision, valleys tend to alluviate, and topographic relief decreases. C. If river incision exceeds hillslope lowering, relief increases and hillslopes steepen. If steepened slopes drive faster hillslope erosion, the coupled system can move toward a steady state.

ones? What drives changes in topographic relief? This chapter does not answer these questions. Rather, it encompasses a rich variety of approaches that help to define the key attributes

that will permit a choice among the competing alternatives.

Rates of erosion and denudation

Definitions

The changing vertical position of a point on the land surface (Fig. 7.2) at any time is a function (England and Molnar, 1990) of: (i) the rate at which bedrock is being carried upward by tectonic or isostatic processes (known as the "bedrock uplift rate"); (ii) the rate at which material in the subsurface is compacting; and (iii) the rate of denudation or deposition at the surface:

surface uplift = bedrock uplift + deposition
– compaction – erosion

Bedrock uplift and deposition contribute to raising the land surface, whereas erosion and compaction serve to lower it. Because compaction and deposition both occur primarily in sedimentary basins near local base level, they are generally ignored in this chapter in favor of denudation and bedrock uplift, which are the main determinants of topography in tectonically active or mountainous landscapes.

Denudation results from the removal of material from a point or region on the Earth's surface and can occur in response to two very different processes. *Geomorphic erosion* (referred to subsequently simply as "erosion") results from mechanical and chemical weathering of rock and removal of loosened or dissolved material by geomorphic agents. Although erosional removal of debris is often incremental, processes like deep-seated, bedrock-involved landsliding can be instantaneous and volumetrically significant. *Tectonic denudation* typically occurs through processes of extension and normal faulting, and such denudation can result in the rapid removal of large volumes of nearly solid rock. A common reference frame for calculating rates of denudation is the geometry of the past land surface. *Surface uplift* or *surface lowering* generally refer to changes in the elevation of a surface, whereas *bedrock uplift* refers to changes in the vertical position of rocks with

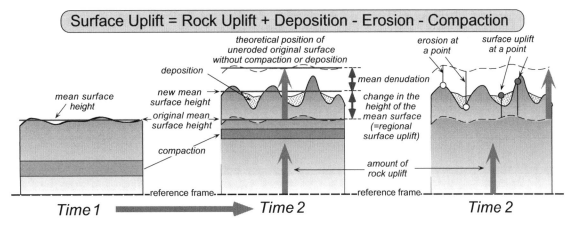

Fig. 7.2   Factors controlling surface uplift.
Between times 1 and 2, a specified amount of rock uplift occurs (vertical arrow). Center panel emphasizes changes in the means of quantities of interest: denudation, surface height, and changes in surface height. Right panel emphasizes changes at a point in the landscape: surface uplift or erosion. Note that erosion is measured *downward* from theoretical position of the uplifted original surface in the absence of erosion, whereas surface uplift is measured *upward* from the original surface at time 1.

respect to a fixed reference frame, such as the geoid (Fig. 7.2).

In most situations, removal of mass from a region will be isostatically compensated by crustal rebound. Given typical crustal ($\rho = 2.75\,\text{g/cm}^3$) and mantle ($\rho = 3.3\,\text{g/cm}^3$) densities, erosion of an average of 100 m of rock across a broad surface will cause the crust to rebound isostatically about 85 m ($\sim 100 \times 2.75/3.3$) and will yield only 15 m of lowering of mean surface elevation. Similarly, crustal thickening will lead to surface uplift, but, owing to the isostatic sinking of the thickened crust, the magnitude of surface uplift will be only about one-sixth of the amount of crustal thickening. Thus, direct measures of the change in elevation of the land surface, which are themselves very difficult to obtain, can only be used to estimate erosion or bedrock uplift rates when other controls, such as changes in mean surface elevation, time scales and lags of isostatic response, volumes of eroded material, and the crustal density structure, are known.

When calculating rates of denudation or surface uplift, a careful distinction should be drawn between local versus regional rates. A local rate may typify a single drainage basin or any point for which a rate can be defined (Fig. 7.2), whereas a regional rate should represent an integrated

rate for a large area, commonly greater than $1000\,\text{km}^2$. It is these regional rates that are geophysically important, because the crust responds isostatically to regional loading and unloading, but is sufficiently rigid to be insensitive to local changes. Locally determined rates can be strongly influenced by the focused action of a geomorphic agent, like a landslide or a rapidly incising river, and such rates are likely to vary in time and space much more rapidly than will rates determined for larger areas. In addition, any given point in the landscape, such as a mountain summit, could be going up or down with respect to the geoid at the same time that the average elevation of the landscape is changing in the opposite direction. For example, erosion that deepens and widens river valleys but does not erode the peaks will reduce the mass of an area and drive isostatic uplift. As a consequence, the altitudes of the peaks could increase at the same time that the mean height of the region is decreasing (Burbank, 1992; England and Molnar, 1990; Small and Anderson, 1995, 1998).

Conceptual framework

Both *mechanical weathering* and *chemical weathering* interact to erode the land surface.

Box 7.1   The role of rock fracture in erosion.

Before rock can be eroded, intact bedrock must be broken into smaller pieces that can be detached from the landscape. This detachment process can occur at the scale of grains, thereby promoting grain-by-grain attrition from the rock surface, or at scales that enable landslides. In all cases, fractures serve to diminish overall rock strength by reducing cohesion and potentially by lessening the effective angle of internal friction. Fracture production results from two major classes of processes: geomorphic and tectonic. Geomorphic processes encompass the broad swath of physical, chemical, and biotic processes that serve to break down rock masses. In almost all cases, these processes are most intense at the surface, and they diminish in effectiveness with depth, typically in some poorly known manner (see figure A). Tectonic fracturing results from myriad stresses within tectonic plates, but, in actively deforming landscapes, the most obvious cause of fracturing is the transport of rocks above irregular fault surfaces (Molnar *et al.*, 2007). Any bend or kink or change in dip of a fault surface causes a concentration of stresses in the rocks of the hanging wall as they are moved and folded above it. The spatial cloud of aftershocks that follows coseismic slip along a fault plane testifies to the dispersed fracturing that occurs as accumulated stress is released in the hanging wall.

Whereas recognition of these fracturing processes and their importance is not new, quantification of their magnitude has been a persistent challenge. How deep is the geomorphically fractured layer? How does fracture density vary with depth? Within any mountain range, how diverse is the degree of tectonic fracturing? How does the degree of fracturing (or lack thereof) influence erosion processes? Our inability to see into the shallow subsurface restricts our insights on these questions.

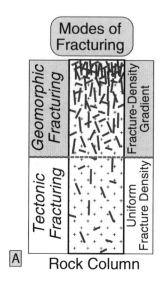

A. Two modes of bedrock fracturing.

Based on data from "backpack-able" portable seismic arrays, new understanding of the upper 10–20 m of the rock column is now emerging. In comparison to the velocity of compressional seismic waves (P waves) in intact bedrock, changes in P-wave velocities in the subsurface can be attributed to fracturing, which impedes the transmission of seismic waves. Recent studies on the South Island of New Zealand (Clarke and Burbank, 2010a, 2011) reveal end-members that correspond to tectonic and geomorphic fracturing models. Geomorphic processes produce an exponential decrease in the magnitude of fracturing with depth (see figure B). In New Zealand, the base of this geomorphic zone ranges from 2 to 18 m, but averages 7 m. Where this geomorphically fractured layer is absent, fracture density is nearly uniform with depth and is attributed to tectonic fracturing. A provocative feature of these data is that the geomorphically fractured upper layer tends to be present only when the rock beneath it has been only weakly fractured by tectonic processes. Where tectonic fracturing is more intense (see figure B), geomorphic fracturing is commonly not seen. Such rocks are interpreted to be sufficiently weak that they fail by bedrock landslides, whose frequency inhibits the development of a weathered layer at the surface.

These observations underpin a simple conceptual model (see figure C) in which geomorphic fracturing propagates downward into a rock column through time, thereby increasing fracture density and reducing rock strength. Such fracturing should affect the erosional efficacy of many surface processes. In sites where landsliding is a major contributor to overall denudation, this fracturing can weaken formerly strong bedrock to the point of failure, such that most landslides are shallow and occur within this geomorphically weakened zone. In contrast, where the underlying bedrock has already been considerably weakened by pervasive tectonic fracturing, landslides of any depth can occur, and a highly fractured, geomorphic surface layer is uncommon.

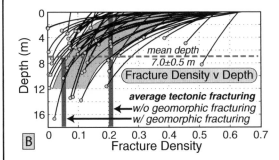

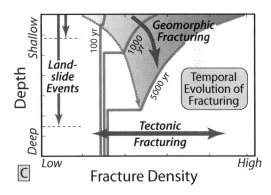

B. Fracture-density profiles for about 70 sites in Fiordland, New Zealand. Where a geomorphically fractured layer is present, the mean fracture density of the underlying, tectonically fractured bedrock (0.05) is four-fold less than the mean (0.2) when no geomorphic fracturing is observed. Strong underlying rock apparently inhibits failure by landslides and promotes geomorphic fracturing.

C. Geomorphic fracturing both propagates a fracture front downward and increases fracture density over time. Tectonic fracturing defines a "starting condition" and, along with lithologic character, determines whether a slope will fail by landsliding without any further fracturing. Whereas deep landslides can rupture beneath any geomorphically weathered zone, shallow landslides may be restricted within it.

For example, chemical weathering can convert micas and feldspars to clays or can dissolve minerals. Such processes weaken rocks and increase their susceptibility to mechanical disaggregation. At the same time, mechanical breakdown of rocks continually exposes unweathered material and provides fresh surfaces for efficient chemical attack. In temperate to alpine and arctic terrains, a key mechanical weathering process is frost cracking. The geomorphic effectiveness of frost cracking is strongly dependent on the presence of water and the time a parcel of rock spends within a certain subzero thermal window (Hallet *et al.*, 1991; Anderson, 1998; Hales and Roering, 2007). Frost cracking can dominate disaggregation processes and facilitate more efficient chemical weathering, even in cool climatic regions. In most tectonically active regions, mechanical weathering appears to dominate locally measured erosion rates (Box 7.1), and rates of dissolution are often uncalibrated.

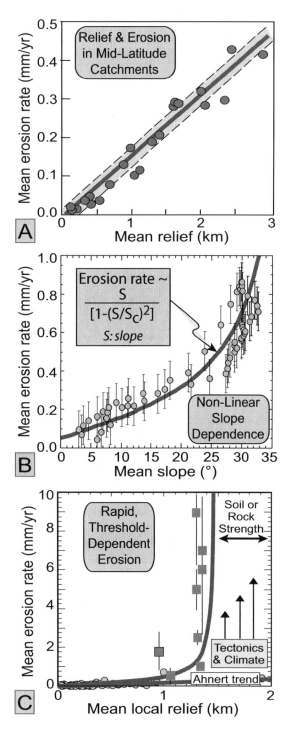

Rates of erosion are commonly defined in terms of lowering of the *bedrock* surface, and have previously been portrayed as a function of the *local topographic relief* (altitudinal difference within an area that is typically smaller than the entire catchment). Owing to the greater potential energy of elevated areas with respect to the local base levels (as represented by adjacent valley bottoms) and owing to the generally steeper slopes that prevail in regions of high relief, the traditional view is that high relief promotes high erosion rates. Measured and estimated sediment fluxes out of drainage basins with slow to moderate erosion rates (<0.5 mm/yr), but lying within a similar temperate climatic regime, display a linear relationship to relief (Ahnert, 1970) (Fig. 7.3A). A similar near-linear relationship of hillslope angle to erosion rate is apparent at slow to moderate rates and lower hillslope angles (Fig. 7.3B). As erosion rates increase, however, they become increasingly independent of slopes or relief (Montgomery and Brandon, 2002). In particular, nonlinear erosion-rate increases can be related to the presence of hillslopes hovering near the threshold angle for landsliding, such that small changes in slope angle are predicted to produce large changes in erosion rates (Fig. 7.3B).

When considering the basin- to range-scale evolution of the land surface, long-term, average rates of erosion, deposition, and uplift are more important than are short-term rates. Nonetheless, given the techniques available, most measured sediment fluxes out of basins sample only short intervals of time. In the context of the well-known glacial–interglacial climatic variations of Quaternary times

---

angle. At rates <0.5 mm/yr, rates are approximately linearly related to slopes, but, as slopes steepen, rates become increasingly independent of slope. This response is consistent with a nonlinear slope dependence in which rates trend toward infinity as a threshold slope is approached. C. In rapidly eroding ranges, relief is a poor predictor of erosion rate. This decoupling is attributed to the role of threshold slopes in accelerating erosion rates. Modified after Ahnert (1970) and Montgomery and Brandon (2002).

Fig. 7.3   Erosion rates as a function of relief and hillslope angles.
A. Topographic relief versus erosion rate, showing a linear correlation for mid-latitude, temperate catchments with rates <0.5 mm/yr. B. Erosion rate as a function of hillslope

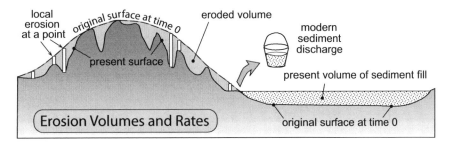

Fig. 7.4   Approaches to estimating erosion rates.
If the original surface topography can be reconstructed, then the missing volume of material can be determined by subtracting the present-day topography from the pre-erosion topography. If the age of that initial surface is known, then a mean erosion rate can be calculated. Similarly, if the volume of sediment in a basin and the age span of its filling are known, and if the size of the area contributing sediment to the basin can be estimated, a mean erosion rate can be approximated. Alternatively, the modern rate of sediment discharge and the contributing area can be used to compute a mean erosion rate.

(Fig. 2.3), sediment fluxes out of basins should be expected to occur in irregularly spaced pulses. Although the timing of major episodes of erosional or sediment discharge may vary between nearby basins, regional climate change more commonly modulates sediment fluxes and creates a coherence among nearby catchments. Such changing sediment pulses could directly relate to precipitation variations (Goodbred and Kuehl, 1999) or to the stability of hillsides influenced by climatically modulated vegetation change (Leeder et al., 1998).

Several approaches can be used to estimate erosion rates (Fig. 7.4). An average denudation rate can be readily computed with three types of data: the volume of eroded material, the area from which that material was derived, and the duration of erosion. The volume of material eroded from a catchment can be estimated using direct measurements of erosion within the catchment itself, gauging the sediment flux out of the catchment, or by assessing the volume of sediment stored in a basin to which that catchment is tributary (Fig. 7.4). These sediment volumes have to be converted to rock volumes with appropriate corrections for density differences. Whereas direct measures of erosion rates in the catchment are informative, rates at individual points must be integrated across the entire catchment. Commonly, basin-

wide estimates with large and unknown uncertainties result from extrapolation of a few point measurements. If, over the span of several years, for example, the total, time-averaged sediment flux, including the dissolved load, out of a catchment could be measured, this flux would provide a good estimate of average, short-term denudation rates in many basins. Similarly, if the volume and age of sediment stored in a basin are known, if it can be shown that significant volumes of sediment did not bypass the basin *en route* to another basin, and if the tributary catchments throughout deposition can be reliably reconstructed, then a long-term mean erosion rate for the tributary catchments can be determined (Métivier et al., 1999). Measurement of both modern fluxes and stored sediment volumes are useful, because they commonly encompass different temporal intervals.

Sedimentary fluxes in rivers

The sediment load of a river can be partitioned into the bedload, suspended load, and dissolved load. If the average contribution of each component over time is known, then mean denudation rates can be calculated. Whereas the highest bedload and suspended-load fluxes typically occur during high water discharge, the highest solute concentrations typically occur during low flows

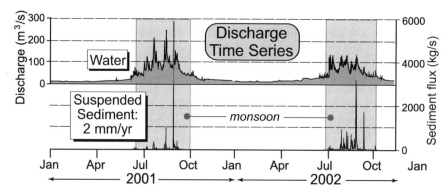

Fig. 7.5 Discharge time series under a monsoonal rainfall regime.
Water discharge during the summer Indian monsoon dominates the annual flow in the Nepalese Himalaya. Storms a few days in length create peaks in the monsoonal discharge. High suspended sediment discharge is typically related to landsliding events that create very peaked sediment fluxes and account for >95% of the total sediment discharge. Integration of the suspended sediment load defines a minimum erosion rate of ~2 mm/yr across the catchment. Modified after Gabet *et al.* (2008).

and can have a strong seasonal dependence. Input of sediment from adjacent hillsides can vary in predictable ways in different climatic regimes. For example, most landslides occur under saturated conditions. In many areas, saturation requires sustained rainfall prior to the triggering storms. In some locales, such as Taiwan, the intensity of typhoon precipitation is often sufficient to trigger landslides without significant prior storms. In parts of the Himalaya, not only is there a daily rainfall threshold, but nearly a meter of rainfall is needed in the prior months before any significant landsliding occurs (Gabet *et al.*, 2004b). Where landslides drive the large-scale sediment flux, sediment discharges tend to be spiky and can be only weakly tied to water discharge (Fig. 7.5). For rivers that are "underloaded" with respect to their transport capacity, sediment discharges may peak early in a storm cycle as material in the bed is mobilized, and then, despite continuing high water discharge levels, diminish as all available sediment has been exported. Similarly, if most sediment input to rivers is landslide-generated, the stochastic nature of landsliding can cause a mismatch between water and sediment discharges (Fuller *et al.*, 2003). Clearly, because of such variations, it is necessary to average over a sufficiently long time to be able to characterize the sediment flux at different river stages and at different times of the year.

Reliable measurement of the contributions from the dissolved load, suspended load, and bedload is only rarely achieved. To estimate suspended loads accurately, both the vertical and horizontal sediment concentration profiles, as well as the spatial distribution of water velocities, have to be either calculated or measured. Bedload sampling almost always consists of isolated measurements at a point in a channel. These measurements then have to be extrapolated to the entire stream bed and over the seasonal or flood cycle. Such extrapolations introduce considerable uncertainty. Although the flux of bedload material is generally greatest during flood stage, bedload sampling during flooding is commonly difficult, if not impossible. All of these complexities contribute to the fact that the true bedload flux is poorly known in most rivers. Bedload is commonly considered to be <10% of the suspended load, but in steep, heavily loaded rivers, bedload can exceed 35% (Pratt-Sitaula *et al.*, 2007).

Despite these difficulties, fluvial sediment fluxes, primarily suspended load only, have been estimated for many rivers near their entry into the ocean (Milliman and Meade, 1983; Milliman and Syvitski, 1992). When the major rivers of the world are considered, those draining the Himalaya and the Tibetan Plateau contribute a disproportionate amount of the sediment discharge to the world's oceans (Table 7.1).

Table 7.1   Sediment fluxes of major rivers of the world, showing catchment areas, sediment load, sediment yield, erosion rates, and runoff for world's rivers with catchments >200000 km^2. Modified after Milliman and Syvitski (1992).

River	Area (×10^6 km^2)	Load (×10^6 t/yr)	Yield (t/km^2 yr)	Erosion rate (mm/yr)	Runoff (mm/yr)
(A) High mountain (>3000 m)					
Magdalena (Colombia)	0.24	220	920	0.341	990
Irrawaddy (Burma)	0.43	260	620	0.230	995
Brahmaputra (Bangladesh)	0.61	540	890	0.330	
Colorado (USA)	0.63	120	190	0.070	32
Indus (Pakistan)	0.97	250	260	0.096	245
Ganges (Bangladesh)	0.98	520	530	0.196	
Orinoco (Venezuela)	0.99	150	150	0.056	1100
Yangtze (China)	1.9	480	250	0.093	460
Parana (Argentina)	2.6	79	30	0.011	165
Mississippi (USA)	3.3	400	120	0.044	150
Amazon (Brazil)	6.1	1200	190	0.070	100
(B) Mountain (1000–3000 m), South Asia/Oceania					
Krishna (India)	0.25	64	260	0.096	140
Godavari (India)	0.31	170	550	0.204	270
Pearl (China)	0.44	69	160	0.059	690
Huanghe (China)	0.77	1100	1400	0.519	77
Mekong (Vietnam)	0.79	160	200	0.074	590
(C) Mountain (1000–3000 m), N/S America, Africa, Alpine Europe, etc.					
Fraser (Canada)	0.22	20	91	0.034	510
Columbia (USA)	0.67	15	22	0.008	375
Limpopo (Mozambique)	0.41	33	80	0.030	13
Rio Grande (USA)	0.67	20	>30	>0.011	
Danube (Romania)	0.81	67	83	0.031	250
Yukon (USA)	0.84	60	71	0.026	230
Orange (South Africa)	0.89	89	100	0.037	100
Tigris–Euphrates (Iraq)	1.05	>53(?)	>52(?)	>0.019	45
Murray (Australia)	1.06	30	29	0.011	21
Zambesi (Mozambique)	1.4	48	35	0.013	390
MacKenzie (Canada)	1.8	42	23	0.009	170
Amur (USSR)	1.8	52	28	0.010	180
Nile (Egypt)	3.0	120	40	0.015	30
Zaire (Zaire)	3.8	43	11	0.004	340
(D) Upland (500–1000 m)					
Vistula (Poland)	0.20	2.5	13	0.005	165
Uruguay (Uruguay)	0.24	11(?)	45(?)	0.017(?)	
Pechora (USSR)	0.25	6.1	25	0.009	415
Hai (China)	0.26	14	55	0.020	
Indagirka (USSR)	0.36	14	39	0.014	150
Volta (Ghana)	0.40	19	48	0.018	91
Don (Ukraine)	0.42	0.77	18	0.007	
Sao Francisco (Brazil)	0.63	6	10	0.004	
Niger (Nigeria)	1.2	40	33	0.012	116
Volga (Russia/Ukraine)	1.4	19	15	0.006	400
Ob (USSR)	2.5	16	6	0.002	130
Lena (Russia)	2.5	12	5	0.002	205
Yenisei (Russia)	2.6	13	5	0.002	220

(continued)

Table 7.1 *(cont'd)*

River	Area (×10⁶ km²)	Load (×10⁶ t/yr)	Yield (t/km² yr)	Erosion rate (mm/yr)	Runoff (mm/yr)
(E) Lowland (100–500 m)					
Yana (USSR)	0.22	3	14	0.005	130
Senegal (Senegal)	0.27	1.9	8	0.003	48
Sevemay Dvina (USSR)	0.35	4.5	13	0.005	330
Dnieper (USSR)	0.38	2.1	5.2	0.002	86
Kolyma (USSR)	0.64	6	9	0.003	140
Sao Francisco (Brazil)	0.64	6	9	0.003	150
St. Lawrence (Canada)	1.1	4	4	0.001	435

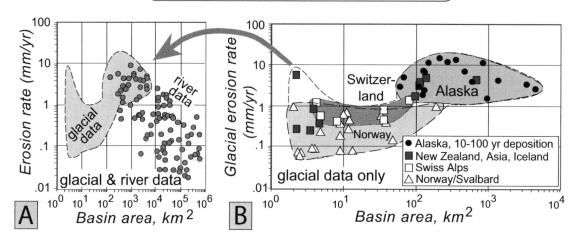

Fig. 7.6   Comparisons of glacial and fluvial erosion rates for basins of different sizes.
A. Fluvial erosion rates (gray dots) from catchments spanning four orders of magnitude in area. Note that rates generally decrease with larger basin size, but show 10- to 100-fold variation for any given basin size. These rates are approximately equivalent to the glacial erosion rates (enclosed dashed field). B. Glacial erosion rates. Alaskan tide-water glacier rates are reduced about three-fold to account for the apparently exceptional rates during rapid glacial recession in the 20th century. Modified after Hallet *et al.* (1996) and Koppes and Hallet (2006).

There, the combination of exceptional topographic relief, ongoing bedrock uplift, highland glaciation, and monsoonal precipitation along the margins of the uplift lead to rapid denudation. Based on the suspended fluxes alone, a mean rate of erosion of 0.3 mm/yr over a region of about 2.6×10^6 km² can be calculated from these data. Moreover, about 20–40% of the drainage basins of the Indus and Ganges are represented by foreland basins, and these have been net sediment sinks during Cenozoic times. If this were still true over the period of measurement, then the mean erosion rate in the hinterland (the tectonically active mountains themselves) would increase to about 0.4–0.5 mm/yr. When compared with mean erosion rates of approximately 0.04 and 0.07 mm/yr for the Mississippi and Amazon Rivers, respectively, the magnitude of modern Himalayan erosion can be appreciated.

Fluvial erosion rates generally decrease with increasing catchment size (Milliman and Meade, 1983; Milliman and Syvitski, 1992), although, at any given catchment size, the observed erosion varies about 10-fold (Fig. 7.6). Even in small

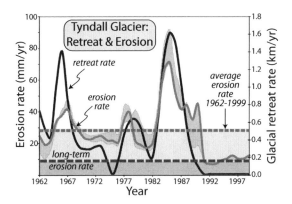

Fig. 7.7   Comparisons of late 20th-century glacial erosion rates and glacier retreat rates.
Rates of ice-margin retreat and glacial erosion (as inferred from sediment volumes) are strongly correlated for the Tyndall tide-water glacier since 1960. Because the glacier is out of equilibrium (the glacier is steeper and ice-flow rates are faster during retreat), the well-constrained erosion rate (~28 mm/yr) is more than three times greater than the estimated long-term equilibrium rate (~9 mm/yr), which is deduced to be equal to the erosion rate during stillstands (1992–99). Modified after Koppes and Hallet (2006).

catchments, the maximum rates of fluvial erosion based on measured sediment loads at the river mouth rarely exceed 5 mm/yr. Some glacial erosion rates have been measured using stored sediment volumes, for which good time control exists. Studies in glaciated basins of sediment fluxes over the past 10–100 years indicate mean, basin-averaged denudation rates that range as high as 20 mm/yr in front of tide-water glaciers in Alaska (Hallet *et al.*, 1996). These high rates have often been cited as indicative of the power and efficiency of glacial erosion. Over the past several decades, however, the Alaskan tide-water glaciers have undergone massive retreat that has caused thinning, steepening, and acceleration of ice-flow rates. Perhaps not surprisingly, sediment delivery and erosion rates correlate with the ice-margin retreat rate (Fig. 7.7). Times of rapid advance or retreat are, therefore, unlikely to reflect steady-state glacial erosion. As a consequence, the mean rate of erosion during retreat is estimated to be three to four times greater than the long-term rate that characterizes glaciers in equilibrium (Koppes

and Hallet, 2006). Despite the resulting downward adjustment of glacial erosion rates, these rates appear to be at least as rapid as the most rapid fluvial erosion rates (Fig. 7.6).

Rather than trying to measure sediment fluxes in a river at high stage, a better estimate of total solid sediment discharge can sometimes be obtained from natural or man-made reservoirs, or from closed depositional basins. Such basins are typically efficient traps of fluvially transported sediment, and they store sediment transported throughout the year. If the shape of the valley prior to damming or the geometry and age of any other stratigraphic horizon in the fill is known, and if the density of the sediment and the volume of the fill can be calculated, then mean rates of sediment delivery and of erosion in the catchment are readily derived. Unfortunately, few of the world's major rivers debouch into reservoirs. Some, like the Columbia River in Washington, are interrupted by numerous low dams. During times of high discharge, however, sediment is flushed out of the reservoirs behind each dam. Of the five large rivers draining the Himalaya and southern Tibet, only the Indus River enters a major reservoir, the Tarbela Reservoir. Although the Tarbela Reservoir was estimated to have a usable lifetime of about 70 years, since the dam's completion in 1974, the toe of the delta has already prograded over 80 km and now abuts the dam. The mismatch between the estimated and the likely usable lifetime of the reservoir demonstrates how poorly the sediment fluxes along this river were known prior to construction, despite the importance of such data to the viability of the project. The rate of sediment infill into the Tarbela Reservoir requires an average erosion rate of 0.4 mm/yr over the upstream catchment, a rate four times higher than that over the entire Indus drainage (Table 7.1). Put another way, the sediment flux into the reservoir reveals that 20% of the Indus catchment area (this mountainous region) accounts for 80% of the eroded material that is measured near its mouth (Milliman and Syvitski, 1992).

Sometimes human intervention provides an opportunity to measure erosion rates quite reliably. For example, along the southern flank of the San Gabriel Mountains of California, many river valleys debouch on to densely populated areas of the Los Angeles basin. Because much of the historical damage from floods has been caused by large debris flows, Los Angeles County engineers have built debris basins at the mouth of each of the canyons. These basins are efficient sediment traps and have been reasonably effective in preventing widespread damage by debris flows to the expensive homes built on the alluvial fans below the mouths of the canyons. When the debris basins approach capacity during flooding, the deposits are excavated and removed by trucks. The amount of debris removed provides an effective measure of the denudation in the catchment (Scott and Williams, 1978) and is recorded in county registers in units of truckloads! Such records provide a rich data set with which to examine the effects of fire, storms, human interference, and revegetation on sediment fluxes and erosion rates (Lavé and Burbank, 2004).

Even when sediment volumes, their ages, and the catchment source area are very well known, however, calculations of meaningful erosion rates can be difficult (Church and Slaymaker, 1989). The data from the tide-water glaciers illustrate some of the pitfalls and challenges of reconstructing erosion rates. Almost all components for an erosion-rate calculation were well constrained: sediment volumes from reliable isopachs in the fiord at the glacier's toe, good time control on a 40-year-long record, and a well-defined source area. Yet, the resulting average erosion rates (Fig. 7.7) appear to greatly overestimate the likely long-term rates (Koppes and Hallet, 2006). Because the glaciers have been strongly out of equilibrium in the past century, their sediment fluxes are probably atypical. It is important to recognize that these caveats do not invalidate the calculated erosion rates: those rates are extremely informative for retreating tide-water glaciers, but they probably should not be interpreted as characteristic of equilibrium glacial erosion.

Rates of erosion based on structural and stratigraphic controls

When it is possible to delineate both the duration of erosion, as well as the volume of rock that was removed during the growth of a fold or slip on a fault, long-term erosion rates can be calculated. In order to do this, several conditions must be met. Both the pre- and the post-deformational geometry of the displaced mass must be known, such that the volume of eroded material can be calculated. The ages of key overlapping and cross-cutting relationships between structures and strata must be known, so that the initiation and termination of deformation and erosion can be determined.

Whereas such volumetric calculations are simple in concept, they can be reliably applied only in unusually well-documented situations, and, even then, they may require a set of commonly untested assumptions. One potentially straightforward geometry to consider is the erosion of the hanging wall of a thrust sheet. The south-vergent Salt Range Thrust in the Himalayan foreland fold-and-thrust belt of Pakistan provides a clear example for application of this methodology (Burbank and Beck, 1991). The hanging wall of this thrust comprises Eocambrian evaporites of variable thickness, a Paleozoic-to-Eocene carbonate and clastic succession about 1 km thick, and a 2.5–3 km thickness of Miocene and younger fluvial strata of the Himalayan foreland basin, the Siwalik molasse (Fig. 7.8). Prior to deformation, the molasse strata formed a predictable wedge-shaped geometry that gradually thinned toward the south. The regional taper and thickness of the foreland strata can be determined from the minimally eroded foreland strata abutting the hinterland edge of the uplifted hanging wall. The known initial thickness and taper are keys to determining how much erosion occurred subsequently. During thrusting, the hanging wall was raised up a 1-km-high footwall ramp and was translated about 20 km to the south (Baker et al., 1988). Stratigraphic evidence indicates that translation took place in two stages: between 6.3 and 5.8 Ma, about 5 km of shortening occurred (Fig. 7.8); and the remainder occurred since 1.5 Ma. During the first stage of thrusting, erosion of the fluvial strata from the leading edge of the hanging wall was great enough to expose Permian rocks at the thrust tip, as evidenced by the presence of distinctive lithologies within the 6 Myr old foreland strata.

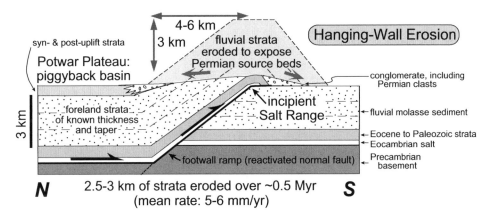

Fig. 7.8   Erosion rates from uplifted hanging-wall strata.
Calculations of erosion rates based on combined structural and stratigraphic data in the Salt Range, northern Pakistan. Three ingredients to calculate rates are available here: thickness and shape of the uplifted hanging wall; timing of initiation of fault slip; and timing of the first appearance of clasts eroded from the Permian rocks beneath the foreland strata. These data define erosion rates that essentially balance rock uplift rates, a common occurrence when poorly lithified rocks are raised above base level. Modified after Burbank and Beck (1991).

This unroofing requires about 3 km of erosion during 0.5 Myr, a rate of approximately 6 mm/yr or 6 km/Myr!

Such rapid erosion rates are not atypical on a global basis for sites where Cenozoic terrestrial foreland strata are uplifted above local base level (Burbank *et al.*, 1999; Dadson *et al.*, 2003). In general, erosion rates of these weakly cemented strata appear to be closely equivalent to rock uplift rates, such that little topographic relief develops within the uplifted foreland strata (Plate 1D). Consider, for example, the Himalayan foreland in Nepal where shortening has exceeded 10 km and rock uplift rates exceed 10 mm/yr (Lavé and Avouac, 2000). Despite these very rapid rates, relief in the hanging wall of active foreland thrust faults (the Main Frontal Thrust) is typically just a few hundred meters, less than 5% of the total rock uplift. Hence, rates of erosion and rock uplift are nearly in balance. In such sites, if the fault geometry and the slip rate along it can be determined and if hanging-wall relief is small, rates of erosion can be readily deduced.

Several challenges confront this type of geometric analysis. In order to reconstruct the eroded mass, its pre-deformational geometry must be assumed or reconstructed. In regions where predictable variations in stratal thicknesses occur, such as in many foreland basins

or passive margins, reconstructions can be done with considerable confidence. Given dated strata that either are cut by or overlie the fault that accommodated the displacement, the duration of deformation can be defined. Commonly, however, one does not know how closely the dates limit the deformation; at numerous sites, either the beginning or end of deformation, but more rarely both, can be defined. In many terrestrial sequences, moreover, it is very difficult to determine accurately the relevant stratigraphic ages. In these sorts of studies, magnetic polarity stratigraphy has provided a powerful tool for creating precise ages for Cenozoic terrestrial strata and in calculating rates of deformational and erosional processes (Burbank and Raynolds, 1984; Fang *et al.*, 2005; Jordan *et al.*, 1988).

Topographically constrained erosion rates

If the geometry and age of a former land surface can be reconstructed, a straightforward differencing with the modern land surface yields a mean erosion rate. For example, individual growing folds experience dissection as they are elevated above local base level. Typically, dissection is focused along river valleys and concavities, such that segments of the pre-deformational land surface may be preserved in a folded, but

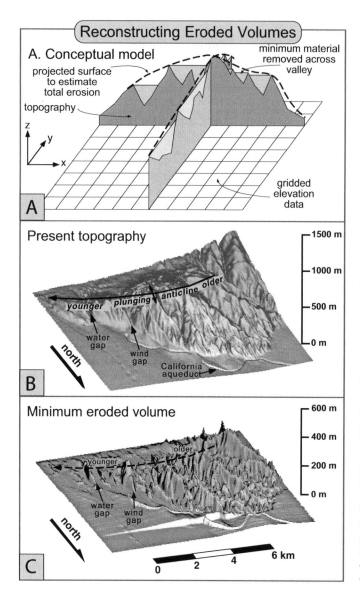

Fig. 7.9   Calculation of volumetric erosion using a digital elevation model.
A. Strategy for calculating minimum eroded volume with orthogonal traverses of a digital elevation model. Subtraction of the present topography from the filled topography yields a minimum volume of eroded material. The actual total mass loss may be better approximated by connecting the high points, but requires a more subjective interpretation. B. Perspective view of the folded and dissected Wheeler Ridge anticline in southern California. Note that the fold plunges to the east where it merges into the plain and that it is much more dissected in its western region. C. Perspective view of the minimum volume of material eroded from the Wheeler Ridge anticline. The material removed from the wind and water gaps is well depicted, but the most erosion has clearly occurred in the western part of the fold. Modified after Brozović *et al.* (1995).

undissected, state, especially in the early stages of fold growth. If the geometry of the undissected surface of the fold can be reconstructed, the modern dissected topography can be subtracted from it in order to define both local and average rates of erosion. The availability of digital elevation models can make such calculations quite straightforward. Minimum eroded volumes can be objectively calculated simply by filling each dissected part of the landscape to the height of the lowest topographic boundary surrounding it (Brozović *et al.*, 1995) and then subtracting the present topography (Fig. 7.9). Alternatively, a reconstructed surface that represents the hypothesized topography in the absence of erosion can be created, and the modern topography (which has experienced erosion) can be subtracted from that surface to estimate spatially variable amounts of erosion (Hilley and Arrowsmith, 2008; Small and Anderson, 1998).

On a considerably larger scale, former planation or low-relief surfaces have been identified in several mountain ranges. These surfaces may be defined by concordant summits, some

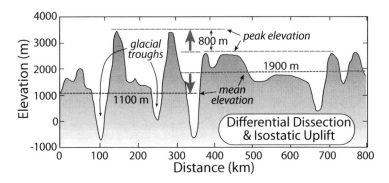

Fig. 7.10 Differential dissection and isostatic uplift of the Transantarctic Mountains.
The summits between 100 and 800 km approximate the position of a pre-rifting erosion surface. Deep glacial incision has lowered the mean elevation about 800 m more between 0 and 350 km than between 350 and 800 km. In turn, the peaks are about 800 m higher where incision is greater. Because 800 m of decrease in mean elevation would be expected to produce only 650 m of peak uplift, the extra 150 m of uplift may be due to some tectonic contribution or to flexure due to three-dimensional differences in erosion. Because the timing of erosion is only loosely constrained as post-Cretaceous, rates of erosion or peak uplift cannot be well defined. Modified after Stern *et al.* (2005).

of which are nearly flat-topped, or by largely undissected, but isolated surfaces (Fig. 7.10). If the former continuity of such surfaces can be demonstrated and the age of the surface can be ascertained, an extensive datum can be defined. Subtraction of the modern topography from this datum can define the volume and mean rate of erosion since the surface began to be incised. Given a knowledge of the eroded mass of rock, the isostatically driven response to its removal can be calculated, and it is therefore possible to determine to what extent the undissected peaks and surfaces would be raised in response to an overall lowering of the mean elevation due to erosion. The uplifted rift flank of the Transantarctic Mountains provides an illustrative example of these concepts (Fig. 7.10). Whereas the high peaks of these mountains are in a perennially frozen zone that experiences little erosion, glaciers in the intervening valleys have created deep troughs and have generated 4 km of relief (Stern *et al.*, 2005). The summits are higher where both incision has been greater and the mean elevation is lower: observations consistent with erosionally driven rock uplift.

Erosion rates based on mixing models

Conservative tracers in fluvial systems can be exploited to deduce spatial variations in erosion rates. A conservative tracer is defined as a measurable quantity that does not change its value due to transport alone, but can change in response to additions or subtractions to the load or discharge of a river. Some examples of generally conservative tracers include isotopic ratios for non-reactive elements, age distributions of detrital grains, and zircon concentrations. These examples are not necessarily truly conservative tracers, because they can change downstream due to chemical reactions or abrasion, even in the absence of any new additions. Such changes, however, are typically slow in comparison to transport times and can, therefore, usually be neglected. For example, the isotopic signature that is contained in water samples or in transported sediments changes as a function of variations in both isotopic ratios and erosion rates as successive tributaries join the trunk stream (Fig. 7.11).

Conservative tracers that can be tied to distinctive source areas can be used to define relative rates of erosion. Consider, for example, a mountain range in which the high peaks near the drainage divide have tracer characteristics (such as cooling ages or isotopic ratios) that are distinct from those in the surrounding terrain. Near the headwaters of the rivers and adjacent to these high peaks, one would expect that the tracer in water or sediment samples would

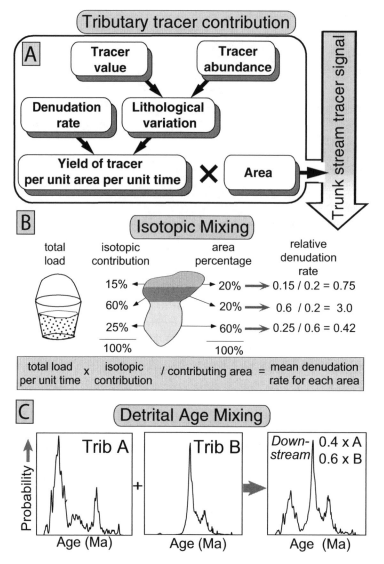

Fig. 7.11   Use of tracers to determine relative and absolute erosion rates.
A. The contribution of a tracer from a tributary catchment depends on lithologic variability, denudation rates, and catchment size. B. Unmixing of the downstream signal derived from source areas with discrete isotopic signatures can reveal relative erosion rates. If the actual erosion rate is known for any of the source areas, then a rate for each source area can be estimated from the isotopic contribution. C. Mixing models for populations of detrital ages from two catchments. The downstream probability distribution of detrital ages (right) can be deconvolved to determine the relative contribution from two tributary catchments, if the detrital age distribution from each catchment is known. Here, the downstream catchment records 40% and 60% of its ages from tributaries A and B, respectively. Modified after Brewer *et al.* (2005) and Amidon *et al.* (2005).

reflect the nearby lithologic characteristics. Now imagine measuring water or sediment in the stream at a point many kilometers downstream from the drainage divide, where the zone containing the high peaks constitutes only a small fraction of the upstream catchment. If the tracer value in the water or sediment in this area is still dominated by contributions from the high peaks, it would indicate that these peaks are the sites of the most rapid denudation.

If you know only the isotopic contributions and the area of each isotopically distinct source area, you can estimate the relative denudation rates between different source areas (Fig. 7.11). If you also know the total sediment load in either the trunk stream or from one of the source areas, then you can also calculate absolute erosion rates for each contributing area (Fig. 7.11). Hence, a simple mass balance based on tracer abundances and source-area ratios can provide a powerful tool for estimating modern erosion rates and their basin-to-basin variation. Keep in mind, however, that any calculated erosion rate is relevant only for the phenomenon being measured. For example, if isotopic compositions of bedload are measured, then one can calculate the rate of mechanical erosion in the source area that produces the observed bedload. If one considers a carbonate terrane versus a granitic one, it is easy to imagine that the relative contributions of these rocks to the dissolved load and the detrital load will vary greatly between these areas, irrespective of the relative erosion rate. In the northwest Himalaya, for example, calcite accounts for less than 1% of the outcrop area, but, because it dissolves much more readily than do the prevalent aluminosilicates, it dominates the Sr isotopic composition of the surface waters (Blum *et al.*, 1998).

Whenever an attempt is made to infer long-term conditions from short-term measurements of conservative tracers, one must ask whether present conditions are representative of the long-term average. Certainly, one year's sediment flux should not be considered to typify even decadal rates. Comparisons of detrital ages between different years or in different grain-size populations commonly display significant year-to-year variation (Ruhl and Hodges, 2005). Typically, the impact of unusual or catastrophic events on tracer distributions is unknown. For example, in several places in the Nepal Himalaya, massive rockfalls and landslides have originated from the slopes of high peaks within the Higher Himalaya (Fort, 1987). The run-out of some of these landslides carried them more than 30 km into the Lesser Himalaya (Yamanaka and Iwata, 1982). Today, the Lesser Himalayan rivers have incised into the landslide debris, leaving behind steep risers sweeping up to the former upper surface of the deposit. As these rivers continue to cut into the bases of these landslide deposits, debris collapses into the river and is washed away. Such landslide deposits can distort the tracer signal in at least two ways. First, because landslide debris is banked against the adjacent slopes, it creates a barrier or buffer that inhibits entry into the river of locally derived bedrock material with its Lesser Himalayan signature. Second, the poorly consolidated Higher Himalayan debris is being actively input into the river due to erosion of the steep risers. In this circumstance, within the Lesser Himalaya, Higher Himalayan contributions to the tracer composition would be expected to be over-represented and inferences based on these tracer studies would be biased and perhaps misleading.

Regolith production rates

In soil-mantled landscapes and in the absence of surface processes that remove unweathered bedrock, the rate of bedrock lowering or erosion is primarily a function of the rate at which bedrock is converted into transportable material (regolith) by chemical and mechanical processes. In the past, these rates have typically been poorly known and difficult to measure. Nonetheless, if the average regolith production rate across a landscape can be determined and if the landscape is in an approximate steady state, then long-term rates of bedrock lowering can be estimated. With the advent of cosmogenic radionuclide techniques, rates of conversion of bedrock to soil can now be assessed more reliably (see the primer on cosmogenic nuclides in Box 3.1).

On a bare bedrock knob that is steadily eroding, the concentration of cosmogenic nuclides [CRN] is dependent primarily on the rate of bedrock lowering (E) (Fig. 7.12A):

$$[\text{CRN}] = \frac{P_0 z^*}{E} \qquad (7.1)$$

where P_0 is the production rate at the surface and z^* is the rock depth (commonly ~60 cm) at which the production rate drops to $1/e$ of P_0, and any decay of the isotopes is ignored. Results

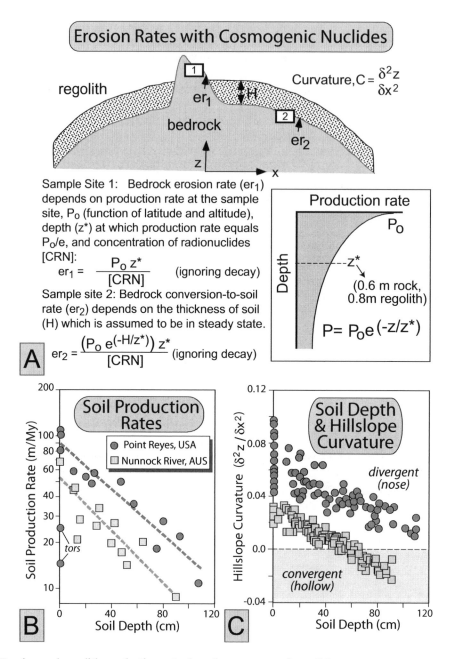

Fig. 7.12 Erosion and regolith production rates based on cosmogenic nuclides.
A. On bedrock knobs (site 1), surface concentrations of cosmogenic radionuclides produced *in situ* yield bare-bedrock long-term erosion rate estimates, whereas those at the bedrock–regolith interface (site 2) yield rock–regolith conversion rates beneath a regolith cover with an assumed steady-state thickness, *H*. If the entire system is in steady state, these rates should agree with each other. The curvature of the hillcrest (*C*) can be used in combination with the regolith production rate, er_2, to yield an estimate of the landscape diffusivity, $\kappa\ (=er_2/C)$, which relates regolith flux to local slope. Inset depicts change in production with depth and defines z^*, where production is 1/e times the surface production rate. B. Soil production rates as a function of soil thickness, showing that rates are fastest when soils are thinnest. C. Soil depth as a function of slope curvature, indicating that soils are thinnest where curvature is highest. Modified after Heimsath *et al.* (2005).

from the employment of this technique have been summarized by Bierman (1994). Studies in alpine areas ranging from the Rockies to Australia have shown that rates of the order of 5–15 μm/yr (5–15 m/Myr) are very common on exposed bedrock surfaces (Small *et al.*, 1999; Quigley *et al.*, 2007).

In regolith-mantled landscapes, regolith shields the bedrock from bombardment by cosmic rays, such that nuclide production rates are lower at the bedrock interface. If the landscape is in a steady state, so that the thickness of regolith (*H*) on a site is constant through time, then the concentrations of nuclides at the regolith–bedrock interface can similarly be used to define the rock-to-regolith conversion rate (Fig. 7.12B):

$$[CRN] = \frac{(P_0\, e^{-H/z^*})z^*}{E} \qquad (7.2)$$

The rates so obtained will be average rates over the time it takes to convert about 0.6 m of rock to regolith under steady-state conditions. This conversion rate is important because landscape denudation by soil-related processes (creep, rain splash, solifluction, shallow landslides) cannot occur faster than the rate at which regolith is produced. Until present, the most rapid known rates of regolith production are ~0.3 mm/yr (Heimsath, 1999). Wherever landscape rates exceed soil production rates, other processes such as glacial erosion or bedrock landslides have to contribute significantly to overall denudation.

At steady state, the rate of regolith production (Heimsath *et al.*, 1997) has been argued to vary as a function of regolith thickness (Fig. 7.12B). The thickness of regolith, in turn, varies as a function of slope ($\delta z/\delta x$), such that thinner soils prevail on steeper slopes. In some relatively slowly eroding landscapes, the thickness variability is a predictable function of the downslope flux of regolith, which can itself be related to the curvature ($C = \delta^2 z/\delta x^2$) of the slope (Fig. 7.12A and C) through the landscape diffusivity constant (κ). In the resultant steady-state diffusive hillslope model, downslope rates of soil movement are fastest where positive curvature is greatest. High-curvature sites are also where soils are thinnest and soil production rates are the highest.

Bedrock incision rates

Rivers, along with glaciers, are the most important geomorphic agents for "setting" the local base level. Incision by rivers determines adjacent hillslope gradients, and the success of a river in removing debris supplied from adjacent slopes influences its ability to incise underlying bedrock. If more material comes off the slopes than the river can transport, it will not incise the underlying bedrock or lower the local base level, and adjacent slope angles may tend to decrease (Fig. 7.1). Thus, a suite of self-regulating feedbacks can develop in which the long-term rate of base-level lowering by river erosion, the sediment transporting power of the river, and the rate of sediment supply from adjacent slopes are in rough equilibrium.

Bedrock incision rates can be calculated directly whenever the geometry and age of a former land surface can be reconstructed. Because of their predictable longitudinal and cross-sectional geometries, river terraces often provide reliable datums for such calculations. One has simply to measure the elevation difference from the terrace tread to the present river level. Owing to their unconsolidated nature, alluvial river terraces are highly susceptible to erosion. Consequently, the history of cut and fill represented by alluvial terraces reveals repeated crossings of the threshold of critical power (Bull, 1991), but tells little about long-term rates of regional denudation or bedrock incision (Box 7.2). In contrast, fluvial terraces etched into bedrock (strath terraces) record previous positions of an actively incising river (Fig. 7.13). Using the height of the strath above the modern stream, if it is assumed that the strath resulted from erosion at or near the low point of the river's bedrock channel, and if the elapsed time since the abandonment of the strath by the incising river is known, then long-term rates of bedrock incision can be calculated (Fig. 7.13).

A recently proposed, alternative model for strath formation (Hancock and Anderson, 2002) posits that straths are created when the bedrock channel is covered with alluvium. This "sediment loading model" (Figs. 2.12 and 7.13) suggests that the alluvium prohibits vertical incision, but provides abundant tools (clasts) that can help the channel etch laterally into the bedrock valley walls. This model is consistent with observations in modern bedrock channels where bedload covers the thalweg during large floods while active lateral erosion occurs on the channel margins several meters above the thalweg (Turowski *et al.*, 2008). Thus, when an aggrading fill achieves a new equilibrium profile, new straths would form on the edges of the fill. In such a scenario, the height of a strath above the modern river channel reveals little about rates of vertical bedrock incision by the channel.

Box 7.2   What are river incision rates measuring?

Changes due to incision in the elevation of a given reach of a river can modify channel gradients and the stability of adjacent hillslopes. For landscapes in quasi-steady state, the rate of river incision is commonly argued to regulate the overall rate of erosion. Hence, many studies focus on defining river incision rates as interpreted from the age and height of reconstructed river channels in the past. Although obtaining reliable ages is commonly a formidable obstacle, a reliable interpretation also depends on understanding the context of the incision history. Several key questions must be answered. Was the incision into bedrock or into alluvial fill? Was the river profile locally perturbed at the study site or do the rates typify adjacent reaches? Is the interval over which incision has been measured long enough to yield a representative long-term rate?

We know, for example, that climate changes commonly drive intervals of aggradation and degradation. Incision of alluvial fill is far easier than incision into bedrock. In most circumstances, therefore, alluvial incision rates are unlikely to place useful limits on bedrock incision or landscape erosion rates. As long as the bedrock is covered with sediment, no bedrock incision can occur (Lavé and Avouac, 2001).

Large landslides can dam big river gorges and cause bedrock incision to cease until the landslide dam is removed. Whereas catastrophic dam breaks can occur shortly after water overtops the dam, more commonly these dams are slowly eroded. Although such dams are sedimentary deposits, they typically comprise large bedrock blocks that erode much more slowly than does river alluvium. As a consequence, such landslides retard the rate of long-term incision even more than do alluvial fills (Ouimet *et al.*, 2007).

Obvious local perturbations of a river's gradient occur when rivers cross growing folds or active faults. A less commonly recognized perturbation can occur when an *epigenetic* gorge has formed. Such gorges are created following episodes of alluviation, landsliding, or glacial advance when a river is subsequently "let down" on to bedrock on the margins of its former valley (see figure A). Deep, narrow, slot canyons can be incised into the underlying bedrock. In comparison to a reach flowing wholly on bedrock, the epigenetic bedrock reach commonly is steepened because the downstream alluvial fill is more readily removed. As a consequence, such steepened bedrock reaches may incise considerably more rapidly. Wherever a bedrock river channel is seen to narrow abruptly in the absence of obvious lithologic or structural controls, the possibility of an epigenetic origin should be assessed.

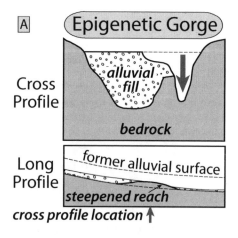

A. Epigenetic gorge in cross-section and profile.

As shown in figure B, alluvial incision rates, instantaneous bedrock incision rates, and long-term rates can exhibit striking contrasts. Rates of alluvial or instantaneous bedrock incision are likely to significantly overestimate long-term rates. Knowledge of each of these rates is useful, because they serve to define the response of the entire fluvial system. But care should be taken to distinguish among them, to define the applicable time frame of incision, and to evaluate what factors influence the observed incision.

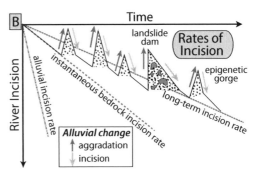

B. Contrasting rates of river incision. Modified after Lavé and Avouac (2001).

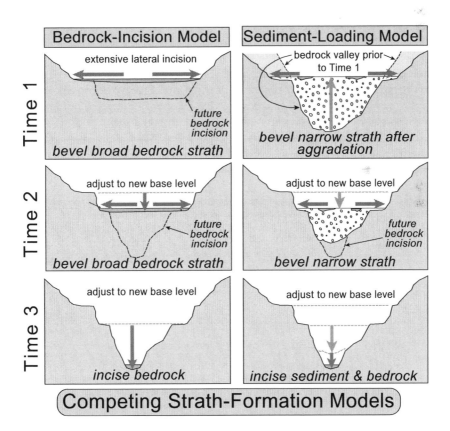

Fig. 7.13   Competing models for the formation of bedrock straths by river incision.
The "bedrock incision model" (left side) is the traditional model in which straths are formed at or near the base of the bedrock channel. When combined with the age of the strath, the amount of subsequent channel incision serves to define the rate of bedrock incision. The "sediment loading model" (right side) predicts that strath formation occurs after aggradation has covered a bedrock channel that was incised previously. The magnitude and timing of bedrock incision is difficult to deduce in this model.

One appealing aspect of this model from the surface-process perspective is that it does not demand that high straths were formed as broad surfaces spanning the entire valley width (Fig. 7.13). Instead, the straths can form as relatively narrow benches on the margins of the fill. Given the sediment loading model, is it still possible to use straths to deduce bedrock incision rates? Yes, but the magnitude of potential alluviation in proportion to strath heights has to be estimated, such that interpretations of bedrock incision histories become more nuanced.

Determining the age of an abandoned strath is typically a challenge to the field geologist. Commonly, a veneer of fluvial gravel mantles the strath surface and may contain organic matter that can be radiocarbon dated (Lavé and Avouac, 2001; Merritts *et al.*, 1994; Weldon, 1986). For straths lacking a gravel cover, calculation of the duration of exposure of the strath surface to cosmic radiation can provide an estimate of the time since abandonment of the strath by a downcutting river. In such cases, it must be assumed or demonstrated that the strath has remained uncovered by alluvial material, landslide debris, or persistent snow cover since abandonment, and that the strath surface itself has not significantly degraded. If the strath displays well-preserved original bedforms, such as potholes, flutes, and polished surfaces, that were created by abrasion in a river channel, then it can be argued that little degradation has occurred. Even so, the strath age might only reflect the time of a modest amount of erosion that occurred when an overlying alluvial fill was removed (Pratt *et al.*, 2002), because only 1–2 m of bedrock removal is required to reset the cosmogenic age to near zero.

Well-preserved straths with associated cosmogenic exposure ages along the Indus River in northern Pakistan have been interpreted to define late Quaternary river incision rates into metamorphic and igneous rocks (Fig. 7.14A and B). This region is argued to contain some of the most rapid bedrock uplift rates in the world (Zeitler *et al.*, 2001b). In the arid Indus environment, some straths are extremely well preserved despite their positions more than 100–200 m above the modern river. Along a 100-km reach of

the Indus River, estimated incision rates vary from about 1 mm/yr to greater than 8 mm/yr (Burbank *et al.*, 1996b). Interpretation of bedrock incision rates using the traditional bedrock-incision model, rather than the sediment-loading model (Fig. 7.13), is supported by three observations: the great height of these straths (some up to 400 m above modern flood levels), the lack of evidence for significant depths of fluvial aggradation, and the sequentially younger strath ages for progressively lower terraces in the most rapidly incising areas. Nonetheless, the very rapid erosion and steep hillslopes in this region would promote removal of any aggradational fill. Consequently, a conservative approach would attach a 20–40% uncertainty to the incision rates. If such high rates were sustained, they would lead to as much as 8 km of incision every million years. The spatial pattern of incision rates of the Indus Rive mimics the pattern of long-term erosion and rock-uplift rate estimates from apatite fission-track dating (Fig. 7.14C and D), suggesting an overall balance. Such near-equilibrium is perhaps not unexpected: if rock uplift and denudation were not in balance, and if rates of more than 5 mm/yr of rock uplift were sustained for very long, rivers that failed to incise rapidly enough to keep pace with the uplift would instead develop very steep profiles. A mismatch of as little as 20% would create 1 km of relief along the river's course in less than 1 Myr!

Rates of denudation by landslides

It has long been recognized that, in actively deforming areas, landsliding commonly provides an important mechanism for delivering material from hillslopes into valley bottoms occupied by rivers or glaciers. We have suggested previously that, wherever overall erosion rates exceed the local conversion rate of rock to regolith, bedrock landslides assume an increasingly important role (Box 7.1). Because landslides are sporadic events, however, calculating their time-averaged contribution to erosion can be difficult. Several different approaches could be used to assess erosion by landslides (Box 7.3). We know that landslides result from interactions between rock strength (cohesion c and the angle of

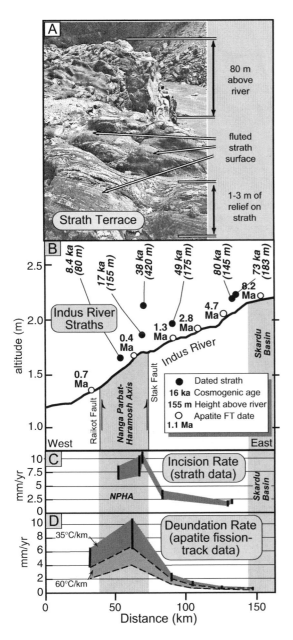

internal friction (ϕ), hillslope gradients (β), relief, pore pressure, and seismic accelerations. In a two-dimensional, steady-state, slope-stability model (Culmann, 1875), and in the absence of pore pressure or seismic shaking, the maximum hillslope height (H_c) can be expressed as

$$H_c = \frac{4c}{\gamma} \frac{\sin\beta\cos\phi}{[1 - \cos(\beta - \phi)]} \qquad (7.3)$$

where γ is the unit weight of the rocks of the hillslope. Recent studies (Schmidt and Montgomery, 1995) have shown that a hillslope's failure characteristics are typically determined by the strength of the weakest strata in a hillslope, rather than by the mean rock strength measured in a laboratory. Furthermore, the orientation of beds or planes of weakness, such as schistocity or foliation, exerts a strong control on overall hillslope strength, because dip slopes have a lower effective angle of internal friction than do anti-dip slopes. Although this formulation predicts the maximum hillslope height in a landscape and suggests that, in an equilibrium condition, there should be predictable limits to local topographic relief, it does not predict a rate of denudation by landsliding, because there is no time dependence in the formulation. Instead, it suggests that, under threshold conditions, peaks will be lowered by landsliding at the same rate that the valley bottoms are lowered, such that this relief is maintained. If the rate of valley lowering or river incision is known, then, in this equilibrium landscape, the time-integrated contribution of landsliding can also be predicted. The rates themselves would be dictated by the rates of valley incision, and the hillslopes basically come along for the ride.

Another approach to quantifying denudation by landsliding exploits a relationship between

Fig. 7.14 Bedrock incision along the Indus River, Pakistan.
A. Annotated photograph of 80-m-high strath above the Indus River near Haramosh, northern Pakistan. B. Strath terraces with their cosmogenic nuclide exposure ages and their heights above the modern river. At river level, apatite fission-track ages are shown. Note that the fission-track ages get progressively older toward the Skardu Basin in the east. C. Incision rates (based on strath ages) versus distance along the Indus. Note that the highest rates are associated with the steepest river gradient. D. Rates of denudation based on apatite fission-track data. Two calculations are made, each with a different geothermal gradient (35 and 60°C/km). Note that the zone of maximum long-term denudation coincides with the zone of most rapid river incision. Modified after Burbank *et al.* (1996b).

Box 7.3   Landslides in landscapes.

Whereas the role of landslides as major agents of erosion is unequivocal in many rapidly eroding mountain ranges, the controls on where landslides occur within the landscape is less well known. Given the critical role played by basal shear stress in triggering landslides, it is not surprising that the preponderance of slides occur on the steepest hillslopes.

The combination of digital topography with extensive GIS databases of landslides underpins new insights on the location of major landslides within a landscape and on differences in their hillslope positions as a function of (i) whether the slides were triggered by earthquakes or by storms (Densmore and Hovius, 2000) and (ii) the "facing direction" of the topography with respect to the position of the epicenter (Meunier *et al.*, 2008). Given a GIS database of landslides, flow lines can be calculated that pass through each landslide and represent the theoretical path of water flowing across the landscape from the ridge crest above the slide to the river channel below it. The position of each slide can then be expressed as a function of its proximity to the crest and channel.

Compilations of data from several active ranges reveal intriguing contrasts in the positions of landslides on hillslopes (see figure A). For example, landslides caused by the 1994 $M_w=6.7$ Northridge, California, earthquake cluster along the ridge crests, a result consistent with the amplification of seismic shaking along ridges. For the Finisterre Range, Papua New Guinea, the 1922 $M_w=6.7$ and the 1994 $M_w=6.9$ earthquakes on the Ramu–Markham Fault prompted landslides clustered near both the crests and the channels. In contrast, for the rapidly eroding Southern Alps, New

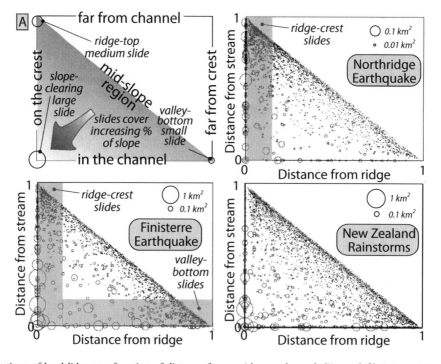

A. Distributions of landslides as a function of distance from a ridge or channel. (Upper left) Interpretational key for other panels. Note that slides spanning an increasing fraction of a hillslope plot closer to the origin. (Upper right) Northridge earthquake landslides occur mostly near ridge crests. (Lower left) Finisterre landslides occur both near crests and near channels. (Lower right) Landslides in the Southern Alps are uniformly distributed.

Zealand, storm-driven landslides were spread rather uniformly across the landscape.

Why does this variability in landslide location occur? Clustering of landslides near ridge crests during earthquakes is expected because of amplification of seismic shaking, which may exceed 10-fold along ridge crests (Buech et al., 2010). At Northridge, for example, mean slope angles are quite uniform throughout the range (see figure B), but landslides are predominant near the ridge crests. Whereas this same phenomenon is seen in the Finisterre Range, secondary landslide clusters occur fairly low on the hillslopes. Topographic data (see figure B) show that inner gorges and the steeper slopes that are associated with such gorges are prevalent in the Finisterre Range. Modeling of seismic waves interacting with topography suggests that the kinks in slopes associated with the crest of an inner gorge also serve to amplify shaking (Meunier et al., 2008). These same models predict that slopes facing away from an epicenter should experience higher accelerations. Hence, local topographic site effects, such as the orientation of slopes, gain in importance. In contrast, storm-driven landslides either may be uniformly distributed along slopes (as in the Southern Alps; see figure A) or may tend to predominate closer to channels. We could predict that higher pore pressures, as well as local steepening of hillslopes by river incision, would promote landsliding proximal to channels (Gabet et al., 2004b). Continued analysis of large spatial data sets, particularly following large storms and earthquakes, promises to shed more insight on to when, where, and why landslides occur within active landscapes.

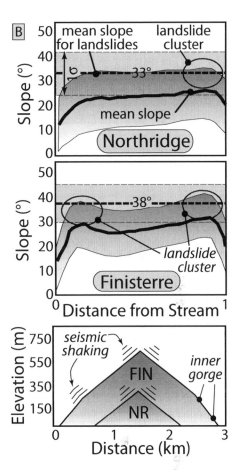

B. Slope distributions for Northridge (top) and Finisterre (middle), with major landslide clusters indicated. Mean slope for observed landslides lies above the average for the entire topography but, within its 1σ distribution (containing 65% of all slides), it includes the steeper hillslopes. (Bottom) Simplified topography for Northridge (NR) and Finisterre (FIN), with the latter showing an inner gorge and amplified seismic shaking.

landslide size and frequency (Hovius et al., 1997). This approach has the advantage of recording a direct time dependence. On the western slopes of the Southern Alps of New Zealand, repeat aerial photographs spanning 60 years have been used to define the distribution and aerial extent of more than 7000 landslides within an area of about 5000 km². These landslides range in size from about 100 m² to 1 km² and define a power-law magnitude–frequency distribution over two orders of landslide size (Fig. 7.15). Stated in cumulative form, the number of slides of magnitude equal to or greater than area A_c, i.e., $n_c(A \geq A_c)$, is

$$n_c(A \geq A_c) = \kappa(A_c/A_r)^{-\beta} A_r \qquad (7.4)$$

where A_r is a reference area, κ is the intercept of the regression when $A_r = 0$, and β is the

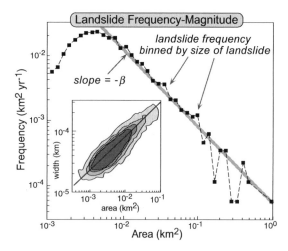

Fig. 7.15 Landslide frequency–magnitude distributions.
Analysis of the area of ~7500 landslides from the Southern Alps of New Zealand reveals a robust frequency–magnitude relationship. Inset shows the area–width relationship for these slides. By relating a landslide's width to its depth, the landslide frequency–area distribution can be converted into a denudation rate. Modified after Hovius *et al.* (1997).

dimensionless scaling exponent equivalent to the best-fit linear regression through the linear region of the log–log plot (Fig. 7.15) of the cumulative distribution. When $\beta > 1.5$, small events are predicted to dominate both the total area covered by landslides and the total landslide denudation, whereas if $\beta < 1.5$, large events are more volumetrically important. In the Southern Alps, $\beta = 1.17$, indicating the statistical importance of events with larger areas (Hovius *et al.*, 1997). Similar studies from the Transverse Ranges of southern California yielded β values of 1.1 ± 0.15 (Lavé and Burbank, 2004).

In order to determine the volume of erosion attributable to landsliding, the aerial extent of a landslide measured on a photograph has to be converted to a rock volume. Although the New Zealand data show a strong relationship between slide area and width (inset in Fig. 7.15), their depth is unknown. Studies elsewhere, however, suggest a linear relationship between landslide depth (d) and either the width (L) of the displaced mass or the square root of its area (A), such that $d = \varepsilon L$ or $d = \varepsilon A^{0.5}$ (Ohmori, 1992; Lavé and Burbank,

2004). In the case of New Zealand, ε is estimated to be 0.05 (Hovius *et al.*, 1997). Integrating over the distribution of landslide sizes, the total volume eroded by landslides is then modeled as

$$V = \frac{2\beta\varepsilon\kappa L_1^{3-2\beta}}{(3 - 2\beta)} \qquad (7.5)$$

where L_1 is the maximum length of a landslide in a particular setting and is limited by the available relief. In the Southern Alps, L_1 is estimated at about 1 km, and the erosion rate that is calculated from the above equation averages 9 mm/yr over the entire study area and ranges from 5 to 12 mm/yr within individual catchments. Although these rates may seem extraordinarily high, they appear to be consistent both with the rates of long-term denudation estimated from fission-track ages in the Southern Alps (Tippett and Kamp, 1993, 1995) and with intermediate-term (10^4–10^5 yr) rates estimated from optically stimulated luminescence (OSL) bedrock ages (Herman *et al.*, 2010). Because the volumetric estimates from landslides are very sensitive to the value assigned to ε (Eqn 7.5), a field-based effort to quantify ε is desirable. Interestingly, the rates of calculated denudation by landsliding, the rates of rock uplift estimated from fission-track ages and OSL dating, and the amount of annual precipitation all co-vary in the Southern Alps, such that areas of high rainfall, high landslide denudation, and high rock uplift spatially coincide. Despite these high rates of hillslope denudation, the rivers in the Southern Alps are "underloaded," such that they export nearly all of the landslide debris delivered to them, and they incise their beds at rates essentially equivalent to the rock uplift rate. Thus, a steady-state landscape can be envisioned here, where balanced rates of river incision and landslide denudation maintain hillslope length and relief through time (Herman *et al.*, 2007; Willett, 1999).

Denudation rates derived from cosmogenic radionuclide techniques at the basin scale

Imagine the power of a technique whereby analysis of a handful of sand in a modern river could reveal the average rate of erosion of the entire upstream catchment! Under some geological circumstances, assessing the cosmogenic radionuclide concentrations in detrital mineral grains can indeed indicate the mean erosion rate across a catchment (Brown *et al.*, 1995; Granger *et al.*, 1996). The rate of accumulation of cosmogenic nuclides in rocks at the Earth's surface is a function of at least two factors: the rate at which the rock surface is being lowered by erosion, and the altitude-dependent rate of nuclide production (Fig. 7.16). Because this altitude dependence is quite well known (a factor of e increase for about each 1700 m of altitude gain), a simple multiplication of the catchment hypsometry by the altitude-dependent production rate can be used to calculate a mean production rate for the entire catchment. If there were no erosion, then the nuclide abundance would increase to a maximum, where the rate of production was balanced by the rate of radioactive decay of the nuclide (i.e., "secular equilibrium"). In the face of erosion, a parcel of rock in the subsurface is gradually brought toward the surface, is exposed to increasing amounts of cosmic-ray bombardment as it approaches the surface, and attains a cosmogenic nuclide abundance that is a function of the rate at which the surface is being eroded (Fig. 7.12). If quartz grains from throughout the catchment contribute to the sediment load in proportion to the local erosion rate, then assessment of the concentration of cosmogenic radionuclides in a sample consisting of many detrital quartz grains can reveal the basin-wide mean erosion rate.

Unlike short-term measurements of sediment fluxes during recent decades, the cosmogenic nuclide signal integrates hundreds to thousands of years of denudation and, therefore, has the potential to yield erosion rates that are largely free of anthropogenic perturbations. In the past 15 years, many dozens of studies have exploited detrital cosmogenic nuclide concentrations to define catchment-wide erosion rates over hundreds to many thousands of years. Such studies have led to a quantum leap in our understanding of changing erosion rates at different time scales (Kirchner *et al.*, 2001), contrasts in rates across diverse tectonic landscapes

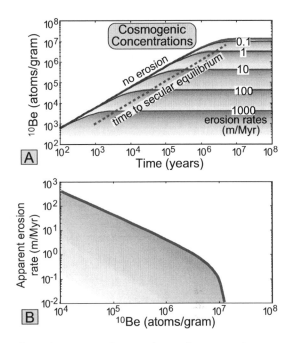

Fig. 7.16 Rock-surface erosion and cosmogenic concentrations.

A. Theoretical [10]Be accumulation based on mid-latitude, sea-level production rates at the upper surface of a rock as a function of time and erosion rate. With higher rates of erosion, [10]Be reaches steady state more rapidly and at lower concentrations. The dashed line indicates the approximate time for an originally pristine surface to reach an equilibrium concentration whereby production and decay are in balance for a given erosion rate. For sustained erosion at 1000 m/Myr, a steady concentration is reached in less than 3000 years, the time required to remove about 2.4 m of rock (equal to four e-folding depths of rock). B. Apparent erosion rates as a function of steady-state [10]Be concentrations. At rates less than 1 m/Myr, radioactive decay accounts for most of the nuclide loss and steady-state concentrations do not indicate mean erosion rates. Modified after Brown *et al.* (1995).

(Vance *et al.*, 2003; von Blanckenburg, 2006), and the influence of climate on erosion rates (Riebe *et al.*, 2001, 2003).

Although attractive in concept, many assumptions lie behind most applications of this cosmogenic technique. These assumptions include: uniform distribution of the target mineral (quartz in most cosmogenic studies); known production rates through time across the catchment; uniform erosion throughout the

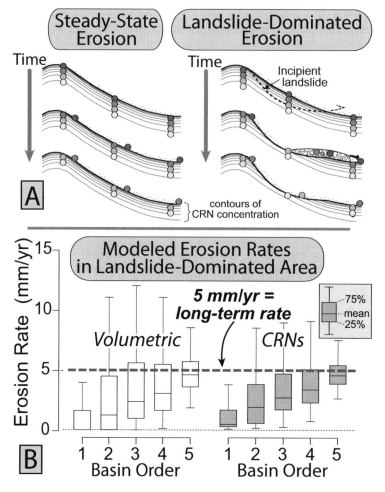

Fig. 7.17   Cosmogenic erosion rates in landslide-dominated catchments.
A. Contrasting models for cosmogenic radionuclides (CRNs) in landscapes dominated by grain-by-grain erosion versus ones where landsliding is prevalent. Whereas in the grain-by-grain case, erosion removes those rocks and soils with the highest CRN concentrations, landslides slice downward to include rocks with low CRN concentrations that will lower the overall detrital CRN concentration. B. Results of a numerical model for a landslide-dominated landscape where the overall long-term erosion rate is 5 mm/yr and where grain-by-grain weathering is 0.1 mm/yr. In the model, landslides are drawn from a magnitude–frequency distribution (Fig. 7.15) and randomly dropped on the landscape to yield a long-term rate of 4.9 mm/yr. The model keeps track of both the volumetric erosion rate (left) and the cosmogenic erosion rate (right). At small catchment sizes (first- and second-order), both volumetric and cosmogenic concentrations will greatly underestimate the long-term erosion rate. The largest catchments come closest to approximating the actual erosion rate. The model results indicate that, in rapidly eroding regions, cosmogenic field studies should target high-order catchments in order to obtain the best estimate of the long-term erosion rate. Modified after Niemi *et al.* (2005).

upstream catchment; known extent of glaciers and annual variations in snow cover; and nuclide concentrations that are independent of grain size or the erosion process. Although some of these assumptions are impossible to test, when they are violated, they can warp or invalidate interpretations. If a large fraction of erosion occurs by bedrock landsliding, for example, cosmogenic concentrations are likely to significantly underestimate the actual erosion rate, especially in low-order catchments (Fig. 7.17). When erosion rates are slow and

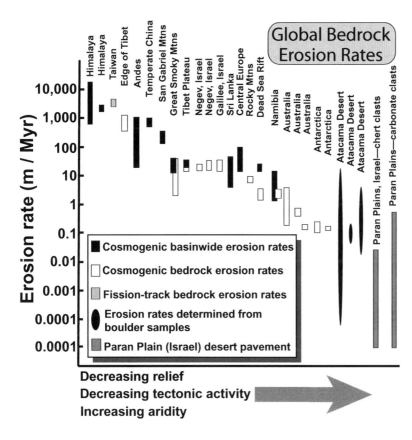

Fig. 7.18 Global span of erosion rates.
Most rates are derived from cosmogenic nuclide concentrations in either detrital sediments or in bedrock. Rates range from 10 km/Myr in very active orogenic belts to as low as 1 mm/Myr – over seven orders of magnitude. Modified after Matmon *et al.* (2009).

much erosion occurs via grain-by-grain weathering, cosmogenic nuclide concentrations can reliably record mean erosion rates. As erosion rates increase above the rock-to-regolith conversion rate and landsliding becomes increasingly important, only the larger catchments (fifth order and above) are likely to approximate the overall erosion rate because only these successfully smooth the stochastic variability of landsliding throughout the catchment. Despite these many caveats, however, when geological conditions permit, measuring nuclide concentrations in sediments in order to determine mean erosion rates has great appeal, because it can integrate an erosional signal from an entire catchment via samples readily collected at its outlet. In combination with estimates of bedrock erosion rates, the results of detrital cosmogenic studies over the past two decades have illuminated the global span to erosion rates and their relationship to climate and tectonics (Fig. 7.18).

Long-term erosion rates based on radiometric ages

Rocks in the crust cool for one of two reasons: either a waning thermal pulse related to a magmatic, hydrothermal, or metamorphic episode; or rocks being moved closer to the Earth's surface due to either tectonism or erosion (Reiners and Brandon, 2006). In response to persistent erosion, deeply buried rocks move upward along the local geotherm toward the surface. During this cooling, various minerals within the rocks will pass through the closure temperatures for different radiometric dating systems (Table 7.2). Notably, the time when minerals pass through their closure temperature can be millions or billions of years after the minerals first crystallized. In contrast to most other approaches described thus far, thermochronology (the study of temperature-dependent mineral ages) examines cooling and erosion histories that commonly span millions of years. These diverse mineral-specific closure temperatures create valuable opportunities for a thermochronologist. Consider a single rock sample from which multiple, datable minerals can be extracted. Based on ^{39}Ar/^{40}Ar dating of hornblende, muscovite, and potassium feldspar, on (U–Th)/He dating of apatite and zircon, and on OSL dating of quartz, a cooling history spanning from 525°C to 35°C might be generated (Table 7.2).

Contrasts in the rate of cooling through time can be interpreted from the radiometric data in order to delineate variations in long-term erosion rates (Fig. 7.19). When cooling rates accelerate toward the present, they are often interpreted to result from enhanced rates of denudation. If no local geological evidence indicates recent normal faulting that could have accelerated cooling, enhanced erosion by surface processes is typically invoked to explain the rapid cooling.

The perennial problem encountered when trying to convert cooling rates to erosion rates is that the local geothermal gradient is almost never reliably defined. Most commonly, a "typical" continental geotherm of 20–30°C/km is assumed and cooling rates are converted to erosion rates on this basis. First, the depth (z) from which the rock came to the surface is calculated:

$$z = c/(\mathrm{d}T/\mathrm{d}z) \qquad (7.6)$$

where c is the closure temperature for the dated mineral, and $\mathrm{d}T/\mathrm{d}z$ is the geothermal gradient. Then, the mean erosion rate (e) is estimated as

$$e = z/a \qquad (7.7)$$

where a is the time of cooling through the closure temperature. Thus, a rock that cooled below 200°C about 2 Ma would be interpreted to have been at 6–10 km depth at that time, assuming a geothermal gradient of 20–30°C/km, and to have been brought to the surface, via erosion, at a rate of 3–5 km/Myr (3–5 mm/yr). Even if the geothermal gradient were known at the start of accelerated denudation, that gradient would not persist during rapid erosion. The "rise" of rocks toward the surface would advect heat upwards, such that the local geotherm would be steepened. Theoretical models of warping of isotherms during rapid erosion and cooling (Craw et al., 1994; Mancktelow and Grasemann, 1997; Stüwe et al., 1994) suggest that gradients as high as

Table 7.2 Radiometric dating systems and closure temperatures for some minerals.

Mineral	Dating system	Closure temperature (°C)
Hornblende	^{40}Ar/^{39}Ar	525±25
Muscovite	^{40}Ar/^{39}Ar	350±25
Biotite	^{40}Ar/^{39}Ar	300±25
K-feldspar	^{40}Ar/^{39}Ar	200±25
Monazite	U–Pb	525±25
Biotite	Rb–Sr	275±25
Sphene	fission-track	275±50
Zircon	fission-track	250±30
Apatite	fission-track	120±20
Zircon	(U–Th)/He	180±20
Apatite	(U–Th)/He	70±15
Apatite	(U–Th)/He: ^{4}He/^{3}He	40±10
Quartz	OSL	35±10

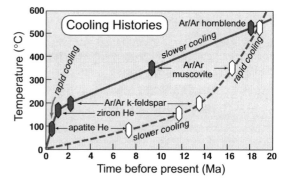

Fig. 7.19 Contrasting cooling histories from thermochronology.
Hypothetic example of cooling histories using two thermochronometers (^{39}Ar/^{40}Ar and (U–Th)/He dating) on five different minerals taken from two rock samples. One sample (solid line) displays rapid cooling at ~150°C/Myr since ~1 Ma, whereas the other (dashed line) shows cooling at a mean rate of ~15°C/Myr for the past 13 Myr. Even with significant uncertainties in the geotherm, these data would suggest rapid Quaternary denudation for the first sample (>2 mm/yr).

60–100°C/km might be achieved with erosion rates greater than 5 mm/yr. Isotherms can be further warped by fluid flow in the crust, which tends to remove heat from peaks and add heat to valleys (Whipp and Ehlers, 2007). The unknown nature of the local geotherm during cooling suggests that large uncertainties should be placed on most erosion rates that are deduced from thermochronological cooling histories. Even with 50% uncertainties, however, statistically significant changes in erosion rates can often be interpreted from thermochronological data derived from multiple mineral systems. Moreover, these thermochronological dates typically provide the best, and commonly the only, means to define average erosion rates at million-year time scales.

Whenever possible, an improved quantification of the local geotherm should be attempted. Sometimes nearby boreholes penetrate deeply enough to provide a calibration of the coolest part of the geotherm. Upper crustal isotherms are sensitive to topographic variations. The ridge-to-valley relief of mountains can be viewed like fins on a radiator that are emitting terrestrial heat to the atmosphere. If the fins are deep and close together, they have little impact on any but the isotherms very close to the surface. At large topographic wavelengths, the isotherms up to 10-km depth or more will mimic the topography. More rapid erosion rates cause a greater sensitivity to smaller-scale topographic variations due to the rapidity with which rocks (and their heat) are advected toward the surface. Beneath high summits, a given isotherm will be at a higher altitude than beneath a valley (Fig. 7.20A), but the near-surface geothermal gradient will be lower beneath the summit than beneath the valley bottom (Stüwe *et al.*, 1994; Braun, 2003).

Consider an example of an alpine landscape with 4 km of local relief and a wavelength of 10 km between summits. If the topography is in steady state, such that erosion and rock uplift are everywhere in balance, a steady-state thermal structure can be modeled as function of the erosion rate (Fig. 7.20). In this particular example, the erosion rate is 2 mm/yr. Despite a spatially uniform erosion rate, the predicted age variations at the surface for several thermochronometers with different annealing or closure temperatures are striking! The ages vary by a factor of 2–4, with the lowest-temperature thermochronometer being most affected (Fig. 7.20C). Cooling rates for helium dates (closure temperature ~60°C in this model) would be estimated to vary from over 150°/Myr in the valley bottoms to less than 40°C/ Myr at the summits. If an "average" geothermal gradient were chosen, say 30°/km, as a basis for

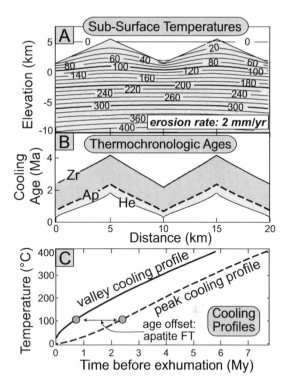

Fig. 7.20 Topographically controlled isotherms, cooling ages, and cooling profiles.
Modeled cooling age pattern of three thermochronometers exhumed at the Earth's surface over two wavelengths of a two-dimensional, V-shaped topography with 4 km of relief. Lapse rate on the surface is set to 6.5°C/km. The erosion rate is spatially uniform at 2 mm/yr, implying the landscape has achieved a steady-state form, and the calculation is carried out over several million years. The chronometers are zircon (assumed fission-track closure temperature of 220°C), apatite (assumed fission-track closure temperature of 110°C), and (U–Th)/He system (assumed apatite closure temperature of 60°C). A. Thermal structure, including upwarped isotherms below summits, but a lower geothermal gradient than beneath valleys. Note the spatial variation in ages despite a spatially uniform erosion rate. B. Expected spatial pattern of the cooling ages at the surface for the three chronometers. C. Cooling histories taken by rocks exhumed in the central valley (solid) and the central peak (dashed) of the topography. As expected from the distance between the deep, flatter isotherms and the surface (in A), the cooling age variation inferred from zircon fission-track ages should tightly mimic the topography. As the closure temperature lowers, the influence of the topography on the thermal structure increases, both lowering the amplitude of the cooling age contrast from peak to valley and imparting a nonlinearity. This influence will be larger in the higher-erosion cases.

calculating erosion rates, the estimated rates would vary from about 5 mm/yr in the valley bottoms to about 1 mm/yr at the summits. Whereas these rates clearly bracket the actual erosion rate (2 mm/yr), the erosion rate associated with any individual age is likely to be misleading.

Given such uncertainties, why bother with this thermochronological approach? The strongest argument is that cooling histories provide a useful long-term perspective that is related to denudation. Such a perspective is difficult to achieve without a means of measuring time over millions of years. Second, the example given here is perhaps not typical. If mean erosion rates or the topography relief were lower, the perturbation of the isotherms and apparent rate variations would be less. Third, even with the valley-to-summit age differences, the data define the limits of past cooling rates and provide useful brackets on acceptable long-term erosion rates. Finally, as long as a reasonable kinematic geometry is known for the deformation, numerical modeling (e.g., Braun, 2003) can now be used to predict both the topography of subsurface isotherms and the three-dimensional array of cooling ages at the surface that would be expected for a given erosion rate (Plate 4). Hence, the effects on ages of the competition of erosion with the lateral and vertical advection of rocks, heat, and fluids in the crust can be modeled (Bollinger et al., 2006; Ehlers and Farley, 2003; Whipp and Ehlers, 2007). Even the effects of changes in relief or erosion rate on the array of surface ages can be explored (Braun, 2002; Braun et al., 2006; Ehlers et al., 2006). Such modeling efforts provide an increasingly robust context for the interpretation of the implications and uncertainties of cooling ages for reconstructing the erosion histories of mountain belts.

To create the best estimates of the past erosion rates and changes in topography, extensive arrays of radiometric ages, reliable kinematic geometries, and considerable computational analysis are needed. Kinematics become increasingly important as the rate of lateral advection increases with respect to rock uplift. The more sophisticated numerical models can incorporate measurable variables, such as heat production and conductivity, as well as computed variables, such as topographically driven fluid flow. All of these factors affect the thermal structure of an orogen and the resultant cooling ages.

In recent decades, tectonic-geomorphic studies have benefited from efforts to develop high-precision thermochronological analyses that are sensitive to increasingly lower temperatures. Because such temperatures occur in the shallow subsurface, these thermochronometers provide rather direct insights on rates of erosion during the most recent intervals of time. For many decades, apatite fission-track dating was the thermochronometer that probed the lowest closure temperature (110°C). Drawbacks with the apatite fission-track approach include the typical depth of the closure isotherm (3–5 km) and its large uncertainties: typically ~10%, but commonly exceeding 50% in regimes of very rapid erosion (Blythe et al., 2007). More recently, the development of the (U–Th)/He thermochronometer has enabled dating of apatite and zircon with closure temperatures of ~60–70°C and ~180°C, respectively, and with typical precisions of <10%, even for young ages (Reiners et al., 2004; Wolf et al., 1996). Thermal histories down to temperatures of 30°C or less have been achieved by stepwise heating of apatite and modeling of concentration gradient of ^{4}He due to thermal diffusion near the outer margins of a grain (Shuster and Farley, 2005). New OSL applications to bedrock samples are interpreted to record closure temperatures of 30–40°C (Herman et al., 2010). Overall, expanding use of these low-temperature thermochronometers will permit much more detailed and more reliable reconstruction of recent erosion at time scales of 10^4–10^6 years.

Wherever you are standing today, somewhere beneath your feet (probably within a few kilometers of the surface) rocks at temperatures above their closure temperature would reveal ages of 0 Ma, if you could date them. Although closure temperatures have been defined for many minerals (Table 7.2), cooling across these temperatures does not represent an on/off switch. For several thermochronological systems, a well-defined thermal zone exists in which some, but not all, of the products of radiometric decay are retained. For apatite fission-track, for example, tracks are annealed at temperatures

Plate 1 A. Abrasion platform exposed during low tide at Kaikoura, New Zealand. B. Beach ridges uplifted by earthquakes at Turakirae Head, Wairarapa, New Zealand. Note that the lowest ridge of sand is the modern storm berm. Image modified from Lloyd Homer, GNS Science. C. Strath terraces in the southern Tien Shan, western China. D. Regional unconformity surface that is beveled across Paleozoic strata and exposed following tilting and subsequent stripping of weakly cemented Cenozoic strata at Gory Baybeiche, Tien Shan, Kyrgystan. E. Glacial moraines cut by range-front normal fault at McGee Creek, Sierra Nevada, California. F. Debris-flow levees in the Tien Shan, Kyrgyzstan.

Plate 2 Paleoseismic indicators. A. Sand volcano caused by the October 15, 1979 Magnitude 7.0 Imperial Valley, Californiaearthquake. B. Cobbles with cylindrical borings (arrows) on raised marine platforms near Sea Ranch, California, are paleo-sea-level indicators because they are found near the high-tide line and provide reliable markers for measuring vertical displacements. C. Open fissure (arrows) formed along a dextral fault during the M 6.7 Superstition Hills earthquake on November 24, 1987. D. Tree trunk ruptured by the M 7.9 Denali Fault earthquake on November 3, 2002. E. Organic-rich layer (arrows) draping faulted fluvial sediments at Pallett Creek, California, on the San Andreas fault. Note the continuity of layers, but changing thickness above the fault tip (circle). Credits: A: National Information Service for Earthquake Engineering – Pacific Earthquake Engineering Research (NISEE – PEER) Library: B, C, E: DW Burbank; D: Patty Craw Burns, Alaskan Division of Geological and Geophysical Surveys.

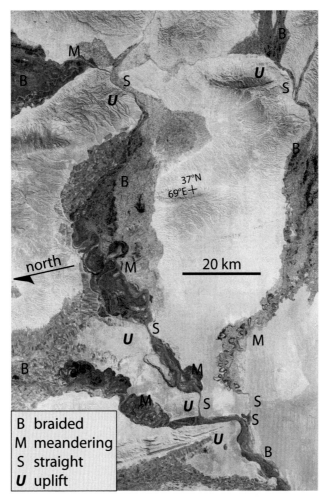

Plate 3 Planform changes in rivers due to interactions with growing folds. Through zones of active rock uplift, rivers become more straight and narrow. Downstream of these constrictions, rivers tend to become braided, but transition to meandering planforms upstream of the next zone of rock uplift, local base level rise, and channel narrowing. Modified after NASA LANDSAT E-30825-05154-7.

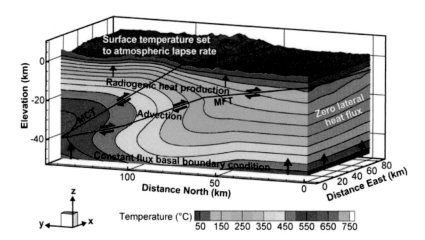

Plate 4 Thermokinematic model of subsurface thermal structure associated with thrust faults. Thermokinematic model designed to mimic the Himalayan orogen as a zone with two major overthrusts (MCT: Main Central Thrust; MFT: Main Frontal Thrust) and over 5 km of topographic relief. The large-scale distortion of subsurface isotherms results from underthrusting of cooler rocks from right to left at a rate of ~20 mm/yr. Note the compression and expansion of isotherms that will modulate the cooling ages expected at the surface as a result of erosion as overthrust rocks are uplifted. Modified after Whipp *et al.* (2007).

Plate 5 Thrust-faulted fluvial terraces in the Tien Shan, Kyrgyz Republic. Offsets of the terrace treads and strata discontinuities visible along the terrace risers reveal the trace of the fault. Note the greater displacement of successively older terrace surfaces and the erosional modification of the thrust-fault scarp on each terrace. Radiocarbon and OSL dates indicate that the T4 terrace dates from ~14 ka and the T2 terrace dates from ~140 ka (Thompson *et al.*, 2002).

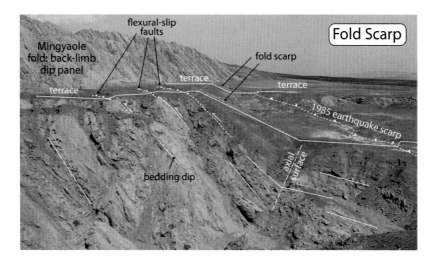

Plate 6 Fold scarp with flexural slip faults. A 30 m high fold scarp on the south limb of the Mingyaole anticline in the western Tarim basin, China. Abrupt dip changes in underlying Neogene strata and folding of a broad terrace delineate an active axial surface. Ongoing flexural slip between beds as they are bent through the axial surface causes flexural slip faults that displace the uplifted terrace surface. Fine-grained sediments (light tan) are ponded against the uphill-facing scarp of the 1985 M_s = 7.4 Wuqia earthquake.

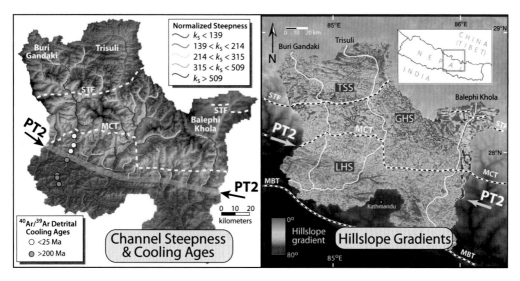

Plate 7 Spatial patterns of channel steepness and hillslope gradients in the Nepalese Himalaya. (Left) Normalized channel steepness in the Burhi Gandaki, Trisuli, and Balephi Khola catchments of central Nepal. Detrital ^{40}Ar/^{39}Ar cooling ages (see Fig. 9.14C) are shown for the Burhi Gandaki. Near-linear zone (PT2: physiographic transition 2) delineates a downstream transition in steepness indices to values less than 215. On opposite sides of this transition, detrital cooling ages show distinctly different populations (<25 Ma to the north; >200 Ma to the south) and delineate domains of rapid and slow cooling, inferred to indicate rapid versus slow erosion, respectively. Note that the PT2 lies significantly south of the Main Central Thrust (MCT). (Right) Hillslope gradients also show an abrupt transition to gentler slopes along the PT2. Together, these landscape indices have been interpreted to be responding to a new, active, surface-breaking fault that lies within the Lesser Himalaya Series (LHS). GHS: Greater Himalayan Series; MBT: Main Boundary Thrust; STF: South Tibetan Fault; TSS: Tethyan Sedimentary Series. Modified after Wobus *et al.* (2003).

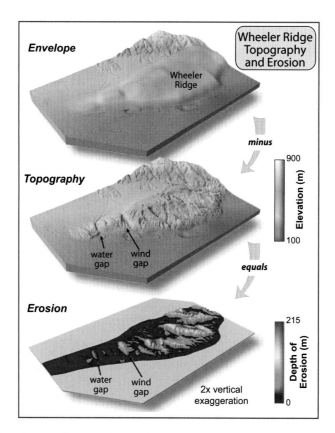

Plate 8 Reconstructed erosion on Wheeler Ridge anticline. Envelope surface is constructed by connecting either ridge crests or remnants of the uplifted, pre-folding depositional surface. After subtraction of the current topography from the envelope surface, the spatial distribution and magnitude of erosion are defined. Enhanced erosion occurs on older parts of the fold, on the forelimb, and in the wind gap. Note that the magnitude of erosion is a minimum because undissected surface remnants are not everywhere present. Graphics courtesy of Scott Miller.

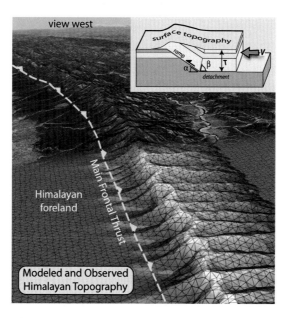

Plate 9 CHILD model of a thrust-generated mountain range. [A black-and-white version of this, along with the full caption, can be found in the main text as Fig. 11.21.]

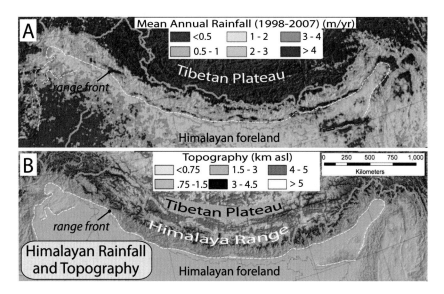

Plate 10 Orographic precipitation and topography in the Himalaya. [A black-and-white version of this, along with the full caption, can be found in the main text as Fig. 11.22.]

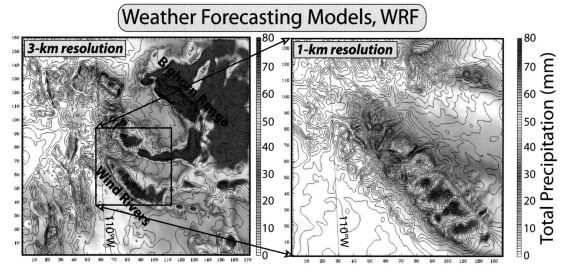

Plate 11 Use of research-grade Weather Research and Forecasting (WRF) models. [A black-and-white version of this, along with the full caption, can be found in the main text as Fig. 11.26.]

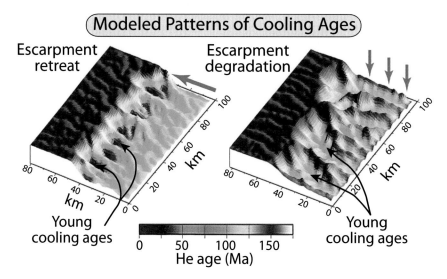

Plate 12 Modeling the generic thermal evolution of an eroding escarpment. [A black-and-white version of this, along with the full caption, can be found in the main text as Fig. 11.28.]

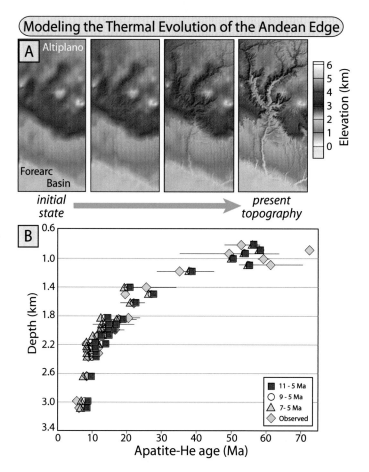

Plate 13 Modeling the thermal evolution of the edge of the Andes. [A black-and-white version of this, along with the full caption, can be found in the main text as Fig. 11.29.]

between about 60 and 110°C, with faster annealing at higher temperatures. This thermal window is termed the "partial annealing zone." Similarly, with (U–Th)/He dating of apatite, a partial retention zone exists between about 40 and 80°C. When the tops and bottoms of these partial retention or annealing zones can be captured in thermochronological sampling transects, they can help to delineate former geothermal gradients and define the magnitude of rock uplift (Fig. 7.21). A recent study from the White Mountains of California used apatite from each sample for both fission-track and (U–Th)/He dating and succeeded in delineating exhumed partial retention and/or annealing zones for each dating technique (Stockli *et al.*, 2000). The ages for samples beneath the base of these zones define the onset of rock uplift and accelerated erosion at 12 Ma. Both techniques capture the penultimate episode of major cooling around 55 Ma, and because the full partial annealing and/or retention zone is encompassed by each, a paleogeothermal gradient and the magnitude of rock uplift since 12 Ma can be estimated for each data set (Fig. 7.21).

Tectonic denudation versus geomorphic erosion

Tectonic denudation occurs when faulting thins the crust and brings formerly buried rocks closer to the land surface. Such thinning can abruptly redistribute large loads from an elevated region on the surface to a topographically lower area, and it promotes an isostatic response that can locally cause bedrock uplift.

It has been suggested that many compressional tectonic regimes, such as those found at convergent plate margins, can be modeled as critically tapered wedges of a Coulomb material (Davis *et al.*, 1983) that is everywhere at the point of failure. Like a wedge of snow driven before a plow, the taper of the wedge is defined as the angle between its upper and lower surfaces and is a function of the frictional coupling at its base, the dip of the basal detachment surface, and the material properties of the wedge itself (Fig. 7.22). Under conditions of constant shortening and consistent material properties and geometries, the wedge will adjust internally to the addition

of new material at its toe (or anywhere else) in order to maintain its taper. If any of the controlling conditions or properties change, the wedge should adjust to a new equilibrium taper.

Geologically, this model implies that an orogenic wedge should undergo constant deformation throughout its mass to accommodate the irregular addition of new material to the wedge, losses of material through erosion, and redistribution of mass through deposition or faulting. Geological materials, however, are not truly Coulomb-like in their behavior. Rather than deforming everywhere, they break and deform along discrete planes (faults). Thus, raising of the surface of the wedge in order to counteract erosion typically occurs through thrust faulting or underplating. When conditions change such that the taper of the wedge becomes too great (supercritical), it can adjust itself by three different mechanisms: propagation of the toe (the leading edge) of the wedge, erosion of the elevated surface, or normal faulting (extension). Normal faulting causes a rapid thinning of the wedge and can promote nearly isothermal uplift and decompression in the underlying rocks.

Large-scale normal faulting under regimes of contraction has been identified in many collisional orogens. Many Cenozoic collisional ranges, including the Himalaya, Alps, and Taiwan's Central Range, have experienced significant extension (Burchfiel *et al.*, 1992; Crespi *et al.*, 1996; Platt *et al.*, 1998; Ratschbacher *et al.*, 1989). In most of these ranges, contraction and extension were coeval, rather than sequential. Commonly, the rate and amount of displacement are difficult to establish, so that tectonic denudation rates are not well constrained. When pressure–temperature–time studies of the rocks beneath a normal fault show rapid depressurization, however, the magnitude, timing, and rate of denudation can be estimated more precisely.

Most tectonicists have traditionally ascribed rapid decompression and rapid cooling to tectonic denudation, that is, normal faulting and extension. Pressure changes on the order of 1 kbar/Myr (~3 km/Myr) were commonly thought only to be possible via normal faulting. Indeed, this interpretation may be correct. However, recent measurements of surface erosion rates by

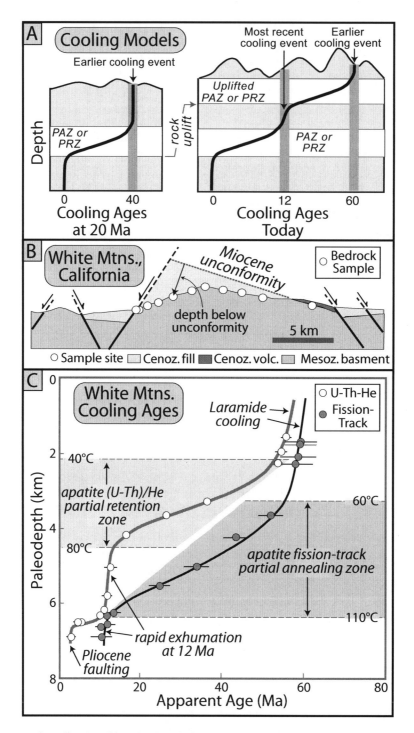

Fig. 7.21 Reconstruction of rock uplift and paleogeothermal gradients with low-temperature thermochronology.
A. Models for cooling ages in the crust. (Left) Ages at 20 Ma. Below the base of the partial retention zone (PRZ) or partial annealing zone (PAZ), cooling ages are zero, whereas above the top of the PAZ or PRZ, cooling ages record cooling due to previous erosion or thermal events. If the ages are all about the same, they indicate rapid cooling of a crustal thickness at least as thick as is preserved. (Right) Ages at present. Ages below the modern PAZ or PRZ

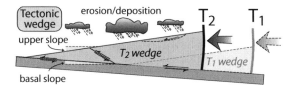

Fig. 7.22 Collisional mountain belt as an orogenic wedge.
Simple model of a growing orogenic wedge at two stages (T_1, T_2). As long as the material properties remain constant, the wedge taper is unchanged, and the wedge grows self-similarly through frontal or basal accretion. Thrusting within the wedge and deposition on its surface can temporarily increase the taper, which will then decrease back to its critical taper via some combination of erosion, extension, or forward propagation of the toe of the wedge.

landsliding (Hovius *et al.*, 1997), glacial erosion (Koppes and Hallet, 2006), and river incision (Burbank *et al.*, 1996b; Finnegan *et al.*, 2008) suggest that geomorphic erosion rates of up to 10 mm/yr can be sustained for perhaps millions of years in rapidly deforming mountains. Thus, it is no longer clear that rapid unloading or decompression, such as that associated with some ultrahigh-pressure rocks (Hacker, 2007), is exclusively caused by lithospheric processes. In fact, documented erosion rates may be fast enough to accommodate nearly all of the estimated rates of unloading. This erosional capability does not, however, discount the likelihood that extension, diapirism, or delamination is implicated in many of the rapidly cooling orogens; it simply means that alternative explanations involving high rates of erosion should be investigated.

Rates of uplift

A distinction has already been drawn between rock uplift, surface uplift at a point, and uplift of the mean surface (Fig. 7.2). Each of these types of uplift is typically referred to some agreed-upon reference frame, such as the geoid or mean sea level. Relative uplift can also be useful to define, particularly at a local scale. By relative uplift, we refer to differential displacement of some point or points relative to others. For example, when an anticline grows, rocks in the core of the fold are uplifted relative to its flanks. Commonly, the scale of a fold is such that there is no geophysical response to localized crustal thickening. Even if the whole area is subsiding and the mean surface is lowering, documentation of relative uplift within a fold remains useful, because it carries implications for the local tectonics and erosion. Within a local reference frame, observations of relative uplift serve to pinpoint patterns of deformation.

Marine terraces

One of the classic means for documenting uplift of rocks comes from studies of marine terraces. As described in Chapter 6 on paleoseismology, a terrace of known age and elevation above present sea level can be used to compute a rate of bedrock uplift (Fig. 6.10), if the position of sea level at the time of its formation is known. Correlation of multiple terraces in a single transect can define changes in rates through time, whereas regional studies of dated terraces can yield two- and three-dimensional bedrock uplift patterns (Chappell, 1974). Where terraces are well preserved, dating of these terraces to define their appropriate correlation to the eustatic record often provides the biggest challenge (Anderson and Menking, 1994; Perg *et al.*, 2001). Where dissection has been extensive, however, simple identification of former terraces within the landscape can be very difficult. On the South Island of New Zealand,

---

Fig. 7.21 (*cont'd*) are still zero. Rock uplift and cooling at 12-Ma have raised the former PAZ or PRZ and are recorded by a zone with nearly homogeneous 12-Ma ages. The previous cooling event (now at 60 Ma) is preserved in the uppermost rocks. B. Structural geometry and sample locations in the White Mountains of California. Key to the interpretation is the linear projection of the Miocene unconformity because the paleodepth of each sample in the Miocene is measured beneath this projected surface. C. Cooling ages for apatite fission-track and (U–Th)/He plotted against paleodepth. Each system reveals a PAZ or PRZ below which is a vertical zone of ~12-Ma ages. Their maximum vertical extent (~2 km for the (U–Th)/He ages) indicates the approximate magnitude of erosion at 12 Ma. A more recent interval of Pliocene faulting, uplift, and cooling is recorded by the lowest helium samples.

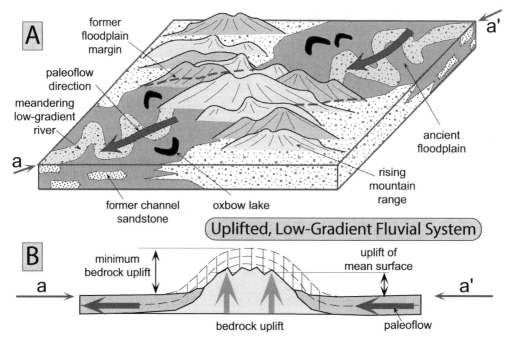

Fig. 7.23   Stratigraphic evidence for surface and bedrock uplift.
A. Stratigraphic evidence for a low-gradient paleoriver system, including meanders, oxbow lakes, and broad floodplains. Paleoflow direction is across the site of the present mountain range. Correlation of stratal units, plus paleocurrent and provenance data, imply that a formerly contiguous, low-gradient fluvial system existed prior to mountain growth.
B. Cross-sectional view of the deformed range. Low-gradient river deposits project into the air above the uplifted rock massif. The amount of surface uplift can be assessed by comparing the mean height of the former depositional surface with the mean height of the present mountain range. The minimum bedrock uplift can be calculated by projecting appropriate stratal thicknesses across the summits of the mountains from which they have been eroded.

where rock uplift and dissection have occurred at rapid rates over the past few million years, Bull and Cooper (1986) used the presence of rounded "beach" pebbles on beveled spurs as a basis for defining remnants of former abrasion platforms more than 1000 m above present sea level.

Stratigraphic constraints

As soon as the terrestrial realm is entered, the sea-level reference frame is removed from direct observation. This absence presents a formidable challenge when trying to interpret rock or surface uplift. Common sense and geological observation, however, tell us that both depositional environments and the characteristics of river systems tend to vary in systematic, predictable ways as one heads upstream from the coast. Except where mountains are present adjacent to the coast,

depositional systems near the continental margins have low gradients. The presence in the stratigraphic record of extensive floodplains, swamps, meandering rivers, estuaries, or fine-grained and far-travelled sediments generally indicates a low-gradient depositional system. Particularly if "low-gradient" terrestrial deposits can be correlated with nearby marine strata of the same age, an inference that the terrestrial system was developed close to sea level is reasonable. If these strata are now incorporated into a mountain range, the rock uplift of individual points and the change in mean surface elevation can be estimated by comparing the mean present topography with the extent of the former low-lying depositional area. More commonly, such "low-elevation" strata are preserved not within the mountains, but along their flanks. If one could demonstrate that strata on opposite sides of the range correlate with each

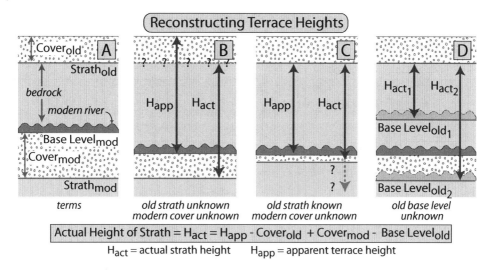

Fig. 7.24 Reconstruction of differential uplift from terrace heights.
Uncertainties in the height of an uplifted bedrock strath above the strath's position when it formed. A. Geomorphic
cross-section with a preserved, uplifted strath that is buried with terrace gravels. The modern river rests on a fill above the
most recently formed strath. B. Uncertainties in the actual strath height due to unknown position of the old strath (the
contact is covered) and unknown depth of fill beneath the modern river. C. If the modern river appears to be flowing on
bedrock, then the apparent and actual strath height may be almost equivalent. D. Additional uncertainty is added if the
base level of the river has changed, such that the strath initially formed above or below the modern river level.

other and contain sedimentological evidence that
former rivers flowed *across* the site of the present
range, thereby connecting these stratigraphic remnants, this implied continuity would be strong evidence for surface uplift (Fig. 7.23). Even if it can
only be shown that a river system on one side of a
range had a distant source area on the far side of
the range, this geometry also demonstrates growth
of the range since deposition. Calculation of the
amount of surface or tectonic uplift will depend
on how reliably the paleoelevation of the ancient
depositional system can be constrained. Correlations of the Neogene stratigraphic record in
Alaska have been used to argue that, only a few
million years ago, low-altitude rivers flowed across
the present site of Mt. McKinley (Denali), the
highest peak in North America (Fitzgerald *et al.*,
1995). Although far from the present coast, these
strata provide a reference frame for assessing
surface uplift.

River profiles

Although commonly far removed from coastal
areas, reconstructed river profiles can be used to

document differential bedrock uplift through
time, if certain assumptions are met. It must be
assumed that the longitudinal gradients of the
modern and ancient valley bottoms were approximately the same and that any differences in
former and present base levels must be small
compared to the magnitude of rock uplift. Both
aggradational and strath terraces can be used to
reconstruct former river profiles. In terms of
reconstructing relative rock uplift, however, measuring between former and current strath surfaces
is preferable, because, in each case, the lowest
level of bedrock erosion is being recorded.
Although surveying a terrace's height above the
modern river is straightforward, determining the
height of a strath above the base level at which it
was formed can be considerably more complex
(Fig. 7.24). Uncertainties can arise due to the
unknown amount of fill beneath the modern river,
the unknown thickness of terrace cover (if the
strath itself is not visible), and the unknown position of base level at the time the strath was formed.

Despite such uncertainties, deformed terraces
clearly record differential uplift. Typically, the
modern river profile is used as a reference frame

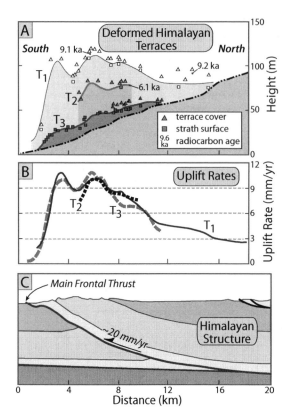

Fig. 7.25 River terraces deformed above the Main Frontal Thrust of the Himalaya.
A. Surveyed Holocene terraces along the Bagmati River, Nepal, in the hanging wall of the Main Frontal Thrust (MFT) within the Himalayan foreland. Note the undulating upper terrace surfaces. B. Relative rock-uplift rates derived from the age of each terrace and its height above the modern river. The spatial coincidence of rates for each terrace results from their kinematic tie to subtle changes in dip of the MFT. The persistence of uplift rates through time argues for steady Holocene slip on the MFT. C. Geometry of the MFT reconstructed from surface dips and slip rate from deformed terraces (B). Modified after Lavé and Avouac (2000).

that is substracted from the observed profiles, such that differential uplift of any portion of the terrace becomes apparent. In a stunning study of deformed terraces within the Himalayan foreland (Fig. 7.25A), Lavé and Avouac (2000) made detailed surveys across a 20 km wide deformed zone that lies above the Main Frontal Thrust, the active fault defining the southern edge of the Himalayan deformation (Fig. 7.25C). Although all of the terraces were of Holocene

age, some had been uplifted more than 100 m! Moreover, a complex pattern of folding in the youngest terrace was progressively amplified in successively older terraces. Each fold appears to reflect subtle changes in dip in the underlying fault that cause spatial variations in rock uplift rates (Fig. 7.25B). Given the dip of the fault and rates of terrace uplift, Lavé and Avouac (2000) were able to argue that, during at least the Holocene, the entire plate convergence that occurs between India and southern Tibet (~20 mm/yr) has been accommodated on the Main Frontal Thrust. This argument is consistent with the megathrust earthquakes that have occurred on the plate-boundary fault descending northward from the Main Frontal Thrust (Fig. 6.31), and it emphasizes again that the currently slow geodetic shortening across this thrust is most likely a response to interseismic elastic strain being stored within the Greater Himalaya (Fig. 5.7) until the next megathrust earthquake (Avouac, 2003).

Usually, as rivers incise into their valleys, terraces along their sides are gradually eroded away by hillslope processes. Thus reconstruction of older river profiles can be nearly impossible in many mountainous regions. Sometimes, however, fortuitous circumstances provide a different means of preserving former valley bottoms. In the Sierra Nevada of California, for example, lava flowed down some alpine valleys during Miocene times and overran the former river bed. Because these lavas were more resistant to erosion than were the surrounding rocks, they have been better preserved and actually stand above the modern valley bottoms, where they provide fine examples of inverted topography. These lavas help define the former profile of a major river flowing west from the Sierra crest. When combined with the modern river gradient in the same valley and the age of the flows, the difference between the profiles constrains the amount and rate of differential bedrock uplift since the lava flow engulfed the valley (Huber, 1981): in essence, these flows are huge tilt-meters. Although the flows seem like robust markers, they are preserved much closer to the edge of the range than to its crest. Hence, their tilt only places loose constraints on the total tilt and differential uplift of the range.

Tectonic and surface uplift rates

Recall that the mean surface is the same as the mean elevation, but that individual points on the surface may increase in elevation at the same time as the mean elevation or surface height is decreasing (Fig. 7.2). Increases in the mean elevation of a region can only occur in response to tectonic processes of crustal thickening, flexural support due to bending of rigid lithosphere, or changes in the density distribution of the crust and underlying mantle. When assessing changes in mean surface elevation, generally we would like to know how much the mean elevation has changed and what role isostatic uplift played in raising parts of the surface. *Tectonic uplift* can be defined as that portion of the total uplift of the mean surface that is not attributable to an isostatic response to unloading. In order to resolve these different components of uplift, we need to know the geometry of the surface at the start of deformation, the present geometry, the volume of material eroded off the original surface, and any changes in the reference frame, such as in sea level, that are used to assess displacement. If we want to define rates, then ages for the beginning and end of deformation are also needed.

Initially, we can note that the total bedrock uplift for a point (Abbott *et al.*, 1997) (Fig. 7.26A) is defined as

$$U_i = (Z_i - Z_{0i}) + E_i + \mathrm{SL}_i \qquad (7.8)$$

where, at any point i, U_i = total bedrock uplift, Z_i = present topographic elevation, Z_{0i} = original topographic elevation, E_i = thickness of eroded material, and SL_i = change in sea level between the beginning of uplift and the present (a sea-level rise is assumed to be positive). The total bedrock uplift can be expressed as the sum of a tectonic and isostatic component:

$$U_i = U_{ti} + U_{ei} \qquad (7.9)$$

where, at point i, U_{ti} = tectonic component of uplift and U_{ei} = isostatic component of uplift. Combining the above equations and adding in

the time of deformation, t, the tectonic uplift rate at a point is

$$U_{ti}/t = [(Z_i - Z_{0i}) + E_i - (U_{ei} + \mathrm{SL}_i)]/t \qquad (7.10)$$

The average of the tectonic uplift rates computed for a series of points across a range then defines a mean rate for the whole range. Note that the data have to be collected across a broad enough area to encompass the region that will respond isostatically to the removal of material by erosion. Thus, the rigidity of the crust and its effective elastic thickness, which together set its flexural wavelength, serve to define the approximate minimum dimensions of a study area. The isostatic response to localized erosion will be compensated and smoothed across this entire area.

Unfortunately, although the approach described is conceptually straightforward, this methodology can be applied to relatively few ranges in the world. The availability of digital topography makes it easy to calculate the present mean surface elevation, but defining a reliable paleoelevation for an ancient surface is commonly impossible. The optimal strategy for making calculations of tectonic uplift rates (Fig. 7.26) appears to require that we study ranges that have experienced large changes in mean elevation, where good dates are available for former surfaces, where enough of an original surface is preserved to permit calculation of the amount removed by erosion, where the amount of sea-level change (SL) is small compared to the change in mean elevation, and where some data can serve to define paleoelevation (Z_{0i}) of the pre-deformational surface.

A successful study in the Finisterre Range of Papua New Guinea (Abbott *et al.*, 1997) exploited a slightly denuded surface of carbonates that had been recently uplifted from considerable depths in the ocean. Foraminiferal assemblages were used to estimate the depositional depth and age of these carbonates. These estimates constrain both the paleoaltitude and the duration of deformation (t). Although the former depositional surface has been uplifted up to altitudes of more than 2 km from water depths of as much as 3 km, much of the original surface is still preserved. By smoothly connecting these preserved surface remnants, an envelope was defined that represents the pre-eroded, but uplifted,

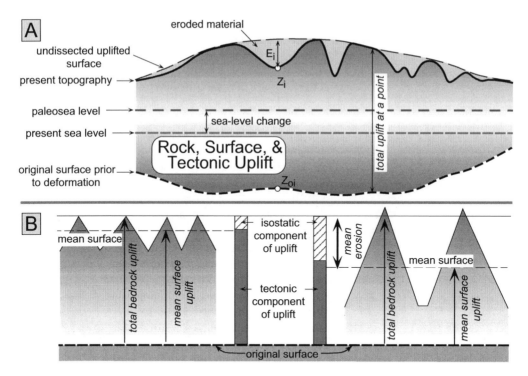

Fig. 7.26 Topographic variables needed to calculate the tectonic component of uplift.
A. Inputs to a calculation of tectonic uplift. Observables are the present topography (Z_i) and present sea level. Variables that are difficult to define can include the depth of eroded material (E_i) and the position of the original surface (Z_{0i}) with respect to a known reference plane (like former sea level), both of which are required to calculate total rock uplift. Modified after Abbott *et al.* (1997). B. Effects of erosion on isostatic uplift. Given an identical amount of bedrock uplift, the relative amount of erosion determines the proportion of the total rock uplift that is due to an isostatic response. In each case, the isostatic contribution is about five-sixths of the mean erosion. The remaining rock uplift is attributable to tectonic effects (crustal thickening or buoyancy effects).

surface. Subtracting the modern topography from this envelope provides an estimate of the amount of eroded material (*E*), and an isostatic response to this unloading can be calculated. In the Finisterre Range, Abbott *et al.* (1997) showed that the uncertainties resulting from erosional isostatic compensation and paleo-sea-level estimates are small (<10%) compared to a tectonic uplift rate that averages 1–2 mm/yr over the past 2–3 Myr. In the Finisterre Range, the small contribution of erosional isostasy to uplift is not surprising, because the raised surface is still only slightly dissected. In contrast, in settings where erosion has removed all but a few remnants of a former surface, isostasy may drive much of the observed rock uplift at isolated points in the landscape (Fig. 7.26B).

Sometimes it can be possible to use large-scale patterns of deposition to discriminate between rock uplift due to tectonic loading versus erosionally driven isostatic uplift. Both of these mechanisms can drive local surface uplift of peaks, but in the former case, uplift results from crustal thickening, whereas in the latter case, it results from erosional thinning of the crust. Collisional mountain belts typically abut basins that receive sediments derived from the adjacent mountains. Two aspects of the basin fill and river systems can distinguish between tectonic loading and erosional unloading (Burbank, 1992). During crustal thickening that is associated with tectonic loading, maximum basin subsidence occurs near the mountain front. This depression is commonly occupied by a longitudinal river flowing parallel to the mountain front in a medial to proximal part of the basin (Fig. 7.27). Transverse rivers debouching from the mountains typically join this trunk

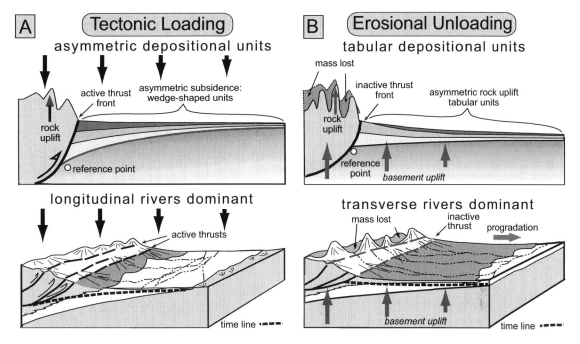

Fig. 7.27 Effects of tectonic loading or erosional unloading on subsidence and depositional patterns in foreland basins.
Either tectonic loading or erosional unloading can cause uplift of individual summits, but, during the latter, mean surface elevations decrease. A. Tectonic loading is typified by asymmetric subsidence in the foreland basin, driven primarily by crustal thickening in the hinterland. In this scenario, transverse rivers are short and join a longitudinal river in a proximal to medial part of the foreland basin. Coarse-grained deposits tend to be restricted to near the mountain front. B. Erosional unloading in the hinterland drives in rock uplift in the proximal foreland. In this scenario, transverse rivers flow across much of the foreland, longitudinal rivers are in a distal position, and coarse facies may prograde far across the basin. Modified after Burbank (1992).

stream a short distance from the mountain front and turn to flow parallel to the mountains. Coarse-grained sediment tends to be trapped in the proximal setting by the rapid subsidence (Heller *et al.*, 1988). In the subsurface, a prism of sediments that thickens toward the mountains provides evidence of persistent tectonic loading in and preferential subsidence toward the hinterland (Flemings and Jordan, 1989).

If the regime switches to one in which erosion prevails over tectonic thickening, such that the crust actually thins, then the reduced mass of the mountains will drive an isostatic uplift within the proximal foreland as well, as a function of the finite flexural rigidity of the crust (Fig. 7.27). Erosion will occur in the proximal foreland, and coarse-grained sediments may prograde as tabular, thin sheets across the medial and distal foreland (Heller *et al.*, 1988).

Transverse rivers can flow across nearly the entire foreland basin before merging with a longitudinal trunk stream, which may itself be pushed toward the far edge of the basin by the proximal rock uplift (Burbank, 1992).

Such a pattern appears to be consistent with the geometry of modern rivers in the Himalayan foreland (Burbank, 1992). Whereas, during Miocene times, coarse strata remained near the mountain front and axial rivers dominated the medial foreland basin, during Quaternary times, tabular sheets of strata extend across the medial and distal foreland, and transverse drainages stretch across the foreland. These changes suggest that the relative balance between crustal thickening and thinning has changed during the past few million years. If true, then uplift of summits observed today would represent an isostatic response to a reduction in the overall thickness of the crust.

Estimates of paleoaltitude

Determining the paleoaltitude of a surface or rock sample represents one of the biggest challenges facing anyone attempting to document uplift of surfaces in inland regions where markers referenced to sea level cannot be readily defined. For example, at an average modern elevation of 5000 m, the Tibetan Plateau is the highest region on Earth and has a major impact on global climate, sediment fluxes, and ocean chemistry. Despite its significance as a huge orographic feature, the timing and magnitude of uplift of Tibet are still highly controversial (Molnar *et al.*, 1993; Murphy *et al.*, 1997; Rowley *et al.*, 2001), with estimates ranging from Cretaceous to Late Pleistocene for when the plateau nearly attained its present altitude. Similar controversy surrounds the timing of uplift or collapse of western North America and the Puna–Altiplano plateau in the Andes.

How does one determine how high some surface was in the past? One means to assess surface uplift is to study the growth of mountain ranges that impact local climatic conditions. The rise of moist air masses as storms approach the flanks of mountains causes orographic precipitation to be focused on the windward side of the range (Bookhagen and Burbank, 2010). In the lee of the range, a rain shadow commonly develops. If you could detect the development of a rain shadow, you might infer that a mountain range had grown in a windward position. For example, the Miocene aridification of the western part of the Basin and Range has been attributed to the rise of the Sierra Nevada in California (Axelrod, 1957) and its interception of moisture from Pacific storms that used to penetrate farther into the continental interior (Smith *et al.*, 1992). Although this interpretation may be correct, it is difficult to place reliable limits on how high the range would have to be in order to produce the observed effect, nor can subsidence of the Basin and Range, as opposed to uplift of the Sierra Nevada, be ruled out as a cause of the observed change. In addition, this approach ignores the impacts of climate changes that occur irrespective of the growth of a given range.

Because atmospheric temperatures generally cool with increasing altitude (a lapse rate of ~6.5°C/km is commonly assumed), as a region rises tectonically and the surface temperature cools, temperature-sensitive plant species and the animals dependent upon them should migrate off the rising surface and move toward lower sites in order to remain in the same temperature range. Another approach to quantifying surface uplift, therefore, begins with correlations between specific assemblages of modern plants and the ambient mean annual temperature. These analyses rely on the observation that the shape or physiognomy of leaves is sensitive to climatic conditions (Wolfe, 1993). A smooth or serrated leaf margin, the presence or absence of a drip tip (an elongate, pointed tip that sheds moisture more readily), leaf size, shape of the base, and spacing of the teeth can each be statistically related to the mean annual temperature at which the plant is growing. For example, a well-documented positive correlation exists between the percentage of entire-margined (non-serrated) species in any assemblage and the mean annual temperature (Wolfe, 1971). Having established these correlations, paleoassemblages of flora can be examined and assigned a mean annual temperature (T_{int}) based on their closest modern analog. Subsequently, these temperatures can be compared with those reconstructed for assemblages of similar age that existed at sea level (T_{sl}). Based on these estimates and an assumed terrestrial lapse rate (TLR), a paleoaltitude (z) can be calculated:

$$z = \frac{T_{sl} - T_{int}}{\text{TLR}} \qquad (7.11)$$

Although fairly straightforward, this approach suffers from at least four limitations. First, because many species have a fairly wide temperature tolerance, their presence as fossils may be only a loose indicator of former temperatures. Second, relatively few fossils are preserved in continental strata. In marine studies, the problem resulting from the wide temperature tolerance of any individual species is overcome by examining many coexisting species to determine a limited temperature range for that assemblage. In terrestrial settings, unfortunately, a paucity of fossil remains commonly precludes estimates with a narrow temperature range. Third, by equating a change in mean annual temperature with a change in altitude of a surface, one ignores the effect of global climate change. The onset of the Ice Ages in the late Cenozoic has lowered mean global temperatures, such that a floral grouping that might have been well adapted at 3000 m in the past is now confined to much lower altitudes. This temperature-dependent response could happen irrespective of surface uplift or subsidence. Without independent evidence of the amount of climate change, inferences of altitudinal changes based on temperature-sensitive assemblages alone are highly suspect. Fourth, correlation of fossil assemblages with modern analogs assumes little or no evolution of these plants in response to environmental change. Finally, the technique also assumes that a terrestrial lapse rate can be reliably assigned to a time typically tens of millions of years ago.

Floristic approaches to estimating paleoaltitudes have recently been improved. One new method utilizes an increase in the stomatal density in leaves that occurs in response to the well-known decrease in the partial pressure of CO_2 at higher altitudes (McElwain, 2004). Analyses of the stomata of fossil leaves could potentially reveal paleoaltitudes with a resolution of a few hundred meters. Leaf-margin analyses have improved owing to two key advances. First, application of multivariate analysis to the physiognomy of leaves has led to a quantitative calibration of modern meteorological environments with physiognomy (Wolfe, 1990). Second, rather than assuming a lapse rate, the moist static energy of an air mass is used to calculate paleoaltitude (Wolfe et al., 1998). Moist static energy (h) is a thermodynamic parameter representing the total energy content of a parcel of air, excluding a negligible amount of kinetic energy. It comprises two components, enthalpy (which consists of both latent heat and thermal energy) and potential energy (due to altitude):

$$h = c_p T + L_v q + gZ = H + gZ \qquad (7.12)$$

where c_p is the specific heat capacity of moist air at constant pressure, T is temperature, L_v is the

latent heat of vaporization, q is specific humidity, g is acceleration due to gravity, and Z is height. On the right-hand side, H is enthalpy and equals $c_p T + L_v q$. Moist static energy is presumed to be conservative, which means that the balance between potential energy, latent heat, and thermal energy may change over time, but their sum remains constant as an air mass moves inland along a trajectory from a coastal site. Multivariate analysis shows that enthalpy is second only to mean annual temperature in terms of its correlation with leaf characteristics. If it is assumed that air masses move zonally (within restricted latitudinal bands) and that the preserved flora are truly isochronous at sites being compared, then changes in enthalpy between sites in the same latitudinal band can be used to calculate changes in potential energy and, therefore, altitude. Owing to likely departures from truly zonal circulation and to uncertainties in the enthalpy calculation, recent paleoaltitude calculations have uncertainties estimated to be 750–900 m (Wolfe et al., 1988). Clearly, this approach is unsuitable for determining small variations of altitude at a given site. Although the uncertainty on an individual estimate is large, studies of multiple related sites allow reduction of the overall uncertainty. The validity of the assumptions (isochronous sites, zonal transport) that underpin parts of this methodology are difficult to evaluate in the past, such that the actual uncertainty may depart significantly from the formal uncertainty.

For many years, a lively controversy has flourished about the paleoaltitude of western North America. Many researchers concluded that this region had risen to its present high elevation during late Cenozoic times (Box 7.4). Application of multivariate analyses on leaves (Gregory, 1994; Gregory and Chase, 1992) and enthalpy calculations (Wolfe et al., 1998) now suggest that many of these areas have been high since early Cenozoic times, and that some, such as the Basin and Range, are considerably lower now than they were 15 million years ago. Clearly, these reconstructions change significantly the way one thinks about the history of western North America!

As storms precipitate on the flanks of mountains, the isotopic composition of their

Box 7.4   Basalt vesicles and paleoaltitude.

The search is ongoing for reliable methodologies that are capable of converting measurable attributes of a pre-Holocene rock sample to the paleoaltitude at which it was formed. One promising strategy utilizes vesicles that are frozen into a lava flow at the time of its eruption (Sahagian and Proussevitch, 2007). Basaltic lava, for example, emerges from a volcano as a bubbly liquid in which bubbles nucleate and grow in the magma due to decompression during rise through the volcanic conduit. These bubbles are well mixed when the magma emerges from the vent and becomes a lava flow. Hence, the bubble population contains equal amounts of gas throughout the flow. As it is emplaced, the flow cools from top and bottom, thereby freezing in vesicles that surround gassy bubbles, while bubbles rise buoyantly through the core of still-molten lava at a speed that depends on their individual sizes. In the end (see figure A), this process results in a lower vesicular zone, a central massive zone (from which all bubbles were able to escape a very slowly rising solidification front), and a highly vesicular upper vesicular zone, the lower part of which includes very large bubbles that coalesced while rising up through the central zone.

Although the bubbles would thus have identical mass distributions at top and base, they are subject to different total pressures due to differences in overburden. At the top of the flow, only the overlying atmosphere exerts pressure, P_{atm}, whereas at the base, the lava itself exerts pressure as a function of its thickness, H. Consequently, two factors control the size of bubbles at the base of the flow: atmospheric pressure and lava weight. Thus, the atmospheric pressure dependence of vesicle size can be expressed by the ratio of vesicle size modes at the top and bottom of a flow:

$$\frac{V_t}{V_b} = \frac{P_{atm} + \rho g H}{P_{atm}}$$

where V_t and V_b are the volumes of the modal bubble sizes at the quenched top and bottom of the flow, respectively, and ρ is lava density ($2650\,kg/m^3$ for basalts). The atmosphere's paleopressure can thus be determined, because all other variables can be measured, and, based on the known relationship of pressure as a function of elevation, a paleoelevation can then be calculated.

In order to make reliable calculations of atmospheric pressure with this approach, however, the thickness of the lava flow measured in the field today must be the same as its thickness at the time of solidification of the upper and lower parts of the flow, as bubbles were "frozen in." Even when the flow top and bottom have been cooled, the thickness of the still-molten flow interior can change due to inflation or deflation. Because evidence for either process can usually be observed by an astute field geologist, much effort is commonly expended on scouting for flows unaffected by inflation or deflation.

After samples have been collected from the top and base of flows with reliable thicknesses, the size distribution of their

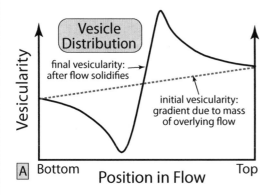

A. Predicted initial and final vesicle distribution in a 3 m thick flow. Modified after Sahagian *et al.* (2002a).

vesicles must be determined in the lab. Several approaches have been used to determine modal bubble size: infusing a plastic into the sample and then dissolving the surrounding rock; making random cuts through a sample and using the statistics of two-dimensional sections to calculate the three-dimensional vesicle distribution; or using X-ray tomography to image distributions directly (see figure B).

used to determine paleoelevation at the time of eruption. Because basalts can be dated radiometrically, the time at which the lava was at that elevation can be determined, and the uplift or subsidence of the site can be determined by subtraction from the current elevation. Recent applications of this method to the Colorado Plateau (Sahagian *et al.*, 2002b) suggest that rates of Cenozoic rock uplift increased about five-fold in the past 5 Myr (see figure C).

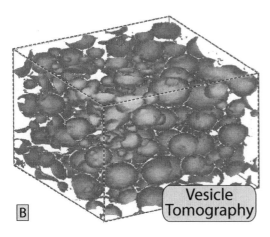

B. Tomographic reconstruction of vesicles. Modified from Sahagian *et al.* (2002a).

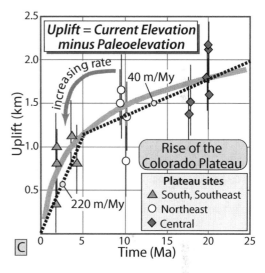

C. Reconstructed net Cenozoic uplift of the Colorado Plateau. Modified from Sahagian *et al.* (2002b).

With a vesicle-size distribution and flow thickness in hand, the equation above can be

rainfall changes. At higher altitudes, the $^{18}O/^{16}O$ ratio is lower than at lower altitudes (Fig. 7.28). Surface waters, as well as plants, animals, and minerals that are in equilibrium with the surface waters, reflect these isotopic compositions of the water in which they grow, and therefore can preserve a record of the variations in isotopic signature (Fig. 7.29). The presence of these isotopic changes suggests some ways in which the growth of mountain ranges could be documented. For example, the isotopic composition of flora or fauna should differ both with altitude and across a range with strong orographic precipitation. Thus, if you were to

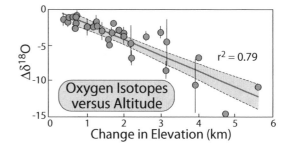

Fig. 7.28 Oxygen isotopic composition as a function of altitude.
Compilation of global data for altitude-dependent changes in oxygen isotopic ratios as expressed in $\delta^{18}O$ ratios. Modified from Chamberlain and Poage (2000).

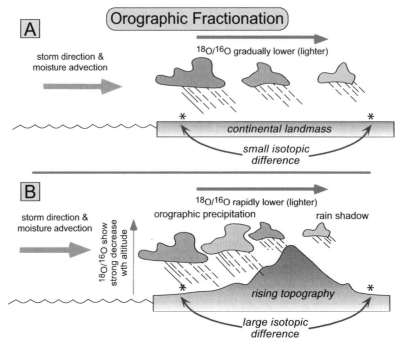

Fig. 7.29   Isotopic fractionation of precipitation and the effects of topography.
A. In the absence of a mountain range, the oxygen $^{18}O/^{16}O$ ratio gradually decreases inland with greater distance from the moisture source. B. In the presence of topography, isotopic fractionation is strongly altitudinally dependent as a result of orographic precipitation. The increased fractionation also causes a stronger gradient in isotopic ratios across the mountain at the same altitude. Stars mark sites that would display weak and strong contrasts in isotopic ratios before and after the mountain range grew, respectively. Importantly, the isotopic ratio at the downwind site reflects not its paleoaltitude, but rather the height of the mountain passes over which the moisture-bearing winds passed.

compare the composition on opposing sides of the range from sites that pre-date and post-date growth of the range, the isotopic difference between the sites should increase significantly as the range rises and increases the fractionation (Fig. 7.29). Because the temperature at the time of precipitation also affects isotopic fractionation, climate change can still affect the isotopic composition of rainfall and needs to be considered. Nonetheless, recent isotopic studies from the South Island of New Zealand clearly demonstrate a strongly increased fractionation from west to east as the Southern Alps rose into the path of storms driven by the prevailing westerly winds during the past 5 Myr (Chamberlain *et al.*, 1999). If many sites representing many time intervals could be sampled, these diverse data could place some robust timing constraints on the topographic growth of the range.

Over the past two decades, numerous studies have exploited $^{18}O/^{16}O$ ratios of authigenic clays, soil carbonates, and volcanic glasses to estimate paleoaltitudes (Cassel *et al.*, 2009; Garzione *et al.*, 2000; Mulch *et al.*, 2006; Rowley and Garzione, 2007). Most such estimates do not account for the fact that growing mountains or plateaus are likely to induce regional climate change in which the distribution of rainfall, the average surface temperatures, and even the major storm tracks and moisture sources, can change (Ehlers and Poulsen, 2009). Each of these changes could add significant uncertainty to paleoaltitude reconstructions based on $^{18}O/^{16}O$ ratios. Some of these issues can be avoided with a new isotopic method, termed the "clumped-isotope thermometer," that uses the bonding characteristics of ^{13}C and ^{18}O to determine the paleotemperature ($\pm 1°C$) of formation of soil carbonates (Eiler, 2007; Ghosh

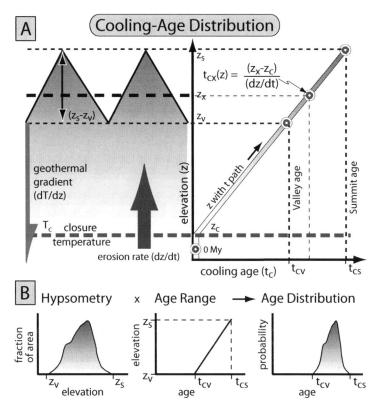

Fig. 7.30 **Cooling ages as a function of relief and hypsometry.**
A. Model for predicting the distribution of cooling ages (t_c) derived from a range with a given relief expressed as the difference between the summit (z_s) and valley (z_v) elevations. A linear age–elevation relationship predicts ages anywhere within the landscape. Closure temperature (T_c) occurs at depth z_c.
B. Probability distribution of detrital cooling ages reflects a convolution of the hypsometry of the range with the age–elevation curve. Modified after Brewer *et al.* (2003).

et al., 2006). This method can also both constrain the $^{18}O/^{16}O$ ratio of the water from which the carbonate grew and underpin a correlation between soil temperature and the isotopic composition of the water. In theory, this correlation enables discrimination between the effects of altitude changes versus those due to climate.

Paleorelief

As rivers cut down into newly rising mountains, the topographic relief commonly grows. Although relief can be independent of elevation, relief has been used as an indicator of erosion rates (Fig. 7.3A). Clearly combinations of relief and drainage spacing control average slope angles, and relief affects the way storms interact with topography to produce orographic rainfall (Bookhagen and Burbank, 2006, 2010). Consequently, reconstructions of paleorelief can provide a valuable window on former landscapes.

In order to define paleorelief, a tracer is needed that is sensitive to relative or absolute elevation

and preserves its signature after it has been eroded from the bedrock. One such tracer relies on cooling ages because many mountain ranges display a generally layer-cake stratigraphy of ages. A sand sample eroded from such a range should contain a suite of individual grain ages that span the exposed topography and its associated ages (Stock and Montgomery, 1996). If a linear gradient of cooling ages within the paleo-topography is assumed (as is typical for many thermochronometers; Fig. 7.20) and erosion was spatially uniform, then the frequency distribution of ages should mimic the hypsometry, i.e., the frequency distribution of elevation, within the source catchment (Fig. 7.30). With the advent of high-precision and relatively inexpensive techniques for dating individual grains, it has become practicable to date hundreds of detrital grains. For any given catchment, if the goal is to capture at least 95% of the paleorelief, then a minimum of about 120 grain ages is required (Vermeesch, 2004). For a linear dependence of age on elevation, changes in relief and/or erosion rate will

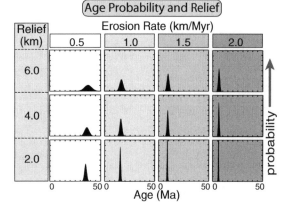

Fig. 7.31   Cooling ages as a function of changes in relief and erosion rate.
Probability distributions of cooling ages for a mineral with a closure temperature of 350°C. A Gaussian-shaped hypsometry is assumed. For any given erosion rate, the span of ages increases with increasing relief, although the mean age remains constant. At increasingly higher erosion rates, the mean age gets systematically younger, whereas the span of ages decreases for any fixed amount of relief. Modified after Brewer *et al.* (2003).

change the expected distribution of ages (Fig. 7.31). The range of ages (oldest–youngest) increases as relief grows, whereas the mean cooling age gets younger as erosion rates increase. Such changes could underpin a research strategy for sampling the stratigraphic record in order to deduce changes in both paleorelief and erosion rate through time. The challenges in applying such a technique, however, are many (Stock and Montgomery, 1996): sampled sediment must have been derived from the same catchment over time; the mineral used for dating must consistently be present at both the top and bottom of the catchment; and any changes in erosion rate must have happened sufficiently long ago and then be sustained at a constant new rate such that the age–elevation relationship in the mountains has been reset throughout the eroding topography. Despite these caveats, in carefully chosen settings, paleorelief can be estimated with more clarity than at any time in the past.

Summary

The balance between erosion and rock uplift lies at the core of tectonic geomorphology. Most landforms in actively deforming areas result from this interplay. New geochronological tools have greatly enhanced our ability to date surfaces, landforms, and events with increasing precision and accuracy. Underpinning any rate calculations, this new temporal control permits both an assessment of conceptual models, such as that of a topographic steady state, and a calibration of rates of erosion, rock and surface uplift, deposition, and deformation – those factors that modulate the evolution of landscapes.

A key realization of recent studies is that mechanical erosion can proceed at rates as high as 5–10 mm/yr. Although the possibility must be assessed in each situation, it is no longer necessary to attribute high rates of cooling (>100°C/Myr) or rapid decompression (>1 kbar/Myr) primarily to tectonic denudation via normal faulting. In some mountain belts, erosion by glaciers, rivers, or landslides has now been measured at rates rapid enough to accommodate nearly all of the documented rates of unloading.

The increasingly widespread availability of digitized topography is permitting rapid characterization of landscapes and facilitates easy calculation of both key topographic variables, such as relief and slope, and geophysical variables, such as mean surface elevation or eroded volumes of rock. Although some of these variables may be statistically tied to erosion, by themselves, they tell us little about the mechanisms of erosion. Key frontiers related to erosion and rock uplift in tectonic-geomorphic studies include the documentation and quantification of erosion processes, discovery of new ways to define key variables, such as paleoaltitude or geothermal gradients, and a fuller exploration of the interplay between deformation and erosion in contrasting climatic and tectonic regimes.

8 Holocene deformation and landscape responses

The record of the most recent 11 000–12 000 years is, on average, considerably more complete than in any interval of similar duration in the past. Even compared with the Pleistocene, we know more during the Holocene about when specific faults ruptured and how much slip occurred in each event, we have much more detail about the tempo and magnitude of climate change, and we have better temporal control on diverse processes, such as rates of erosion, the strength of the monsoon, or earthquake-induced sediment pulses. Better than any time in the geological past, the Holocene allows us to examine interactions, cause and effect, and positive and negative feedbacks. Hence, Holocene studies provide a unique opportunity for tectonic geomorphologists to gain robust insights on how various parts of the geomorphic system work and interact. It is important to remember that specific rates and processes in the Holocene (especially the late Holocene) are unlikely to characterize many previous Quaternary intervals: sea level has been quite steady for the past 7000 years; the Asian monsoon is weak; glaciers are highly retracted during the current interglaciation; and human society is having an unprecedented impact on the Earth's surface and atmosphere. Despite such differences that impede use of the Holocene as a modern analog for most of the Quaternary, the quantitative understanding that can be gained on interacting processes during the Holocene can be aptly applied to the same processes in the Quaternary. Thus, knowledge of how sediment fluxes change in response to increased precipitation, vegetation changes, seismic activity, or glacial flow rates provides an invaluable foundation for assessing impacts of changes in these variables in the past. Similarly, in an effort to understand how landscapes evolve, we can study Holocene *transients* (intervals of change from one state to another) that enable us to measure many of the relevant variables. Consider, for example, a modern river whose channel profile has been offset by a large earthquake or that has been abruptly diverted by a landslide. In either case, the river probably has to begin incising a new channel into solid bedrock. Given our ability to measure discharge, channel slope, channel cross-sections, sediment loads, rock strength, and differential bedrock incision, measurements of these variables and the temporal evolution of the channel can provide insights on the controls, rates, and interactions of incision processes that would be impossible to glean in past geological intervals.

Although paleoseismic studies tend to focus on the intense deformation expressed in rather narrow fault zones, interferometric geodetic data (Fig. 5.17) clearly demonstrate that large earthquakes can cause significant displacement more than 100 km away from a fault. Geomorphic responses at Holocene time scales to such regional deformation, as well as to localized

Tectonic Geomorphology, Second Edition. Douglas W. Burbank and Robert S. Anderson.
© 2012 Douglas W. Burbank and Robert S. Anderson. Published 2012 by Blackwell Publishing Ltd.

faulting, form one focus of this chapter. Despite recognition that many elements of the geomorphic system in alluvial settings are linked to each other, each element does not have a similar response time or sensitivity to changes imposed on it (Whipple and Tucker, 1999). Reconsider, for example, a drainage basin (Fig. 1.3) and the hierarchy of sensitivity of its landscape elements to imposed changes. In order to change the catchment area, the interfluves have to be shifted laterally, but in order to shift the interfluves, the hillslopes that define where the interfluves are located have to migrate in space. Changes in the shape of hillslopes and removal of material from them are typically most sensitive to slope angle, with higher angles leading to less slope stability and more rapid material transport (Fig. 7.3). Changes in the slope angle itself are related to changes in the altitudinal differences between the interfluves and the river channels. If the river channel is lowered by incision erosion, adjacent hillslope angles are steepened, rates of creep should increase, and the hillslope may become unstable and prone to landsliding. Thus, the erosion or aggradation of river channels will strongly influence hillslope responses. In a drainage basin, a hierarchy of sensitivity to most tectonically imposed changes would look like this: catchment area (least sensitive) to interfluves to hillslopes to channels (most sensitive).

Another way to envision the sensitivity of a landscape is to consider the impact on it that a relatively small, tectonically induced, change might have. For example, if folding causes a region to be tilted 1°, what difference would this make to various elements in the drainage-basin hierarchy? The catchment area and interfluves would be insensitive to such changes at short time scales. If hillslopes were poised at maximum stable slope angles, then some of them could be destabilized by such tilting, but, under most circumstances, they, too, would be largely unaffected. River channels, on the other hand, typically have equilibrium gradients of considerably less than 1°. Our hypothesized slope increase of 1°, therefore, would greatly increase the stream power due to its proportionality to

slope (Box 2.2). A river could respond very quickly to this increased power by starting to erode its bed. In a sense, landscape elements like catchment areas and interfluves can be characterized as having considerable geomorphic *inertia*, such that they tend to change slowly, whereas the dynamic nature of rivers and their rapid responses to changing external controls indicate that they have little inertia. In terms of examining the impact of tectonism on Holocene landscapes, those elements that have little geomorphic inertia provide the clearest responses.

It is useful to keep in mind the magnitude of typical deformation rates. Most vertical deformation occurs at rates of a fraction of a millimeter per year. Uplift at 1 mm/yr yields 10 m of uplift during the Holocene. Very rapid and sustained bedrock uplift rates are seen in ranges like the Himalaya (Burbank *et al.*, 1996b; Zeitler *et al.*, 2001b), the Southern Alps of New Zealand (Tippett and Hovius, 2000), and the Central Range of Taiwan (Dadson *et al.*, 2003; Willett *et al.*, 2003), where rates as high as 10 mm/yr have been described. We have already seen that such rates in the Himalayan foreland have driven more than 100 m of vertical displacement of Holocene river terraces (Fig. 7.25). Rates of horizontal displacement on strike-slip faults can be much more rapid (>30 mm/yr) and might be expected to perturb geomorphic systems even more. Indeed, such faults cause significant lateral displacements of features at Holocene time scales (Liu *et al.*, 2004; Sieh and Jahns, 1984). Because strike-slip faults may induce little differential vertical movement, however, the millennial-scale stability of hillslopes and the profiles of streams along strike-slip faults are typically less affected than in settings dominated by dip-slip faulting.

Given (i) the hierarchy of sensitivity to change within terrestrial geomorphic systems, (ii) the magnitude of deformation during Holocene tectonism, and (iii) the relative impacts of vertical versus horizontal displacements, an examination of Holocene landscape responses to deformation in terrestrial settings focuses naturally on fluvial systems. River channels typically extend

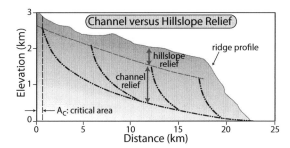

Fig. 8.1   Relief of rivers and hillslopes in nonglacial landscapes.
River channels occupy most of the relief in many landscapes. Their incision rates determine local base level for all bounding slopes, and rivers are the most geomorphically sensitive element to tectonic perturbations at Holocene time scales. The critical area, A_c, comprises the catchment area above the head of the fluvial channel. Modified after Whipple and Tucker (1999).

through 70–90% of the topographic relief of non-glacial landscapes (Fig. 8.1) (Whipple, 2004), and, although fluvial channels occupy only a tiny fraction of any catchment area, their behavior is a key control on adjacent slopes. We will, therefore, examine some theoretical, experimental, and natural responses of rivers to deformation at the scale of a few meters to tens of meters over periods of decades to thousands of years.

Base level

John Wesley Powell (1875) introduced the concept of *base level*: the lower limit of the landscape below which rivers cannot erode. In most cases, the ultimate base level is sea level, although in closed tectonic depressions, such as Death Valley or the Dead Sea, it can be lower. *Local base level* refers to the lowest topographic point in any particular area. A lake, for example, represents the lowest level to which upstream rivers and hillsides can erode for the duration of the lake's existence. The *base level of erosion* is a concept that involves all reaches of a fluvial system. It is the "equilibrium (graded) longitudinal

profile below which a stream can not degrade and at which neither net erosion nor deposition occurs" (Bull, 1991). When a river is at the base level of erosion, its equilibrium longitudinal profile can be regarded as an infinite succession of adjacent local base levels that will be unchanging until the controlling variables on the system change (Gilbert, 1879). As the river adjusts to imposed changes over time, it may re-establish an equilibrium profile at several positions within the landscape.

Theoretical perspective on fluvial erosion and river profiles

Lowering of a river's profile through time requires incision of its bed. A river flowing in alluvium can typically lower its bed much more readily than one flowing on bedrock. In the former case, incision depends on mobilizing sediment on the river bed, whereas in the latter case, the underlying bedrock has to be loosened and then mobilized. Hence, these two classes of rivers are typically defined as *transport-limited* or *detachment-limited* rivers. In Chapter 2, we briefly examined the concept of stream power (Box 2.2): a representation of the energy expenditure that results from changes in potential energy as water flows downstream. Stream power can be estimated in the context of a river's entire discharge at a point:

$$\Omega = \rho_w g Q_w S \tag{8.1}$$

where Ω is stream power, ρ_w is the density of water, g is gravitational acceleration, and Q_w and S are total discharge and average channel slope at that point, respectively. Specific stream power, or the power per unit area of the bed, is equal to Ω/w, where w is the bed width. Specific stream power can also be expressed as the product of bed shear stress, τ, and mean velocity, $\bar{v}$, as follows:

$$\Omega / w = \rho_w g d S \, \delta x / \delta t = \tau \bar{v} \tag{8.2}$$

where d is the water depth, x is bed length and t is time.

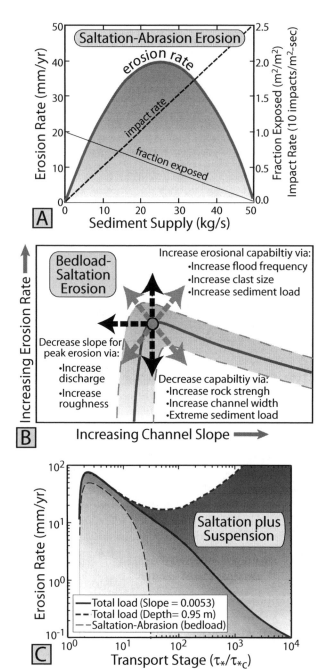

Fig. 8.2 Erosion by saltating bedload and suspension.

A. Erosion by saltating bedload alone is predicted to vary as a function of the rate at which clasts impact the bed, the energy per impact, and the fraction of the underlying bedrock exposed. Because both high impact rates and abundant bedrock exposure favor erosion, the most rapid erosion occurs where these competing factors are roughly balanced. Note that the cited erosion rate represents the annual erosion that would occur if flood conditions persisted all year. The actual erosionrate, therefore, depends on the frequency of such floods. Modified after Sklar and Dietrich (2004). B. Effects on maximum saltation–abrasion erosion rate due to changes in a single variable, when others are held constant. Rates increase with greater clast size, sediment load, and flood frequency, because each tends to increase the total energy delivered to the bed. Changes in discharge or roughness, while other variables are constant, shift the slope at which maximum erosion occurs, but not its magnitude. Modified after Goode and Burbank (2009). C. Predicted erosion due to impacts by both bedload and suspended load. The solid line represents erosion when the channel slope is fixed (0.0053), such that depth increases with higher transport stages. Alternatively, if depth is fixed (0.95 m), variations in slope can drive accelerated erosion at high transport stages (thick dashed line). Erosion by bedload only (thin dashed line) is shown for reference. In all models, 60-mm-diameter clasts are assumed, but sand is also included in the saltation–suspension model. Note that transport stages of ≥100 times the critical shear stress are considered exceptional flows. Modified after Lamb et al. (2008).

Intuitively, the incision of the bed of a river is more related to specific stream power than to total stream power, both because specific stream power relates to energy expenditure per area of the bed and because shear stress exerts the force along the bed that moves sediments and removes bedrock. Nonetheless, neither total nor specific stream power actually provides much insight on the processes that cause a river to erode its bed. Many numerical formulations for river incision incorporate a threshold shear stress, τ_c, that has to be exceeded before any erosion can occur, because such stress is required to get bedload moving. In fact, some recent models focus on the erosion caused exclusively by saltating bedload, in which erosion is driven by the energy of clast impacts against the underlying bedrock (Sklar and Dietrich, 1998, 2004). Both theory and laboratory experiments on saltating bedload suggest that, if too little bedload is available, erosion is ineffective, whereas if too much bedload is present, a high proportion of the bedrock is covered and, thereby, protected from erosion (Fig. 8.2A). These results predict that a sweet spot exists in which erosion rates will be maximized. This maximum rate is modulated by interactions among many variables in some predictable, but also in some surprising, ways (Fig. 8.2B). For example, the maximum erosion rate generally increases with larger clast sizes, more frequent floods, and greater channel depth (Goode and Burbank, 2009). In contrast, with larger discharge, the maximum rate does not change, but it occurs at a lower channel slope. In the original bedload-saltation model, erosion went to near zero as the transport stage greatly increased because more and more particles went into suspension. A newer and perhaps more realistic model includes the predicted erosion due to impact of the suspended load as well (Lamb *et al.*, 2008). This inclusion creates a much more extended tail or even an acceleration of erosion at high transport stages (Fig. 8.2C).

To test any of the erosion models, a direct measure of the magnitude of erosion and its location within the channel would be very valuable. Given the very high precision that is required for annual or storm-by-storm measurements, relatively few such studies have been undertaken (Hancock *et al.*, 1998; Hartshorn *et al.*, 2002). With the growing availability of terrestrial laser scanners that can create high-precision surveys of exposed river beds at low flow, the opportunity now exists to provide more robust documentation of event-driven erosion.

Some existing studies reinforce arguments for the importance of saltation or suspended loads. For example, detailed surveys before and after supertyphoons in Taiwan revealed gravel aggradation across the deeper parts of the bedrock channel during the typhoons and focused lateral erosion on the bedrock walls just above the top of the gravel fill (Turowski *et al.*, 2008). On the other hand, the presence of a hierarchy of bedrock flutes of different spatial scales on the downstream flank of large boulders and bed irregularities argues for effective erosion by suspended sand loads in turbulent Himalayan rivers (Whipple *et al.*, 2000).

Whereas discharge is almost always difficult to measure, the proliferation of digital elevation models (DEMs) in the past decade makes extraction of stream networks and catchment areas from desktop computers straightforward. Because discharge is typically considered to be proportional to upstream area (Snyder *et al.*, 2003), area is commonly substituted for discharge in calculations related to stream power. This assumed proportionality is not strictly true: it deviates wherever topography exerts a strong control on rainfall (Roe, 2005). For many catchments of a few tens to thousands of square kilometers, however, area can be reasonably substituted for discharge, such that

$$Q_w = k_a A \tag{8.3}$$

where A is catchment area and k_a relates the mean rainfall to the discharge. Although the actual process of erosion (abrasion, plucking, cavitation, bedload impacts) is commonly unknown and undoubtedly varies in space and time on any river, erosion (E) is now commonly expressed in a stream-power formulation as

$$E = KA^m S^n \tag{8.4}$$

where K is a dimensional coefficient of erosion, while m and n are positive constants (Howard and Kerby, 1983; Whipple and Tucker, 1999). The exponent m can be varied to reflect whether erosion is considered proportional to total stream power, in which case $m \sim 1$ because $Q \sim A$, or to specific stream power, in which case

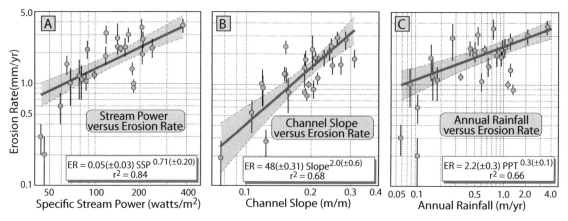

Fig. 8.3   Himalayan erosion rates as a function of stream power, channel slope, and rainfall.
Average catchment-wide erosion rates derived from analysis of detrital cosmogenic nuclide concentrations in the Himalaya and Tibet are compared with variables expected to influence erosion rates. A. Specific stream power incorporates spatial variability in rainfall distributions and provides the strongest correlation ($r^2=0.84$) with erosion rates. B,C: As separate variables, either channel slope (B) or rainfall (C) predict two-thirds of the variability in erosion rates, but explain about 25% less of the variability than does specific stream power (A). Compilation courtesy of Bodo Bookhagen.

$m \sim 0.5$ because $E \sim Q$/width, and width is typically considered to vary as a function of $Q^{0.5}$ (Whipple, 2004).

The proposition that greater discharge, steeper channel slopes, or greater stream power would lead to higher rates of erosion makes intuitive sense. Until recently, however, few data were available to provide a clear demonstration of a linkage between erosion rates and these quantities. As discussed in the previous chapter (Fig. 7.18), compilations of erosion rates at time scales of centuries to a few millennia have recently emerged from studies of cosmogenic nuclide concentrations in detrital sediments (Granger *et al.*, 1996; von Blanckenburg, 2006). Because these concentrations are interpreted to represent the average erosion rate in the catchment upstream of the sampling site, the rates are expected to be responsive to the catchment-averaged rainfall, channel slope, or specific stream power. Detrital cosmogenic data from the Himalaya and Tibet, for example, show quite strong correlations of erosion with each of these variables (Fig. 8.3). Variations in specific stream power predict over 80% of the variance in erosion rates, whereas variations in rainfall or channel slope predict about two-thirds of the

variance. In this Himalayan region, remotely sensed rainfall data with a high spatial resolution (5×5 km) show very strong orographic control: topography modulates rainfall distributions, such that five- to 10-fold gradients in rainfall over a few tens of kilometers are not uncommon (Bookhagen and Burbank, 2010). Hence, the summed rainfall over a catchment is a much more reliable predictor of discharge than is catchment area in the Himalaya. Not surprisingly, then, for these Himalayan data, discharge based on area is a considerably weaker predictor of erosion ($r^2=0.48$) than is rainfall ($r^2=0.66$). In assessing these correlations (Fig. 8.3), keep in mind two caveats: such correlations provide little insight on the actual process of erosion; and correlation does not necessarily indicate causation. It is possible that all of these factors (slopes, rainfall, erosion rates) are driven by another variable, such as rates of rock uplift or variations in rock strength. Nonetheless, if the correlation of specific stream power with erosion rates could be shown to be broadly applicable, the increasing availability of digital topography and remotely sensed rainfall would underpin far more efficient comparisons of spatial variations in erosion across tectonically active regions.

At any point along a river profile, the change in height with time, $\delta z/\delta t$, is the difference between rock uplift, U, and erosion (Eqn 8.4), such that

$$\delta z/\delta t = U - E = U - KA^m S^n \qquad (8.5)$$

In this context, K amalgamates many different variables that control erosional efficiency, including rock erodibility, sediment load, climate, erosion process, hydraulic geometry, and the return period for effective discharges. Even though many of these variables are usually poorly known, the effects of changes in some of them are commonly predictable: weaker rocks, higher rainfall, and shorter return periods should all lead to more rapid erosion for any given area or channel slope.

For a steady-state profile in which channel elevation at a point does not change, $\delta z/\delta t$ equals zero, and Eqn 8.5 can be rearranged in terms of the equilibrium slope, S_e, as follows:

$$S_e = (U/K)^{1/n} A^{-m/n} \qquad (8.6)$$

If U and K are constant, then the ratio m/n defines the rate of change of channel gradient as a function of drainage area. In other words, the long profile of the river is a function of m/n, which is defined as the *concavity* and is typically designated as θ. A channel with no concavity represents a linear gradient, whereas a channel with high concavity will have steep headwaters and gentle downstream slopes (Fig. 8.4).

For the purposes of analyzing river networks using DEMs, Eqn 8.6 is commonly recast as

$$S = k_s A^{-\theta} \qquad (8.7)$$

where k_s is the *steepness index* and equals $(U/K)^{1/n}$ (Whipple *et al.*, 1999; Wobus *et al.*, 2006c). The steepness index in relation to U/K makes intuitive sense: for any given catchment, we would expect channel steepness to increase for higher uplift rates (Fig. 8.3B) or to decrease for weaker rocks or more rainfall. Our intuition, however, can sometimes fool us. In looking at the profiles in Fig. 8.4B, many people would deduce that profile B has higher concavity, whereas in fact both profiles have identical concavities, but different steepness indices.

The utility of channel indices like concavity or steepness becomes obvious when applied to

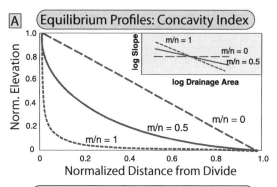

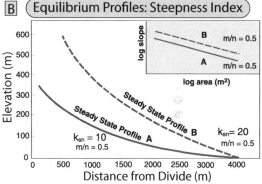

Fig. 8.4   Channel concavity and steepness.
A. Channel concavity, expressed as the ratio of m/n in the formulation for channel slope (Eqn 8.6). Inset shows how channel slope changes as a function of drainage area for different concavities. B. Despite identical concavity values ($m/n=0.5$; see inset), these profiles have different gradients as reflected by their steepness index, k_s, in Eqn 8.7. The steepness index is defined by the y intercept in log slope–log area space (inset). Modified after Duvall *et al.* (2004).

diverse DEMs. When data for channel slope versus upstream catchment area are systematically plotted in log–log space, several key channel characteristics typically emerge (Fig. 8.5). First, at small catchment areas, slope may be largely independent of area. This region is considered to be either hillslopes or channels where processes such as debris flows, rather than river incision, dominate erosion (Stock and Dietrich, 2006). Second, at a certain critical area that is equivalent to about $1\,km^2$ (or $10^6\,m^2$) in many studies (A_c in Fig. 8.5), the fluvial channel head is recognized by the start of a progressive decrease in channel slope. Third, in many analyses, this decrease can be fit by a line with a slope that is equivalent to the channel concavity, θ or m/n.

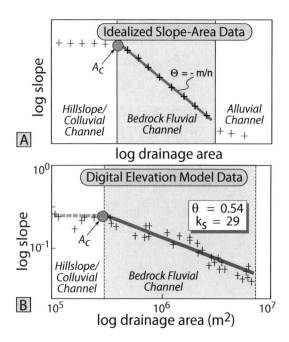

Fig. 8.5   Channel characteristics derived from DEMs.
A. In idealized log slope–log area space, channels with constant concavity define a straight line with slope θ, equal to $-m/n$. The fluvial channel begins at a critical catchment area, A_c. B. Example of slope–area data extracted from a 25-m DEM. This catchment is experiencing rock uplift of ~4 mm/yr, but shows average concavity. Despite rapid uplift, steepness is modest, at least in part due to very weak local rocks, which also promote a smaller-than-average critical area, A_c. Modified after Duvall *et al.* (2004).

For many catchments, concavities of 0.5 ± 0.2 are typical. Interestingly, except for rivers in a transient state, this concavity is largely independent of uplift rate, rock type, or precipitation regime, unless these factors exhibit strong variability within a catchment. Hence, departures from normal concavity, spatial variations in steepness, and perturbations to the linear data array in log area–log slope space can steer us to parts of a catchment where tectonic processes of interest may be disrupting normal channel profiles.

Caveats on river profiles

In rapidly eroding ranges such as the Southern Alps or the Himalaya, river channels are typically conceptualized as primarily bedrock channels that are being systematically lowered during large discharges. Similarly, river profiles are assumed to display sustained, predictable decreases in channel slope in a downstream direction (Fig. 8.5) in the face of steady rock uplift and invariant lithology. Although the presence of alluvial fills (commonly resulting from climate change) is recognized as impeding vertical bedrock incision along a river channel, such fills can be remobilized and rapidly incised by the same rivers that deposited them. As a consequence, the downstream, bedrock profile of such rivers is expected to be reoccupied following incision of any alluvial fill.

In contrast to fluvially aggraded fills that incrementally accumulate grain by grain, landslides can deliver large volumes of disaggregated bedrock to the river channel in a geological instant. The resultant mass commonly dams the river and a lake develops as the river ponds upstream of the dam. Whereas, for a few major rivers, landslide dams are removed rapidly by powerful outburst floods (Burbank, 1983), for most rivers, some blocks within any large landslide will exceed the river's transport capacity. As a consequence, the landslide dam cannot be readily eroded and instead remains as an obstruction for centuries or millennia (Korup, 2006). Upstream of the dam, fluvial and lacustrine sediments commonly aggrade to the height of the outlet across the dam. Within the dam itself, a narrow, steep rapid that is lined with remnant, untransportable boulders is developed (Fig. 8.6A). As long as the landslide dam persists, some of the former bedrock channel will be protected from erosion. Along some rivers in eastern Tibet, for example, over 80% of the bedrock profile is shielded from erosion by a succession of landslide dams and associated upstream aggradation (Fig. 8.6B and C). Thus, the persistence of landslide dams, their recurrence intervals, and their spacing along a river network determine key aspects of the long profile of the river, the amount of bedrock that is exposed to erosion, and the long-term rate of bedrock erosion (Ouimet *et al.*, 2007). The potential imprint of landslides instills a cautionary note on the unfettered interpretation of river profiles derived from DEMs: high

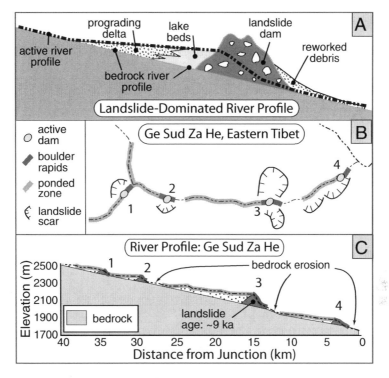

Fig. 8.6 Impact of large landslides on river profiles.
A. Conceptual model for the geometry of deposition and the active river profile after a major landslide dam event. Fluvial–lacustrine aggradation upstream of the dam occurs concurrently with downcutting through the crest of the landslide. Bedrock is shielded from erosion both upstream and downstream from the slide. B. Geomorphic map of four landslides, ponded zones, and boulder rapids along a river (Ge Sud Za He) in eastern Tibet. C. Long profile of the Ge Sud Za He illustrating the lateral extent (>80%) of bedrock cover by landslide-related sediment. The largest landslide (number 3) dates from ~9 ka, yet is still >150 m high. Modified after Ouimet *et al.* (2007).

steepness indices, abrupt changes in concavity, and high calculated stream power can each be related to landsliding, rather than to incision along a predominantly bedrock profile.

Knickpoints

When a reach of a stream is steepened with respect to the adjoining reaches, it defines a topographic *knickpoint* (Fig. 8.7). Because the river typically is no longer in equilibrium, the channel will begin to adjust its longitudinal profile. Across the steepened knickpoint, stream power will typically increase due to the increased slope, and erosion of the stream bed will be enhanced. The common effect of this erosion is to cause the knickpoint to migrate or propagate upstream. Such migration is a key process in tectonic geomorphology because it serves to communicate changes from downstream to the upstream parts of the network. A steepened reach could develop in the absence of tectonism (or eustatic change) simply due to a difference in the erodibility of the bedrock: more resistant rocks will tend to underlie steeper reaches of a stream (Fig. 8.9B). We are more interested in tectonically generated knickpoints that form through differential folding or faulting of a reach of a river. Normal faulting in which the downthrown (hanging-wall) block is in the downstream direction causes a step (knickpoint) to develop in the longitudinal profile of the river. Similarly, thrust faulting that differentially uplifts an upstream reach with respect to a downstream reach will create a knickpoint. In either case, the base level

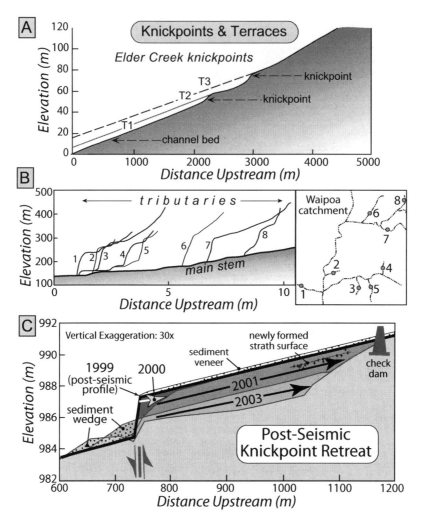

Fig. 8.7   Fluvial knickpoints in California, New Zealand, and Taiwan.
A. Longitudinal river profile of Elder Creek, California, showing two knickpoints that are carved in bedrock. T_1 and T_2 represent strath (bedrock) terraces that appear graded to the crest of individual knickpoints. The terrace–knickpoint geometry suggests that the knickpoint migrated progressively upstream and caused abandonment of its previous bed, which now forms the associated strath terrace. Modified after Seidl and Dietrich (1992). B. Fluvial knickpoints in the Waipaoa catchment, New Zealand. River profiles extracted from a 25-m DEM show knickpoints 50–100 m high on small tributary channels. A wave of incision is interpreted to have swept up the Waipaoa catchment beginning ~18 ka. Whereas rapid knickpoint retreat occurred in the main channels, knickpoint migration stalled near the mouth of many smaller tributaries, as shown in the map. Modified after Crosby and Whipple (2006). C. Knickpoint retreat after the 1999 Chi-Chi M_s=7.6 earthquake. Following ~3–4 m of vertical, coseismic displacement, repeated channel surveys document rapid knickpoint retreat across the shales in the hanging wall. In 2001, the average retreat rate was ~1 m/day, whereas from 2001 to 2003, the channel profile was lowered about 2 m (probably by waves of knickpoint migration originating at the fault scarp), despite little migration of the upstream extent of the knickpoint. By 2001, extensive sediment accumulated adjacent to the fault in its footwall. Modified after Sklar et al., 2005).

of erosion with respect to the upstream segment of the river has been lowered across the fault.

The abrupt changes in channel slope represented by a knickpoint are readily discernible in river profiles extracted from relatively high-resolution DEMs. When projected in log slope–log area space (Fig. 8.8), the steepened channel slopes of the knickpoint zone rise

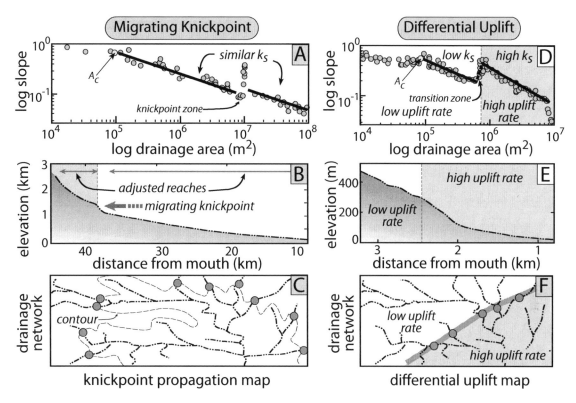

Fig. 8.8 Migrating knickpoints versus zones of differential uplift.
A. In slope–area space, a knickpoint appears as an anomalous steepened reach separating reaches with similar concavity and steepness indices, k_s. B. In profile, the steepened reach separates two adjusted reaches. C. In map view, knickpoints may be located at similar elevations in any subregion within a larger catchment; and for nearby catchments of similar sizes, knickpoints may be located a similar distance from the catchment mouth. Gray circles (in both C and F) indicate steepened reaches. D. In slope–area space, steepened reaches can bridge between zones with different steepness indices (indicative of different uplift rates). Note that, if the upstream area had a higher uplift rate, the upstream reach should have the high k_s values, and a knickpoint (with steeper slopes, as in A) could separate the reaches. E. In profile, the steepened reach may be broader and less well defined. F. In map view, the steepened reaches may align or create a systematic pattern that is associated with a boundary between zones with different uplift rates. Modified after Wobus *et al.* (2006c).

above the broader trend that defines overall channel concavity. Based on several contrasts, the signature of a migrating knickpoint can sometimes be distinguished from a similarly steepened reach that is due to differential uplift (Wobus *et al.*, 2006c). In slope–area space, a knickpoint stands out as an anomaly that separates reaches with similar steepness indices (Fig. 8.8A), whereas for differential uplift, a steepened reach can bridge between adjacent reaches that have different steepness indices (Fig. 8.8D). In a map view, knickpoints within a given part of a larger catchment may occur at

similar elevations and be distributed in analogous positions with respect to the mouth of various tributaries (Fig. 8.8C). This pattern is similar to that seen in the Waipaoa catchment (Fig. 8.7B). In contrast, differential uplift is more likely to create steepened reaches that are aligned with each other, have weaker associations with a particular position within a tributary catchment, and are less associated with a given elevation (Fig. 8.8F).

The mechanism of knickpoint formation and rates of upstream migration in natural settings have not been well documented, especially in

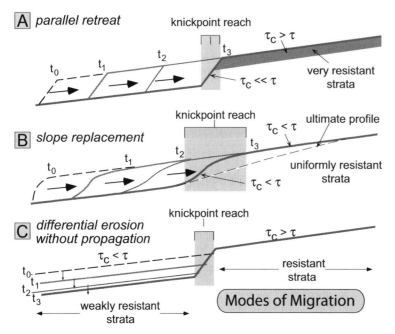

Fig. 8.9 Models for knickpoint migration.
Models for knickpoint migration or formation based on stream-table analyses. Here τ_c is the threshold stress to initiate erosion of the bed; and τ is the actual bed shear stress. A. Parallel retreat of the knickpoint through time with the initial shape being maintained. A resistant rock unit caps the units that are being incised. B. Knickpoint propagation in uniformly resistant rocks with slope replacement during migration. C. Knickpoint formation due to differential erosion of strata with contrasting resistance to erosion. No propagation is required here. Modified after Gardner (1983).

relation to faulting events. Rates of knickpoint propagation in bedrock-floored rivers are generally poorly known, but recent studies of knickpoint migration in channels carved into Hawaiian basalts suggest long-term upstream propagation rates of more than 2 mm/yr (Seidl *et al.*, 1994), whereas migration rates a thousand times faster (~2 m/yr) are estimated for knickpoints in the weak forearc mudstones and siltstones of northeast New Zealand (Crosby and Whipple, 2006) (Fig. 8.7). Rates as high as 300 m/yr have been observed for the knickpoint caused by the Chi-Chi earthquake in 1999 (Fig. 8.7C).

Calculations of knickpoint migration rates depend both on interpretations of the geomorphic history and on dating the sequential migration. The study in Hawaii relied on cosmogenic nuclide exposure dating of bedrock exposed along channels to determine how rapidly incision and translation have occurred in the past, whereas the New Zealand study exploited dated volcanic ashes preserved in terraces that pre- and post-date knickpoint migration. Without such dating, rates of knickpoint migration in most geomorphic settings remain speculative. Moreover, in natural settings, the original shape of the knickpoint or how that shape has been modified as it propa-

gated is commonly impossible to document. Is the original shape simply translated through the landscape or is it smoothed and flattened during translation? What controls the speed of migration? Under circumstances where natural settings reveal little about the geomorphic processes of interest, sometimes useful insights on these processes can be obtained through small-scale experiments and modeling that attempt to reproduce key aspects of the natural setting.

Experimental responses to base-level lowering

Using a long, narrow flume with a steady input flow of water to a pre-carved shallow channel, Gardner (1983) explored temporal and spatial changes in shape and flow parameters across a knickpoint. A comparison of the threshold shear stress (τ_c) needed to initiate erosion with the actual bed shear stress (τ) serves to predict how and where erosion may occur. Above a knickpoint's lip, the channel was observed to become steeper, narrower, and deeper, such that the bed shear stress (as predicted by Eqn 8.2) steadily increased to a maximum at the lip and promoted rapid erosion at this point.

In uniformly resistant bed materials, the originally steep knickpoint evolved through slope replacement into a reach of nearly uniform slope without a clear knickpoint (Fig. 8.9B). On the basis of his experiments, Gardner (1983) concluded that, in homogeneous bedrock, parallel knickpoint retreat would occur only if the rock were pervasively jointed (Box 7.1). Alternatively, he observed that parallel retreat could occur in layered rocks of variable resistance when a highly resistant layer overlies less coherent rocks (Fig. 8.9A). In this situation, undercutting of the lip and its collapse would cause the knickpoint to migrate upstream with little modification to its shape. Even in the stream-table experiments in which slope replacement caused the knickpoint to be transformed into a uniform slope, this transformation took many days during which a clear knickpoint migrated upstream through the channel.

When channel networks, rather than single channels, are the experimental focus, some stream tables utilize sprinkler systems that "rain" on a rectangular, unchannelized, gently inclined, sediment surface. The resulting discharge eventually forms a self-organized channel network. To mimic base-level lowering with such a set-up, the outlet of a network can be artificially lowered in order to create a knickpoint (Parker, 1977). Following base-level lowering, the knickpoint propagates up the main stem and, as the junction with each tributary is encountered, a new knickpoint tends to form at the mouth of the tributary and then begins to migrate up it. In such experiments, the rate of knickpoint migration has been found to be proportional to the discharge (or upstream drainage area), such that, as the headwaters are approached, the rate of migration dramatically slows (Fig. 8.10A).

As a knickpoint migrates through a reach, one can readily envision that the base of the upstream channel is lowered by erosion and that the new, lowered channel may be graded to its downstream continuation. These stream-table experiments showed, however, that this geometry is only the first stage in a *complex response* of the fluvial system that is modulated by sediment fluxes (Fig. 8.10B). Imagine what happens to the river bottom at a particular

point. Prior to the migration of the knickpoint, it "feels" no difference in its base level of erosion and maintains its former geometry. As the knickpoint migrates through, however, the reach is steepened, stream power increases, more sediment is eroded than is deposited, and the bed is lowered until it is approximately graded to the adjacent downstream reach (although it may not have attained the base level of erosion at this time). As the knickpoint migrates farther upstream, the sediment flux to that downstream site increases due to the enhanced upstream erosion. Without a corresponding increase in discharge or slope, the increase in sediment load can cause deposition within the previously scoured channel, so the base of the channel rises. Ultimately, as the knickpoints near tributary headwaters, sediment production decreases, and the stream incises into its bed once again (Fig. 8.10B). The multiple events of incision and aggradation highlight the multi-faceted response to simple base-level lowering. The fact that similar responses may be recorded by Holocene fluvial strata suggests that considerable caution may be needed when interpreting phases of aggradation and degradation.

In other stream-table experiments, the sediment yield was measured as the stream network adjusted to multiple events of base-level lowering (Schumm *et al.*, 1987). Following each lowering, a striking initial increase in sediment yield subsequently decayed exponentially. This decay was not smooth, however, but was punctuated by secondary peaks and troughs (Fig. 8.10C) and by high variance. The secondary peaks in sediment yield were interpreted to result from sporadic collapses of valley walls that suddenly shunt extra sediment into the channel. In recent years, experimental knickpoint formation, rate of migration, and role of knickpoints in inducing collapse of adjacent hillslopes have been further documented with stereo digital cameras and video on stream tables with circular sides (Bigi *et al.*, 2006; Hasbargen and Paola, 2000). These experiments show that knickpoints form even when base level is steadily lowered. The cumulative height of successively formed knickpoints suggests that their upstream

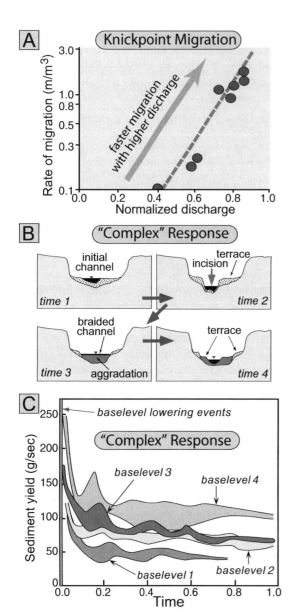

Fig. 8.10   Stream-table experiments on the effects of base-level lowering.
A. Rate of knickpoint migration in a stream-table experiment as a function of discharge (or drainage area) contributing to a site. These data suggest that knickpoints migrate 20 times faster for a catchment about twice as large (twice the discharge). B. Schematic cross-sections of a channel responding to a knickpoint migrating past it. Alternating intervals of incision and deposition are sometimes termed "complex response." Time 1: prior to knickpoint passing, small alluvial terraces flank the channel. Time 2: immediately after knickpoint passed, channel has incised through the

migration could account for up to 30% of the total erosion. Hillslope failures are more likely following knickpoint passage, and, despite steady base-level fall, sediment fluxes vary cyclically by 20–30%. The dynamism of these artificial landscapes is striking!

Numerical experiments

The apparent conflict between Gardner's (1983) observation that knickpoints degenerate into uniform slopes during stream-table experiments using homogeneous material and field observation that at least some knickpoints in homogeneous bedrock appear to be rather long-lived (Wobus *et al.*, 2006a) suggests that other approaches, including numerical models, may be useful in investigating knickpoint migration and persistence. Some recent models (Crosby *et al.*, 2007) exploit a stream-power formulation of the form of Eqn 8.5 in which $m=0.5$ and $n=1$ to calculate the change in height at each point along a channel profile (Fig. 8.11). Initially, the model is run forward to allow development of a detachment-limited profile that is adjusted to base-level fall at a steady rate of 1 mm/yr. Once an equilibrium profile is attained, the rate of base-level fall is accelerated 10-fold to 10 mm/yr for 5 kyr (equivalent to 50 m of base-level change) and then is restored to 1 mm/yr (inset in Fig. 8.11A). Following restoration to the background rate of base-level fall, this model predicts that the knickpoint that formed in response to rapid base-level fall will migrate upstream as a kinematic wave with a speed that is proportional to the area (and, therefore, discharge) upstream of the knickpoint. Notably, the

alluvium and the underlying bedrock; lateral migration causes it to widen its valley. Time 3: increased sediment load from upstream causes the bed to aggrade. Time 4: decreased sediment load causes incision of the bed.
C. Sediment yield resulting from multiple events of base-level lowering on a stream table. Note the high variance in yield, the secondary peak that occurs after the initial rapid drop in yield, and the gradual damping of the oscillations in sediment fluxes. Modified after Schumm *et al.* (1987).

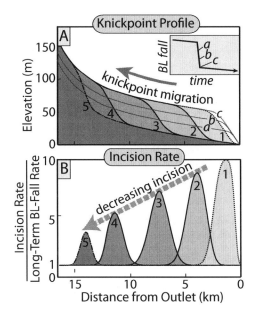

Fig. 8.11   Simulation of knickpoint migration.
A. Knickpoint migration using a detachment-limited,
stream-power model in response to an accelerated fall in
relative base level. The knickpoint initially forms when
base-level fall accelerates 10-fold (shown by profiles a, b,
and c for successively younger stages). Note that, in this
model, the base-level fall can be conceptualized as a rise
in the land surface (profiles a, b, and c) with respect to a
fixed base level. Return to the long-term rate of base-
level fall allows upstream migration of the knickpoint
(profiles 1 to 5). Note that the profiles are not evenly
spaced in time: migration is fastest in the initial stages
when the upstream catchment area is largest. History of
base-level fall is shown in the inset. B. Incision rate in
comparison to the long-term rate of base-level fall. Both
rates and volumes of erosion decrease upstream and
with time. Modified after Crosby *et al.* (2007).

values of *m* and *n* modulate the way in which
the height and shape of the knickpoint evolve as
it migrates. With the values used in this model,
the knickpoint maintains its height and basic
shape as it migrates. The bedrock incision rate is
maximized at the steepest part of the knickpoint
and returns to the background incision rate both
above and below the knickpoint (Fig. 8.11B). As
the knickpoint migrates, the catchment upstream
of it shrinks, so that both the incision and
migration rates decrease. These predictions are
broadly consistent with the observations from
the Waipaoa catchment (Fig. 8.7), where knick-

point heights are quite similar, and knickpoint
migration seems to have stalled near the mouth
of small tributaries.

Channel patterns and characteristics

Many modern rivers display contrasting channel
patterns – including meandering, braided, anastomosing, and straight channels – along their
courses, and some of these rivers are known to
cross sites of ongoing deformation. Wouldn't it
be useful to be able to examine the channel
pattern of a river in order to get insights on the
nature of active deformation along its length? To
succeed in this, we would need to know how
changing channel patterns of the rivers correlate
with differences in deformation rates and
with patterns of relative uplift or subsidence.
Moreover, we would need to know how to
distinguish between changes in patterns that
result from non-tectonic versus tectonic causes.
Such distinctions may sometimes be made with
little ambiguity when planform changes coincide
with clearly contrasting tectonic domains. Before
discussing some of the common controls on
planform changes that may be independent of
tectonics, we examine three examples from
fold-and-thrust belts.

First, let us consider a channel that crosses an
actively growing fold. Upstream of the fold axis,
the channel has to respond to a rising local base
level, especially following coseismic uplift.
We might expect the channel to aggrade as its
slope decreases in response to rising base level.
Some studies have shown that anastomosing
channels can form under these conditions
(Jerolmack and Mohrig, 2007; Smith and Smith,
1980). (Note that anastomosing channels are
multi-threaded, like braided channels, but the
channels tend to be stable, relatively narrower
and deeper, and flanked by vegetated, rapidly
aggrading floodplains that help to suppress
avulsion.) On the downstream side of the fold
axis, the channel gradient will tend to be
steepened. In response to differential uplift,
the channel is likely to be single-threaded,
more incised, and potentially either more
sinuous or straight.

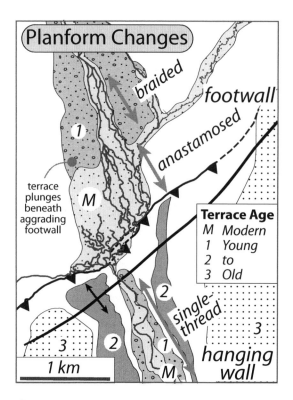

Fig. 8.12   Planform river changes due to differential uplift.
Irishman Creek in New Zealand's Southern Alps changes from braided to anastomosed as it aggrades in the realm of active footwall subsidence, as evidenced by the terrace that plunges beneath the active floodplain. The active floodplain narrows, and the channel becomes single-threaded, deeper, and more sinuous as it crosses the hanging-wall fold. Modified after Amos *et al.* (2007).

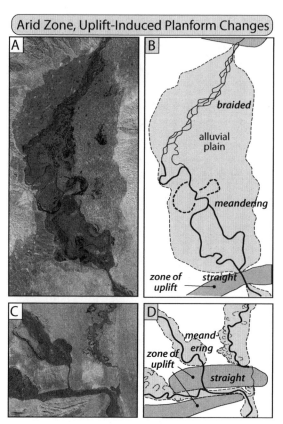

Fig. 8.13   Planform changes related to active uplifts, western Pamir foreland.
Satellite imagery and interpretations of rivers and structures along the Amu Darya (Oxus River) bounding the arid, Afghanistan–Tadzhikistan border. See Plate 3 for color imagery. Flow is toward the lower right in all images. Changes in planform river patterns appear intimately associated with tectonic controls exerted by folding, tilting, and possibly faulting. A, B. Image and interpretation of upstream reach of the Amu Darya showing transitions from braided to meandering to straight channels as an active uplift is approached. C, D. Image and interpretation of downstream reach of the Amu Darya. Note the abrupt narrowing of the right-hand tributary channel (C) as it crosses the zone of uplift.

Most of these characteristics are seen in the Mackenzie basin, New Zealand, where Irishman Creek crosses a thrust fault and its associated hanging-wall anticline (Fig. 8.12). Active footwall aggradation is denoted by a change from braided to anastomosed channels. Late Pleistocene river terraces on the footwall plunge gently toward the hanging wall and then disappear beneath the active floodplain nearly a kilometer from the fault itself. Presumably, footwall subsidence, driven by the load of the hanging wall, caused downward deflection of the river terrace. As the river crosses into the hanging wall, it becomes single-threaded, narrower, more incised, and more sinuous.

Second, in satellite images of the foreland fold belt to the west of the Pamir Range (Fig. 8.13 and Plate 3), a repetitive planform pattern is apparent. Upstream of channel constrictions, broad meander belts are present and are typically associated with braided channels farther upstream. This transition from braided to meandering in the middle of large basins suggests that fluvial gradients progressively decrease downstream of the basin constriction. Several meander belts are consistently positioned to one side of the basin mid-line and suggest broad tilting in the direction of offset. The channel constrictions at the basin exits correspond with abrupt transitions from meandering to straight and typically much narrower river courses. Several such straight channel reaches correspond with folds that are visible on the satellite images and are suggestive of active uplift. Rivers are diverted toward or around the end of some of the growing folds, whereas water gaps have developed across others. Between growing folds or active thrusts, strike valleys with little apparent perturbation of the river courses are present. The repeated sequence of braided to meandering to straight channels suggests that these planform transitions represent a predictable sequence that can be used to help recognize localized zones of active deformation (Attal *et al.*, 2008).

Third, evidence for contrasting tectonic regimes can sometimes be inferred through comparisons of adjacent drainages that display different channel patterns. A satellite image of northern Pakistan (Fig. 8.14) shows the Indus and Kabul Rivers merging and flowing through a water gap in the Attock Range. At the point of their merger near where they enter the water gap, both rivers are responding to the same local base level. The Indus, now incorporating the Kabul's discharge, is flowing on or near bedrock through the Attock Range, which is experiencing active bedrock uplift (Burbank and Tahirkheli, 1985). Despite this common base level, the planform river patterns for each river upstream from the water gap contrast strongly with each other: the Indus displays a strongly braided channel pattern, whereas, at a comparable distance upstream, the Kabul shows a nearly straight channel that suggests active incision of its bed. The braided Indus may result from aggradation in response to rock uplift in the Attock Range, but this river also drains rapidly eroding regions of the Himalaya, such that its planform pattern may be attributable to its high sediment load. Why does this contrast exist between aggradation versus incision on adjacent rivers that have a

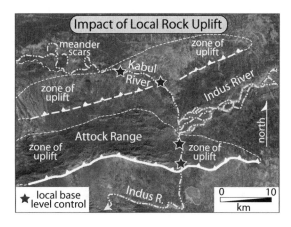

Fig. 8.14 Indus and Kabul Rivers, channel patterns. Satellite image of the Indus and Kabul Rivers in northern Pakistan as they join and pass through a water gap in the Attock Range. Note (i) common base level at the water gap through the Attock Range, (ii) contrasting planform river patterns, (iii) zone of inferred deformation along the Kabul River, and (iv) ponding (high-sinuosity pattern) upstream of the uplift.

common base level? The likely cause can also be inferred from the satellite image, which suggests that both upstream and downstream controls are important. Linear trends both east and west of the Kabul River delineate active transpressional faults (Fig. 8.14). The zone of uplift, delineated by lighter colored, less vegetated areas, extends across the Kabul River toward the east-northeast and lies north of the Indus River. Throughout this uplifting reach, the Kabul River is incising. Just upstream of this zone, however, the river displays meanders, cutoffs, and swampy regions indicative of low gradients. These features suggest that the Kabul system is ponded upstream of the uplift, whereas, when crossing the zone of active rock uplift, the channel narrows and straightens, thereby increasing its gradient and specific stream power.

It is remarkable how much tectonic information can be inferred from planform river patterns such as these! Such interpretations can be reinforced when analysis of planform patterns is combined with structural interpretations of stratal geometries or with local drainage patterns that are indicative of growing folds and faults. The widespread availability and ever-increasing quality of remotely sensed imagery allows for

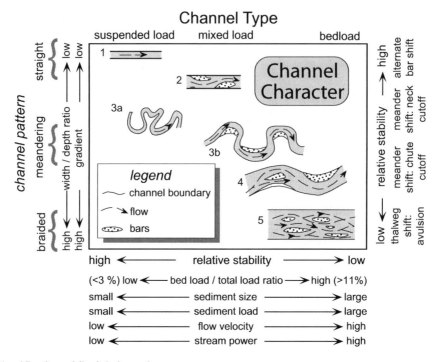

Fig. 8.15   Classification of fluvial channels.
Channel patterns and the persistence of channel shapes are conceptualized as a function of nature of the sediment load, velocity, and stream power. Modified after Schumm (1986).

armchair exploration of tectonic environments across the globe. Viewing such imagery with programs such as Google Earth brings the added advantage of perspective views of the landscape – you can check whether visible landforms are consistent with inferences about deformation. Simple observations of downstream changes in planform patterns, correlation of planform changes with topographic and geomorphic forms along the margins of a river, and quantification of the channel position with respect to valley margins, all serve to delineate potential deformation regimes.

Numerous previous studies of channel patterns (Fig. 8.15) indicate that they respond to several competing controls, including sediment size and load, flow velocity, and stream power (Schumm, 1986). In general, braided rivers are favored by high sediment fluxes, coarse bedload, weak bank material, and high variability in water discharge, whereas meandering rivers are typified by lower sediment fluxes, finer bedload,

more cohesive bank material, lower flow velocity, less variable discharge, and lower stream power. Braided rivers quite commonly change downstream into meandering rivers as slope and mean grain size diminish (Fig. 8.13). Each of the variables affecting channel patterns can clearly be independent of tectonic activity. On the other hand, for a given grain size, specific stream power (Eqn 8.2) during bankfull discharge appears to distinguish between braided and meandering rivers (Fig. 8.16) (Van den Berg, 1995). Given the sensitivity of stream power to channel slope, subtle tilting due to tectonic processes could certainly affect whether a river was braided or meandering.

Stream-table experiments

Stream-table experiments designed to investigate possible responses of channel patterns to changing slopes commonly take one of two approaches: tilting of the entire stream table, or creating

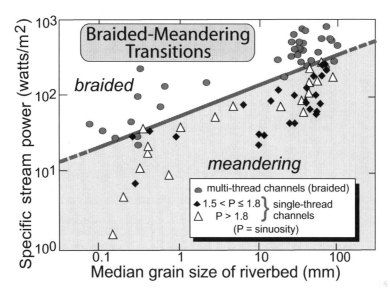

Fig. 8.16   Channel pattern as a function of specific stream power during bankfull discharge versus mean grain size of the river bed.
Specific stream power (W/m²), calculated from river slope and channel cross-section at bankfull stage. Sinuosity (*P*) is the ratio of the length of the thalweg to the valley length. High sinuosity (*P* > 1.5), single-thread (meandering) rivers define a separate field from that of braided rivers. Based on data from more than 200 rivers, this plot predicts that, for a given grain size, the channel pattern is a function of specific stream power and that changes in water depth or slope can cause the river to cross the threshold defining the different channel patterns. Modified after Van den Berg (1995).

localized uplift by folding the substrate beneath a reach of the channel. Both approaches show potentially diagnostic transitions in channel forms. As slope increases while discharge and sediment supply are held constant, initially straight channels become more sinuous (Fig. 8.17). Sinuosity continues to increase until a threshold slope is attained, after which any additional steepening of slopes leads to a braided river pattern. Further experiments indicate that, for different discharges, but with a constant grain size, the transition to braided occurs at different stream power, but at approximately the same slope (~0.012–0.013) for each discharge (Edgar, 1973). In these experimental results, increased sinuosity should not be equated with meandering channel forms, because truly meandering, single-thread channels have only recently been created on stream tables (Peakall *et al.*, 2007), well after these experiments were conducted. Nonetheless, the tendency toward higher sinuousity as channel slopes steepen suggests that planform

changes represent one way in which rivers could respond to tectonic deformation.

Although the reasons that rivers meander are not completely understood, meandering rivers tend to dissipate energy (through localized bank erosion and sediment transport) in about equal amounts along equal lengths of channel and to minimize the total energy expended along the river course. At the regional scale, sinuosity has been shown to be a function of both rock strength and the frequency of high-intensity discharge (Stark *et al.*, 2010). Weaker rocks and more frequent storms cause higher sinuosity. At the local scale, when a channel is steepened, stream power initially increases, such that a stream formerly in equilibrium will be pushed toward a non-equilibrium condition. By increasing sinuosity, the channel is lengthened, and the slope along the thalweg decreases, thereby counteracting the increase in the valley gradient. As a consequence, the rate of work done per unit length of channel is

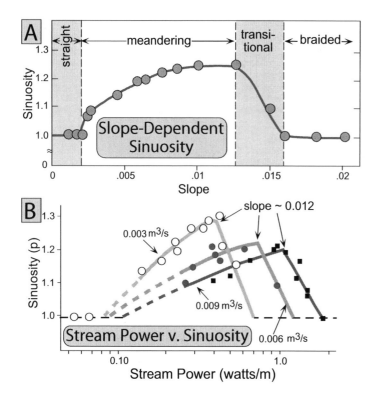

Fig. 8.17 Slope-driven changes in river sinuosity: stream-table results.
A. Channel sinuosity on a stream table as a function of slope. Sinuosity increases until a threshold is crossed at a slope of ~0.013. Modified after Schumm and Khan (1972). B. Sinuosity as a function of stream power for different discharges. Note that the transition from higher to lower sinuosity occurs at the same average slope for all three discharges. Modified after Edgar (1973).

lessened in comparison to a shorter, steeper, and straighter channel.

In another stream-table experiment, a kink fold was created across an established channel midway along its length (Ouchi, 1985). The fold directly deformed only the middle third of the stream table, and fold growth was mimicked by incrementally increasing the height of the fold crest. Such growth caused an increase in the valley slope on the downstream side of the fold axis, whereas it diminished the slope on the upstream flank of the fold. When folding was superimposed on a pre-existing moderately sinuous river pattern, the primary response to the uplift was an increase in the thalweg sinuosity in the downstream part of the uplift, where the slope was steepened (Fig. 8.18A). The elevation of the thalweg itself increased in this uplifted area, but the increased sinuosity compensated for the steeper slope of the "valley" bottom. Thus, a convexity developed in the valley slope profile, but little significant aggradation or degradation occurred at the fold crest or

on its downstream limb. Upstream of the uplift, deposition occurred along the reduced upstream channel gradient as multiple stable channels were established (an anastomosing pattern). With continued uplift, a cutoff formed downstream of the axis of uplift, and the channel steepened, straightened, and assumed an "island–braided" pattern. Folding experiments with a multi-threaded river caused erosion across the fold crest, where a deepened, single-thread channel bounded by river "terraces" was created (Fig. 8.18B). Downstream of the fold axis, the steepened slope and increased sediment flux (due to erosion of the fold crest) augmented the pre-existing braiding. Upstream, however, the decrease in slope and sediment flux generated alternate bars with a tendency toward a more sinuous, single thalweg.

Given the simple observation that sinuosity tends to increase with increased slope (up to a threshold), it might be expected that active deformation could be revealed by map patterns of meandering rivers showing localized changes

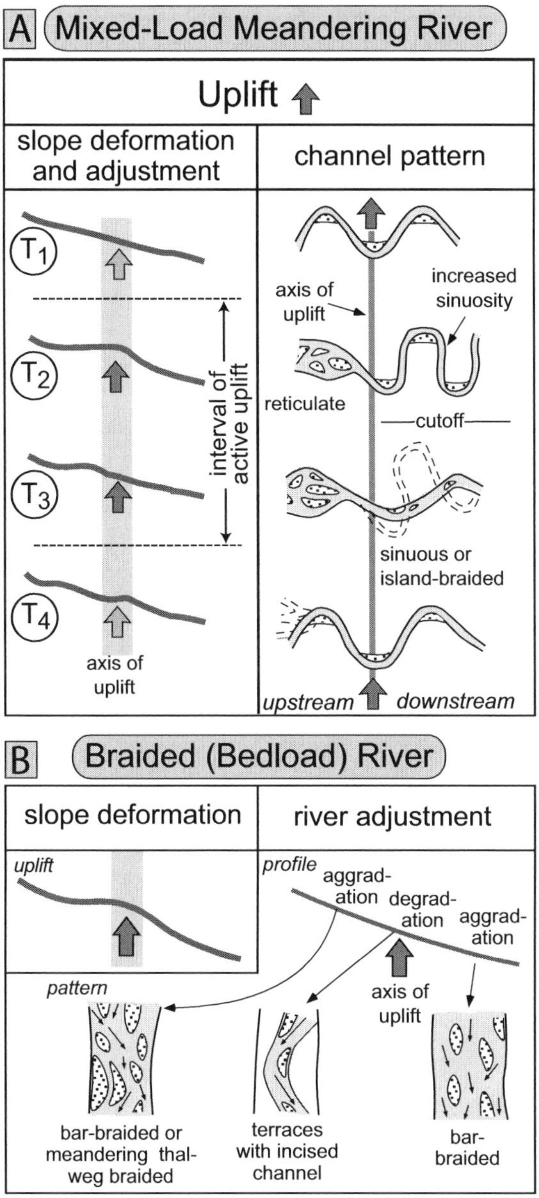

Fig. 8.18 Effect of localized uplift on stream-table channels.
A. Schematic response of a mixed-load meandering river to the growth of an anticline. (Right) Change in valley profile with time as uplift is initiated and then wanes. (Left) Planform response to uplift. Initial growth causes increased sinuosity on the downstream fold limb and increased deposition on the upstream limb. When the slope threshold is crossed, a braided channel develops, incision occurs across the fold crest, and the locus of erosion migrates upstream. As uplift wanes, sinuous pattern

in sinuosity along the river course or indicating a temporal change in sinuosity along a given reach. Similarly, along generally braided rivers, zones of incised, single-thread channels bounded by terraces would be likely candidates for zones of active uplift.

Numerical experiments

Numerical models have also been developed to predict the effects on deposition and erosion of a channel flowing across a domal uplift (Snow and Slingerland, 1990). Although conceptually similar to the experimental stream-table set-up of Ouchi (1985), a sinusoidal fold, rather than a kink fold, was modeled. The model specified a straight channel with a fixed cross-section flowing across non-cohesive bed material, i.e., a transport-limited channel. Maximum rates of uplift were 5 mm/yr, and the folding affected 100 km of the stream course (Fig. 8.19A). Results of the modeling show major erosion across the axis of uplift and centered on it. Upstream deposition occurs primarily due to growth of the fold and "ponding," whereas very limited downstream deposition results from the enhanced sediment flux and the need to maintain continuity with the uplifted river course (Fig. 8.19B). Through time, the zone of erosion widens, and the point where elevation increases the most migrates upstream well beyond the axis of the uplift (Fig. 8.19 C). Deposition rates are the same magnitude as erosion rates during the first few hundred years, then deposition stabilizes, whereas erosion keeps pace with uplift. With time, deposition upstream of the axis of uplift causes the profile to become increasingly asymmetrical and convex. Perhaps most importantly, despite ongoing uplift, the profile approaches a steady state by approximately 5 kyr that could be viewed as representing a dynamic equilibrium

is re-established. B. Schematic response of a braided river channel to folding. Changing gradients and elevations promote erosion across the fold crest, increased sediment fluxes downstream, and aggradation on both flanks of the fold. Modified after Ouchi (1985).

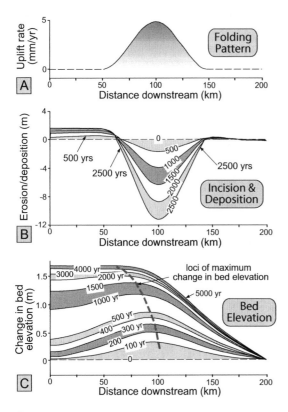

Fig. 8.19 Modeled channel profile in response to anticlinal uplift.
A. Pattern of uplift, reaching a maximum of 5 mm/yr across a zone 100 km wide. B. Erosional and depositional responses to uplift. Note that the erosion generally mimics the uplift pattern, but that deposition occurs upstream of the anticline. C. Change in bed elevation. Note that the point of maximum change migrates upstream through time and that the system approaches a steady state (dynamic equilibrium) in 3–5 kyr. Modified after Snow and Slingerland (1990).

in which each increment of uplift is matched by equivalent erosion.

It is instructive to note that, after 5 kyr and 25 m of "rock" uplift, the elevation of the channel has only increased ~1.5 m, indicating that 95% of the uplifted material has been eroded away (Fig. 8.19). Given the non-cohesive character of the material beneath the channel, this erosion is perhaps not surprising. If the model were run with more resistant bedrock, a much greater change in elevation would be expected, because the channel would require greater steepness

and stream power to erode the bedrock. Nonetheless, the basic geometric relationships would remain similar: a steepened reach extending all the way across the zone of uplift, the steepest channel slope localized near the zone of highest uplift, and a likely approach to steady state, such that, in the long term, erosion and uplift would balance.

Rivers crossing folds

These stream-table and numerical experiments make predictions about fluvial responses to deformation. Tilting and uplift or subsidence due to folding along a trend lying perpendicular to a river course will modify down-valley gradients. Tilting along an axis that lies parallel to a river course could deflect or displace a river system. Discrete offsets related to faults will affect the local base level, as well as causing tilting.

Several examples illustrate these different tectonic scenarios. Simple monoclinal folding along the post-Vicksburg flexure in southeast Texas appears to have caused a considerably steepened valley gradient along the San Antonio River (Fig. 8.20A). Through this reach, both the gradient of the valley floor and the sinuosity double (Ouchi, 1985), whereas the gradient of the channel itself only increases by about 10% (Fig. 8.20B). Thus, the increased sinuosity significantly moderates any channel steepening, a response consistent with the predictions from the stream-table experiments (Fig. 8.18).

If a river is going to sustain its course across an actively growing fold, the river typically has to erode fast enough to counterbalance the rate of rock uplift, which has the potential to continuously raise local base level. The previous discussion suggests several adjustments that would enhance the erosive capability of a river: increased channel slope and decreased channel widths can lead to higher shear stress, specific stream power, and erosive potential (Whittaker et al., 2007). In their meticulous study of side-by-side rivers crossing rapidly uplifting folds, Lavé and Avouac (2000) observed intriguing responses

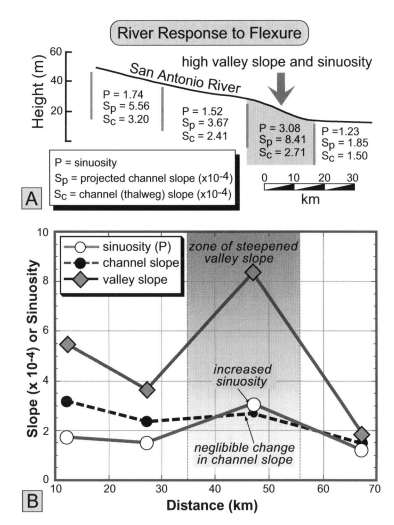

Fig. 8.20 **Fluvial response to localized uplift.**
A. Longitudinal profile of the San Antonio River and associated changes in valley slope (S_p), river slope (S_c), and sinuosity (P). Modified after Ouchi (1985). B. Plot of valley gradient, channel slope, and sinuosity along the San Antonio River. Through the reach where the valley slope more than doubles, the sinuosity also doubles, whereas the channel slope is almost unchanged.

in each river (Fig. 8.21). Both rivers narrow significantly across the zone of rock uplift. In fact, a tight spatial correspondence exists between the rate of rock uplift (as inferred from deformed terraces; Fig. 7.25) and the magnitude of channel narrowing. Incision rates as high as 12 mm/yr broadly mimic the magnitude of narrowing. In contrast to their similar response of narrowing, the Bakeya River becomes steeper across the zone of most rapid uplift, whereas the Bagmati River shows no significant steepening across this same zone. Although the cause of this contrast is unknown, the Bagmati has a far larger upstream source area and carries a considerably greater sediment load. Possibly, the

channel slope is adjusted primarily to carry the sediment load. Alternatively, because the maximum incision rate along the Bagmati is 30% lower than that along the Bakeya, perhaps the Bagmati's erosive capability is increased sufficiently via channel narrowing alone. This interpretation would be consistent with the observation that the Bakeya channel steepens only where its incision rates exceed about 10 mm/yr (Fig. 8.21).

Given that both narrowing and steepening of the channel can enhance incision rates, does one take precedence over the other, or do both typically occur synchronously? The data from the Nepalese foreland (Fig. 8.21) suggest that narrowing can occur without steepening. Such

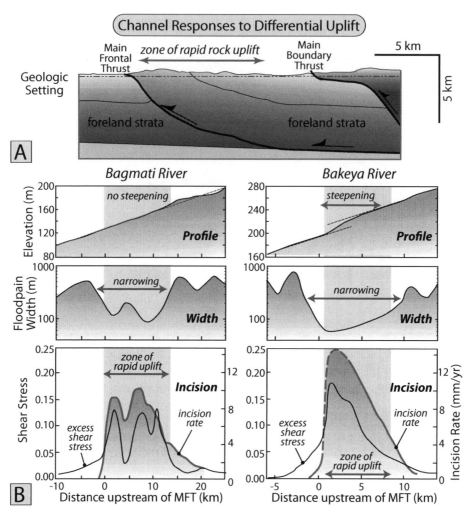

Fig. 8.21   Antecedent channel responses to rapid, differential uplift.
A. Channel setting. Slip along the Main Frontal Thrust (MFT) in the Himalayan foreland of central Nepal drives hanging-wall uplift at rates exceeding 10 mm/yr (Fig. 7.25). B. Channel profile, floodplain width, incision rate, and calculated excess shear stress for the Bagmati and Bakeya Rivers as they cross the MFT hanging wall. The Bagmati narrows without steepening. In the zone of the highest incision rate, the Bakeya both narrows and steepens. Patterns of excess shear stress are closely coupled to channel narrowing. Modified after Lavé and Avouac (2001).

sequencing is also borne out by small fluvial channels that were cut across short-wavelength folds in New Zealand (Amos and Burbank, 2007). Detailed geodetic surveys document channel morphology across and along channels that cross late Quaternary folds with wavelengths of hundreds of meters and differential uplift of a few meters (Fig. 8.22A and B). These data reveal abrupt channel narrowing as the magnitude of uplift and incision increases (Fig. 8.22C).

Moreover, the data suggest that narrowing typically occurs earlier and can occur independently of any channel steepening (Fig. 8.23). In particular, if the magnitude of differential uplift is sufficiently small, narrowing appears adequate to increase the channel erosion rate in order to balance uplift rates. At higher rates or magnitudes of uplift, channels are interpreted to initially narrow and then to steepen in order to achieve a greater overall rate of incision. Studies of river width,

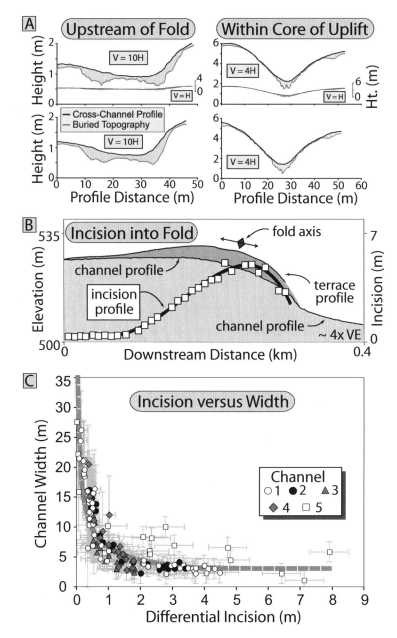

Fig. 8.22   Width changes in antecedent channels crossing small growing folds.
A. Surveys of abandoned channels in the Mackenzie Basin, New Zealand, reveal three- to eight-fold decreases in width as channels traverse a fold. Note that, because loess and soils have partially filled the abandoned channels, their former geometry is reconstructed using closely spaced soil probes. B. Channel incision is measured by comparing the channel bottom to the uplifted former terrace surface across the asymmetric fold. C. Compilation of data from five channels showing that 1–2 m of uplift causes a five- to 10-fold channel narrowing. Modified after Amos and Burbank (2007).

slope, and incision rates for bedrock rivers in Taiwan reach very similar conclusions (Yanites *et al.*, 2010): in response to increasing rates of differential uplift, rivers become progressively narrower until a width/depth ratio of ~10 is attained, after which they also become steeper.

The very detailed documentation of rock uplift rates above the Main Frontal Thrust in central Nepal provides a robust framework for examining how the profiles of rivers respond to uplift gradients. Given that rivers that are not tectonically perturbed have average concavities of ~0.5, we can predict three end-members of concavity for rivers crossing growing anticlines (Fig. 8.24): (i) rivers originating in zones of high uplift rates and flowing to low-uplift zones will have high concavity because their headwaters will be differentially uplifted with respect to

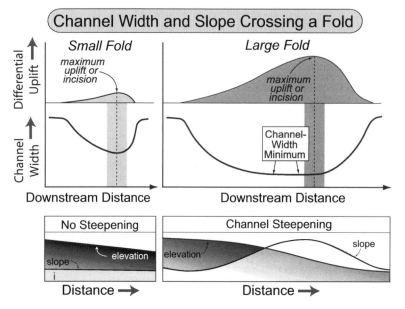

Fig. 8.23 **Dependence of antecedent channel response on scale and rate of folding.** For sufficiently small folds, narrowing of the channel width can accelerate erosion enough to keep pace with uplift. For larger folds, narrowing occurs first, but subsequently the channel also steepens in order to amplify erosion. Modified after Amos and Burbank (2007).

their outlets; (ii) rivers flowing parallel to the uplift gradient should show normal concavity (~0.5), even if the uplift rate is high, because no differential uplift is occurring along the river's course; and (iii) antecedent rivers that cross the fold are likely to have low or even negative concavity because they tend to steepen as they cross the zone of high uplift (see Bakeya River; Fig. 8.21). Studies of river profiles in the foreland where Lavé and Avouac (2000) completed their study show precisely this behavior (Fig. 8.24). The relatively weak rocks in these rapidly uplifting folds permit rivers to attain apparently equilibrium profiles, even at Holocene time scales (Kirby and Whipple, 2001). Such systematic and tectonically related changes in concavity (and channel steepness) suggest these fluvial indices, which can be readily extracted from DEMs, could be used in a predictive way to identify areas of strong differential uplift and even to estimate erosion rates (Kirby and Whipple, 2001).

Most of the examples discussed thus far result from deformation in which the trend of the fault or fold axis is approximately perpendicular to the river. What happens when the tilting axis is more parallel to the river? At least two different scenarios can be envisioned. In the case of coseismic deformation, instantaneous tilting occurs across the affected region, such that

part of the former floodplain may experience considerable differential subsidence (Fig. 4.26). The magnitude of displacement can be large (several meters), and the tilting may cause an immediate avulsion of rivers into new low points in the landscape. Alternatively, tilting may occur incrementally through largely aseismic deformation or in small coseismic steps, in which case rivers would respond over longer time scales.

An example of fluvial responses to incremental tilting is found in northern California, where a resurgent dome within Long Valley caldera has inflated and deflated periodically. Between 1979 and 1983 and between 1988 and 1992, the crest of the dome rose about 40 cm (Fig. 8.25A) and 20 cm, respectively (Castle *et al.*, 1984; Langbein *et al.*, 1995). The Owens River flows along the margin of the dome, and, through much of its course, the river is nearly parallel to the elliptical contours of the recent domal deformation (Fig. 8.25A). The recent uplift has not caused any dramatic shift in the present-day river, but, as the river flows around the flank of the dome, it displays two, parallel meander belts (Fig. 8.25B) separated by 200–300 m (Reid, 1992). Although both belts show similar meander wavelengths and amplitudes, the inner belt (closer to the dome) now contains an underfit stream and is both older and approximately 60 cm higher in elevation than

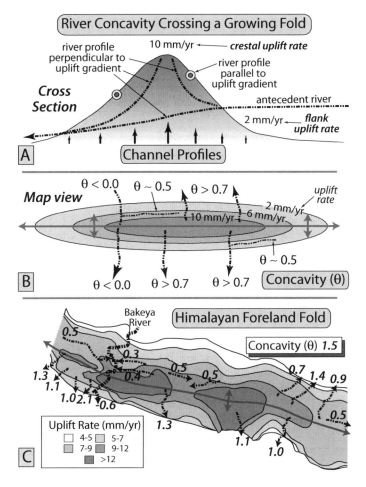

Fig. 8.24   Effects of active fold growth on river concavity.
A. Schematic cross-section of an active fold with a strong uplift gradient (2 mm/yr at the flanks to 10 mm/yr at the crest). Three classes of rivers are depicted: (i) high-concavity rivers flowing from high to low uplift rates; (ii) normal-concavity rivers flowing parallel to fold axis; and (iii) antecedent rivers. B. Schematic map view illustrating typical concavity for the three classes of rivers above. C. Example of rivers crossing a very rapidly deforming fold in the Himalayan foreland of central Nepal. Uplift data from Lavé and Avouac (2000). Uplift map modified after Hurtrez *et al.* (1999). Concavity values from Kirby and Whipple (2001).

the outer belt (Fig. 8.25C). This configuration has been interpreted to result from systematic, outward avulsion of the Owens River in response to outward tilting of the volcanic dome (Reid, 1992). The most recent avulsion event occurred between 1856 and 1879. Because doming progressively lowers the outer part of the floodplain with respect to the inner part, the river will tend to avulse into this lower, outer region during flooding events. The inner belt is not fully abandoned by the Owens River, but it retains only a fraction of its previous discharge. Consequently, the remnant channel appears underfit (Fig. 8.25C). If the region between the belts is considered to have been nearly horizontal prior to deformation, then the present slope between the two channel belts can be projected to the center of the dome to estimate the cumulative recent uplift of the dome.

It is instructive to think about the likely differences in river patterns that would result from slow versus abrupt tilting. If slow tilting occurs, the river channel will preferentially shift towards the outer, lower part of its meander belt. One way to test whether such migration has occurred is to measure the position of the active channel with respect to the entire width of the meander belt. By compiling the relative channel position on many transects across the meander belt, the mean position of the channel with respect to the mid-line of the meander belt can be determined (Reid, 1992). A significant displacement from the statistical center can be interpreted to indicate ongoing tilting that is forcing continued river migration.

At a more regional scale, comparing the position of a meander belt with respect to the mid-line of a drainage basin (the axis of

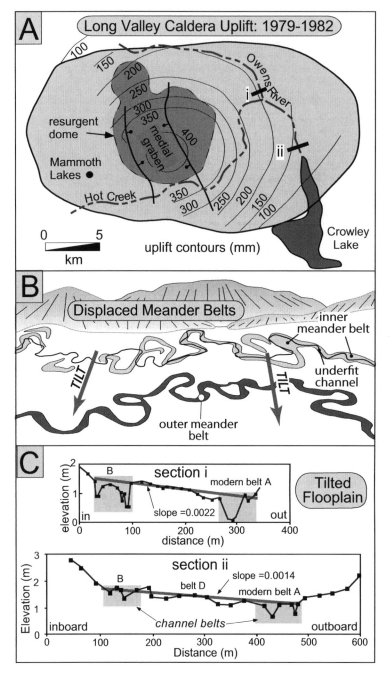

Fig. 8.25 Owens River dome.
A. Map of recent deformation (1979–83) of the resurgent dome of Long Valley caldera and the course of the Owens River, showing locations of cross-sections (i) and (ii).
B. Photographic sketch showing parallel meander belts of the Owens River, but with an underfit stream in the inner belt that is closer to center of the uplifting resurgent dome.
C. Topographic cross-sections showing the mean slope between the inner and outer meander belts. Projections of these slopes toward the dome's center suggest the cumulative magnitude of recent uplift (15–35 m). Modified after Reid (1992).

symmetry along the basin center) can be used to infer tilting (Cox, 1994). A consistent bias of the position of the active meander belt with respect to the basin's mid-line can indicate tilting in the direction of bias. The relative position of the meander belt can be quantified as

$$T = dO/dL \qquad (8.8)$$

where dO equals the distance of offset of the mean river course or meander belt from the mid-line and dL equals the distance to the mid-line from the drainage divide defining the basin

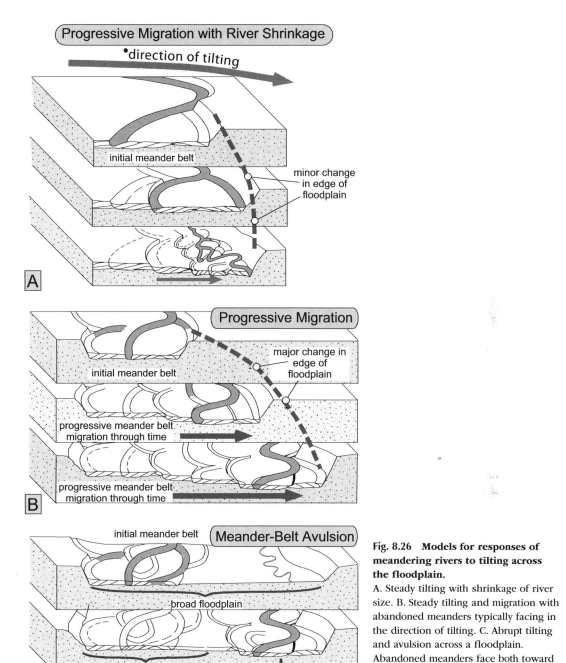

Fig. 8.26 Models for responses of meandering rivers to tilting across the floodplain.
A. Steady tilting with shrinkage of river size. B. Steady tilting and migration with abandoned meanders typically facing in the direction of tilting. C. Abrupt tilting and avulsion across a floodplain. Abandoned meanders face both toward and away from tilt direction. Modified after Alexander *et al.* (1994).

margin. When $T=0$, the river is located near the mid-line of the basin, whereas, as $T \rightarrow 1$, the river is shifted increasingly close to the basin's margin (Cox, 1994). These ratios can be determined from topographic maps and can be easily used to provide a quick overview of potential zones and directions of tilting within a drainage basin.

In the process of shifting, a meandering river will abandon oxbows and meander bends, which then remain as visible scars on the landscape.

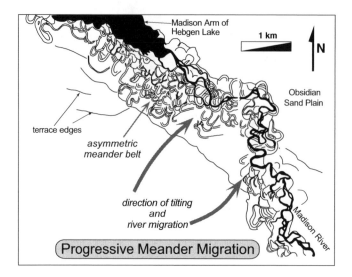

Fig. 8.27   Pattern of meander scars of the Madison River near the 1959 Hebgen Lake rupture.
Normal faulting during the Hebgen Lake M=7.3 earthquake caused large displacement and tilting that was down to the north-northeast (see Fig. 4.26). Note the position of the river channel on the northern and eastern margin of its meander belt. Partial drowning of the meander belt due to coseismic damming of the lake is obvious from the truncated meanders along the southwest shore of the lake. In the region depicted here, the coseismic offset was only 0–1 m down to the north. The series of meander scars facing toward the present river and the scale of those abandoned meanders suggest that the river maintained its size, but migrated laterally in response to tilting, probably as the cumulative result of repeated earthquakes on the Hebgen Lake fault. Modified after Alexander *et al.* (1994).

Under conditions of steady tilt, but without major avulsions that completely relocate the meander belt, one might expect to see a suite of meander scars all facing in the same direction and indicating the generally unidirectional migration of the river across its floodplain toward a low point (Alexander *et al.*, 1994) (Fig. 8.26). If tilting was abrupt and caused a major avulsion and relocation of the meander belt, the channel pattern should reflect that jump in average position (Fig. 8.26C). In the area that has been abandoned, meander scars should face in both directions, as is typical of a modern meandering river, rather than facing one direction – which indicates progressive, unidirectional migration. Measurements of the relative position of the old channel with respect to the width of the meander belt can indicate whether the channel position was symmetrical within the meander belt and untilted prior to avulsion, whether it had begun to migrate in the direction of tilt prior to avulsion, or whether it had been off-center in the opposite direction prior to avulsion, which would suggest that the direction of tilt was suddenly reversed

(Reid, 1992). In the Owens River setting (Fig. 8.25), for example, this last condition would be analogous to having the dome deflate prior to inflation, whereas, with a folded structure, it could be analogous to deformation during a seismic cycle in which the interseismic and coseismic displacements are roughly equal in magnitude, but in opposing directions.

In contrast to such oscillating uplift and subsidence, regions with persistent coseismic faulting and interseismic sediment loading commonly produce a unidirectional and sustained tilt (e.g., Figs 4.25 and 4.30). In the area affected by the 1959 Hebgen Lake normal-fault earthquake (Fig. 4.26), such tilt is manifested by meander patterns displaying a consistent migration direction toward both the fault trace and the zone of maximum displacement (Fig. 8.27).

Summary

At Holocene time scales, deformation tends to occur in one of two modes: episodic coseismic displacements; or steady, incremental folding, tilting, or rotation. Both modes can trigger geomorphic responses as deformation modulates base level, differential uplift, and tilting. Abrupt changes in base level due to seismic events commonly initiate waves of erosion, such as represented by knickpoints, or waves of deposition that propagate outward from the active fault. Perhaps surprisingly, inherent thresholds related to geomorphic processes can also create propagating waves of erosion in response to steady changes in base level.

Because river systems are the most widespread geomorphic features that are strongly affected by small changes in slope, much analysis of Holocene landscapes focuses on fluvial responses to deformation. Moreover, aggradation or erosion by rivers sets the local base level for adjacent hillslopes, such that fluvial behavior typically modulates the entire landscape response to tectonism.

River planform patterns may provide insight on the distribution and nature of deformation on both historic and prehistoric time scales. Given their sensitivity to subtle changes in gradient, rivers commonly respond rapidly to tectonically imposed changes. Whereas tilting of fluvial terraces of a few degrees can usually be revealed only through careful surveying, changes in river patterns along a river's course can be readily observed from maps, aerial photographs, and satellite images. Because rivers respond to numerous controls (water and sediment discharge, slope, roughness, sediment caliber), observed planform changes commonly cannot be uniquely attributed to a tectonic cause. Nonetheless, the changes in fluvial planform patterns serve to pinpoint areas where deformation may be occurring. Moreover, the type of the planform change, such as from meandering to straight or from lower to higher sinuosity, suggests what the nature of the deformation may be: increasing slope, lateral tilting, or changes in erosion rates.

Three different aspects of geomorphic studies have significantly improved our understanding of tectonically active, Holocene landscapes. First, field studies have focused on detailed measurements that provide insights on interactions within geomorphic systems, so that we now have much better understanding of why rivers narrow or steepen in response to differential uplift or what modulates rates of knickpoint migration. Second, the recent proliferation of digital elevation models has provided new tools for studying modern fluvial systems. Not only can channel slopes and contributing areas be more reliably obtained, but also specific fluvial indices, such as channel steepness, channel concavity, and slope–area relationships, can be readily calculated. Such indices enable holistic exploration of landscapes in search of both common characteristics and anomalies. These indices also serve to pinpoint rivers that are being perturbed by active deformation because their indices depart from normative values in non-deforming sites. Third, advances in dating techniques have underpinned increasingly successful attempts to quantify rates of tectonic and geomorphic processes and interactions.

Clearly, geomorphic systems other than rivers will also initiate responses to tectonic deformation at Holocene time scales. As fluvial knickpoints migrate past the toes of hillslopes, steepened hillslope gradients will initiate changes in the flux of material off the hillside into the channels and may also influence the rates, efficiency, or nature of the dominant hillslope processes that modify the landscape. In many rapidly eroding orogens (erosion at more than several mm/yr), the entire landscape is predicted, on average, to have experienced multiple landslides during the Holocene, and many tens of meters of bedrock should be removed during that time. Hence, such landscapes have the potential to approach steady-state forms during the Holocene. Most active orogens, however, are eroding at slower rates (<2mm/yr). As a consequence, most non-fluvial landscape responses may be difficult to discern at Holocene time scales. In more slowly deforming orogens, the scales and rates of tectonic deformation, base-level lowering, and tilting are sufficiently small that hillslopes may not be pushed over a threshold that would provoke a dramatic or readily recognizable response. It is only as one examines longer time scales (>10kyr) that such landscape responses become clear. In the following chapter, deformation and landscape responses at intermediate time scales (12ka to 300ka) are examined.

Deformation and geomorphology at intermediate time scales

9

Studies of tectonic geomorphology at intermediate time scales provide opportunities to gather a different suite of insights on both tectonic deformation and landscape responses than can be obtained from investigations confined to the Holocene. By "intermediate" time scales, we refer to intervals extending from the Holocene–Pleistocene boundary at 11.6 ka (Steffensen *et al.*, 2008) to about 300–400 ka. Over such time scales, the landscape becomes a summation of both episodic and continuous tectonic and geomorphic processes. Whereas many active faults will have experienced only a few earthquakes during the Holocene, these faults may sustain dozens to hundreds of earthquakes over 300 000 years. These repetitive ruptures smooth the variations in displacements that frequently occur between any two successive earthquakes (Fig. 4.4), such as might be seen during the Holocene, and they cumulatively allow better, more confident definition of long-term mean rates of deformation.

Knowledge of such longer-term rates is invaluable when trying either to understand the implications of a shorter-term record or to assess its relevance to other time scales. For example, geodetic measurements collected over a few years are now being used by a growing number of researchers to characterize long-term rates of regional deformation. But, are the rates measured at decadal scales truly typical of tectonic rates over 100 000 years or more

(Fig. 5.1)? Only knowledge of rates at these longer time scales can answer that question. Similarly, the persistence of deformation rates and calculated recurrence intervals based on trenching studies of Holocene faulting can only be assessed using a longer-term record. Thus, by considering intermediate time scales, more representative averages can be obtained for deformation rates.

At time scales of more than 10 kyr, erosion becomes an increasingly important factor in modifying the landscape. Pristine tectonic forms become degraded over these longer intervals. Sufficient time elapses for interactions between discrete deformation events and ongoing surface processes to shape the landscape into characteristic forms, such that river patterns, dissected limbs of folds, and modified fault scarps can be viewed as recording these interactions. At time scales extending well beyond the Holocene, interpretation of the stratigraphic record becomes increasingly important as an approach to reconstructing past deformation: as mountains are eroded, their history can be recorded by the sediments eroded from them. Fortunately, at intermediate time scales, many tectonically produced topographic features are still sufficiently well preserved that they provide clear evidence of the summation of individual tectonic events that determined their shape.

At scales of hundreds of millenia, intermediate time scales smooth the intrinsic variability of

---

weather and provide a better long-term average of the fluxes of water, temperature, and winds that drive most surface processes. Superimposed on any such average conditions, however, are major glacial–interglacial cycles that create a strong, complex, and highly variable climate signal, which, in turn, modulates surface processes as they shape the landscape. Across such cycles, we should expect: (i) sea level to fluctuate by more than 100 m; (ii) major changes in sediment fluxes, discharge, and erosive capacity in rivers; and (iii) variations in hillslope stability as moisture content, soils, vegetation, and hillslope diffusivities change in response to climate. All such changes commonly complicate the interpretation of the landscape.

Despite the geomorphic changes wrought by changing climate, such climatic forcing also creates additional evidence in the landscape that allows better determination of the tectonic rates and processes. Geomorphic markers, such as marine or fluvial terraces, which form in response to climatic variability, can provide multiple reference surfaces with which to define the geometry of deformation. If the ages of these markers can be determined, then intermediate-term rates of deformation can be calculated. At these time scales, it is sometimes also possible to distinguish between the climatic and the tectonic imprint on the landscape. Consider the growth of a simple fault-bend fold. At the scale of individual seismic events, its growth is highly episodic, such that instantaneous spurts of growth are followed by extended quiescent intervals. Over the span of numerous faulting events, however, the fold will be seen to grow continuously. In contrast, climatically controlled processes will typically be relatively steady at the scale of a single or a few seismic events, whereas they become highly variable at those intermediate time scales over which tectonic processes may appear steady. Such contrasts between the time scales of episodicity and steadiness in tectonic versus climatic processes provide an excellent opportunity to try to separate their relative influence on both evolving landforms and sedimentary successions related to those landforms. Given a desire to understand how tectonically influenced landscapes evolve

through time or to reconstruct depositional or erosional responses that are likely to be preserved in the stratigraphic record, this more time-averaged landscape offers an important geological perspective.

Unresolved problems at intermediate time scales

Certain classes of problems are best attacked at intermediate time scales – those in which the signals of interest need to be integrated across many millennia. Most commonly, those same signals need to record significant change at these same time scales. For example, when snapshots of a fault or landscape can be captured at multiple moments over many tens to hundreds of millennia, diverse models for fault behavior and landscape evolution can be tested. Because faults commonly grow at time scales of 10^4–10^6 years, observations of well-dated geomorphic markers can enable us to test competing models for fault growth. For example, do faults obey the classic concept of progressive lengthening with each successive earthquake? Or, do they rapidly achieve an approximately fixed length early in their history if they encounter barriers that restrain their tips (Fig. 9.1)? The latter model predicts faults dominated by "characteristic earthquakes" (Fig. 4.7), whereas the former model predicts continuous change in fault length with time. These models can be tested if data are available over sufficiently long time intervals (many millennia), because both the position of successive fault tips and the shape of the displacement profiles change with time in distinctive ways for each model.

Climatic changes at intermediate time scales have clearly led to the waxing and waning of glaciers, as well as to the growth and shrinkage of lakes. Because both expanding glaciers and lakes impose new loads on the crust, we might wonder whether their growth and decay could influence the behavior of faults. If changes in the balance between maximum and minimum stresses on fault surfaces were sufficiently great, then rates of fault slip might be expected to vary in concert with the climatically induced changes in crustal loads (Fig. 9.2). In fact, depending on

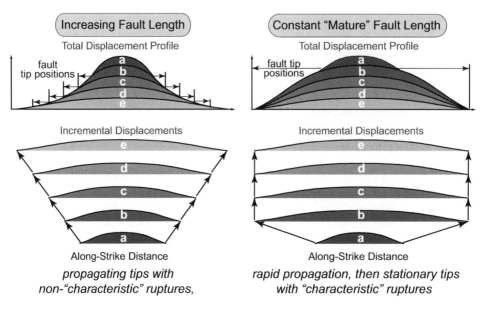

Fig. 9.1   Competing models for accumulation of displacement on faults.
(Left) The "traditional" model suggests faults lengthen in each rupture as they accumulate displacement. (Right) The "fixed length" model suggests that faults propagate rapidly in early stages of growth, stop lengthening as fault tips encounter strong barriers, and then simply accumulate displacement at a nearly fixed length. Modified after Amos *et al.* (2010).

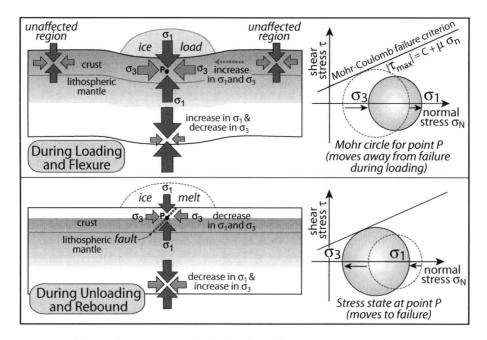

Fig. 9.2   Conceptual linkages between crustal loads and fault slip.
Suppression of fault slip during temporary loading by glaciers or lakes. (Top) Ice (or water) loads and associated crustal flexure can increase crustal stresses such that a fault is moved farther away from failure (as represented by the Mohr–Coulomb failure criterion). (Bottom) Removal of the load and crustal rebound can then drive the fault to the point of failure. Modified after Hampel and Hetzel (2006).

the sensitivity of faults to small changes in loads, spatial and temporal changes in the locus and/ or rates of bedrock erosion in a mountain belt due to climate variability might also modulate fault slip rates. Given the striking correlation of seasonal cycles of monsoon-modulated water loads in the Himalayan foreland (see Box 5.2) with both Himalayan seismicity and convergence rates (Bettinelli *et al.*, 2008), these proposed linkages between climate and fault slip rates at intermediate time scales make conceptual sense. Nonetheless, more high-resolution data are needed at these time scales to explore the characteristics of such potential linkages.

Because plate convergence rates can be as high as 100 km/Myr, unsteadiness in convergent processes should be well expressed at intermediate times scales. For example, geomorphic studies along subduction zones have shown that, when seamounts or a rough sea floor impinge on a trench, the forearc deforms (Fisher *et al.*, 1998). But, can the timing, wavelength, and magnitude of that deformation provide insight into the tectonic processes and kinematics that govern the collision? Again, competing models (Taylor *et al.*, 2005) make contrasting predictions for the pattern and rates of deformation at intermediate time scales (Fig. 9.3). If a seamount that is rafted along by a subducting plate steadily slips beneath the forearc, it should generate a migrating wave of forearc uplift and subsidence at spatial and temporal wavelengths that scale with the size of the seamount and the convergence rate. In contrast, if the seamount gets stuck against the forearc, collisional stresses could be abruptly transferred to the outer forearc and might drive rapid uplift above a subduction zone that had become temporarily locked. If the plate interface eventually ruptures through the seamount, the release of those stresses could drive rapid subsidence. These different forearc behaviors can be assessed with data gathered at intermediate time scales.

In this chapter, two different aspects of tectonic geomorphology at intermediate time scales will be examined. First, techniques for calibration of rates of deformation are illustrated by focusing on the use of various types of

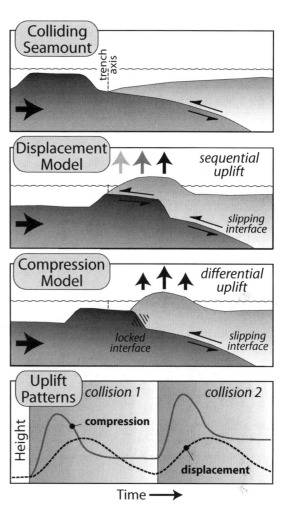

Fig. 9.3 Models of forearc deformation for a subducting seamount.
Colliding seamount (top), representing thickened crust, approaches the forearc, where its collision could cause distinctive patterns of deformation. In the "displacement model," the seamount slides beneath the forearc, thereby driving accelerated uplift that migrates landward and gradually decelerates in its wake. In the "compression model," the seamount gets stuck as it impinges on the trench and forearc. The locked interface translates the collisional stresses inland and initially causes abrupt and rapid uplift. If the locked interface then ruptures and dismembers the seamount, subsidence will also be rapid. Each model predicts a different temporal and spatial pattern of uplift (bottom). Modified after Taylor *et al.* (2005).

geomorphic markers. Second, some landscape responses are described, using examples of growing faults and folds, deforming forearcs, and tilted mountain ranges.

Calibrating rates of deformation

Marine terraces

Several approaches can be employed to define the vertical deformation pattern along a coastline using either abrasion platforms or coral terraces (Figs 2.1 and 2.2). A rising coastal landmass is like a strip chart that records and uplifts the geological record of each successive sea-level highstand. As in all rate studies, one needs knowledge of the initial and final elevation of a marker, as well as the time it took to traverse that vertical distance. The key here is that the landmass is assumed to be rising steadily (with good enough data, this assumption can be tested), whereas sea level is varying widely over glacial–interglacial cycles that have periods on the order of tens of thousands to one hundred thousand years. The abrasional or constructional platform that is created at time zero is a nearly planar, gently sloping surface. Subsequent emergence of a platform implies that the rock mass has moved upward relative to sea level, and submergence implies the opposite. In order to know which has moved, the sea or the land, other information (e.g., eustatic records) must be brought to bear. On the other hand, any warping of this surface can be used directly and unambiguously to document relative movement of one part of the coastal landmass with respect to another. Depending on what is already known about the landscape, one can either use the terraces to deduce a sea-level curve, or use the terraces to deduce the rate of uplift of the landmass. In any case, in order to document rates of deformation, one must be able to date the platforms.

The determination of the absolute ages for former sea levels has been something of an industry for a large number of researchers for several decades. Unfortunately, most of the sea-level highstands that we would like to date are older (>40 ka) than can be dated using [14]C

approaches. As described in Chapter 3, uranium-series (U–Th) dates on aragonitic material have provided ages of surfaces several hundred thousand years old. With improved techniques and measurement capabilities, uranium-series dates have become increasingly precise, such that dates of 100 ka may have measurement uncertainties of <1% (Edwards et al., 1997), and chronologies of sea-level changes have been extended back to greater than 200 ka (Andersen et al., 2010; Edwards et al., 1997).

In the ideal case, we have a reliable global (eustatic) sea-level curve (sea level relative to present versus time; Fig. 2.5) and a flight of dated terraces. The age and the present elevation of each terrace (relative sea level) with respect to the position of the correlative eustatic sea level at the time of terrace formation can be used to define amounts and rates of rock uplift through time. But, what if the terraces are undated, or only one of them has an approximate age? Can they still be used to define the uplift rate? In this situation, one typically assumes a steady uplift rate (or assigns one, if a single terrace is dated), and then determines graphically (or numerically, on the computer) how well the observed heights of terraces correlate with the predicted position of terraces based on the eustatic sea-level curve and the constant apparent sea-level change (Fig. 9.4A).

In this graphical matching technique, a terrace should exist at an appropriate height above sea level for each of the high sea-level peaks defined in the eustatic curve. If the elevation of each sea-level highstand matches with that of a preserved terrace, then the assumption of constant uplift appears warranted. If the match is poor, then a different uplift rate can be tried. If the match is still unsatisfactory, the rate of uplift might be varied through time in order to obtain a satisfactory match (Fig. 9.4B). If one permits the uplift rate to vary without constraint, however, a perfect, but probably meaningless, match can always be obtained. The simplest and often most convincing approach, therefore, assumes a constant uplift rate. If rates are varied through time to obtain a match, then a geologically reasonable rationale for the proposed

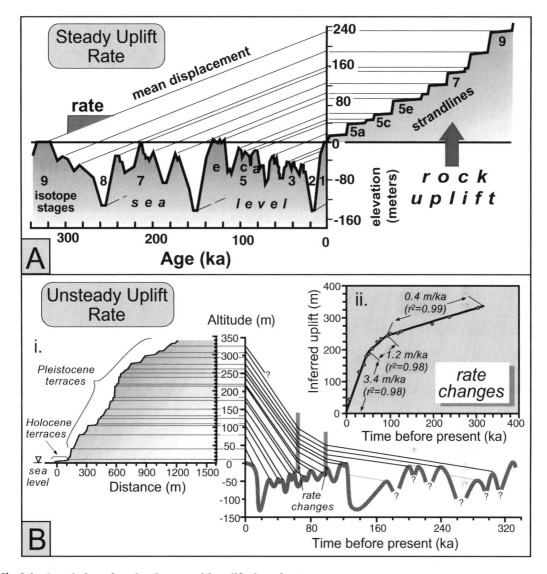

Fig. 9.4   Correlation of sea-level curve with uplifted marine terraces.
A. Graphical correlation of sea-level variations with coastal terrace record based on a steady rock-uplift rate. Note that not all highstands older than 130 ka or younger than 50 ka are represented in the terrace record. Some older ones are obscured by subsequent higher sea levels, whereas some younger ones are still below sea level. Modified after Lajoie (1986). B. i. Example of terrace correlation in northern California based on the assumption of abrupt accelerations in the rate of uplift. This coastal area has been strongly affected by the passage of the Mendocino triple junction during the past 100 kyr, such that accelerated uplift is not unreasonable. ii. In order to correlate each of the observed terraces with a sea-level highstand, the Middle Pleistocene rate of bedrock uplift is inferred to have tripled at ~100 ka and then to have tripled again at ~60 ka. Modified after Merritts and Bull (1989).

changes should be offered for this variation. One observation that could support an assumed acceleration in the rate of uplift would be the presence of older, higher terraces that are more closely vertically spaced than younger, lower terraces (Fig. 9.4B). Given the nature of the sea-level curve in which the frequency of high sea-level stands appears lower prior to 125 ka,

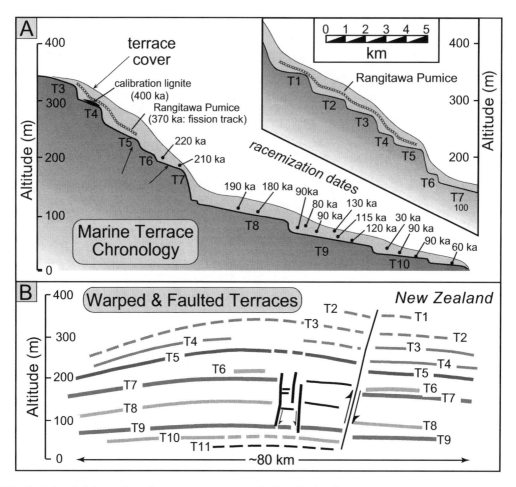

Fig. 9.5　Dated and deformed marine terrace sequences in New Zealand.
A. Terrace chronology developed using fission-track dating on pumice and amino acid racemization dating on organic material overlying the terraces. These dates span more than 300 kyr and provide minimum ages for each terrace. Correlation among sections is based on pumice, soil stratigraphy, and geomorphic position. B. Same terraces warped and offset across faults along 80 km of coast. Tracing and correct identification of terraces is key to discovering any offset across faults. The progressively greater offsets of older terraces indicate that the fault has been periodically active throughout the period of uplift. Note the difference from a faulted trench stratigraphy (Fig. 6.3) in which the strata at lower elevations record more displacement than those higher in the trench walls. Modified after Pillans (1983).

such a spatial distribution of terraces argues for an increase in uplift rates. Be aware, however, that the younger sea-level record is better known and that its complex and rapid changes have emerged from recent success with high-precision dating. Similar sea-level complexity (and multiple, closely spaced highstands) may also be documented in the pre-125-ka record when we develop a similar chronological precision.

If flights of marine terraces can be correlated with the sea-level curve at numerous sites along a coastline, or even if they can only be laterally correlated from one site to the next, these data can be used to define patterns of differential vertical uplift. Abrupt offsets in terrace heights with increasingly larger separations for successively older terraces on the one side will indicate the presence of persistently active faults (Fig. 9.5A and B), whereas steady convergence or divergence between individual terraces of a given age will exemplify broad warping of an area.

The deformed patterns of uplifted Pleistocene terraces are an obvious and readily observed consequence of variable bedrock uplift rates in coastal regions. Moreover, comparisons of the shapes of deformed terraces of different ages allow one to assess spatial and temporal variability in rock uplift rates. Commonly, the faults that are responsible for uplift of the terraces are either blind faults or break the Earth's surface below sea level. In such circumstances, these warped terraces take on added significance, because they may permit testing of hypotheses about the behavior of unexposed faults or folds that are thought to be responsible for their uplift. For example, comparisons of coseismic coastal uplift patterns with patterns of warped terraces can be used to assess whether a long succession of characteristic earthquakes on one or two local faults could have generated the observed terrace pattern. Such a situation was previously described (see Fig. 4.8) on the California coast in the vicinity of Santa Cruz, where uplifted and broadly warped marine terraces are well preserved. Alternatively, the geometry of warped terraces on the flanks of a growing anticline can be used to deduce the orientation and typical slip direction along the buried fault(s) responsible for terrace uplift (Ward and Valensise, 1994). The terraces are like bathtub rings around a growing fold: each one originally formed a horizontal surface (see Box 9.1). But now, in their deformed positions, their geometry can be inverted to estimate the depth, dip, and slip on underlying faults.

Most marine abrasion platforms vary from 100 to 500 m wide. The depth to wave base and the requirement that the seaward slope of the platform permits removal of debris from the shoreface appears to control the maximum platform width. In some areas, however, platforms greater than or equal to 1 km wide are preserved (e.g., Fig. 9.5A). Unusually wide terraces are likely to result from one of two conditions: the presence of very weak, readily eroded rocks; or successive reoccupations of a given terrace level. If the amount of bedrock uplift between two highstands is just a bit less in magnitude than, but in the same direction as, the vertical difference in sea level between the two highstands, then a previously formed terrace can be reoccupied by a slightly higher sea level and laterally extended (Kelsey and Bockheim, 1994).

It is important here to note that there exists a "terrace survival problem" analogous to the glacial moraine survival problem (see Box 2.3). A sequence of elevations corresponding to a terrace flight at one location along a coastline might be different from another nearby sequence in that one or more terraces might be missing (Anderson et al., 1999). This mismatch results from the fact that: (i) platform width is dependent on very local variables such as lithology, structure, and orientation of the coastline relative to the dominant wave energy (Adams et al., 2005); and (ii) a younger terrace platform grows in width at the expense of older platforms, whose outer edges are progressively nibbled away by the cliff at the back of the younger terrace. Hence, younger terraces can locally eliminate older ones. In addition, this geomorphic reality raises a cautionary flag against using platform width as a correlation tool.

Until recently, only emergent platforms were used in defining uplift patterns. This limitation has changed with the advent of increasingly available and detailed bathymetry and with new drilling methods that allow collection of geological materials from the sea floor or from beneath younger sediments. An example of the use of sub-sea-level platforms utilizes corals dredged from about 2 km deep on a submerged platform in the Huon Gulf of the Solomon Sea (Galewsky et al., 1996). These ancient platforms were first visualized using sidescan sonar and detailed bathymetry, where both anomalously flat surfaces and striking spires (interpreted to be coral pinnacles) were identified as potential targets for dredging. Samples retrieved from these surfaces were dated at approximately 340 ka using uranium-series dating and, thereby, revealed a long-term subsidence rate of about 6 mm/yr, one of the first documentations of such sustained rates. With the explosion of high-resolution sea-floor data, the use of submerged terrace platforms promises to become more routine.

Box 9.1   Marine terraces on a growing fold.

In addition to the effects of differential warping or fault offsets on terrace patterns, one can wonder how a changing sea level would interact with a three-dimensional deforming shape, such as a growing fold that is increasing in amplitude through time. Some simple models of a growing anticline (Ward and Valensise, 1994) predict patterns of strand lines that are not intuitively obvious. In the model (see figure), a fold grows above a buried thrust fault in such a manner that the crest of the fold is uplifted at a steady rate through time. Every 100 kyr a new terrace is cut into the margins of the fold during a highstand, referred to here as a reference sea level. Thus, one might expect that, given a uniformly growing fold and terrace-cutting episodes regularly spaced in time, the terraces would show a uniform spatial distribution. Such spacing is not borne out in the model, however, because the emergent part of the fold becomes broader through time.

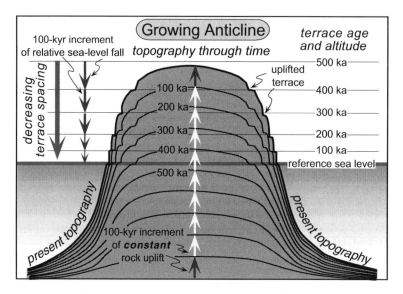

Model of marine terraces etched into the flanks of a growing anticline. Modified after Ward and Valensise (1994).

Although uplift is steady at any point along the fold, it decreases from the crest of the fold toward its flanks. Therefore, younger terraces that are formed on the ever-widening fold as it emerges above sea level are spaced more closely than older terraces formed when the emergent fold was narrower. Whereas the changing vertical spacing of the terraces, such as those depicted here, would often be interpreted as indicating a steady deceleration in the rate of uplift, the variable spacing is in fact a simple consequence of the shape of the steadily growing fold. This model would predict a uniform vertical spacing of terraces only in situations where the coastal region is experiencing a block-like uplift. Thus, in analyzing any coastal terrace sequence, the cause of the oceanward slope of the land surface should be considered. Is it a result of folding such that interior regions have been uplifted more rapidly than the coastal area? Or is it a geomorphic surface created in the absence of differential tilting? Given the model predictions described here, the distinction between a tectonically created slope and a geomorphic slope is fundamental to making a reliable interpretation of a terrace succession.

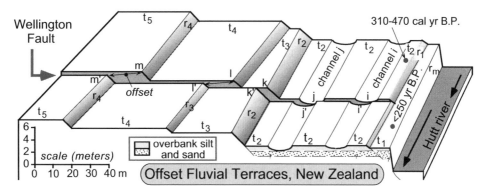

Fig. 9.6   Displaced fluvial terraces along the Wellington Fault, New Zealand.
The lowest terrace, dated at <250 yr, is not displaced by this strike-slip fault, whereas the next highest terrace (t_2) and the channels cut across its surface show a displacement of ~4 m attributable to the last major earthquake. Note that any fault displacement of the riser (r_1) between t_2 and t_1 was beveled off during creation of t_1. The amount of offset of the risers increases systematically with each older terrace. Important controls on terrace width are exerted by the geometry of the river channel prior to abandonment (note the triangular shape of t_3). The vertical throw across the fault is small (~10%) compared to the horizontal displacement. Modified after Van Dissen *et al.* (1992).

Fluvial terraces

When a river course flanked by flights of fluvial terraces is oriented at a high angle to a strike-slip or dip-slip fault, the terraces displaced by the fault can provide an excellent record of progressive offsets. If some or all of the terraces can also be dated, then the rates of fault displacement over the duration of the dated sequence can be derived. In regions where climatically controlled terraces are widespread, it may not be necessary to date the terraces directly adjacent to the fault, because sequences of soils, loess, or volcanic ashes that overlie the terrace treads may permit correlation with other, better dated, terraces in the same region.

The assessment of offset fluvial terraces requires several steps. First, the correlation of terrace treads and risers across the fault must be determined. Because of changes in the river course through time, the height of risers between terraces along strike-slip faults is generally a better guide to correlation than is the width of the terrace tread, which is sensitive to patterns of lateral migration by the former channel. If, however, significant vertical (dip-slip) displacement has occurred along the fault, then riser height will vary as well. In any case, the

most reliable correlation will usually result from consideration of the entire suite of treads and risers and any relative or absolute dating of their surfaces. Second, the offset of formerly continuous risers is measured across the fault for each terrace level. In addition, any other linear features, such as channels or gullies, located on the terrace treads and trending across the fault are also measured (Fig. 9.6). As described earlier (Box 6.1), careful attention should be paid to whether risers have been modified since the upper tread was abandoned and whether slip rates should be based on ages for the upper or lower treads that bracket the riser. Finally, to the extent permitted by available dates, a history of cumulative offset and rates of offset through time is developed.

Although a similar methodology could be applied to marine terraces that have been cut by strike-slip or dip-slip faults, fluvial terraces have a considerable advantage in terms of reconstructing the record of the past 10–50 kyr because the available terrestrial record in this interval is commonly more complete. Whereas climatically driven changes in water and sediment discharge between 10 and 50 ka have generated multiple fluvial terraces in many sites, sea level was below present for this entire

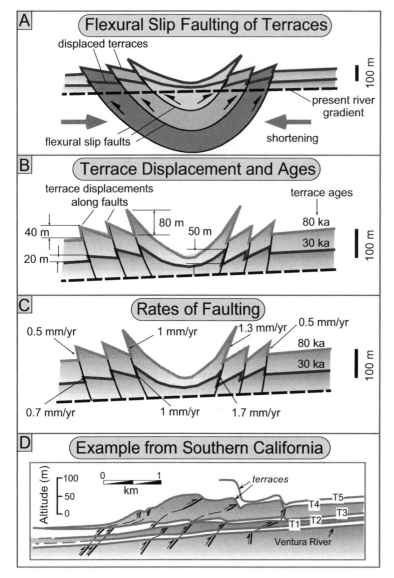

Fig. 9.7   Terraces displaced across flexural slip faults.
A. Geometry of terrace displacement defines greatest displacement near the core of the fold. B. Terrace ages and displacements of terraces across different faults. Note that offsets are not symmetrical across the synclinal axis and that younger terraces display lesser offsets compared to the older terraces. C. Rates of fault displacement based on offset of dated terraces. Note that rates are not necessarily constant along the same fault through time. D. Displacement of terraces by flexural slip faults along the Ventura River, California. Ages assigned to the terraces are 92 ka (T5), 54 ka (T4), 38 ka (T3), 30 ka (T2), and 17 ka (T1). Modified after Rockwell *et al.* (1984).

interval (Fig. 2.5), such that, with the exception of rapidly rising coasts, marine terraces younger than 80 ka are still commonly submerged.

Where rivers cross active folds or dip-slip faults, fluvial terraces can also record progressive displacement. Where faults rupture the surface, it is expected that the age and height of the terrace will generally correlate with the magnitude of cumulative displacement (Plate 5). Flights of fluvial terraces can overcome some of the limitations imposed on paleoseismological studies by the fact that trench walls in alluvium are often unstable, so that deeply excavated trenches are uncommon. Trenches in any of the terrace risers may reveal paleoseismological data on the last few ruptures that display perhaps a few meters of displacement in each rupture. In contrast, the entire vertical suite of terraces can serve to define the long-term displacement history, such that several tens to hundreds of meters of displacement can be recorded by the higher terraces (Plate 5). As with any study seeking to define rates, key constraints are provided by dating of the formation and abandonment of each terrace.

Across actively deforming zones in which multiple faults are closely spaced, a single terrace may be displaced by several faults. Even without knowledge of the terrace age, the variable displacement of the marker surface by each fault will indicate how deformation has been partitioned among the active structures, and relative rates of displacement can be defined. Because flexural slip faults (Plate 6) exploit weak interbeds as slip surfaces, the limbs of tightening synclines sometimes display fairly closely spaced faults (Figs 4.34 and 9.7). Studies along the Ventura River in southern California of terraces offset across flexural slip faults in the Canada Larga syncline (Rockwell et al., 1984) provide well-calibrated examples of both progressive terrace displacement and differential partitioning of displacement among several faults (Fig. 9.7D).

In the case of growing folds, warped fluvial terraces can provide unique insights into the two-dimensional geometry of the fold and its rate of growth. Antecedent streams that maintain their courses across growing folds will often produce strath terraces that may or may not be mantled with a veneer of alluvial debris. The terraces develop during intervals when lateral abrasion dominates over vertical incision (Figs 2.12 and 7.13). In cases in which (i) terraces have extensive, down-valley continuity and (ii) the deformed treads within a rising structure, such as a fold, appear to grade into undeformed treads beyond the structure, it is likely that climatic fluctuations controlled the periods of major terrace formation. Alternatively, if the growth of the structure itself was tectonically pulsed, then terraces may have formed during intervals of reduced deformation rates (e.g., Lu et al., 2010). Most published work (Medwedeff, 1992; Suppe et al., 1992; Vergés et al., 1996; Hubert-Ferrari et al., 2007) in which rates of fold growth are well calibrated, however, is inconsistent with a pulsed deformation model.

It is worth stressing that a crucial component in the analysis of both marine and fluvial terraces is the correlations that are drawn between physically isolated terrace remnants.

Often, erosion makes it impossible to trace terrace surfaces confidently, even along smoothly folded structures. Whenever faults are encountered, correlation of terraces across the fault becomes even more difficult. Because interpretations of offsets are entirely dependent on such correlations (Fig. 2.17), characterization of the terrace surface and its subsurface stratigraphy is often a major element in any such study (Merritts et al., 1994). Soil development, loess stratigraphy, tephra layers, and relative and absolute dating techniques can all be used to distinguish between and correlate among terraces.

Landscape responses at intermediate time scales

We distinguish here between landscape features that permit a direct calibration of deformation rates, such as terraces, and features that represent part of the landscape response to deformation. Calibration features primarily comprise displaced geomorphic markers whose initial shape is quite well known. The initial geometries of most other elements in the landscape, ranging from stream channels to hillslopes, are less easily traced backward in time, because these geometries represent an integrated response to ongoing deformation, base-level variation, and climate change. These features, therefore, only indirectly calibrate rates of deformation.

Stream gradients

River networks represent a hierarchical organization of tributary streams (lower order) routing flow into trunk streams (higher order). For a graded river flowing across uniform rock types and experiencing uniform uplift, the downstream channel gradient gets systematically gentler as a function of increasing discharge, which itself tends to vary as a function of catchment area (Fig. 8.5). Departures of the river gradient from this idealized, smooth shape may reflect variations in the rock strength of the river bed or variations in rock-uplift rate. Numerical models of tectonically perturbed rivers predict

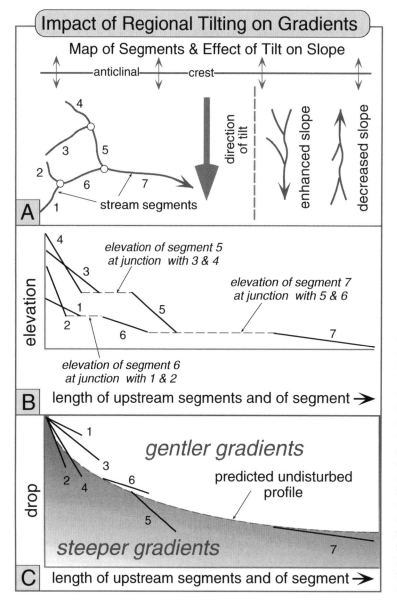

Fig. 9.8 Gradient analysis of tilted stream segments.
A. Stream network is divided into segments based on positions of stream junctions. The model assumes that, if tilting is occurring, streams flowing in the direction of tilting will have increased gradients, whereas those flowing in the opposite direction will have gentler gradients. B. Depiction of stream gradients. Each stream segment is plotted according to its length and to the elevation of the upper and lower ends of the segment. The upstream end of the segment (on the *x* axis) is determined by the length of segments that are upstream of it and tributary to it. Thus, segments 1 and 2 flow into segment 6, segments 3 and 4 flow into segment 5, and segments 5 and 6 flow into segment 7. Segments 1 to 4 all represent first-order tributaries that have no upstream segments – their upper ends plot on the left-hand margin. The sum of the lengths of segments 1 and 2 determines the horizontal position of the upstream end of segment 6 into which they flow. C. Departure from ideal gradient. When compared to the idealized gradient, steeper segments suggest tilting in the downstream direction. Compare the relative sense of tilting with the hypothesized fold orientation in the top panel. Modified after Merritts and Hesterberg (1994).

that they will approach a graded profile rather rapidly (Snow and Slingerland, 1987) once the perturbation ceases. Thus, anomalously steep or gentle river profiles, especially when not correlated to lithologic contrasts, may be interpreted as responses to ongoing tectonism.

Several approaches use data extracted from topographic maps to identify areas of active deformation, such as zones where stream gradients depart from expected longitudinal profiles. Consider a region experiencing tilting toward the south. Rivers flowing south will tend to have their gradients steepened, whereas rivers flowing north will have their gradient lessened with respect to untilted regions. One approach to the analysis of tilted channel networks (Merritts and Hesterberg, 1994) begins by segmenting a drainage network into its component tributaries (Fig. 9.8A) and measuring (i) the length of each segment and

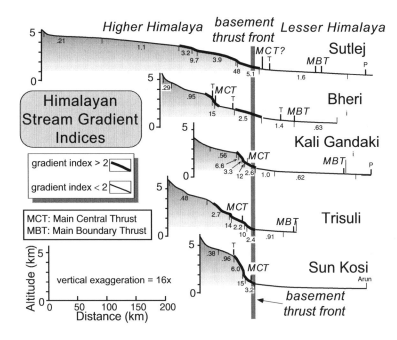

Fig. 9.9 Longitudinal profiles and stream-gradient indices for Himalayan rivers. Thicker segments of the profile indicate reaches where the local gradient index (SL) is more than twice the index (k) for the entire profile: $SL/k \geq 2$. The steepest gradients are not associated with the Main Boundary Thrust or active deformation to the south. Rather, they occur near the Main Central Thrust and appear to result from upward ramping of the overthrusting Himalaya above a deep-seated basement thrust. Modified after Seeber and Gornitz (1983).

(ii) the elevation of its upper and lower ends. The elevation range of each segment (y axis) can then be plotted against the sum of the lengths of the upstream segments and the length of the segment itself (x axis) (Fig. 9.8B). The lower elevation of each feeder or tributary segment is matched to the upper elevation of each segment into which it flows. Subsequently, using the x axis position as dictated by the length of the upstream segments, each segment is compared with an idealized logarithmic longitudinal profile (Fig. 9.8C). Significant departures from the ideal profile serve to identify segments that could be interpreted to indicate increased or decreased gradients over time due to tilting. When coherent areas are located in which all or most of the streams flowing in a given direction show the same tendency toward steepening or flattening, regional patterns of warping can be deduced (Merritts and Hesterberg, 1994). One must, however, exercise considerable caution in the application of such a technique, as it rests on the assumption of an "ideal" profile, and all effects of the variations in lithology or grain size of the material involved in the fluvial system must be assumed to be small compared to the tectonically induced changes in slope.

Before digital topographic data became widely available, departures from expected channel gradients were sometimes identified on the basis of changes in the stream-gradient index (SL), which compares the slope of a local reach with the distance to the drainage divide (Hack, 1973). For a short reach, the stream-gradient index can usually be approximated by

$$SL = (\Delta H / \Delta L) L \qquad (9.1)$$

where L is the distance measured from the drainage divide to the mid-point of the reach, and the slope of the short reach ($\Delta H / \Delta L$) is considered constant. For a well-adjusted channel profile, the stream-gradient index will remain nearly constant or change only slowly. Abrupt increases in the index typify oversteepened reaches.

A pioneering study of major Himalayan rivers by Seeber and Gornitz (1983) used stream-gradient indices to identify those river reaches that were anomalously steep (Fig. 9.9). Their analysis clearly showed that the steep reaches were not associated with what were considered the younger, active faults, such as the Main Boundary Thrust (MBT; Fig. 9.9), but instead were localized either above a deeply buried thrust ramp in the basement or near the trace

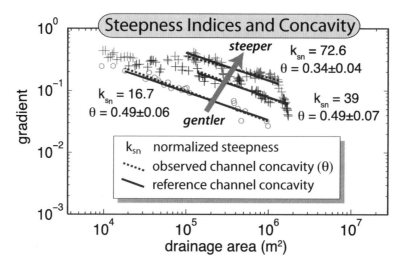

Fig. 9.10   River channel steepness indices and concavity.
Forced regressions through slope–area data using a reference concavity (0.45 in this example) define differences in normalized channel steepness (k_{sn}). For these channels, the observed concavity (θ) is similar to the reference concavity, but normalized steepness varies five-fold. Note the difference in channel slope between the channels for any given drainage area. Modified after Kirby *et al.* (2007).

of the Main Central Thrust (MCT; Fig. 9.9). This study helped promote a new view of this collisional orogen in which rates of rock uplift, large-scale crustal structure, and regional topography became inextricably linked.

With the widespread availability of digital topography and of programs based on geographic information systems (GIS) that are designed to analyze such topography, the use of stream-gradient indices has been largely replaced by use of a steepness index, k_s (Wobus *et al.*, 2006a). Recall from the previous chapter that $S = k_s A^{-\theta}$ (Eqn 8.7), where S is slope, k_s is the steepness index, A is upstream catchment area, and θ is concavity (see Fig. 8.4). Although concavity varies among rivers, it typically ranges between 0.4 and 0.7 (Whipple, 2004). For the sake of comparison among different rivers, the steepness index can be normalized, k_{sn}, by using the same reference concavity, θ_{ref}, for all channels being analyzed, such that

$$k_{sn} = k_s A_{cent}^{-(\theta_{ref} - \theta)} \qquad (9.2)$$

where A_{cent} is the area upstream of the mid-point of the reach being analyzed in the DEM (Wobus *et al.*, 2006a). In practice, this approach finds

the normalized steepness, k_{sn}, when a regression with a fixed reference concavity (commonly 0.45) is forced through channel slope versus area data to yield the best fit (Fig. 9.10). Combined Matlab and GIS programs that analyze a DEM and make a spatial map of variations in normalized steepness can currently be downloaded from http://www.geomorphtools.org. Such maps can permit ready identification of river reaches or whole regions characterized by anomalous steepness.

Analyses of river profiles on a rapidly deforming fold (Kirby and Whipple, 2001) have shown that high concavities characterize rivers whose headwaters are uplifting more rapidly than more downstream reaches (Fig. 8.24), but that concavities are commonly normal for a channel exposed to a uniform uplift rate, irrespective of whether it is rapid or slow. We might then wonder what happens to concavity and steepness when the rate of uplift changes at a given site. From a theoretical perspective, we can predict that, if the relative uplift rate doubles (for example, by accelerated fault slip at the channel's outlet), a knickpoint will develop that propagates upstream and that, downstream of the knickpoint, the channel will become steeper in order to erode at a rate that balances the new uplift rate (Fig. 9.11).

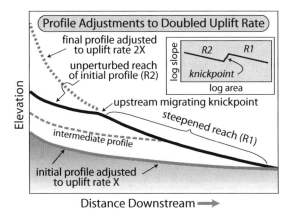

Fig. 9.11 River-channel adjustments to accelerated uplift.
Channel profile in a transient state due to doubling of the uplift rate with respect to the outlet. As a knickpoint sweeps upstream, the lower channel profile (R1) steepens, whereas the upper profile (R2) retains its initial gradient, one that was adjusted to the original uplift rate. Once the knickpoint sweeps to the headwaters, the entire profile will be steepened. Inset shows lower and upper reaches in slope–area space. Note the different steepnesses (vertical position of the line with respect to area), but same concavities (slope of the line). Modified after Whipple and Tucker (1999).

Upstream of the knickpoint, the pre-acceleration gradient will persist, because that part of the channel will not "feel" the acceleration at the outlet until its effects are translated upstream by the knickpoint (Whipple and Tucker, 1999).

One well-documented region where both spatial and temporal changes in rock uplift rate have occurred is the northern Californian coastal region. Here, over the past few million years, the Mendocino triple junction has swept northward across the region and created a wave of accelerated uplift followed by gradual restoration of rates in the wake of the triple junction (Merritts and Bull, 1989). Studies of coastal terraces, such as that depicted in Fig. 9.4, reveal more than six-fold increases in uplift rates over the past 100 kyr for some sites, thereby providing a striking context in which to examine channel responses to these differences in rates. Whereas the southern part of the region has sustained uplift rates <0.5 mm/yr for the past 300 ka, the northern region accelerated to 3–4 mm/yr at ~100 ka. Despite these strong contrasts in uplift rates, channel concavity, θ, remains nearly constant (Fig. 9.12) throughout the study area (Snyder et al., 2000). Such constancy suggests that (i) the channels are in rough equilibrium with

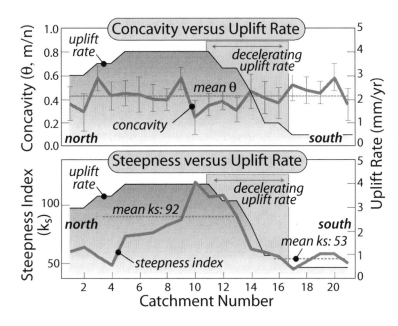

Fig. 9.12 Channel concavity and steepness in different uplift regimes.
Rates of rock uplift deduced from marine terraces define rates range from <0.5 mm/yr to 4 mm/yr along the northern California coast (Merritts and Bull, 1989). (Top) Concavity is largely insensitive to uplift-rate variations. Mean concavity is ~0.45, consistent with expectations for adjusted channels. (Bottom) Steepness shows strong contrasts, with higher average steepness in the high-uplift-rate zone. Note correlation of steepness index with uplift rate across the zone of decelerating rates. Modified from Snyder et al. (2000).

the uplift rate and (ii) channel-profile adjustments to the increased rates have occurred in less than 100 kyr, the time when rates accelerated. In contrast to the constant concavity, channel steepness shows strong regional variability (Fig. 9.12). The steepness index, k_s, is nearly twice as large in the high-uplift-rate region as in the low-rate region. Notably, the uplift rate and steepness index co-vary in the zone of changing uplift rates (catchments 12 to 16; Fig. 9.12). In the northernmost catchments, however, the steepness index decreases despite the rather high uplift rate. Although the cause for this decoupling is not known, orographic rainfall is about twice as great in the north than in the south (Snyder et al., 2000). To the extent that the efficiency of erosion is linked to the mean annual discharge, higher rainfall could drive more erosion and promote less steep channels for a given erosion or uplift rate.

From landscapes to faults

Sometimes one of the biggest challenges for field geologists is simply recognizing the existence of major faults. Such faults may be blind, they may rupture the surface where few geomorphic markers exist with which to recognize differential displacement, or dense vegetation may obscure clear views of offset features (e.g., Fig. 1.5). The identification of faults under these conditions commonly requires a quantification of landscape attributes that respond to the uplift or subsidence caused by active faults.

A recent study of the region surrounding Mount Tamalpais along the northern San Andreas Fault (Kirby et al., 2007) exploits changes in mean topography, channel steepness indices, hillslope angles, and topographic relief to delineate a spatial gradient that is interpreted as a response to a blind thrust (Fig. 9.13). Along a transect southwards toward Mount Tamalpais, both the mean and maximum elevations rise, as does the average steepness of hillslopes and the fraction of slopes considered to be at a threshold angle for failure. Steepness indices for a suite of similarly sized catchments along this north–south transect show an exponential

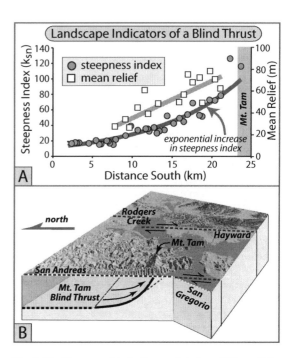

Fig. 9.13   Topographic indices related to differential uplift above a blind thrust.
A. Normalized steepness index (reference concavity = 0.45) and relief in the inner gorge as a function of distance. Both parameters increase towards the south to where Mount Tamalpais is located (~24 km) and suggest a concomitant increase in rock uplift rates. B. Schematic model for an uplift gradient above a listric blind thrust. Vertically exaggerated topography and major dextral strike-slip faults are shown. Modified after Kirby et al. (2007).

increase toward the south (Fig. 9.13A). Channels in all these catchments display inner gorges whose walls are at threshold angles, suggesting that these walls are eroding by bedrock landsliding, presumably because the channels are incising at rates faster than soil-mantled hillslopes can erode. The relief from the channel bottom to the top of the inner gorge also progressively increases toward the south (Fig. 9.13A) and is interpreted to indicate increasingly rapid rates of channel incision. The presence of an inner gorge, rather than a gradually steepening hillslope rising above each channel, suggests that this landscape is in a transient state. This suite of observations suggests that rock uplift rates increase

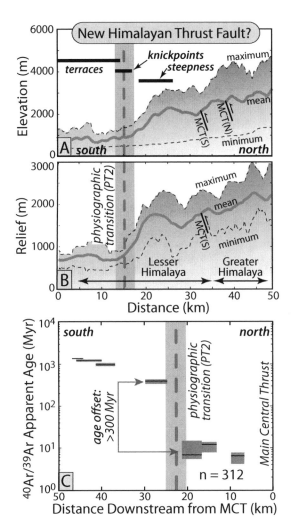

Fig. 9.14 Topographic indices and detrital ages related to a proposed Himalayan fault.
A. Maximum, mean, and minimum elevation along a north–south swath focused on the Burhi Gandaki river in central Nepal (see Plate 7). The northern and southern traces of the Main Central Thrust (MCT) and the north–south extent of fluvial terraces, river knickpoints, and normalized channel steepness indices greater than 450 are shown. Major topographic changes occur >20 km south of the MCT at the "physiographic transition 2" (PT2). Note the upward deflection of the mean elevation toward the maximum, indicating a transient state north of the PT2.
B. Maximum, mean, and minimum of local relief show an abrupt increase at the PT2. Transition between Greater and Lesser Himalaya is taken as the MCT(S).
C. Detrital muscovite cooling ages ($n=312$) from seven tributary catchments indicating a >10-fold offset of ages across the PT2. Box widths correspond

systematically toward the south and toward Mount Tamalpais, a scenario consistent with slip above a blind listric thrust fault (Fig. 9.13B).

During efforts to locate previously unknown active structures, both regional maps and swath profiles of topographic characteristics can be very useful, particularly for identifying anomalies or trends. In a swath profile, spatial data on transects perpendicular to the long axis of the swath are assessed at successive steps (typically equivalent to the pixel size of the data) along the swath. The data from each transect can be examined statistically to define their attributes: commonly maximum, mean, and minimum of some characteristic. For a typical swath profile of elevation in most landscapes, the mean elevation will lie closer to the minimum than to the maximum. If the mean elevation approaches the maximum, this upward deflection of the mean is likely to indicate a transient state of adjustment to an accelerated uplift rate in which rates of hillslope lowering are not keeping pace with the rate of rock uplift (Fig. 9.14). Similarly, changes in the average hillslope angle or topographic relief can be a signal that the landscape has adjusted (or is adjusting) to changes in patterns or rates of rock uplift.

Recent studies in the Nepalese Himalaya have combined diverse topographic attributes with bedrock cooling ages to identify what may prove to be a major, active, and previously unrecognized fault (Wobus *et al.*, 2003, 2006b). The location of this fault was initially deduced from maps of hillslope angles and channel steepness indices that show an abrupt discontinuity that lies south of the Main Central Thrust and within the Lesser Himalaya (Plate 7). Swath profiles across this region show concurrent increases in both mean elevation and mean relief (Fig. 9.14A and B) and led to the naming of this topographic break as the "physiographic transition 2" (or PT2). (The PT1 lies at the break between the Tibetan Plateau and the Greater Himalaya.)

to the widths of sampled catchments orthogonal to the swath. Box top and bottom show the 25th and 75th percentiles, with horizontal lines indicating the median age. Modified after Wobus *et al.* (2003, 2006b).

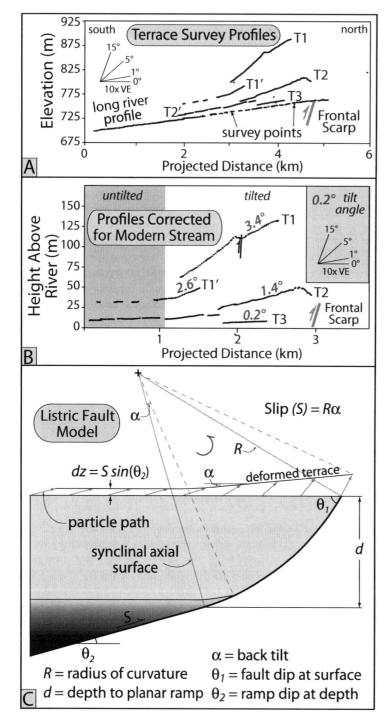

Fig. 9.15   **Tilted backlimb terraces and listric thrust faulting.**
A. Surveyed terrace and river profiles deformed above the Irishman Creek Fault, Mackenzie Basin, New Zealand. Terraces T1 to T3 span ages ranging from about 140 to 20 ka. B. Heights of terrace treads with respect to the modern river gradient. Note the progressively greater magnitude of tilting for older (higher) terraces, but the consistent position of the hinge representing the transition to the untilted domain (~1.2 km). C. Model for a listric fault linking to a planar ramp. When applied to the Irishman Creek data in panels A and B, the model predicts 190 m of slip on the T1 terrace, the listric-to-planar fault transition at 1.4 km depth, and a planar ramp dip of 17°. Modified after Amos et al. (2007).

Other changes observed at the PT2 include the abrupt northern end of Lesser Himalaya river terraces, commonly observed knickpoints in river profiles, and greatly increased channel steepness to the immediate north (Fig. 9.14A). The totality of these topographic changes points to a higher rock uplift rate to the north of the PT2, but left open the question of whether this rate change was due to a surface rupturing fault or a change in the ramp angle of a deeply buried fault surface. Wobus *et al.* (2003, 2005) also showed that a major discontinuity in the ^{40}Ar/^{39}Ar cooling ages of detrital muscovite that was collected from small tributary catchments straddles the PT2 (Fig. 9.14C). North of the PT2, cooling ages average 10 Ma or less, whereas south of the PT2, cooling ages range from an average of 300 Ma to >1000 Ma. These data imply that the northern catchments have cooled at rates above 30°C/Myr, whereas the southern catchments experienced rates averaging below 1°C/Myr. This simplest, although not unique, explanation for the abrupt change in cooling ages is a previously unrecognized, surface-rupturing fault at the PT2.

Given that the cooling ages north of the PT2 imply rapid cooling since 10 Ma, the reader might justifiably wonder why this example is included under "intermediate time scales." The reason is that erosion rates north of the PT2 are sufficiently high that many hundreds of meters of erosion would occur during 500 kyr and, as a consequence, the topographic indices examined here (hillslope gradients and channel steepness) would be sensitive to erosion of that magnitude.

Where terraces of different ages are preserved, patterns of deformation across tens to hundreds of thousands of years can be deduced, and sometimes constraints can be placed on the underlying fault geometry. Consider, for example, a suite of terraces in which progressively older terraces are increasingly back-tilted above a thrust fault (Fig. 9.15A). After subtracting the gradient of the modern channel, the magnitude of tilting can be defined, as well as a narrow zone in which each terrace transforms from tilted to simple planar uplift (Fig. 9.15B). Based on deformed terraces on New Zealand's South Island, Amos *et al.* (2007) argued that this configuration of terrace deformation is most consistent with an underlying, near-surface listric thrust fault that transitions to a planar fault at depth (Fig. 9.15C). For this model, if the width of the tilted backlimb and its tilt magnitude, as well as the position and dip of the frontal fault, are known, then the radius of curvature of the listric fault, the magnitude of fault slip, the depth

to the transition to a planar fault, and the dip of the planar ramp can be calculated (Amos *et al.*, 2007). If the vertical offset (dz in Fig. 9.15C) of the non-tilted terrace treads is known, this offset provides a check of the dip prediction for the planar fault at depth. It is important to note, however, that listric faulting is not the only way to produce progressive terrace tilting: both detachment folding (Scharer *et al.*, 2006; Suppe *et al.*, 2004) and simple-shear fault-bend folding (Suppe *et al.*, 2004) can produce similar tilting. Sometimes the local geology can rule out alternative models, but commonly subsurface imaging is needed to test among them.

Fault-bend fold theory (Fig. 4.36A) (Suppe, 1983) predicts that, whenever a planar marker is transported through an active axial surface, the marker will deform in a geometrically predictable way. Consequently, planar geomorphic features, such as fluvial terraces, that are beveled or deposited across an active axial surface can be folded following the rules for fault-bend folds. This deformation is particularly obvious and useful on the backlimbs of folds where structural advection of a terrace through an active axial surface commonly forms a fold scarp (Fig. 9.16A and Plate 6). The dip of the fold scarp is a predictable function of (i) the dip of the original terrace surface and (ii) the angular difference in dip of the underlying fault plane on either side of axial surface (Hubert-Ferrari *et al.*, 2007; Chen *et al.*, 2007). From a tectonic perspective, the key aspect of the fold scarp is that its length approximates the slip on the fault since the terrace was created (Fig. 9.16B). Clearly, if the age of the terrace is known, then a fault slip rate can also be determined. Where multiple dated terraces are present, temporal changes in slip rates can also be assessed.

A spectacular example of a fold scarp was recently described by Hubert-Ferrari *et al.* (2007) from the southern margin of the Tien Shan in western China (Box 9.2). In this arid setting, erosional modification of many geomorphic features is modest, such that large triangular facets representing extensive remnants of fold scarps are well preserved in the landscape. In the cross-section at Quilitak fold (Fig. 9.17A), a remarkable, planar facet

Box 9.2 Geological versus geodetic rates of deformation.

The proliferation of GPS data on relative velocities across orogens and plate boundaries has provided a far more complete view of the modern patterns of crustal deformation. Although it is tempting to use these decadal rates as representative of rates applicable to hundreds of thousands of years, this equivalence has relatively rarely been tested. Along some well-studied strike-slip fault zones, such as the southern San Andreas Fault, where numerous paleoseismic studies define rates of slip on multiple faults at time scales of many thousands of years, geodetic and geological rates are quite closely matched.

Across contractional orogens, however, this match of rates is less clear. For example, in the Himalaya, geomorphic studies (Lavé and Avouac, 2000) clearly show that, at Holocene time scales, ~20 mm/yr of slip occurs at the Main Frontal Thrust (see Fig. 7.25). Yet, the geodetic shortening across the same zone is only a few millimeters per year. This large mismatch is interpreted to result from the presence of a locked megathrust, such that elastic strain is accumulating in the Greater Himalaya and is episodically released in large earthquakes that translate slip to the frontal thrust (Avouac, 2003). The Tien Shan, which span from China across Kyrgyzstan to Kazakhstan, are not underlain by a single megathrust, but instead comprise a series of fault-bounded ranges that are separated by narrow basins (see figure A). Recent earthquakes in the Tien Shan suggest that most fault planes dip quite steeply (~45°) and extend to depths of ~20 km (Ghose *et al.*, 1997, 1998), which is near the brittle–ductile transition. Hence, these faults are unlikely to be linked by a relatively shallow crustal detachment.

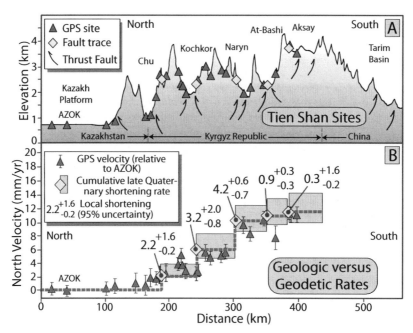

A. Topographic profile of the Tien Shan with major thrust faults, GPS sites, and studied fault indicated. B. Comparison of GPS north velocities with late Quaternary shortening rates. Modified after Thompson *et al.* (2002).

In an effort to define fault-slip rates across the Tien Shan, Thompson *et al.* (2002) analyzed numerous fold scarps, fault scarps, trenches across faults, and deformed terraces on major faults within each of the Tien Shan basins. Most of the calibrated slip rates span the past 14–15 kyr (see Fig. 2.13), but some extend to >100 ka. One striking result of Thompson *et al.*'s study is that the cumulative geological rates, when plotted versus the geodetic data across the same area (Abdrakhmatov *et al.*, 1996), provide a good match to the geodetic rates (see figure B). This match suggests that, at least at time scales of several millennia, geodetic and geological rates are equivalent in the Tien Shan. This distributed deformation in the Tien Shan at geological time scales may be relevant for other orogens that comprise multiple fault-bounded ranges and lack an underlying megathrust. When Thompson *et al.*'s (2002) study was published, it was the first to provide calibrated slip rates across an entire contractional orogen that encompasses complex fault patterns.

stands ~700 m high. Based on a fault-bend fold model, the facet records ~1.1 km of slip (about equal to the length of the fold scarp) on the underlying thrust fault (Fig. 9.17B). Along strike on the Quilitak fold, multiple facets rise to parallel linear ridge crests that display accordant heights and are nearly 4 km long (Fig. 9.17C and D). A surface fit to these ridge crests is interpreted to represent the uplifted erosion surface.

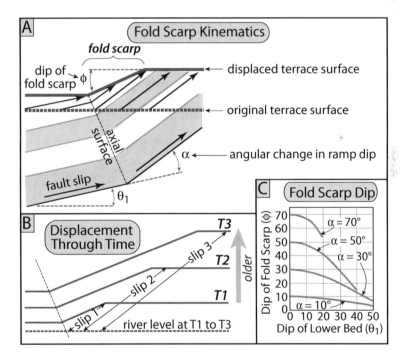

Fig. 9.16 Fold scarp kinematics and progressive terrace deformation.
A. Model for formation of a fold scarp in a fluvial terrace. The underlying fault has two planar segments that are separated by an active axial surface. Displacement is parallel to the underlying fault for material on either side of the axial surface (slip vectors). Terrace created at the dashed line will form a fold scarp whose length approximates (but underestimates) the fault slip since the terrace was formed. Modified after Thompson *et al.* (2002). B. Cartoon depicting three terraces that formed at different times at the same river height (dashed line). Length of the fold scarp approximates the slip for each terrace. Displacement pattern is based on the fault geometry of part A. C. Fold scarp dip (ϕ) as a function of the dip of the gentler dipping fault plane (θ_1) and the angle (α) between the fault planes on either side of the axial surface. Modified after Chen *et al.* (2007).

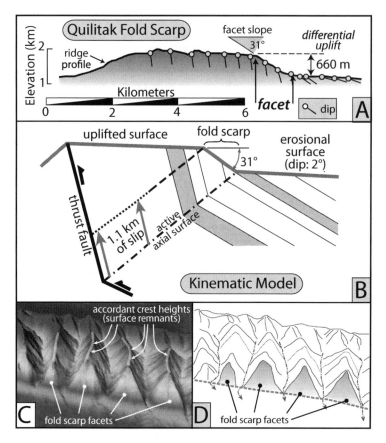

Fig. 9.17 Fold scarps in the southern Tien Shan.
A. Topographic profile and bedding dips along a ridge crest in Quilitak fold. Profile passes through a triangular facet with a dip of 31°, a height of 660 m, and a length of ~1 km. The facet is interpreted as a remnant of a fold scarp.
B. Formation of a fold scarp using a kinematic geometry that is consistent with the observed structure and geomorphology. C, D: Perspective views of fold-scarp facets and accordant ridge heights that define the approximate position of the uplifted strath or pediment. Modified after Hubert-Ferrari *et al.* (2007).

Given that Hubert-Ferrari *et al.*'s (2007) study suggests that the Quilitak fold has been deforming for several million years, why is the current morphology dominated by a single folded and moderately eroded paleoerosion surface? They suggest that, for most of Quilitak's growth, the sum of erosion of the fold plus sediment accumulation in the surrounding basin roughly balanced rock uplift, such that very little topography developed within the fold, despite its accommodating several kilometers of shortening. Within the past 200 kyr, however, they estimate that shortening rates across Quilitak increased about six- to eight-fold. As a consequence of this acceleration, the fold emerged as a prominent topographic entity that is dominated by the uplifted erosion surface that was present just prior to the accelerated growth (Hubert-Ferrari *et al.*, 2007).

In most sites of active folding, suites of deformed geomorphic markers that span a wide age range are absent, such that the temporal evolution of the shape of a fold becomes difficult to discern. A tilted fold limb could have attained its current geometry by maintaining a constant dip and by lengthening through the toe of the fold or by lengthening through the fold's crest, or the limb could have rotated with a fixed length as in the listric fault model (Fig. 9.18A). Perhaps surprisingly, the long profiles of rivers incised into the fold limb may provide clues about the kinematic pathway of fold growth (Goode and Burbank, 2011). Under the assumption that channel incision rates are linearly proportional to stream power and that stream power is a function of upstream catchment area and channel slope, the temporal evolution of a river's long profile can be modeled as a function of the limb kinematics (Fig. 9.18B). A channel on a limb that is lengthening through its toe, for example, will always have an instantaneous incision rate that is maximized at its

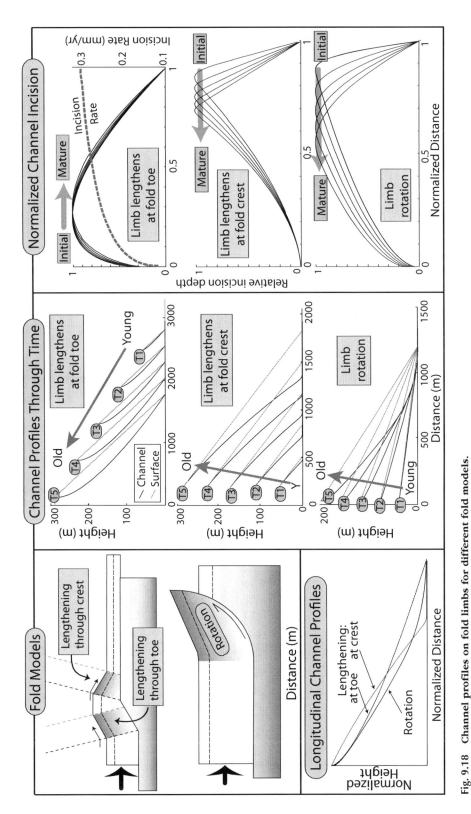

Fig. 9.18 **Channel profiles on fold limbs for different fold models.**

(Left, top) Different modes of limb growth: lengthening through the toe or crest at a constant angle, or rotation of a limb of constant length. (Center) River incision patterns through five time steps for different fold models. Both the pristine fold surface and the channel profile are depicted for each step. (Right) Spatial patterns of incision for an undissected surface for five time steps (as in part B). Note the migration of the point of deepest incision through time (arrows). Top figure also shows the instantaneous incision rate versus distance (dashed line). (Left, bottom) Normalized long profiles for each fold model at time step 5. Note the strong contrasts in concavity and overall shape. Modified after Goode and Burbank (2011).

toe. But, because pristine material continuously migrates through the axial surface at the toe and, thereby, adds new uneroded material to the base of the limb, the maximum amount of net incision occurs closer to the top of the fold (Fig. 9.18C). In contrast, a limb that is lengthening through the fold crest or that has a constant length but is steepening via rotation is predicted to have the most channel incision near the lower part of the fold limb. Whereas the boundary conditions for these models will certainly not all be matched by many field settings, the modeled patterns of incision provide a means to infer fold kinematics when other means may be lacking.

Transient landscapes

Because climate changes occur at rates that commonly outpace a landscape's tendency to reach equilibrium with the current climate, most landscapes are technically in a transient state all the time. Here, transient landscapes refer instead to landscape changes that occur, for example, in response to changes in the rate of base-level fall, rock uplift, or precipitation. Stepwise changes in any of these boundary conditions will force a landscape toward a new configuration. When a landscape is captured in this transitional interval, insights on both the nature and rates of change, as well as on key controlling variables, can commonly be obtained.

Knickpoints

Landscapes that are in transition from one state to another can provide insights on how surface and tectonic processes respond to changing boundary conditions. The migration of a knickpoint, a deceleration in the rate of rock uplift, a change in climate, or the growth of a fold, will each induce changes that will affect the rates of surface processes and ultimately modify the shape of the landscape. Such transitional landscapes commonly provide opportunities to examine landscape regions that have been affected by the change or ones that have not yet been affected. The contrasts between such areas can serve to highlight the nature and rates of landscape change and perhaps can provide quantitative insights on controlling variables.

Consider the impact of a knickpoint on a trunk stream as it sweeps progressively upstream past numerous tributary channels. Each tributary will experience a relative drop in base level at the junction with the main stem. This drop is equivalent to increasing the rock uplift rate of the tributary catchment, such that we expect a knickpoint then to migrate up each tributary as well (Fig. 9.11). The Yellow River on the northern Tibetan Plateau provides a spectacular example of the impact of knickpoint migration on tributary channels and on rates of erosion (Harkins et al., 2007). In the upstream end of the study area, the Yellow River progressively steepens, producing an upward convexity, as it descends ~900 m over the next 350 km (Fig. 9.19A). Two basins at the downstream end of the study area contain Plio–Pleistocene fills that rise ~600 m above the current level of the Yellow River. The tops of these fills are aligned and project upstream approximately to the level of the top of the modern Yellow River knickpoint (Fig. 9.19A). Along this projected gradient, several strath terraces are preserved, suggesting that the Yellow River previously flowed along this more even, gentler gradient. Recent magnetostratigraphic studies (Craddock et al., 2010) show that the Tongde Basin was actively aggrading until ~0.5 Ma. Hence, if we assume that the accelerated incision of the Yellow River began at 0.5 Ma, the knickpoint migrated at a mean rate of ~50 cm/yr as it swept more than 250 km upstream! Along dozens of tributary channels, Harkins et al. (2007) used DEM analysis to document pronounced knickpoints that separate steep lower reaches from gentler upper reaches (Fig. 9.19B). In order to estimate the pre-knickpoint gradient of each tributary, Harkins et al. (2007) calculated the normalized steepness index, k_{sn}, (Eqn 9.2) for the river reach above the knickpoint. They then projected a channel with the same steepness downstream in order to estimate the former elevation of the tributary mouth (Fig. 9.19B). When the reconstructed elevations of the former mouths of numerous tributaries are plotted along the

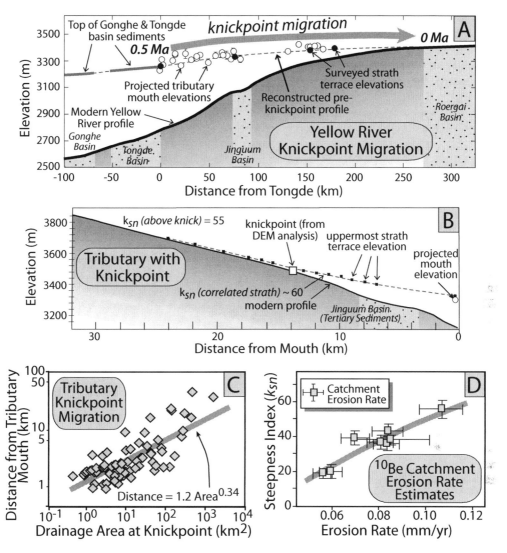

Fig. 9.19 Knickpoint migration along the Yellow River.
A. Long profile of the Yellow River, showing a pronounced knickpoint and the position of late Cenozoic basins. The tops of the downstream basins define a gradient that projects to the top of the knickpoint 250 km upstream. Along this gradient lie both preserved straths (black circles) and the projected former elevations of tributary mouths (open circles). B. Example of a tributary with a knickpoint ~14 km from the mouth. Observed normalized channel steepness above the knickpoint is used to project the pre-knickpoint channel gradient to the mouth (dashed line). Multiple strath terraces (small black squares) lie along this projected profile. C. The distance of the knickpoint from a tributary mouth scales with the drainage area upstream of the knickpoint. D. Catchment-wide erosion rates above knickpoints scale with the normalized steepness index. Modified after Harkins *et al.* (2007).

Yellow River, they align both with the observed straths and with the upstream projection from the downstream Gonghe and Tongde Basins to the top of the main-stem knickpoint (Fig. 9.19A).

Through their analysis of many dozens of tributaries, Harkins *et al.* (2007) developed an

extensive database with which to assess both proposed controls of knickpoint migration and landscape responses to falling local base level. For example, these data show a robust relationship between upstream drainage area and the distance each knickpoint has migrated

(Fig. 9.19C). Given that discharge is roughly proportional to catchment area, this relationship is consistent with observations from stream-table experiments in which migration rates scaled with discharge (Fig. 8.10A). This relationship further suggests that an erosion rule based on stream power can adequately describe the observed profiles. Harkins *et al.* (2007) also used detrital [10]Be concentrations in channel sediments upstream of knickpoints to determine catchment-wide erosion rates. Not only are these rates about an order of magnitude lower than incision rates below the knickpoints, but they also scale with the normalized steepness index, k_{sn} (Fig. 9.19D). Notably, most tributaries display a convex upward profile downstream of the knickpoint (Fig. 9.19B). This convexity suggests that these tributaries are still in a transient state of incomplete adjustment to their new base level. Note, for example, the contrast with the concave-up profiles above and below a knickpoint in a theoretical model for a migrating knickpoint (Fig. 9.11). Overall, this study of the Yellow River exemplifies the wealth of data and insights that can be gleaned from combinations of DEM analysis, chronological control on key events, calibrations of erosion rates, and both usage and tests of numerical models of rivers.

Pressure ridges

When strike-slip faults depart from verticality and when the fault trace bends into the path of the fault-slip vector, the resultant fault-normal stresses cause contraction and uplift, thereby forming a pressure ridge (Fig. 4.19). If the structural anomaly is persistently attached to a block on one side of the fault, the anomaly acts as a point source that drives uplift of the opposing block as it slides by. Once a given segment of the opposing block moves past the anomaly, uplift ceases. Thus, a very discrete spatial and temporal window exists in which the opposing block transitions, first, to experiencing the accelerated rock uplift and, second, to exiting the zone of uplift. The spatial extent of the window depends on the length of the structural anomaly parallel to the fault trace, whereas the pattern of uplift depends on the

subsurface shape of the structural anomaly. The temporal extent of the window depends on the fault slip rate and the anomaly's length. When a pressure ridge formed under these conditions can be identified, an illuminating opportunity exists to make a robust space-for-time (ergodic) substitution and to examine how the land surface responds to this transient pulse of uplift.

Along the San Andreas Fault in southern California, the Dragon's Back pressure ridge (Fig. 9.20) provides just such an opportunity (Hilley and Arrowsmith, 2008). Subsurface imaging at the Dragon's Back site (Unsworth *et al.*, 1999) defines a 2-km-long, structural knuckle attached to the North American Plate that juts beneath the Pacific Plate and drives rock uplift. The slip rate on the San Andreas Fault in this area is ~33 mm/yr (Sieh and Jahns, 1984), such that each kilometer of Dragon's Back's length represents ~30 kyr. Recent acquisition of high-resolution (1-m pixel) lidar topography provides a high-resolution spatial database with which to quantify how the pressure ridge evolves in time and space as a geomorphic entity.

By mapping flat-lying rock formations that become uplifted in the pressure ridge, Hilley and Arrowsmith (2008) show that, over the first 2 km (to the southeast), the rock-uplift rate ranges as high as 2.3 mm/yr (Figs 9.20C and 9.21A). During the ~70 kyr that it takes for any point on the Pacific block to pass across the structural anomaly, the total rock uplift is ~80 m along the crest of the fold (Figs 9.20B and 9.21B).

Dragon's Back ridge is underlain by weakly consolidated Quaternary sediments that can be readily eroded. Consequently, the response time of various geomorphic processes to the tectonic forcing is expected to be quite rapid. For each drainage basin along the pressure ridge, Hilley and Arrowsmith (2008) measured several topographic metrics, including basin width and area, channel concavity and normalized steepness, relief within a radius of 50 m, density of landslide scars, and hillslope gradients. Their results show that local relief, normalized channel steepness, landslide density, and hillslope gradients all broadly track the uplift rate (Fig. 9.21). Despite the small size (<0.5 km[2]) of the drainage

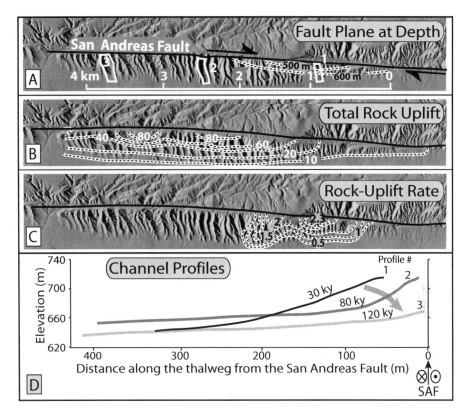

Fig. 9.20   Deformation patterns and channel profiles at Dragon's Back pressure ridge.
A. Shaded relief of Dragon's Back and the surrounding terrain. Note the leftward bend of the fault trend
which drives compression. Contours indicate depth to the fault surface and show that the San Andreas Fault
is offset in the subsurface toward the southwest. Numbers indicate channel locations in part D lidar DEM
from www opentopography.org. B. Total rock uplift within the pressure ridge. C. Rock uplift rate in mm/yr.
D. Channel profiles for three catchments (shown in A). Ages are inferred based on the position of the
basin along the pressure ridge and the assumed slip rate. Note the high concavity at 80 kyr and the much lower
channel relief by 120 kyr. Modified after Hilley and Arrowsmith (2008).

basins that developed on the pressure ridge, channel and hillslope processes are shown to be intimately linked. For example, as uplift ceases, channel concavity abruptly increases (profile 2 at 80 kyr; Fig. 9.20D), thereby driving rapid channel incision. This incision undercuts adjacent hillslopes and, thereby, promotes increased rates of landsliding (Fig. 9.21E). Based on the space-for-time substitution, the response time of the channel is only ~7 kyr, whereas rates of landsliding only slowly diminish: more than 70 kyr is required to transition from hillslopes dominated by mass wasting to ones dominated by diffusive processes that typify the nearby landscape. This order-of-magnitude difference

in response times indicates a fundamental lag between channels versus hillslopes. Channels are the drivers that determine the rate of local base-level lowering and short-term rates of hillslope steepening; hillslope processes then respond to that steepening.

For Hilley and Arrowsmith's (2008) study, the combination of high-resolution topography, superb time control, and a well-constrained structural setting underpins their ability to closely examine both the pattern of tectonic forcing and the diverse geomorphic responses to this forcing. Investigations of different, but similarly well-constrained, tectonic settings, as well as analogous settings but with contrasting

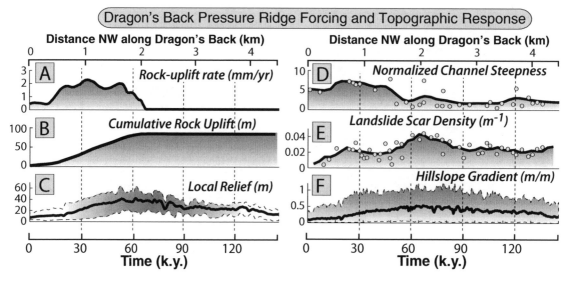

Fig. 9.21   Rock uplift and geomorphic metrics at Dragon's Back pressure ridge.
Distance is measured toward the northwest and starts at the inception of uplift. Time is based on an assumed
San Andreas slip rate of 33 mm/yr or 1 km in 30 kyr. Topographic data were extracted from a 1-m DEM.
A. Rock-uplift rate along the crest of the pressure ridge. B. Cumulative rock uplift along the crest. C. Local topographic
relief measured in a 50 m radius: average relief (solid line) and 95% bounds (shaded area). D. Normalized channel
steepness based on a reference concavity of 0.68 (mean of the entire data set). Only catchments for which a linear
log area–slope trend could be identified that spanned more than an order of magnitude of catchment area were
included. E. Landslide scar density. F. Hillslope gradients. Modified after Hilley and Arrowsmith (2008).

climate, lithology, or vegetation, promise to shed
more insight on tectonic–geomorphic linkages.

Growing folds

Whether they are laterally propagating or have
rather fixed tips, growing folds represent a
special type of transient landscape, because,
along their length, progressive changes occur in
their geometric attributes, such as the magnitude
of differential uplift or the slope of their limbs.
Perhaps most importantly, laterally propagating
folds can provide a robust space-for-time substitution, and, when the ages of uplifting surfaces
are known, rates of processes can be defined.

As described previously, as faults accumulate
displacement, they commonly increase the size
and length of their rupture surface. The
plunging noses of hanging-wall anticlines and
footwall synclines that develop in conjunction
with either emergent or blind thrusts would,
therefore, be expected to propagate laterally as

the tip of the underlying fault migrated with
each successive rupture event. Simultaneously,
especially in the early stages of development,
these growing folds progressively increase in
amplitude and breadth with each successive
earthquake. As their flanks rise above local base
level, they are attacked by erosional surface
processes, and their pristine surfaces begin to
be modified. Meanwhile, as their noses propagate laterally, the resultant uplift influences the
gradients and geometries of nearby fluvial
systems. Thus, two rather different responses to
growing folds are potentially recorded in the
geomorphological record: in one, the shape of
the fold is modified by surface processes; in the
other, the actual growth of the fold modifies
the surface processes (Keller *et al.*, 1999). The
degree of surface modification may be an indication
of the age of various parts of a fold and the local
rate of deformation, whereas river patterns with
respect to the fold may reveal their interactions
over time. Analysis of these interactions can

provide insights on a history of fold growth that are otherwise unattainable (Burbank *et al.*, 1996c, 1999; Keller *et al.*, 1998). Several key data sets can contribute to such a history, including: topographic data on the fold crests, limbs, and nose; dates on geomorphic surfaces and features along the fold; detailed analysis of the underlying structural geometry; and variations of structure along the length of the fold.

Consider first the ways in which a fluvial system that was previously flowing across a relatively low-relief landscape may interact with a growing fold or a suite of growing folds. For any individual fold, as it emerges above the adjacent land, a new drainage divide is defined along the fold crest and new catchments are formed along its flanks: it subdivides formerly continuous drainage systems, and the new catchment configuration is closely tied to the fold geometry. Asymmetrical folds with steep forelimbs will have short, steep catchments on their forelimbs and elongate, gentler catchments on their backlimbs. This asymmetry is readily visible in map patterns of river courses (Talling and Sowter, 1999; Talling *et al.*, 1997).

Streams that had formerly flowed across the site of the growing fold either (i) are diverted parallel to the fold axis and around the nose of the fold, (ii) become entrenched as antecedent streams that incise across the uplifting fold, or (iii) bevel off the top of the emerging fold so that it has little or no topographic expression. In order to maintain its course through a water gap across a rising fold, an antecedent stream must maintain a basinward-dipping gradient across the fold. Otherwise, the stream will be diverted along the upstream margin of the fold. A folding event will cause relative uplift of the channel reach within the folding domain (Fig. 9.22A), causing a likely instantaneous reversal of the channel gradient immediately upstream of the fold. Sediment will tend to be ponded in this depression and begin to fill it. At the same time, a knickpoint will begin to propagate upstream from the newly steepened zone on the downstream end of the uplifted reach (Fig. 9.22B). A competition, therefore, exists between the rate of differential rock uplift in the fold and the rate of aggradation upstream of the fold.

If the sediment load of an antecedent stream is insufficient to aggrade as fast as the fold is rising, the stream is likely to be diverted (Humphrey and Konrad, 2000). Several other factors can affect whether a channel can sustain its course. First, if a layer of alluvium that is a few meters thick is part of the longer-term transport load of the channel floor, then, in response to uplift that is less than the thickness of the alluvium, the channel will simply incise its bed rapidly through the alluvium and sustain its course. Second, if rock strength is low across the core of the fold, then a knickpoint will more rapidly propagate from the downstream to the upstream end of the uplift and, thereby, restore a downstream gradient. Third, as the fold widens during continued growth (Fig. 9.22C), the gradient of antecedent channels across the fold will decrease and cause a concomitant decrease in the erosive power of the stream (Burbank *et al.*, 1996c). Whereas the first two conditions promote maintenance of an antecedent channel, the last condition promotes its diversion.

When analyzing the map pattern of stream valleys associated with growing, laterally propagating folds, the expectation should be that a series of wind gaps along the fold crest will record the progressive defeat of older, antecedent streams and that, where a stream is still antecedent, the fold has not broadened and uplifted sufficiently to defeat the stream. As successive antecedent streams are defeated, the remaining streams will capture the discharge of the defeated streams and will augment their discharge (and probably stream power) as they cross the anticlinal crest. Thus, the most likely location for an antecedent stream is near the propagating nose of a fold (Burbank *et al.*, 1996c; Jackson *et al.*, 1996).

An illustrative example of both antecedent and diverted channels is well displayed in the southern Tien Shan of western China (Hubert-Ferrari *et al.*, 2007). Here, as deformation encroaches on the adjacent foreland, a new, elongate fold that is 200–300 m high has grown in front of an older fold that is cut by antecedent streams. Prior to the growth of the new fold, those antecedent streams fed alluvial fans about 10 km long whose apices lay on the downstream

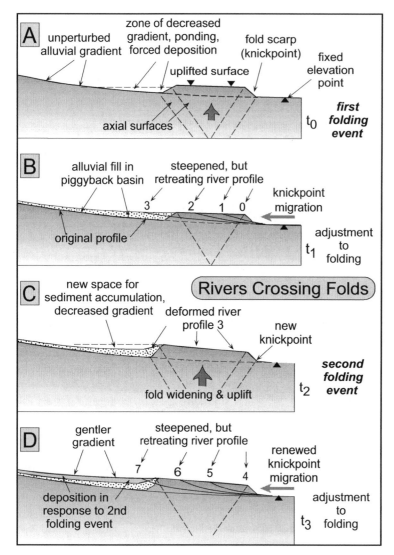

Fig. 9.22 Interactions of an antecedent stream with a growing fold.
Fold uplift creates accommodation space for sediment in the piggyback basin. Deposition in the piggyback basin allows the stream gradient to steepen across the fold, but also raises the river above the surrounding plain. Such deposition promotes avulsion, often away from the former water gap. With each increment of uplift, a knickpoint forms at the downstream end of the fold and propagates upstream. Modified from Burbank *et al.* (1996c).

edge of the older fold (Fig. 9.23). As is typical with fans, a suite of radial channels emanated from each fan apex. During growth of the new fold, most of these channels were defeated, leaving elongate linear valleys and numerous wind gaps that now decorate the fan crest. Owing both to the lateral continuity of the new fold and to the fact that most of the modern fans have beveled laterally several kilometers into the upstream flank of the fold, most defeated streams do not appear to have been diverted around the fold tip. Instead, a few of the antecedent channels have

gathered the flow from defeated ones and sustained their courses across the new fold (Fig. 9.23). The relationships between the fans, channels, and the new fold change along strike. Beyond the fold tip, the lateral extents of the old and modern fans are the same. At the tip, the modern fan margin has been deflected ~5 km away from the tip. Farther along the continuous fold, modern fans have radii that are 3–5 km shorter than the older fans that preceded them. The piggyback basin is aggrading with fan sediments and, depending on that rate of aggradation

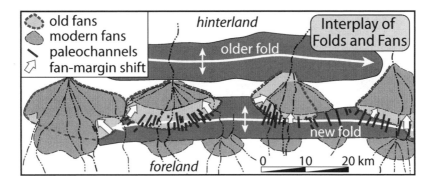

Fig. 9.23 Growing folds, antecedent rivers, and abandoned paleochannels.
Interactions of an older and younger fold with drainage systems in the southern Tien Shan. The older fold is the Quilitak fold previously described (Fig. 9.17). North of the newly emergent fold, fan margins have retreated, whereas along the new fold's crest, abandoned paleochannels record the radial drainages of alluvial fans that pre-date the younger fold. Only a few of the paleochannels persist as modern antecedent streams. Modified after Hubert-Ferrari *et al.* (2007).

versus the rate of tip propagation and vertical fold growth, more of the channel flow may someday be diverted around the new fold's nose.

In order to determine rates of deformation and geomorphological modification of growing folds, dates are needed to define when the fold propagated into a given position and how rapidly it grew vertically and laterally. One such dated structure is Wheeler Ridge anticline near the southern end of the San Joaquin Valley of California (Fig. 4.39). Because Wheeler Ridge encompasses an actively producing oil field, numerous wells with accompanying electric logs provide a basis for correlation among the wells, help to define the stratigraphic boundaries in the subsurface, and serve to delineate the overall structure (Medwedeff, 1992). The anticline is strongly asymmetric, with a relatively steep (~45°) forelimb and much gentler (17°) backlimb. The fault geometry underlying the anticline has been interpreted as a wedge thrust (Fig. 9.24). The Wheeler Ridge anticline plunges to the east and dies in an actively aggrading alluvial plain. From the fold tip, the differential uplift gradient is ~200 m/km across the easternmost 3 km of the anticline (Fig. 9.25) (Medwedeff, 1992). The topographic relief of the fold is considerably less than the structural uplift, because synfolding aggradation has raised the surface of the surrounding depositional basin.

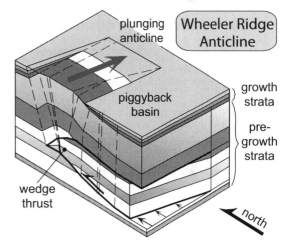

Fig. 9.24 Block diagram of structure of Wheeler Ridge.
Structural interpretation of a wedge thrust with the anticline being underlain by a north-vergent thrust fault and a passive roof thrust. Displacement and uplift decline eastwards. Modified after Medwedeff (1992).

As the fold propagates, formerly actively aggrading surfaces adjacent to the fold are uplifted and incorporated into it. Following initial uplift, these surfaces accumulate few new sediments (except some loess) along the crest of the fold. If the timing of the end of active alluvial deposition or the age of the soil that developed on the uplifted surface can be determined at several points along the fold, these

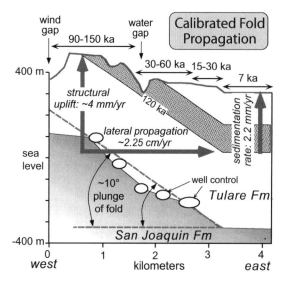

Fig. 9.25   Calibrated rates of fold propagation, crestal uplift, and sediment aggradation for Wheeler Ridge.
Schematic cross-section along the plunging axis of Wheeler Ridge anticline, showing ages, relief, and propagation rates based on well data. Note the importance of surface ages for defining the rates of uplift, propagation, and sedimentation. Modified after Medwedeff (1992).

ages will constrain the initiation of fold growth at those points. Based on soil stratigraphy and both radiocarbon and uranium-series dates at Wheeler Ridge (Fig. 9.25), the youngest uplifted area at the emergent eastern end of the anticline is dated at about 7 ka, whereas the alluvial surfaces just east of the wind gap are estimated to date from 90–150 ka (Keller *et al.*, 1998; Medwedeff, 1992; Zepeda, 1993). When combined with the structural and stratigraphic data, the ages indicate that, during the past 100 kyr, sediment accumulated in the adjacent basin at a rate of about 2 mm/yr, and the crest of the fold uplifted at a rate of about 4 mm/yr, while the nose of the fold propagated eastward at a rate of about 25 mm/yr.

If one were to assume that a significant earthquake occurs on the underlying fault once every 400 years, these rates imply that, with each rupture, the fold crest would be uplifted approximately 1–2 m, and the fault would extend its eastern tip about 10 m. The rates of growth of

the vast majority of folds identified within the entire geological record are essentially unknown. By developing a chronology that spans 100 kyr, these dates at Wheeler Ridge provide unique insights on the mean rates of anticlinal growth and facilitate development of more detailed and reliable kinematic models of fold growth and geomorphic dissection.

The drainage geometry and topography in the vicinity of Wheeler Ridge reveal interactions between fold growth and river responses (Figs 4.39 and 9.25). Whereas the floor of the wind gap has been uplifted ~100 m above the expected fluvial gradient across the current fold, the crest of the fold lies 300 m above the floor of the paleovalley. Hence, an antecedent river had successfully incised across the fold during about three-fourths of its vertical growth. An asymmetric catchment funnels runoff from near the current wind gap to the present water gap (Fig. 4.39). This asymmetry suggests the direction of fold propagation and river diversion. At the current water gap, the fold is narrower than at the wind gap and the upstream catchment is larger, such that the discharge and potential stream power (dependent on channel slope) would also be greater. Given the approximately 200 m of relief bounding the water gap (Figs 4.39 and 9.25), it, too, has persisted for a considerable interval of fold growth. Drainage on the fan to the east of the fold swings around the fold nose and converges toward the northwest, where finer-grained sediment accumulates in a depositional "shadow zone" in the lee of Wheeler Ridge (Fig. 4.39B).

In the context of the reconstructed history of fold growth, the geomorphic responses to folding and the modification of the fold by surface processes can also be examined in more detail than is commonly possible. The apparently systematic eastward propagation of the Wheeler Ridge anticline permits the ergodic hypothesis to be applied. Visual inspection of Wheeler Ridge (Fig. 4.39A) clearly shows significant changes in the character of the land surface along the length of the fold. Initially, as the nose of the fold propagates eastward, a planar region of the alluvial surface is uplifted and gently folded. At this point, the geomorphic surface should precisely mimic the structural geometry

Table 9.1   Effects of slope steepening and lengthening on geomorphology and surface processes.

Feature/process	Effect of steepening and slope lengthening	Reason behind effect
head of first-order streams	channels begin higher upslope	channel initiation is function of (slope × area)
ruggedness (mean relief × drainage density)	increases with steepening and lengthening	more first-order channels lower on slope and deeper channels to match relative base level
creep and landslides	increases with steepening and lengthening	shear stress is a function of slope: $\sigma = \rho g h \sin \alpha$
exported material	increases with steepening and lengthening	flux proportional to slope (dy/dx)
balance of uplift versus incision	uplift tends to outpace incision for antecedent streams	width of structure increases with time; discharge stays constant, but river slope and mean stream power decrease

of folding. Following uplift, however, surface processes begin to modify the surface in several ways. Channel heads appear on slopes exceeding about 5° (Keller *et al.*, 1998; Talling and Sowter, 1999). Antecedent streams incise into the growing fold and, in the particular situation at Wheeler Ridge, are commonly localized by transverse tear faults (Mueller and Talling, 1997). As gullies deepen, the pristine, but uplifted, alluvial surface becomes more dissected, and landsliding begins to occur on the sides of the larger gullies. As the steep forelimb lengthens, creep and shallow landsliding become increasingly important, and gullies extend their heads toward the crest of the fold. If fold widening and vertical uplift are sufficiently rapid, antecedent streams are defeated and diverted, leaving behind wind gaps (Fig. 4.39A).

Modification of the fold's geomorphic surface can be analyzed from at least two perspectives: changes along the length of the anticline; and contrasts between the forelimb and the backlimb. The western parts of the anticline are older, have greater topographic relief, and commonly are steeper than the younger, eastern parts of the structure. In any transverse cross-section of this fold, the forelimb is consistently steeper than the backlimb (Fig. 9.24). All slope-dependent processes would, therefore, be expected to attack the forelimb more vigorously than the backlimb. Similarly, catchment areas, discharge, and relief generally increase toward the west. The net results are that the fold becomes

increasingly dissected toward the west and that the forelimb is more dissected than the backlimb along any transverse section. Qualitative predictions of changes along the fold (Table 9.1) suggest how various surface processes will be influenced by the growing relief and both lengthening and steepening of the fold limbs.

Analysis of a 30-m digital elevation model (DEM) of Wheeler Ridge permits further quantification of several aspects of the geomorphology. Slopes and relief were calculated within a sliding 150 × 150 m window along the fold limbs, excluding the wind and water gaps (Brozović *et al.*, 1995) (Fig. 9.26). The resulting distributions show that both slope and relief (Fig. 9.26) are consistently higher (i) on the forelimb than on the backlimb and (ii) in the older parts of the fold. These data support the hypothesis that the enhancement of slope-dependent processes promotes greater dissection of the fold limbs (Table 9.1).

As shown earlier (Fig. 7.6), digital topography can also be used to calculate minimum volumes of eroded material along an anticline. At Wheeler Ridge, the pre-dissection topographic surface can be reconstructed by connecting undissected remnants of the uplifted surface of the fold. Subtraction of the current topography from that pre-dissection surface defines the magnitude of erosion throughout the fold (Plate 8). These calculations indicate that the most extensive erosion has occurred in the older, more strongly uplifted segments of the fold, whereas very little erosion

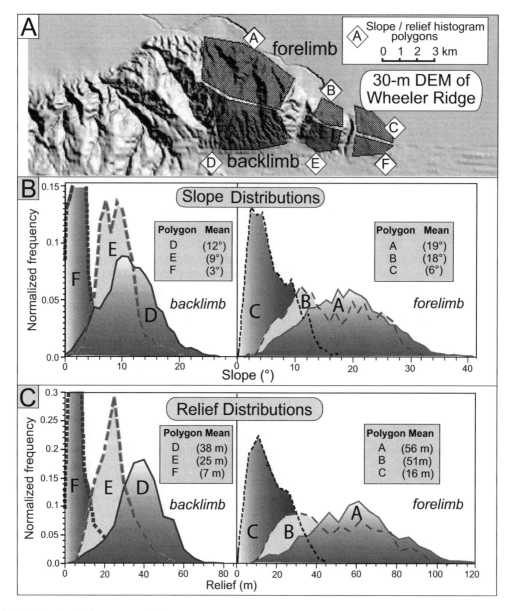

Fig. 9.26 Wheeler Ridge geomorphology.
A. Polygonal areas represent calculation areas on either flank of the Wheeler Ridge anticline. Magnitude of rock uplift and dissection increase toward the west. B. Slope distributions (150×150 m window) for polygons shown above. Note progressive changes from old (A) to young (C) and the persistent difference between the forelimb (A–C) and backlimb (D–F). C. Relief distributions for the same polygons. Modified after Brozović *et al.* (1995).

has occurred near the eastern nose of the fold. The magnitude of erosion (eroded volume/source-area size) is always higher on the forelimb rather than the backlimb (Fig. 9.27A); this contrast supports the concept of erosion rates being slope dependent. The rates are also higher on the older parts of the fold, where greater relief is present and more time for dissection has elapsed (Fig. 9.26). A plot of the rates of dissection versus the mean slope (Fig. 9.27A) shows

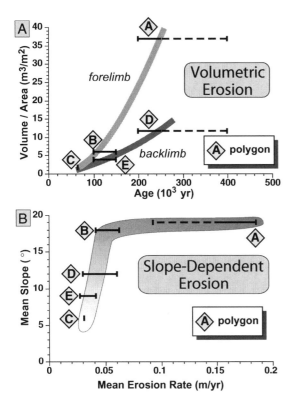

Fig. 9.27 Quantified erosion on Wheeler Ridge anticline.
A. Eroded volume per unit area plotted versus the age of the dissected surface. Note increasing rates with age and persistently higher rates on the forelimb for each age grouping. Polygons are the same as those shown in Fig. 9.26. B. Mean erosion rate versus slope angle in each polygon. Modified after Brozović *et al.* (1995).

predictable increases in rates for increasingly steep slopes. The abrupt increase in erosion rate for slopes >17° probably does not indicate some threshold slope angle, but rather that the longer slopes, greater relief, and more areally extensive gully headwalls in the older part of the fold have promoted more rapid erosion there.

This analysis of Wheeler Ridge highlights some of the ways in which ages on uplifted surfaces, recognition of structural geometries and geomorphic patterns, and quantification using digital topography can be used to develop a fuller understanding of the rates of fold growth and dissection over thousands of years. If combined with field measurements

of geomorphic processes and with paleoseismic analysis, it may be possible to develop more realistic models of fold development and erosional modification that span from decades to many thousands of years.

The fault-bend fold developed above the Main Frontal Thrust in central Nepal (Fig. 7.25C) provides a well-calibrated setting for examining how rocks that experience rapid lateral advection (~20 mm/yr) are eroded as they are uplifted. One intriguing prediction is that topographic features can also be advected across a fold (Miller and Slingerland, 2006; Miller *et al.*, 2007). Observations that support this prediction include the fact that a high proportion of valleys and saddles in the topography are aligned from one side of the fold to the other across the fold crest (Fig. 9.28A and B). This alignment suggests that northward-draining river valleys that formed on the hinterland flank of the fold are advected across the fold crest and become southward-draining valleys on the distal (foreland) flank. Numerical modeling of landscape evolution in the context of rapid lateral advection of rocks (Miller and Slingerland, 2006; Miller *et al.*, 2007) predicts that such alignment of valleys should be common for major topographic elements, such as ridges and valleys, when advection rates are rapid compared to erosion rates (Fig. 9.28C and D). Whereas topographic advection on these frontal folds is predicted to occur at scales of a few kilometers and ~10^5 years, larger features such as deep river gorges may also be advected at longer time scales (Koons, 1995).

Fault behavior

Propagating versus fixed faults

Age-calibrated data on displacement profiles gathered at intermediate scales can serve to discriminate between faults whose tips are propagating and those whose tips are fixed (Fig. 9.29). In a broad sense, the shape of the overlying fold might show little difference as a function of whether the fault tips are fixed or propagating (Fig. 9.1), although fixed-tip folds are expected to have steeper displacement gradients near their tips than do steadily

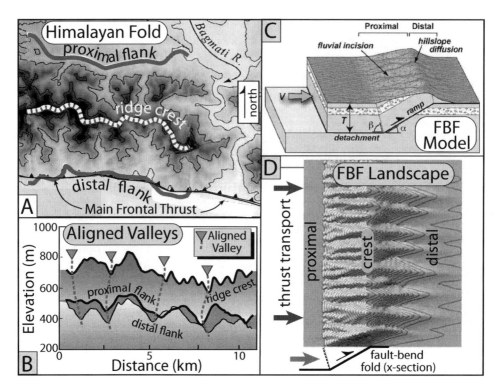

Fig. 9.28   Topographic advection on fault-bend folds.
A. Contour map of topography on a fault-bend fold developed above the Main Frontal Thrust in central Nepal (see Fig. 8.21 for cross-section). Lines along the crest and the distal and proximal flanks depict the location of topographic profiles in B. B. Topographic profiles show an overall alignment of major valleys on both fold flanks and the ridge crest. C. Set-up for numerical model of eroding topography above a rapidly advecting fault-bend fold (FBF). D. Map view of predicted model topography depicting clear valley alignment on either flank of the fold. Cross-section at bottom shows relationship of kinks in the fault surface to the topography. Modified after Miller and Slingerland (2006) and Miller *et al.* (2007).

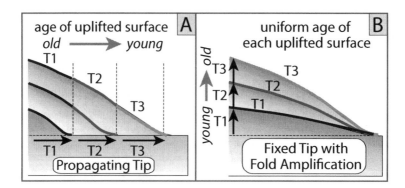

Fig. 9.29   Propagating versus fixed-tip folds.
A. Propagating fold tip in which the geomorphic surface beyond the fold tip is progressively incorporated into the propagating tip, causing surface ages to get younger toward the tip. T1 to T3 represent both time increments of fold growth and surface ages along the fold's crest. B. Fixed-tip fold in which continued displacement leads to amplification of the fold. The upper surface has the same age, irrespective of position with respect to the tip.

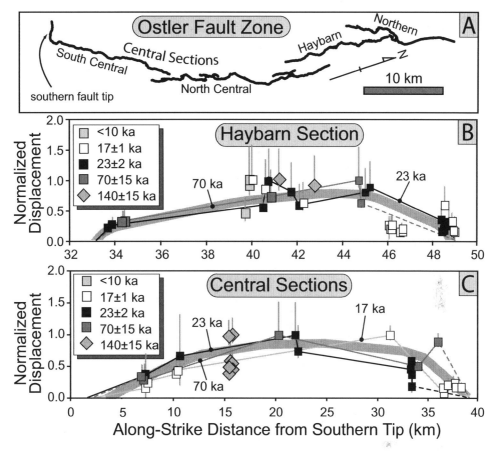

Fig. 9.30   Fixed-tip faults in the Ostler Fault zone.
A. Map of the surface trace of the Ostler Fault, an east-vergent thrust fault. Normalized displacement versus length on the Haybarn (B) and Central (C) sections. Similarities of the normalized profiles at 17, 23, and 70 ka argue for fixed tips on each of these segments. Modified after Amos *et al.* (2010).

propagating folds (Fig. 4.10). The use of normalized displacement gradients and of ages of geomorphic surfaces can permit unambiguous discrimination among these fault-growth models. For a fold whose tip is propagating, ages along the upper surface of the fold should get progressively younger toward the tip (Fig. 9.29A). In contrast, two diagnostic features help distinguish fixed-tip folds. First, deformed geomorphic surfaces should have the same age along their length (Fig. 9.29B). Second, normalized length–displacement profiles should remain similar irrespective of age, i.e., ongoing differential uplift amplifies the fold in a self-similar way.

Given the progression of ages along its crest, Wheeler Ridge anticline provides an excellent

example of a propagating fold (Fig. 9.25). In contrast, a succession of deformed and dated terraces along the Ostler Fault in New Zealand suggests that the tips of current fault segments have been fixed for many millennia (Fig. 9.30). A suite of fluvial terraces ranging in age from Holocene to the penultimate glaciation (140 ka) have been deformed along the length of the Ostler Fault (Amos *et al.*, 2010). Although older terraces have been significantly more deformed than younger terraces, the patterns of normalized displacement versus length appear indistinguishable from one terrace to the next, despite a seven-fold range of ages (Fig. 9.30B and C). This similarity in shape and the absence of lengthening of the fault

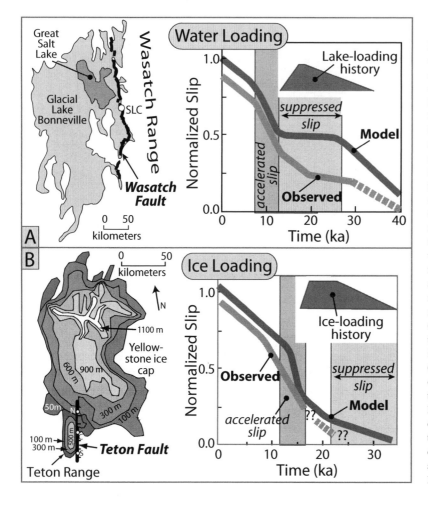

Fig. 9.31 Climatic modulation of fault slip rates.
A. Water loading. (Left) Areal extent of glacial Lake Bonneville in proximity to the Wasatch Fault. (Right) Numerical model of normalized slip and slip-rate changes in response to lake-loading history that extends from 34 to 13 ka (dark trapezoid) in comparison to observed slip (see Fig. 6.30). Note that the model predicts a prolonged interval of suppressed slip when the lake is present, followed by brief, but significantly accelerated, slip rates when the lake load is removed. B. Ice loading. (Left) Reconstructed ice thickness for the Yellowstone ice cap and the Teton Range. (Right) Modeled slip history of the Teton Fault in response to the waxing and waning ice load compared to the fault's observed slip history. Note that the observed history only extends to ~16 ka. Modified after Hampel *et al.* (2007) and Hetzel and Hampel (2006).

provide strong evidence for a fixed fault tip, at least for the past 140 kyr.

Climatically modulated fault slip

The observed temporal coincidence of summertime water loading in the Himalayan foreland and suppression of both seismicity and convergence rates in the hinterland suggest climate–tectonic linkages at seasonal time scales (Box 5.2). Uncertainty remains, however, about whether variations in loads at intermediate time scales also impact rates of deformation. From a theoretical perspective, we expect that changing water, ice, or rock loads could influence fault behavior (Fig. 9.2), and several recent studies provide data supportive of such linkages. In each

case, climate change has induced the expansion and subsequent contraction of either a large lake or an ice cap during late Quaternary times. Whereas the growing load of the lake or glacier is predicted to have moved nearby normal faults farther from failure and, thereby, to have reduced their slip rates, waning ice or water loading should have caused accelerated slip rates (Fig. 9.2). For example, at its maximum extent at ~18 ka, glacial Lake Bonneville covered more than 50 000 km² (Fig. 9.31A) and was over 300 m deep (Gilbert, 1890). During most of Lake Bonneville's existence, numerical models predict that slip rates on the nearby Wasatch Fault should be suppressed (Hetzel and Hampel, 2006). As the lake shrank at ~12.5 ka, slip rates are predicted to have rapidly accelerated for several thousand years. Although the observed slip history on the Wasatch Fault (Fig. 6.30) does not perfectly match the modeled slip-rate changes, the shape of the observed curve mimics the inflections of the model curve and supports the hypothesis of slip modulated by water loading (Fig. 9.31A).

The Teton normal fault borders the eastern flank of the Teton Range in Wyoming (Byrd *et al.*, 1994). During the last glaciation, an ice cap up to 1 km thick inundated the highlands of Yellowstone to the north of the Tetons (Fig. 9.31B), while alpine glaciers expanded within the Tetons themselves (Foster *et al.*, 2010; Hampel *et al.*, 2007). Modeling predicts a deceleration of slip rates as the ice loads grew, and a rapid acceleration when the ice loads disappeared. Notably, the load of the Yellowstone ice cap is predicted to have a much greater effect on slip rates on the Teton Fault than do the more local, but much smaller, Teton glaciers. Given the currently known slip history of the Teton Fault, slip rates were about twice as rapid between 16–8 ka as they were between 8–0 ka: changes consistent with predicted changes as the ice disappears (Fig. 9.31B) (Hampel *et al.*, 2007).

The general match between predicted and observed slip on the Teton and Wasatch Faults is provocative and suggests that climate changes that create changes in stresses on faults can modulate their slip rates. A more convincing test of this proposed climate–tectonic linkage, however, awaits the development of slip histories that extend through an entire glacial or lake cycle and show rate changes consistent with a time-calibrated loading–unloading history.

Catch and release: seamounts and forearcs

Earlier we described the challenge of discriminating among different forearc behaviors when thickened crust, such as that represented by a seamount, is rafted into a subduction zone and collides with the forearc (Fig. 9.3). Contrasting uplift–subsidence scenarios are proposed depending on whether compression dominates as a seamount gets pinned against the front of the forearc, thereby inducing widespread and rapid uplift, or whether displacement dominates as a seamount slides under the forearc, thereby

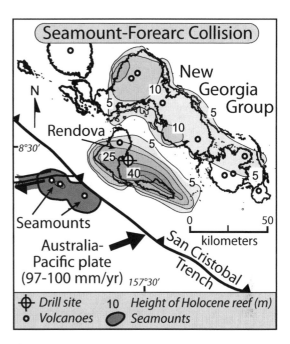

Fig. 9.32 Seamount–forearc collision in the tropical Pacific.
The New Georgia group of forearc islands sit above the rapidly converging Australia–Pacific subduction zone. Contours of Holocene reef heights indicate the pattern of recent uplift. Seamounts impinging on the subduction zone, volcanic centers, and the drill hole on Rendova island are shown. Modified after Taylor *et al.* (2005).

causing a progressive, more localized wave of uplift. Tropical coastal regions can provide fertile sites for testing these competing models, because the general history of sea-level change (both magnitude and age) is well known (Figs 2.5 and 9.4) and many coral-rich marine terraces can be dated. With these data, local terrace elevations and ages can be compared with the sea-level curve to deduce patterns and rates of uplift and subsidence.

Recent studies in the tropical Pacific have illuminated rich histories of rapid uplift and subsidence (Taylor *et al.*, 2005). In the New Georgia group (islands on the forearc of the Pacific Plate where the Australia Plate underthrusts at a rate of ~100 mm/yr), several adjacent seamounts have entered into the subduction zone (Fig. 9.32). Holocene terraces have been differentially uplifted as much as 40 m above sea level, and they provide unambiguous evidence

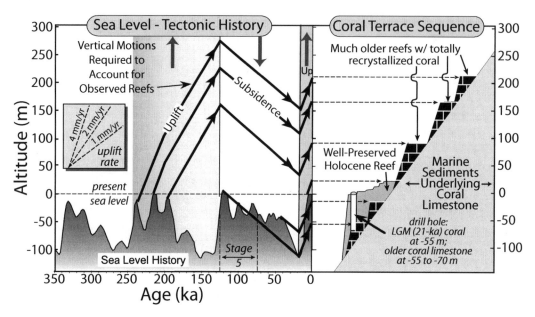

Fig. 9.33 Coral terrace sequence on the New Georgia group forearc and inferred tectonic history.
(Right) Uplifted Holocene reefs on Rendoza island (Fig. 9.32) are spatially bracketed by much older, recrystallized reefs. Drill hole shows Last Glacial Maximum (LGM) corals at −55 m, indicating ~65 m of uplift since 20 ka. No Marine Isotope Stage 5 terraces are recognized above sea level. (Left) Correlation (black lines) of sea-level history with observed and dated terraces defines abrupt changes in uplift and subsidence over the past 250 kyr. Modified after Taylor *et al.* (2005).

for recent uplift at rates as high as 6 mm/yr (Taylor *et al.*, 2005). Similarly, drill-hole data reveal that Marine Isotope Stage 2 terrace deposits that were formed at a depth of ~120 m are currently only at −55 m (Fig. 9.33) and define an average uplift rate of ~3 mm/yr since 20 ka. But these rates have not been sustained for long: no Marine Isotope Stage 5 terraces are currently exposed above sea level, and terraces immediately above the Holocene terraces are completely recrystallized, indicating their overall antiquity (Fig. 9.33). These observations provide persuasive evidence for multiple and abrupt changes in forearc uplift and subsidence over the past 20–300 kyr. Taylor *et al.* (2005) argue that these changes could not be induced by a displacement (underthrusting) mechanism (Fig. 9.3). For example, over the past 40 kyr, as rapid subsidence was replaced by rapid uplift, an impinging seamount would have advanced only a few kilometers, not enough to cause such an abrupt transition. Instead, Taylor *et al.* (2005) suggest that subsidence–uplift history is consistent with impinging seamounts that jam up the subduction

process, cause temporary locking of the upper parts of the subduction zone, and drive elastic-like compression of the nearby forearc. Seismic rupturing of the seamount is speculated to cause collapse of the forearc and rapid subsidence, as is seen in the New Georgia islands. This striking example of abruptly changing patterns of uplift highlights the potent insights that can be gleaned at intermediate time scales when well-dated geomorphic markers with known initial positions can be tracked as they deform.

Summary

Tectonic deformation and interactions with surface processes over intervals of hundreds of thousands of years produce the landscapes that we see today in many tectonically active areas. At vertical uplift rates of about 1 mm/yr and horizontal displacement rates of about 1 cm/yr, hundreds of meters of uplift and several kilometers of lateral displacement occur over these "intermediate" time spans. In combination with climatic and lithologic variations, such movements exert a fundamental control on landscape development.

Given the slope dependence of many surface processes and the sensitivity of fluvial systems to small variations in surface gradients, geomorphic responses to specific tectonic perturbations are commonly predictable. At intermediate time scales, powerful insights are obtainable when the growth of structures can be quantified. Not only does this quantification reveal mean rates of deformation and structural propagation, but it also provides a reliable context for interpreting geomorphic responses to deformation and testing predictions concerning those responses.

Numerous difficulties, however, can thwart successful landscape analysis at intermediate time scales. Determination of tectonic rates typically depends on dating of displaced geomorphic surfaces. In many geomorphic settings, dating of surfaces that exceed the range of radiocarbon dating (>40 kyr) is difficult or impossible. Soil chronologies, uranium-series dates, cosmogenic exposure ages, or luminescence ages can sometimes bridge the gap between radiocarbon and argon–argon dating. Because each of these dating approaches may lack accuracy, reliability is improved by (i) the use of several techniques in conjunction with each other whenever possible, (ii) a focus on settings in which redundant ages can be determined for the same surface, or (iii) use of sites for which multiple rate calculations can be made from a succession of offset markers.

In tectonically active coastal settings, ages of terraces are often inferred through correlation with a dated sea-level curve. Numerous uncertainties still exist in this curve with respect to both the magnitude of past sea-level variations and the actual timing of those changes. The interval prior to the last interglaciation (>130 ka) has only been partially calibrated. To the extent that the reliability of the sea-level curve and resulting correlations can be enhanced, the accuracy of deformation rates in coastal domains will improve.

The timing of geomorphic responses to climate variations in terrestrial settings is even less well understood than those along the coast. Development of some geomorphic markers,

such as fluvial terraces, have both direct and indirect climate controls. For example, even if climatic conditions are conducive to terrace building, if insufficient sediment is available for transport within the upstream drainage, aggradation will not occur. Nonetheless, geomorphic markers provide a critical basis for gauging deformational and denudational processes within many landscapes. Therefore, understanding of the character and rates of response of markers to climatic changes, as well as the temporal lags in those responses and the potential impact of autocyclic processes, is of paramount importance in reconstructing tectonic and geomorphic histories at intermediate time scales.

The punctuated production of geomorphic markers by climatic variations provides a means both to delineate changes in tectonic rates and deformation patterns through time, and to create a robust framework in which to examine landscape evolution. At intermediate time scales, accumulated displacements can be sufficiently large to allow incisive testing of conceptual models for fault and fold propagation through time, fault linkage, steadiness of fault-slip rates, and stick–slip behavior in subduction zones at multi-millennial time scales. Where rates can be shown to change, transient landscapes can be used to explore how the surface responds to changing boundary conditions.

Individual folds commonly grow over an interval of a few hundred thousand years in tectonically active settings. During this period, stream patterns and preserved surfaces commonly provide unambiguous information on the amount and geometry of deformation. If fold growth were to persist at rates of 1 mm/yr for a million years, 1 km of displacement would occur. In most cases, structural relief of this magnitude would cause erosion to obliterate all but the most persistent geomorphic markers, such that the details of fold evolution would be difficult to reconstruct from geomorphology alone. It is at these longer time scales, however, that mountain ranges develop and orogens evolve. In the succeeding chapter, we examine landscapes that represent an integration of geomorphic and tectonic processes at time scales of a million years or more.

Tectonic geomorphology at late Cenozoic time scales

10

Actively deforming ranges and orogens that evolve over hundreds of thousands to millions of years result from a long-term interplay between constructional and erosional processes. Mean rates of fault slip and folding create spatial patterns of bedrock uplift or subsidence that drive the evolving shape of unmodified tectonic landforms, whereas the summation of all active erosional and depositional processes determines how those pristine geometric forms are modified over time. Currently observable deformation patterns, climate, and erosion rates may have only tangential relevance to a range's overall evolution. At such long time scales, climate will have varied between glacial and interglacial conditions dozens of times, and nearly all geomorphic thresholds will potentially have been crossed, perhaps numerous times. Faults that are currently active are unlikely to be the faults that were predominantly responsible for the building of the mountain range, especially in contractional mountain belts. Consequently, the rates of active tectonic and geomorphic processes that we can examine during our present interglacial conditions are unlikely to be representative of average rates over millions of years. Whereas we can use our knowledge of tectonic and erosional processes at short and intermediate time scales to understand the phenomena that build and tear down mountains, a more integrative perspective is commonly needed to synthesize the growth and decay of orogens at long time scales.

Some landscape responses are best viewed at long time scales. For example, dynamic topography can be represented by long-wavelength crustal tilting that is driven by mantle or deep crustal processes. Such subtle tilting only expresses itself geomorphically at long time scales. Migration of drainage divides at long time scales may result from sustained gradients in shortening across a landscape, asymmetric rock uplift, different base levels on opposite sides of a range, or storm tracks that are persistently followed. Such divide migration in bedrock massifs is commonly difficult to discern at intermediate or shorter time scales, but it may be inferred at long time scales either where a pristine landform shape that is attributable solely to deformation can be reconstructed, where along-strike ergodic substitutions can be made in laterally propagating ranges, or where comparisons can be made among analogous ranges that differ in only a few major variables.

Over these prolonged intervals of tectonic activity, the record of detailed interactions of short-term deformation and surface processes is often obscured owing to their incremental impact on the present-day landscape. In the analysis of long-term deformation, the appropriate suite of questions to pose and the approaches to solve them differ from those applied to processes and landforms at shorter time scales. In general, a larger spatial framework (hundreds to thousands of square kilometers) is appropriate for addressing the products of long-term deformation. Bedrock cooling histories, stratigraphic data from bounding basins, the topographic character of entire ranges, and the geographic character of large river catchments can each provide useful clues to the interpretation of the tectonic-geomorphic history (Allen, 2008). We will see, for example, that increasingly available digital data sets, such as digital elevation models (DEMs), allow rapid statistical characterization of landscapes at these scales, that catchment geometries can serve to delineate long-term interactions among faults, and that patterns of bedrock cooling ages and hillslope angles can sometimes be used to assess orogenic steady state.

Hot topics and unresolved questions

Over the past decade, several "hot topics" have motivated numerous tectonic-geomorphic studies at long time scales. For example, the appealing concept of a topographic steady state has been difficult to document with field data. Given the many difficulties of reconstructing paleoelevations, particularly in mountain ranges experiencing active erosion (see Chapter 7), we cannot hope to be able to compare present and former topographic characteristics in most mountain ranges. Consequently, different approaches are needed, such as the use of along-strike ergodic substitutions in propagating ranges for which the tectonic forcing through time is well known.

Similarly, the common correlation of higher stream power with more rapid erosion rates suggests that climate could affect tectonic deformation: where erosion is more intense, the loss in mass from an orogen might be expected to induce a tectonic response whereby the influx of rock (via faulting) into the eroded region is enhanced (Whipple, 2009). But, does this interaction occur? How can we demonstrate it?

In the context of global topography, orogenic plateaus, such as the Tibetan Plateau in the Himalayan orogen or the Puna–Altiplano Plateau in the Andes, represent some of the largest topographic features on Earth. These plateaus influence climate by redirecting the jet stream and guiding storm tracks. Their steep topographic margins represent huge gradients in potential energy. What can tectonic geomorphology tell us about how these plateaus grow and decay? Do they grow steadily or in pulses? What controls their upward and outward growth? We introduce several of these "long-time-scale" problems here and then delve into them in more depth later.

Steady-state topography

The concept of a dynamic equilibrium was developed many decades ago (Hack, 1975) and is now commonly described as a "steady state" when applied to mountain ranges (Fig. 1.2). Steady state is an appealing concept because it suggests a dynamic balance whereby rock uplift is counterbalanced by erosion, such that the topographic character of a range remains steady through time despite ongoing deformation. For example, numerical models of the topography of a range subjected to sustained deformation predict that the range will evolve toward a persistent topographic shape. The resulting topographic shape is commonly modeled as a function of three variables: the ratio of the horizontal to vertical velocity (lateral versus vertical advection of rock, or frontal accretion versus underplating); the erosional efficiency of rivers; and the diffusivity of hillslopes (Willett et al., 2001). Although these models typically have an accretionary flux from one side (Fig. 10.1), the inbound material can be added to the orogenic wedge in two different ways: frontal accretion or underplating. In models characterized by rapid lateral advection, the predicted steady-state topography is typically asymmetric (Fig. 10.1). Similarly, lateral gradients in rainfall tend to push the drainage divide away from the moisture source (Willett, 1999), whereas either underplating, more efficient erosion, or more rapid diffusion are predicted to engender more symmetrical topography. The challenge for tectonic geomorphologists is to examine actual mountain ranges and to use their characteristics along with model predictions to gain insights into the controls on topographic characteristics and patterns of erosion and deformation. Whereas model predictions may seem straightforward, many questions remain to be answered with respect to actual landscapes: What characteristics

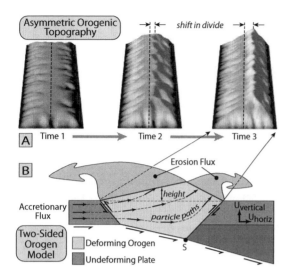

Fig. 10.1 Predicted topographic evolution in a collisional orogen.
A. Orogenic topography in which fluvial incision and hillslope diffusion drive erosion that competes with vertical uplift and horizontal advection. The progressive lateral shift in the drainage divide results from a high ratio of horizontal to vertical velocity. The "Time 3" panel represents a topographic steady state that has become independent of the duration of accretion or erosion. B. Key geometric elements of a numerical model dominated by frontal accretion (little to no underplating). At steady state, the accretionary influx is balanced by the erosional efflux. Note the predicted particle pathways through the orogen and the spatial variation in the ratio of horizontal to vertical velocity. Point S is a fixed point where incoming continental crust detaches from the incoming mantle and obducts on to the "backstop." Modified after Willett *et al.* (2001).

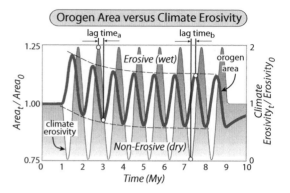

Fig. 10.2 Interplay between climate erosivity and orogenic cross-sectional area.
In this numerical model of a critically tapered wedge, the erosivity of climate regularly varies on a million-year time scale and the orogenic cross-sectional area responds to changes in erosion rate by shrinking when erosion rates are high during wet phases and expanding during dry phases. Note the lag time for the orogenic area to respond to the climate change. Subscript "0" denotes initial conditions. Modified after Meade and Conrad (2008).

2009; Roe, 2005; Smith, 1979). For many decades, the prevalence of young cooling ages in many of the world's large mountain belts suggested to geologists that late Cenozoic uplift of the mountains had helped trigger widespread alpine glaciation. This idea was turned on its head by Molnar and England (1990), who argued that accelerated erosion by glaciers caused young cooling ages and drove isostatic uplift of ranges. Since that time, numerous studies have sought linkages among climate change and the tectonic evolution of orogens. The conceptual rationale behind such linkages is generally straightforward, especially in the context of critically tapered orogenic wedges (Dahlen and Suppe, 1988): spatial variations in precipitation should drive correlative variations in erosion; and the resulting removal of mass from localized regions of an orogen should induce a restorative tectonic influx (a return to critical taper), commonly by spatially focused strain. This rationale has been used to predict that the zones of most active thrusting in an orogen will depend on the side from which storms approach the orogen (Willett, 1999). At even larger scales, surface erosion has been predicted to "draw" channelized flow of lower crust toward the surface (see Fig. 1.9),

serve to define a steady-state landscape? At what temporal and spatial scales is it relevant to assess steady state? How do we tease out the effects of asymmetric climate forcing versus asymmetric tectonic forcing? And how do we account for all the fluxes into and out of orogens?

Climate–tectonic interactions

High topography resulting from tectonic deformation can be observed to influence climate. Orographic precipitation tends to focus rainfall on the windward side of ranges and to create rain shadows on their lee sides (Galewsky, thereby localizing deformation (Beaumont *et al.*, 2001, 2004). Several modeling studies of climate–orogen coupling (Meade and Conrad, 2008; Whipple and Meade, 2004) suggest that the width and cross-sectional area of an orogen should vary in concert with the effective erosivity of a given climate (Fig. 10.2). Such studies suggest diverse observations that could serve to test such models. For example, a switch to a more erosive climate should cause an increased erosional efflux. If the convergence rate is held steady across an orogen that can be represented as a critically tapered wedge, the increased rate of erosion should be balanced by a narrowing of the width and cross-sectional area of the deforming orogen. Hence, we should observe coeval abandonment of distal faults, acceleration of deformation in the hinterland, an increase in the exhumation rate concomitant with decreased cooling ages in the orogen's core, and a slowing of the rate of subsidence in adjacent forelands as the size of the orogenic load decreases (Whipple, 2009). Because collection of such a diverse array of data is uncommon and requires a truly interdisciplinary suite of observations, testing a proposed climate–tectonic coupling provides a formidable challenge.

Growth of orogenic plateaus

Thousands of kilometers of interplate convergence over millions of years can create orogenic plateaus, such as those associated with the Himalaya and the Andes. Significant unresolved questions remain about how these plateaus grow upward and outward. Do plateaus grow steadily upward as the lithosphere systematically thickens beneath them, or do they rise in more abrupt bursts, such as those hypothesized to occur when part of a thickened and densified mantle detaches, sinks, and is replaced by warmer, buoyant asthenospheric material (Molnar *et al.*, 1993)? If surface uplift is pulsed, is there a geomorphic signature that records the change from slow to more rapid uplift?

Plateaus also grow outward, as well as upward. Again, we commonly do not know whether that growth is steady or pulsed, and what tectonic style modulates outward growth (Fig. 10.3). The traditional model for outward growth posits propagation of thrust faults toward the foreland that gradually extend the margin of the plateau. In contrast, some recent observations suggest that the lower crust beneath at least part of the Tibetan Plateau is partially melted (Nelson *et al.*, 1996). If so, then gravitational gradients might be expected to drive this flow outward and could extend the plateau laterally, a phenomenon that could occur with little or no faulting at the surface (Fig. 10.3B). A third mode of proposed plateau growth is a variant of outward thrust propagation. In this model (Fig. 10.3C), newly formed thrust faults at the leading edge of a plateau create piggyback basins that fill with sediment as the fault accumulates slip. Over time, the increasing load of the basin fill and the resultant isostatic subsidence make it more difficult to continue slip on the existing fault and, thereby, promote formation of a new thrust closer to the foreland (Hilley *et al.*, 2005). A corollary of this model is

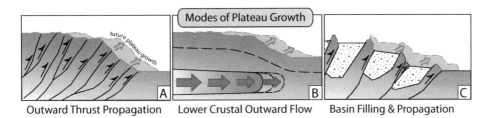

Fig. 10.3 Modes of growth of orogenic plateaus.
A. Traditional model of progressive outward propagation of bedrock thrust faulting. B. Partial melting of the lower crust beneath a thickened plateau and outward flow in response to gravitational gradients across the plateau margin drive expansion without significant surface faulting. C. Sediment fills in piggyback basins create loads and subsidence that favor formation of more distal thrust faults.

that, if the piggyback basin is breached and drained of sediments, deformation is likely to step back toward the hinterland. Such a scenario suggests a climate sensitivity: a more erosive climate would lead to a retraction of the leading edge of deformation. Each of these scenarios poses questions that can be at least partially answered with tectonic-geomorphic studies that assess which faults have been active and when, how basin filling, basin emptying, and faulting interrelate, and whether the plateau margin is progressively extending in the absence of faulting.

Range fronts, basins, and normal faults

Commonly, ruptures on active normal faults can be characterized as a series of linear fault segments separated by transfer zones with more complex geometries (Figs 4.9 and 4.16). The linear trends result from the fact that most normal faults intersect the surface at high angles (~60°), such that the trace of the fault is only slightly deflected by surface topography. In many cases, normal faults also approximately define a boundary between an erosional domain in the uplifted footwall and a depositional, nearly horizontal, domain above the downthrown hanging wall. As seen earlier, a predictable pattern of co- and interseismic vertical motions exists (Figs 4.25 and 4.26), with the greatest magnitude of footwall uplift and hanging-wall subsidence proximal to the fault. Enhanced hanging-wall subsidence near the fault tends to steer regional river systems toward the fault, and, subsequently, their deposits tend to fill the space available to accommodate sediment to the extent that the sediment supply permits. Despite such deposition, coseismic footwall uplift commonly results in positive topography that bounds the depositional basin along the fault. One would, therefore, expect that recurrent normal faulting would produce a relatively linear mountain front delineating the footwall–hanging wall boundary.

Facets and drainage spacing

The topographic evolution of normal-faulted mountain fronts depends strongly on the relative rates of faulting, erosion, and deposition (Ellis *et al.*, 1999). Rivers flowing across the fault from the footwall uplift will tend to dissect and embay the mountain front, whereas active faulting will tend to restore its linear character (Fig. 10.4). Hence, both facet geometry and the linearity of the range front are clues to the activity of the bounding fault. Facets initially develop as footwall scarps in bedrock. Given the steep dip of most normal faults (~60°), however, the facets degrade and lay back over time due to weathering and erosion. Consequently, steep, high facets are signatures of rapidly slipping faults, whereas gentle, low facets are typical of

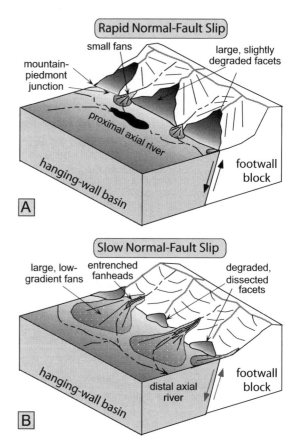

Fig. 10.4 Range-front fans and facets in normal faulted mountain ranges.
A. Rapid footwall uplift and hanging-wall subsidence create a linear range front, large facets, small piedmont fans, and a proximal axial river. B. Slower deformation leads to large, low-gradient fans, small facets, entrenched fanheads, and distal axial rivers.

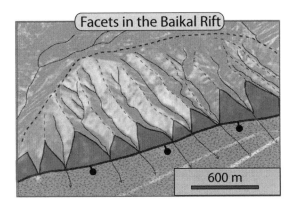

Fig. 10.5 Facet spacing and shape along an active normal fault in the Baikal rift.
Active range-front faulting in the Barguzin Basin produces triangular facets. These facets have an average slope of ~35°, heights up to 900 m, and estimated slip rates of ~1 mm/yr. Modified after Petit *et al.* (2009b).

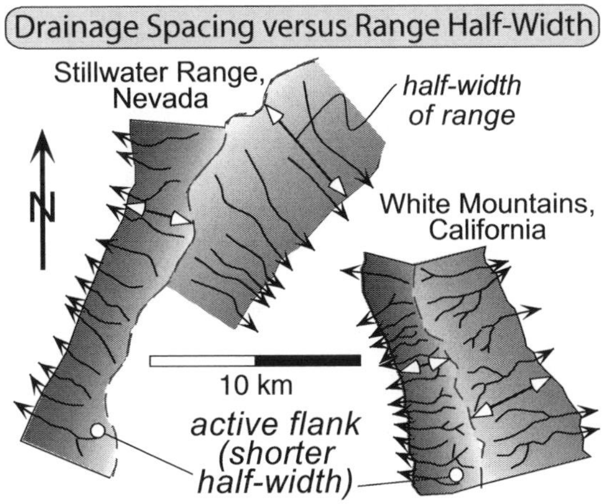

Fig. 10.6 Asymmetric flank widths on fault-block mountains in the Basin and Range.
Note that the average spacing of the outlets scales with the average length of the drainages, such that more closely spaced outlets are associated both with shorter tributary basins and with the more rapidly uplifting flank of the range. Modified after Talling *et al.* (1997).

slower slip rates (Petit *et al.*, 2009a). Although facet steepness correlates with slip rate, facet widths tend to scale with drainage-basin spacing (Figs 10.5 and 10.6). Spacing can be measured as the ratio between the mean length of the basins, i.e., the mean distance from the main drainage divide to the mountain front, and the

mean spacing of the mouths of the basins along the range front (Wallace, 1978). For tectonically active, extensional fault blocks, length-to-outlet spacing ratios are typically between 1.8 and 4 (Hovius, 1996; Wallace, 1978) with a mean of about 2.5 (Talling *et al.*, 1997). In "well-behaved" ranges bounded by half-grabens, when opposing flanks are compared, differences in the distance to the main divide are clearly reflected in differences in the spacing of the major rivers draining each flank, such that shorter drainages are associated with more closely spaced rivers (Fig. 10.6). In terms of trying to decipher the tectonic regime from the drainage patterns, the key attribute of these patterns is that shorter and steeper drainages are commonly found on the more tectonically active flank of the range. In more complex geological uplifts and in older mountains, where drainages have had more time to integrate or cannibalize other nearby drainages, basin shapes are more irregular and river-outlet spacing can be wider, yielding spacing ratios as low as 1.2 (Wallace, 1978).

Geomorphology and quantitative assessment of normal-faulted range fronts

A normal-faulted range front that has experienced persistent rock uplift typically displays a related suite of geomorphic characteristics. Mountains in the footwall block may rise more than 1000 m above the adjacent basins. Rapid footwall uplift tends to result in a linear mountain front characterized by high relief and deeply incised streams (Fig. 10.5). Within the footwall uplift, these V-shaped valleys have steep gradients and small to non-existent floodplains. Within the hanging-wall basin, the length and gradient of transverse streams or fans, as well as the relative distance from the depocenter to the mountain front in comparison to the width of the basin, provide indicators of the balance between tectonically driven subsidence and sediment supply. Lower transverse river gradients, longer fan lengths, and increasingly distal depocenters should indicate lower rates of active deformation and subsidence along the basin-bounding fault; whereas short, relatively steep alluvial fans, the presence of clear fault scarps, and proximal depocenters typify regimes of more rapid faulting (Fig. 10.4). The amount of sediment accommodation space in the adjacent basin is highly sensitive to the rate of faulting, given the tendency of hanging-wall subsidence to be several-fold greater than footwall uplift (Fig. 4.25). For any given sediment flux from the footwall, slower subsidence reduces accommodation space and promotes fan progradation. In humid settings associated with rapid normal-fault slip, short transverse streams sourced in the footwall can be tributary to axially flowing master rivers that are located near the mountain front. In drier climates, closed-basin lakes could be located at the toes of the short tributary fans.

Whereas the slopes on short, range-front fans may approach 15° near their apices, in general, when more than 0.5 km from the fault, fans typically have slopes of 1–3° (Wallace, 1978). If the mean slope of a group of fans or a bajada is more than 3–4° over distances exceeding 2 km, such slopes suggest that this portion of the fan is underlain by tilted bedrock and that faulting has propagated out into the basin (Wallace, 1978).

In order to deduce the current state of activity on a normal fault, several different attributes of the mountain front and of the associated depositional and erosional systems can be examined. The amount of incision at the apex of fans aligned along the mountain front–piedmont junction is determined by (i) the balance between the rock uplift rate in the footwall, (ii) rates of fluvial erosion in the footwall, and (iii) variations in sediment supply. If active uplift is elevating the stream within the mountains more rapidly than it is incising, active deposition will occur on the apex of the fan (Bull and McFadden, 1977). To the extent that incision outpaces footwall uplift, the fanhead will tend to become entrenched. The degree of entrenchment and the age of the dissected remnants of the former fan apex provide a general indicator of the amount and timing of the change from active to less active footwall uplift. It must also be realized, however, that sediment fluxes can vary rapidly as the climate changes.

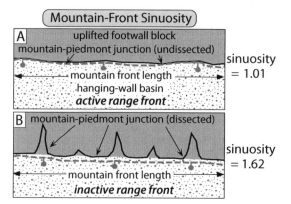

Fig. 10.7 Sinuosity of faulted mountain fronts.
A. Active normal faulting with linear mountain–piedmont junction leading to low sinuosity.
B. Embayment of mountain front along a less active fault creates higher sinuosity.

Changes in sediment and water discharge can lead to fanhead incision or aggradation that is independent of any tectonic variations. Therefore, entrenchment that persists through several climate cycles is more likely to be a response to tectonic forcing than entrenchment occurring solely within one cycle.

Several simple numerical measures of the mountain front and its related fluvial system have been used to classify the state of long-term tectonic activity. Because a range front above an active normal fault is generally straight, whereas an inactive one becomes increasingly embayed, the *sinuosity* of the mountain front is a potential indicator of the level of long-term tectonic activity (Bull and McFadden, 1977). Sinuosity, S, is determined by dividing the length of the mountain–piedmont junction, L_{mp}, by the length of the associated range, L_r (Fig. 10.7):

$$S = L_{mp}/L_r \qquad (10.1)$$

A sinuosity near 1 is usually interpreted to characterize an actively deforming range, whereas as the sinuosity increases to 2 or more, it indicates a highly embayed range front with relatively little active faulting. These data can be readily derived from topographic maps and DEMs, although some subjectivity is involved when designating the mountain–piedmont junction. This technique can be relatively insensitive to reactivation of old faults. Furthermore, if renewed faulting has been insufficient to restore a linear range front, even a presently active mountain front will appear embayed on a topographic map.

If rates of tectonic activity decrease or stop, then surface processes will dominate further evolution of the landscape. Lateral planation by transverse rivers will widen their valleys within the mountains, and gradual back-wearing will decrease their gradients. As valley widening extends into the footwall uplift, lateral and frontal slopes of triangular facets will be further reduced, and the mountain front will become increasingly embayed. Within the depositional basin, decreasing subsidence will decrease the space available for deposition, leading to progradation of the transverse fans, fanhead entrenchment, and basinward displacement of axial rivers or lacustrine depocenters (Fig. 10.4B).

In order to evaluate mountain-front sinuosity along an extensive range front, criteria for sub-dividing the range front into segments can serve to limit the extent to which non-tectonic variability influences topographic measures of relative fault activity. Such criteria could include: major changes in lithologic resistance, major changes in the orientation of the range front or steps in the bounding faults (Fig. 4.16), changes in footwall lithology, cross-cutting river valleys that are large in proportion to the range, and significant changes in the geomorphic character of the range (Wells *et al.*, 1988).

Topographic cross-sections parallel to the mountain front, but within the footwall block, can be examined to determine the ratio of valley width to the height, R, of the adjacent drainage divides (Fig. 10.8) in order to evaluate the effect of tectonism on valley geometries (Bull and McFadden, 1977). Low width-to-height ratios suggest prolonged incision and uplift, whereas high ratios indicate that degradational processes, such as valley widening and ridge-crest lowering, have dominated the recent landscape. Because it takes time for base-level effects to propagate upstream into the footwall block, measurements are usually made close to the range front, e.g., 1 km up-valley. The width of

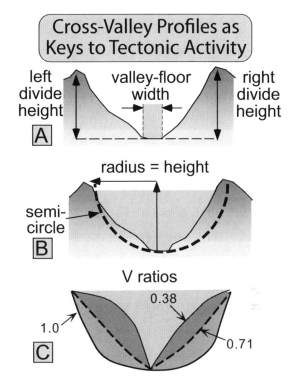

Fig. 10.8 **Measurements of valley shapes in footwall blocks.**
A. Valley-floor width-to-height ratio. B. Parameters for calculating a "V ratio." C. Examples of V ratios for several valley shapes.

the valley floor, w_{vf}, is divided by the mean relief represented by the height of the right and left drainage divides, e_{ddr} and e_{ddl}, respectively, above the valley bottom, e_{vb}:

$$R = \frac{w_{vf}}{\frac{1}{2}[(e_{ddr} - e_{vb}) + (e_{ddl} - e_{vb})]} \quad (10.2)$$

Another way to evaluate the cross-valley profiles is to determine the ratio of the area of the actual valley cross-section (in other words, that region that has been eroded away) to the area of a semicircle with a radius equal to the height of the adjacent drainage divide (Mayer, 1986) (Fig. 10.8). Valley ratios (V ratios) near to or greater than 1.0 signify U-shaped valleys that are typically broad and shallow, whereas valleys with V ratios less than 0.5 are usually steeply incised (Fig. 10.8). Interpretation of the valley-floor width-to-height ratios can be

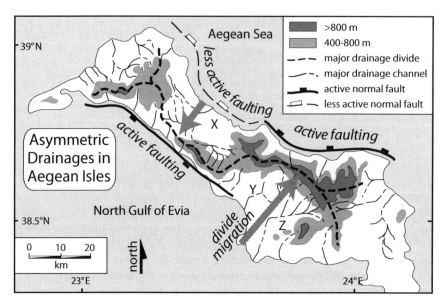

Fig. 10.9 Asymmetric drainage basins responding to differences in the activity of range-bounding faults in the Aegean Sea.
Active normal faults show a clear spatial association with short, steep footwall drainages. Large rivers (X, Y, Z) flow down the backslopes of these blocks. Drainage divides are strongly shifted from a medial position toward the active range front. Modified after Leeder and Jackson (1993).

complicated by variations in both stream size and lithologic resistance. Weaker lithologies and larger rivers will lead to broader valley floors than will more resistant rocks and smaller rivers in the same uplift regime. Consequently, owing to the complexity imposed by geological conditions, "quantitative" measures, such as sinuosity or V ratios, should be regarded as qualitative guides to the status of tectonic activity over multiple climate cycles.

Drainage basins in extensional ranges

As opposed to individual fault scarps, which may be strongly modified over short time intervals, drainage basins have considerable persistence as geomorphic landscape elements. Such inertia suggests that drainage basins can provide insights on the long-term evolution of the landscape. Basin geometries develop in response to the nature and distribution of uplift and subsidence, the spatial arrangement of faults, the relative resistance of different rock types, and climatically influenced hydrologic parameters.

Because of the nearly linear character of many normal faults and quite readily predicted pattern of subsidence and uplift associated with them (King and Ellis, 1990; King et al., 1988), it is reasonable to develop a conceptual scheme for drainage basins in extensional terrains based on modern examples and theories.

Because many extensional faults bound half-grabens, asymmetric footwall uplift typically produces back-tilted blocks with long, relatively gentle basins on the "back" side of the block and with steep, short drainages on the side abutting the fault scarp. Thus, one criterion to distinguish the relative activity of range-bounding normal faults is to examine the size and steepness of basins on either flank of a range: the flank lying directly above the more active fault will tend to have smaller, shorter, and steeper drainage basins (Figs 10.9 and 10.10) (Leeder and Jackson, 1993). In some fault-bounded grabens, a master fault delineates one boundary of the graben and one or more antithetic faults cut the hanging wall and define the opposite boundary. In this circumstance, the relative size and steepness of the opposing drainage basins developed on the foot-

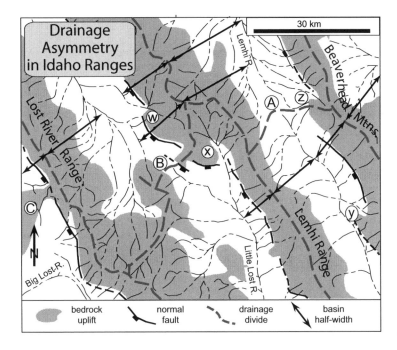

Fig. 10.10   Extensional mountain ranges in the Basin and Range, Idaho.
Asymmetry of transverse basins with short basin half-widths associated with active extensional faults. Axial drainages flowing to the northwest or southeast are separated by basement ridges at A, B, and C. Oblique drainages that develop near fault terminations, step-overs, or segment boundaries occur at W, X, Y, and Z. Modified after Leeder and Jackson (1993).

wall uplifts of each fault can sometimes indicate which is the more active fault (Leeder and Jackson, 1993). Similarly, if an axial river flows through the basin, it will tend to be displaced from the center of the basin toward the more active fault, where subsidence is generally the greatest.

Fault terminations and offsets or step-overs along range fronts, where displacement is transferred from one major fault strand to another, often have distinctive relationships with drainage basins. If we accept that most faults lengthen as they accumulate more displacement (Cowie and Scholz, 1992), it would be expected that rivers would tend to be displaced away from the zone of maximum uplift and toward or around the growing tip. As fault tips propagate toward or past each other, deflected rivers will tend to be localized into the space between the tips, which may eventually be considered a "transfer" zone. The reason these catchments become large is that, as the fault propagates, the trapped catchment incrementally expands toward and beyond the fault tip. Commonly, rather large drainage basins that are oriented obliquely, rather than orthogonally, to the range front develop near fault terminations or transfer zones (Figs 10.10 and 10.11).

As faults grow and their tips overlap, one fault may predominate over the other and accommodate some of the displacement that was previously taken up on the other fault (Fig. 10.12). Evidence that a propagating fault is accumulating displacement at the expense of another fault (Fig. 10.12) may be derived from a combination of the following: (i) The diminution of topographic relief along the footwall of the propagating fault – despite active faulting along the propagating fault, the amount of footwall uplift and topographic relief is minimal along these young faults. (ii) Changes in the size of alluvial fans and their proximity to the range front – as faulting steps forward into the hanging wall, the proximal part of the former hanging wall becomes part of the footwall of the new fault. Footwall uplift drives incision (fanhead entrenchment) of the old fans and displaces their apices away from the topographic range front. The abandoned fan surface may be tilted steeper than depositional gradients as the former hanging wall becomes a newly activated footwall. (iii) Changes in drainage basin size and orientation in the newly formed footwall – large, oblique catchments that marked the termination of the older fault may be incorporated into the footwall of the new fault,

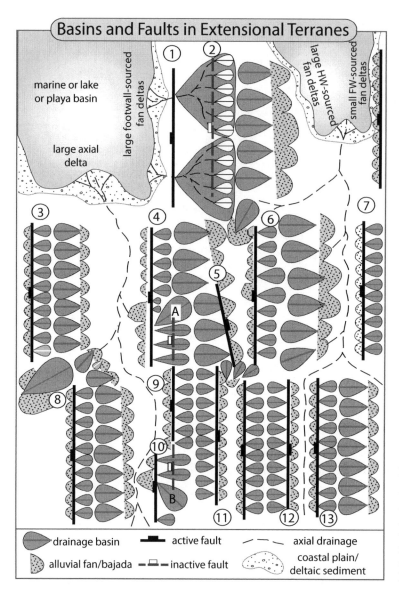

Fig. 10.11 Summary of drainage geometries in extensional basins. Active normal faults (3, 8, and 13) have short, steep drainages adjacent to the fault and longer, gentler drainages on the back-tilted flank, which are commonly hanging-wall basins. Large catchments oriented oblique to the range front occur near fault terminations and step-overs, such as between faults 3 and 8 or between 2 and 6. Large oblique drainages (A and B) mark the terminations of fault 9, which have become inactive due to the propagation of faults 4 and 10. Abandonment of fault 2 as fault 1 becomes active causes large catchments to form in the sediments of the footwall to fault 1. Axial rivers tend to be deflected toward the more active basin margin. Closed basins develop near faults 5 and 11. Modified after Leeder and Jackson (1993).

where most of the catchments are small and orthogonal to the fault (Figs 10.11 and 10.12).

In the Basin and Range, comparison of the overlapping Pearce and Tobin Faults in Pleasant Valley, Nevada (Fig. 10.12), shows that the size of the fans, the amount of topographic relief, and the size of the footwall catchments all decrease toward the tip of the propagating young fault and toward the zone of overlap, whereas the amount of entrenchment and the distance of the active fanhead from the fault trace increase in the same direction.

Displacement gradients and fault linkage

Theoretical calculations indicate that hanging-wall subsidence and footwall uplift should be maximized where fault displacement is the greatest (King and Ellis, 1990). For faults that propagate symmetrically from an initial central rupture, this theory implies that maximum subsidence and fault displacement should occur near the mid-point of the fault trace (Fig. 4.10A). At the same time, we know that, as multiple faults propagate along a given trend, either they

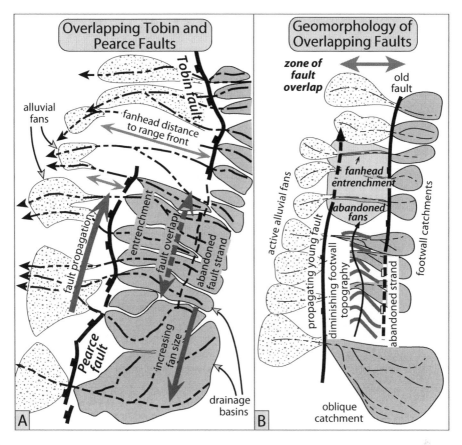

Fig. 10.12   Large-scale geomorphology where a young fault tip propagates past an older normal fault.
A. Interpretation of the overlap zone of active normal faults in Pleasant Valley, Nevada, where the Pearce Fault is propagating northward past the tip of the Tobin Fault. B. Cartoon of major geomorphic features in such zones of overlap. Note diminishing footwall topography toward the propagating fault tip, incorporation into the young footwall of oblique catchment marking the end of the old fault, and entrenchment of the fans that sat in the hanging wall of the older fault and are in the footwall of the new fault. Modified after Leeder and Jackson (1993).

can link up to form a continuous displacement surface, their tips can propagate past each other, or they might encounter barriers that inhibit their propagation (Fig. 4.10).

These options raise the issue of how one determines whether a given normal fault resulted from a linking up of several faults or from the growth of a single rupture. One starting point is to delineate competing models for the fault evolution, footwall topography, and basin subsidence (Fig. 10.13). In the case of a single fault that simply increases its length and displacement through time, the "bow-and-arrow" rule should apply, such that a systematic decrease in displacement occurs toward the fault tips (Fig.

10.13A). In the case where a fault consists of segments that have linked together, the footwall topography should show displacement "deficits" near the segment boundaries, such that there are topographic saddles and basement highs at the segment boundaries (Fig. 10.13B). This geometry would be consistent with the contention (Schwartz and Coppersmith, 1984) that the segments rupture with characteristic earthquakes, such that strain should accumulate in a predictable fashion, with more strain in mid-segment and less at the segment boundaries. In the case where segmented faults link together and then compensate for the displacement deficits at the segment boundaries (Fig. 4.12), the geometry could be

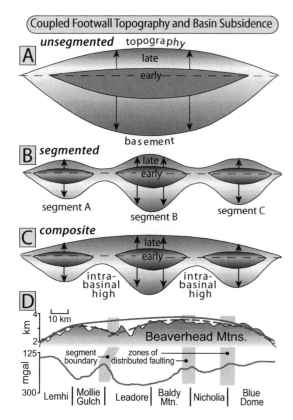

Fig. 10.13   Development of segmented and unsegmented normal faults and basins.

Models for displacement showing topography of the crest of the uplifted footwall and of the downthrown hanging wall. Darker shaded areas show early displacement on fault(s). A. Unsegmented fault with monotonic decrease in displacement toward the tips of the fault. B. Segmented fault with more displacement in the center of each segment than at the tips. Older faults have joined, but there is a persistent displacement deficit at the segment boundaries. Both the footwall topography and basement ridges reflect the fault segmentation. C. Following linkage of segmented early faults, displacement deficits at the boundaries are erased in the footwall topography. Intrabasinal high-standing basement ridges are sites where deformation is distributed among several smaller normal faults. D. Footwall topography and gravity anomalies of the Beaverhead Range, Idaho. The Mollie Gulch–Leadore segment boundary is associated with a gravity high, complexly faulted basement ridge, and a topographic saddle. The Nicholia segment is bounded on each end by modest gravity highs. In each case, displacement due to distributed faulting nearly matches that in the center of the segment. The Beaverhead Range appears to show aspects of models B and C. Modified after Anders and Schlische (1994).

indistinguishable from that of a single growing fault (Fig. 10.13A). Alternatively, the footwall topography could show smooth displacement gradients, whereas intrabasinal high points could mark the former segment boundaries (Fig. 10.13C) (Anders and Schlische, 1994).

Next, the character of the landscape in question should be examined. Does the range display a smooth topographic profile or do saddles separate major crests? Does the bounding fault display a clear segmentation with discrete regions where faults step over, overlap, or break into numerous smaller faults? Do regional gravity anomalies show a gravity low above a hanging-wall basin that extends the full length of the range and is systematically deeper toward its center, or do the gravity anomalies indicate the presence of intrabasinal basement ridges? If there are intrabasinal ridges, do they correlate spatially with segment boundaries or with topographic saddles in the footwall?

Using digital elevation data, it may be possible to test whether segment boundaries are associated with slip deficits (Simpson and Anders, 1992). For example, consider some range fronts that have been interpreted as being bounded by segmented normal faults and in which some visual correspondence exists between the segment boundaries and topographic lows in the footwall. If topography is controlled by segment boundaries and characteristic offsets on each segment, then the topography should slope downwards from the mid-points to the end-points in each segment. If, on the other hand, the entire range can be treated as being controlled by a single fault, then, for half of each segment, the topography should increase toward the end-point, rather than decrease.

In the Beaverhead Mountains in Idaho (Fig. 10.13D), segment boundaries have been identified based on map patterns of faults (e.g., Fig. 4.16), but only some of these appear to represent zones of linkage between older, smaller faults. For example, at the Nicholia–Blue Dome boundary, no slip deficit is apparent in the footwall topography, and mapping across the basement high shows that six smaller faults have together accommodated nearly as much displacement as the single bounding fault does in the middle of the adjacent segment (Anders and Schlische, 1994). The gravity highs here result from the distributed faulting in the absence of a significant slip deficit. Whereas the topographic and basin pattern appears analogous with model C (Fig. 10.13C), the total slip accumulation is consistent with model A (Fig. 10.13A). In contrast, the Mollie Gulch–Leadore segment boundary is associated with an apparent slip deficit, as indicated by a topographic saddle, such that the linkage between segments appears to have happened more recently. This geometry is consistent with model B, in which segment boundaries represent zones of persistent slip deficits (Fig. 10.13B).

Do all normal faults that link together tend toward a state in which displacement gradients change smoothly from end to end along the entire composite fault, eventually eliminating deficits associated with former segment boundaries? This behavior seems to be clearly true for some ranges, but we do not know whether slip deficits observed at some present-day segment boundaries will persist through time or be smoothed and eliminated. The position of any given range on this spectrum from segmented to unsegmented can be evaluated by comparing model predictions for these end-members (Fig. 10.14) to surface and subsurface data on fault geometries and displacement, footwall topography, stratal thicknesses, and stratal tilt in the hanging-wall basins.

Contractions, folds, and drainage networks

The Himalaya, Tibetan Plateau, Caucasus, Alps, Zagros, Taiwan, and many of the major, non-volcanic mountain ranges of the world have developed over spans of millions of years in response to compressional stresses exerted between converging plates. Contrasting landscape and tectonic characteristics emerge in this setting at differing spatial and temporal scales. In the previous chapter, we described the growth of simple folds, such as Wheeler Ridge or Dragon's Back, over a span of 10–200 kyr (Figs 9.20 and 9.26). Here we examine what happens to such folds over longer time spans, and how multiple growing folds and faults may interact in the landscape.

Topography of folds

Consider the growth of contractional folds under these basic assumptions: both blind thrusts and ones that cut the surface drive differential rock uplift and folding; faults propagate laterally as they accumulate strain; and erosive forces attack hanging walls when they are uplifted above local base level. What governs similarities or differences in the geomorphic evolution of individual folds? Certainly, the geometry of the causative fault is important. For example, the hanging wall of a purely fault-bend fold will never be uplifted more than the height of the footwall ramp, whereas uplift in a displacement-gradient fold can greatly exceed the ramp height (see Fig. 4.36). Variations in the stratigraphy and erosional resistance of the hanging wall also affect its evolution. During the development of the fold-and-thrust belts that are commonly found associated with collisional mountain ranges, thrust faults propagate beneath and into the foreland basins bounding the range. When the hanging wall consists only of the sediments that have previously filled the foreland, their relatively homogeneous and typically low resistance to erosion will yield a more spatially uniform and predictable pattern of dissection (see Fig. 9.26) than when resistant bedrock is also uplifted in the hanging wall. When both foreland-basin strata and bedrock are uplifted and exposed to erosion, the large contrasts in erodibility between the less indurated sediments and the bedrock promote a distinct erosional and topographic pattern. During initial stages of uplift (which could amount to 1–5 km, depending on the stratal thicknesses in the foreland), relatively rapid dissection and stripping can occur of the weakly cemented sediments. Thus, the fold might develop limited surface topography with only a few hundred meters of relief, representing a small fraction of the total rock uplift (Fig. 10.15). As resistant bedrock is elevated above base level, rates of surface erosion can drop abruptly and a much higher fraction of the rock uplift may begin to be reflected in surface uplift. In semi-arid to arid landscapes, these bedrock ridges may be almost undissected despite several kilometers of uplift. Such is the case in the Tien Shan of

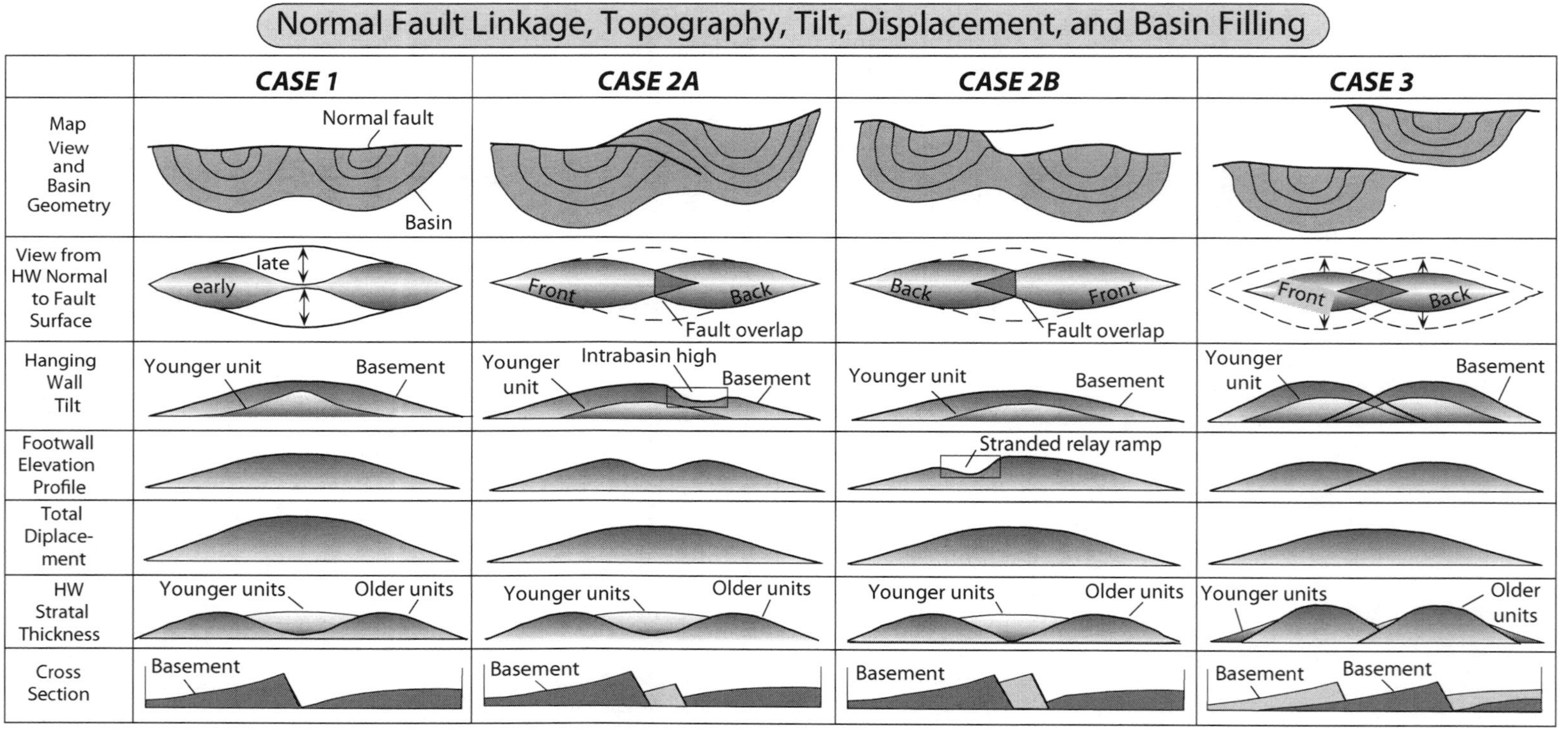

Normal Fault Linkage, Topography, Tilt, Displacement, and Basin Filling

	CASE 1	CASE 2A	CASE 2B	CASE 3
Map View and Basin Geometry				
View from HW Normal to Fault Surface				
Hanging Wall Tilt				
Footwall Elevation Profile				
Total Displace-ment				
HW Stratal Thickness				
Cross Section				

Fig. 10.14 Displacement and basin characteristics for linking or overlapping normal faults.

Four scenarios for linkage geometry and associated structures and basins. In the map view (top row), the shaded area represents the hanging-wall basin with contours on the sediment thickness. In the view from the hanging wall, the dashed line (cases 2 and 3) represents the predicted geometry following more displacement. Note that all cases have the same total displacement and displacement gradient, but that differences in stratal tilting and footwall topography distinguish among them. Case 1: faults link along a plane. Case 2A: rearward fault intersects the frontal, basinward fault. Case 2B: frontal, basinward fault intersects rearward fault. Case 3: overlapping, but non-intersecting faults. Modified after Anders and Schlische (1994).

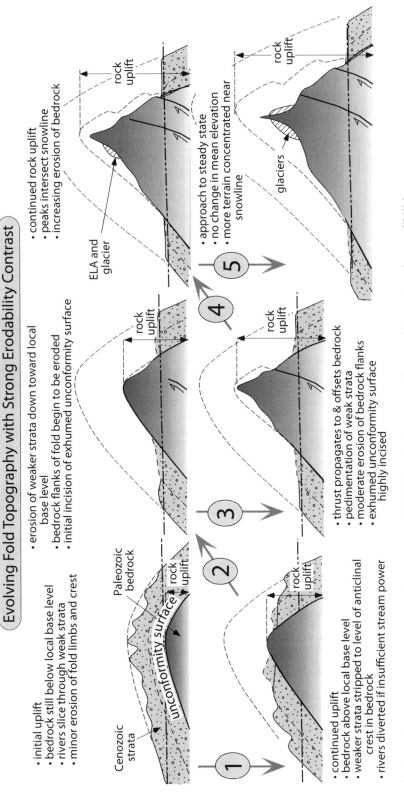

Evolving Fold Topography with Strong Erodability Contrast

1
- initial uplift
- bedrock still below local base level
- rivers slice through weak strata
- minor erosion of fold limbs and crest

2
- continued uplift
- bedrock above local base level
- weaker strata stripped to level of anticlinal crest in bedrock
- rivers diverted if insufficient stream power

3
- erosion of weaker strata down toward local base level
- bedrock flanks of fold begin to be eroded
- initial incision of exhumed unconformity surface

4
- thrust propagates to & offsets bedrock
- pedimentation of weak strata
- moderate erosion of bedrock flanks
- exhumed unconformity surface highly incised

5
- continued rock uplift
- peaks intersect snowline
- increasing erosion of bedrock

- approach to steady state
- no change in mean elevation
- more terrain concentrated near snowline

Cenozoic strata

Paleozoic bedrock

unconformity surface

rock uplift

ELA and glacier

glaciers

Fig. 10.15 Structural and geomorphic development of a fold in which pre-growth strata have contrasting erodibilities.
In the uplifted hanging wall, strong contrasts in erodibility favor rapid removal of overlying, weakly cemented sedimentary strata (typical of a foreland basin), whereas well-lithified basement rocks resist erosion. The unconformity between these strata provides a reliable marker with which to track the geometry of deformation. With sufficient rock uplift and steepening of slopes, the bedrock also begins to be significantly modified by erosion, especially if the range crest rises above the snowline and glaciers begin to grow. Modified after Burbank *et al.* (1999).

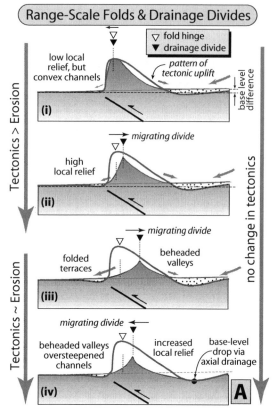

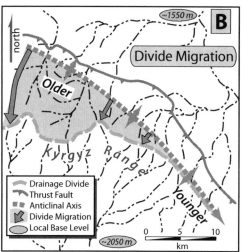

Kyrgyzstan (Burbank *et al.*, 1999; Sobel *et al.*, 2006), where folded and faulted bedrock ridges have been lifted more than 2 km above local base level. Prior to uplift, the bedrock was overlain by about 5 km of Cenozoic sediment. Today, these sediments have been stripped from nearly all fold flanks rising more than 300–500 m above local base level. These exhumed bedrock ridges display generally bow-shaped topographic profiles, suggestive of more recent and lesser displacement near the fold tips, whereas, along the central part of the forelimbs of many of the larger folds, faults have cut the surface.

With the common exception of detachment folds, most folds are asymmetric, with steeper forelimbs than backlimbs (Fig. 4.36). If rainfall is uniform or at least symmetric across the fold crest, then the channels on the forelimb would be expected to erode faster than backlimb channels. As a consequence, the drainage divide should migrate toward the backlimb until equilibrium slopes are developed whereby rock uplift is balanced by erosion on each flank of the fold (Fig. 10.16). Now consider the impact

outpaces erosion, whereas in more equilibrium phases (iii) and (iv), erosion and rock uplift are in balance. (i) Initially, incipient erosion is much slower than rock uplift, such that the structural crest of the fold and the drainage divide coincide. Erosion may be greater on the backlimb because it has bigger catchment areas. (ii) As fold growth continues, the divide migrates toward the backlimb owing to both gentler slopes and its higher base level. (iii) Over time, relief diminishes with respect to rock uplift, and the divide occupies an equilibrium position between drainages with equivalent concavity. (iv) Base-level fall near the backlimb steepens the local channel gradients and drives migration of the divide to a new equilibrium position closer to the forelimb. Such migration could behead channels that formerly drained to the forelimb. Modified after Ellis and Densmore (2006). B. Example of divide migration with respect to structural crest of a range-scale fold, eastern Kyrgyz Range, Kyrgyzstan. The range has propagated eastward over time, such that a space-for-time substitution can be made. More retreat of the drainage divide toward the backlimb has occurred in the western, older part of the range, whereas the divide and the structural crest coincide in the eastern, younger part of the range. In this range, divide migration has been enhanced by headwall erosion by glaciers. Modified after Oskin and Burbank (2005).

Fig. 10.16 Range-scale folds and migration of drainage divides.
A. Conceptual model for divide migration in response to the differential dip of fold limbs and to different local base levels on opposite flanks of the fold. Pattern of tectonic rock uplift remains constant in each panel. In the topographic building stages (i) and (ii), tectonics

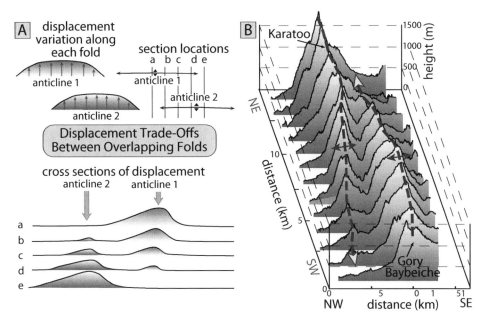

Fig. 10.17   Transfer of displacement among *en echelon* folds.
A. Models of displacement transfer between overlapping folds. As displacement decreases toward the nose of one fold, it increases in the adjacent fold. Shortening recorded by bed-length changes in each fold shows displacement compensation, which smoothes the integrated displacement variation between the two folds. B. Topographic profiles across the oppositely plunging and overlapping termini of the Gory Baybeiche and Karatoo folds, Tien Shan, Kyrgyzstan. The lines of the topographic profiles are oriented approximately perpendicular to the fold axes. Except low on the flanks where Cenozoic strata remain, the topographic surface of both folds is defined primarily by the only slightly dissected regional unconformity that represents the upper limit of the bedrock, such that a similar structural level is being measured in each fold. Note the similarity of the topographic profiles to that predicted when deformation is transferred between two oppositely growing structures (A): as one dies, the other grows, and the total shortening remains fairly constant along strike.

of different base-level elevations on either flank of the fold (Ellis and Densmore, 2006). Assuming equilibrium slopes and equal channel concavity (Figs 8.4 and 8.5) on both flanks, the drainage divide will also be displaced toward the flank with the higher base level. If the base level subsequently changes, for example, via drainage capture that incises an intramontane basin, the divide will shift again in order to achieve a balance with the new base levels and the pattern of rock uplift (Fig. 10.16A(iv)).

The Tien Shan not only provides excellent examples of bedrock-cored folds that grew over the past few million years, but also illustrates how displacement may be transferred between multiple structures. Imagine a convergence between two semi-rigid blocks with growing folds accommodating the shortening

between them. In zones of overlap of the fold tips, one would expect a trade-off in the amount of shortening: as the shortening that one fold accommodated decreased toward its tip, the shortening in the overlapping fold should increase (Fig. 10.17). Based on analysis of an approximately 90-m DEM, topographic profiles oriented perpendicular to the fold axes and across the overlapping tips of two folds illustrate this accommodation of shortening between the two folds (Burbank *et al.*, 1999). DEM-based topographic profiles can serve to illustrate such shortening accurately only when the topographic surface approximately coincides with a structural marker. In the Tien Shan, the exhumed unconformity that cuts across the bedrock and that underlies the weak Cenozoic strata serves this purpose.

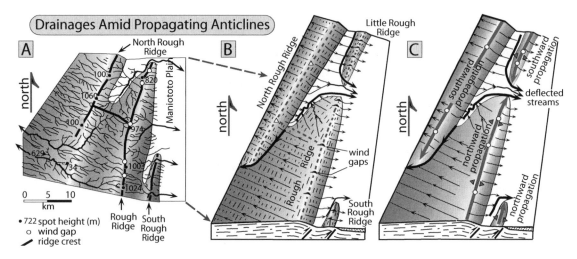

Fig. 10.18   Stream patterns in interfering bedrock folds.
A. Drainage network, fold crests, plunging fold noses, water gaps, wind gaps, and their elevations in the vicinity of
Rough Ridge, central Otago, New Zealand. Note that the elevation of the wind gaps decreases as the fold nose is
approached and that, within the piggyback basins, highly asymmetrical drainages extend upstream of the water gaps.
Cosmogenic radionuclide exposure ages along the nose of South Rough Ridge (lower right) indicate that it has been
propagating north at a rate of ~1 km/Myr (Bennett *et al.*, 2000). B. Cartoon of the drainage patterns in the vicinity of
Rough Ridge. Note the wind gaps along the fold crests and the clear diversion of drainages around the nose of each
growing fold. C. Interpretation of the history of growth of the four anticlines and drainage development along their
flanks. Modified after Jackson *et al.* (1996).

Drainage development from folds to orogens

If several folds are growing simultaneously and
their noses are propagating toward or past each
other, complex river patterns and topography
can sometimes reveal the history of lateral fold
propagation and vertical growth. Envision the
propagation of a young fold adjacent and parallel to an older fold. Steep, short drainages typically develop on the forelimbs of such folds
prior to any interference between them. In the
direction of fold propagation, the crest of the
younger fold is likely to decline in height, and
the width of the fold commonly narrows
(Fig. 10.18). Streams that formerly flowed unim-
peded from the older fold into the basin will be
diverted parallel to the young fold, if they have
insufficient erosive capacity to maintain their
course across the growing anticline. Such diver-
sions will leave wind gaps along the crest of the
young fold as channels are sequentially defeated.
The diverted drainage systems are typically

asymmetrical because the defeated rivers are
usually diverted toward the nose of the fold, that
is, in the direction of fold propagation. Growth
of an asymmetric upstream catchment that
encompasses the diverted streams continues
until the discharge and stream power through a
water gap closer to the fold's nose is sufficient
to balance the uplift rate.

In central Otago, New Zealand (Jackson *et al.*,
1996), excellent examples are displayed of wind
gaps along plunging fold crests, asymmetric
drainages in piggyback basins, diverted drainages,
and persistent water gaps that developed in
response to growth of a suite of thrust-related
folds (Fig. 10.18). The asymmetry of the
underlying folds, the relative timing of fold
growth, and the propagation directions of the
multiple folds can be readily deduced from these
folds. Whereas the full growth history of these
folds remains unclear, recent cosmogenic dating
of "sarsen" stones (residual quartzose monoliths
that sit atop the ancient erosion surface that
serves as a structural marker defining the

magnitude of differential uplift and folding) shows that the fold noses propagate and amplify vertically at rates of about 1 km/Myr and 100 m/Myr, respectively (Bennett *et al.*, 2000). Although these rather slow rates imply that these Otago folds are several million years old, the prevalent semi-arid climate has limited overall downwearing of the fold surface to about 30 m, and drainages are fairly lightly etched into the schist that underlies each fold.

When the drainages develop in folded ranges with weakly lithified sediments draping over bedrock (Fig. 10.15), another interesting uncertainty emerges concerning the development of the current fluvial network. Did today's drainage patterns develop due to interactions of rivers with the mantle of overlying sediments or with the resistant bedrock once it became exposed during deformation (Oberlander, 1985; Tucker and Slingerland, 1996)? Were rivers that pre-existed folding "let down" or superposed on to the deforming landscape, such that they are independent of the growing structures? Have rivers that currently cut across folds shifted laterally during fold growth (Fig. 10.18) or did they maintain an antecedent course? Commonly, the answer to this must be gleaned from the stratigraphic record where changes in provenance and river systems can be documented. In response to prolonged convergence, imbricated folds and thrust faults build entire mountain belts. The drainages that develop on these ranges might be expected to differ in their geometric properties in response to contrasts in climate, lithology, rock uplift rate, and/or erosion rates in different settings. In fact, such variation is remarkably limited among many ranges. Instead, studies of the spacing (S) of major drainages and comparisons to the half-width (W) of the ranges on which they developed demonstrate a strikingly consistent relationship, with a spacing ratio (W/S) of about 2. This consistency suggests that some of the early-formed irregularities in river courses that developed in response to individual growing folds or faults are smoothed during the drainage expansion and competition among drainages that occur as mountains grow (Talling *et al.*, 1997). This consistent ratio has another intriguing implication: as a range widens, some

laterally adjacent drainages must be captured to maintain the drainage spacing.

Past stream captures are typically difficult to document with the preserved geomorphic record. Wind gaps are defined as topographic saddles through which a now-defeated river formerly flowed. But, how do we distinguish such a saddle from a pass on a ridge that never had a river flow across it? Most typically, little evidence is preserved of a former river. Some of the best indicators of a pre-existing, through-flowing river include the following: (i) preservation of sediments that indicate long-distance fluvial transport – well-rounded clasts, well-sorted sandstones, mature sediment shapes and compositions, and non-local lithologies; (ii) fluvial terrace remnants that are now folded across the water gap; and (iii) remnant fluvial channels that do not scale with the current catchment geometry – channel widths or the depths of channel incision do not decrease toward the current drainage divide. In the Otago folds, for example, quite broad channels etched into the bedrock can be traced through a wind gap, up and over the fold crest.

True wind gaps where former rivers have been defeated clearly indicate that some catchment has gained drainage area and discharge at the expense of another, that is, a drainage capture has occurred. But which river and how do we recognize it? In their study of the Otago folds, Jackson *et al.* (1996) use both verifiable and assumed wind gaps and the known direction of fold propagation to make logical deductions about how the drainage network progressively evolved as the folds grew. Other geomorphic criteria that can be used to deduce river capture and reversal of drainage directions include the following: (i) river terraces that tilt in the opposite direction to the modern river gradient and are unrelated to any local structural deformation; (ii) "barbed" rivers whereby river junctions define atypical angles, for example, by pointing upstream or entering at nearly right angles even though the converging rivers have similar discharge and low gradients; (iii) provenance indicators that are inconsistent with the current upstream source area; and (iv) convoluted catchment geometries.

If the stratigraphic record of sediments derived from the drainage can be examined, then changes in provenance and fluvial character, such as in paleodischarge calculations, can also indicate past river capture events. In rapidly eroding orogens (>1 km/Myr), the geomorphic signal of river captures has a short residence time. Within 100 kyr after a capture, many tens of meters of bedrock will have been eroded, most older river terraces will have been obliterated, and channels will have adjusted to new gradients and discharge. In more slowly eroding orogens, histories of drainage evolution and river capture can sometimes be reconstructed.

One of the more spectacular examples of orogen-scale catchment expansion through successive capture events has been proposed for the Yangtze River in western China (Clark *et al.*, 2004). Today, the Yangtze is about 6300 km long, its catchment encompasses nearly 2×10^6 km², and it spans from the Tibetan Plateau to the East China Sea (Fig. 10.19). The river's map pattern reveals many barbed junctions and a highly convoluted river course. In a few sites, reversals of drainage are inferred from upstream-sloping river terraces. Five major capture events have been proposed to have more than doubled the Yangtze's catchment area (Clark *et al.*, 2004). The timing of these events is not well defined, but they are inferred to have migrated from east to west as each successive capture increased the paleo-Yangtze's discharge and would be expected to have driven a knickpoint up the newly captured river, thereby promoting enhanced headwater erosion and subsequent capture events. Although the timing of capture events remains elusive, future analyses of drill cores from near the Yangtze's outlet in the East China Sea and the Red River's outlet in the South China Sea (Fig. 10.19) will likely reveal the chronology and provenance data needed to test and calibrate the overall river-capture model.

Whereas geomorphic evidence currently provides the primary support for the Yangtze River synthesis, offshore isotopic and seismic data have been used to deduce other continental-scale river capture events. In the western Himalaya, for example, the five major rivers of the Punjab (which means the "country of five rivers" in Sanskrit), which now flow southwest into the Indus River, have been interpreted on

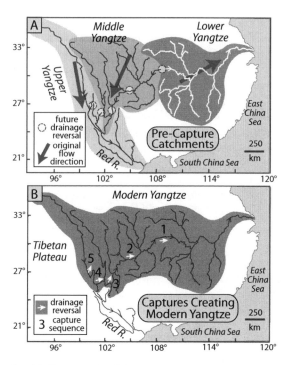

Fig. 10.19 Growth of the Yangtze catchment through river capture.
A. Proposed pre-capture network geometry in which rivers in the western half of the Yangtze's current drainage flowed into the Red River. B. Sequence of drainage reversals and captures that is proposed to progress from east to west through time. Modified after Clark *et al.* (2004).

the basis of Miocene paleocurrents to have flowed southeast into the Ganges River (Burbank *et al.*, 1996a). New evidence to support this apparent switch in drainage from the east side of the Indian subcontinent to the west side has emerged from the sedimentary record in the northern Arabian Sea near the mouth of the Indus River. There, a major increase in sediment discharge from the Indus after the Miocene, as deduced from seismic data, is coeval with a change in neodymium isotopes that reveals a change in source areas (Clift and Blusztajn, 2005). These isotopes suggest an increase in contributions from the Greater Himalaya (where the headwaters of the Punjab rivers currently reside) at the expense of contributions from the Indus suture zone (which the Indus River seems to have persistently drained during Neogene time). In theory, the loss of contributions from the Punjab catchments should be detectable in records from the Bay of Bengal where the Ganges debouches its sediment load. But, the addition of new, isotopically distinctive source areas (as happens when the Punjab rivers are captured by the Indus) is much easier to detect than the incremental decrease in an existing and more isotopically homogeneous source area, as would be the case for the Ganges catchment.

As is clear from the Yangtze example (Fig. 10.19), river-network anomalies that represent departures from expected geometries may point to sites where former catchments have been significantly modified. In addition to features that may connote capture events, such departures include anomalous spacing of drainages or downstream narrowing of catchments. Whereas no specific tectonic events are known to be associated with the capture sequence on the Yangtze or in the Punjab, clear tectonic inferences can be drawn from some river patterns. Consider, for example, the southeastern Tibetan Plateau, where three of the region's great rivers (the Mekong, Salween, and Yangtze) rotate from easterly to southerly courses as they swing around the eastern tip of the Himalaya and exit the plateau (Fig. 10.20A). The most curious aspect of their courses through this bend is that they converge, such that they flow parallel to each other and are spaced only a few tens of kilometers apart (Hallet and Molnar, 2001). This region of closely spaced, parallel channels represents a major width anomaly in which the catchments are about 10-fold narrower than would be expected at these downstream positions (Fig. 10.20B, C, and D). So, what causes such striking narrowing of these catchments? Hallet and Molnar (2001) propose a tectonic model (Fig. 10.20E) in which crustal-scale shear occurs around the northeast corner of the Indian craton as Eurasia (and the Tibetan Plateau) converges with it. North of the rigid indentor (India), crustal shortening in Tibet would be expected to cause narrowing of catchments oriented east–west. East of the indentor, large-scale dextral shearing would be expected. Simple geometric assessments suggest that the shear strain that is required to cause both the proximity and the parallelism of the rivers is about 6 ± 1. Shear strain is defined here as the ratio of shear displace-

ment to the width of the shear zone, and, in this setting, such strain is equivalent to about 900 km of dextral shear across a 150-km-wide shear zone where the anomalous drainages occur (Fig. 10.20E). Such strain is consistent with modern geodesy (see Fig. 5.15A) which depicts a striking turning and flow of crustal material around the eastern syntaxis where the three rivers described here are in close proximity.

Steady state and pre-steady state

Because convergence between two plates can be sustained for millions of years, rock uplift in contractional mountain belts can persist for similarly long time spans. But the height of mountains cannot grow forever. At some point, rates of erosion or tectonic extension ought to balance the rates of rock uplift, such that the range comes into a topographic steady state. Prior to that time, the mean surface elevation can increase through time, reflecting rates of rock uplift that outpace erosion rates, whereas in post-steady-state conditions, erosion outstrips rock uplift. Even in steady-state conditions, rock uplift will not be balanced by erosion at every point in the landscape at all time scales. For example, we expect that changes in erosion rates are coupled to changes in climate at time scales of decades to many millennia, but, in the general absence of evidence indicating similarly paced changes in tectonic rates, we deduce that tectonic rates are steadier and that short-term imbalances between erosion and uplift are likely. On average, however, under steady-state conditions, a balance should exist at time scales that span several climate cycles. Thus, we examine here how a balance between rock uplift and erosion is attained in such circumstances and how we may recognize whether or not this balance has been achieved.

Categories of steady state

With respect to mountain ranges or orogens, at least four categories of steady state can be defined: flux steady state, thermal steady state, exhumational steady state, and topographic steady state (Willett and Brandon, 2002).

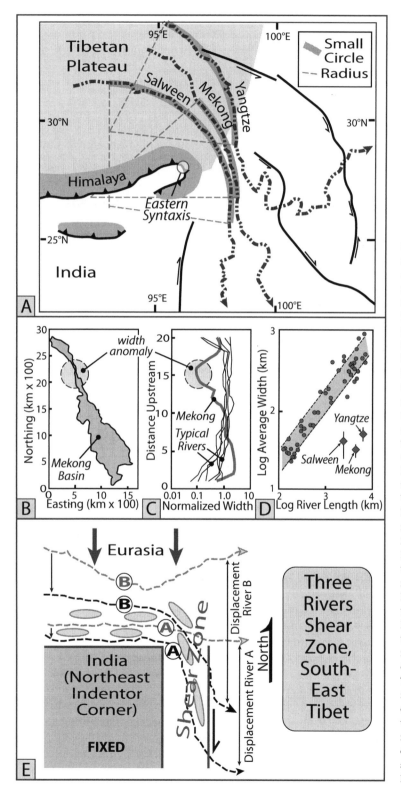

Fig. 10.20 Three Rivers shear zone in southeastern Tibet.
A. Map of the anomalous upper reaches of the Salween, Mekong, and Yangtze Rivers where they flow in close proximity and parallel to each other as they exit the Tibetan Plateau. B. Map of the Mekong catchment that highlights the region of unusually narrow catchment width. C. Comparison of upstream changes in normalized river width between the Mekong and six "typical" rivers. Basin widths were measured at 20 equally spaced points spanning the length of each river, normalized to unity, and plotted on a log scale. Whereas significant narrowing at basin outlets is common, the mid-course narrowing of the Mekong is clearly anomalous. D. River length versus basin width for the world's 50 longest rivers compared to the Three Rivers. The narrowness of the three river basins attests to the large-scale crustal shortening in this region. E. Schematic tectonic model of the impact of the corner of a rigid indentor (India) with Tibet and Eurasia. Convergence north of the indentor drives catchments closer together, whereas dextral shear along the right-hand margin of the indentor causes both necking of the basins and river alignment. Modified after Hallet and Molnar (2001).

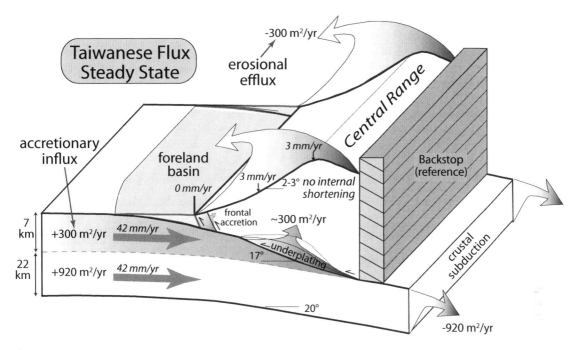

Fig. 10.21   Flux steady state for the Central Range of Taiwan.
The accretionary influx is divided between the upper 7 km, which is underplated (and later eroded from the Central Range), and the lower 22 km, which is subducted. The accretionary flux equals the layer thickness times the convergence rate. Note that the 300 m²/yr flux is underplated beneath the Central Range and then is balanced by erosion from the range. The crustal subduction is not balanced within the Central Range, but, because it subducts, it does not formally contribute to the mass of the orogen. Modified after Simoes and Avouac (2006).

In a *flux steady state*, the influx of rock into a range is balanced by the efflux out of the range (Fig. 10.21). Consequently, the volume or cross-sectional area of the range remains constant through time. Most commonly, the efflux is simply due to geomorphic erosion, but any type of tectonic removal of rock from the range will also add to the efflux. An assessment of a flux steady state requires both geophysical determinations of the rate of tectonic additions and losses from a range, as well as geomorphic measures of volumetric erosion rates.

A *thermal steady state* exists when the spatial distribution of rock temperatures with respect to the surface is unchanging. Such a state occurs when the conduction of heat and the advection of rocks (which carry their heat with them) balance the cooling at the surface that occurs due to erosion (or heating in response to deposition), such that the depth-dependent

thermal structure of a range is time-invariant. In order to attain a thermal steady state, the velocity field must also be in steady state, such that heat advection is time-invariant. At present, few data exist to assess a thermal steady state: present-day subsurface temperatures are poorly defined, and reconstruction of past thermal patterns is equally fragmentary.

A thermal steady state is a necessary precondition for an *exhumational steady state*, whereby rates of erosion are spatially invariant. Commonly, long-term erosion rates are assessed using low-temperature thermochronometric techniques, such as fission-track or (U–Th)/He dating, in which a cooling age, a known closure temperature, and a geothermal gradient are combined to estimate the average erosion rate since a mineral passed through its closure temperature. Minerals used to assess an exhumational steady state must have had their ages reset due to heating above their closure temperature. Unreset mineral ages will reveal pre-orogenic ages. If particle pathways through a range and the thermal structure of the range are time-invariant, then an exhumational steady state likely exists (Fig. 10.22A), such that the cooling ages at any site within the range will be constant through time. For a contractional range in which rock is accreted from the side, numerical models predict nested zones of cooling ages for each thermochronometer (Willett and Brandon, 2002). In this case (Fig. 10.22B), the mineral that has the lowest closure temperature, such that its cooling age is locked in closest to the surface, will occupy the broadest zone of reset ages, and this zone will become increasingly narrow for minerals with higher closure temperatures. During growth of a range and prior to steady state, the zone of reset ages should gradually expand, but it should reach a persistent extent at steady state (Fig. 10.22C).

Until we acquire an ability to go backward or forward in time, a strict test of an exhumational steady state will remain elusive, because we cannot assess, for example, what the actual pattern of cooling ages was two million years ago. Two quite different approaches, however, can be used to argue in favor of an exhumational steady state. First, the presence of nested zones of cooling ages that have a distribution consistent with the known kinematics of a range of orogens support an inference of steady state, because this geometry suggests a sustained and consistent thermal history for rocks across an orogen. In several well-dated orogens, such as the Olympic Mountains in Washington (Batt *et al.*, 2001) or the Central Range of Taiwan (Fuller *et al.*, 2006), nested zones of cooling ages are observed (Fig. 10.23).

An alternative approach to assess an exhumational steady state relies on changes in the cooling ages of detrital minerals in well-dated stratigraphic records (Bernet *et al.*, 2001, 2004; Bullen *et al.*, 2001) (Fig. 10.24). These ages record the time it took for each mineral to travel, first, to the surface from its passage through its closure temperature at some depth and, second, from the surface to the basin where it became part of the depositional record. Typically, this last transport phase is far shorter than the earlier

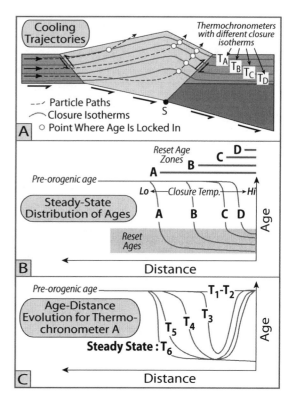

Fig. 10.22   Cooling ages in a steady-state orogen.
A. Kinematic model and particle pathways for a convergent orogen. The positions of closure isotherms for thermochronometers with different closure temperatures ($T_A < T_B < T_C < T_D$) are warped by lateral and vertical advection of rock within the orogen.
B. Predicted distribution of ages across an orogen at steady state as a function of closure temperature (lower closure temperatures are reset across a broader zone).
C. Temporal evolution of the spatial distribution of cooling ages from initial (T_1) to steady state (T_6). Modified after Willett and Brandon (2002).

phase and is ignored. The difference between the cooling age of the mineral and its depositional age is defined as the "lag time." Short lag times indicate that the depth-to-surface transect was rapid and imply high rates of erosion at that time. The typical strategy in stratigraphic lag-time studies is, first, to date 50–100 individual grains, such as apatite or zircon, from a succession of well-dated stratigraphic horizons. Next, the distribution of ages at each horizon is analyzed statistically to pick out the youngest population of ages (Brandon, 2002), which is

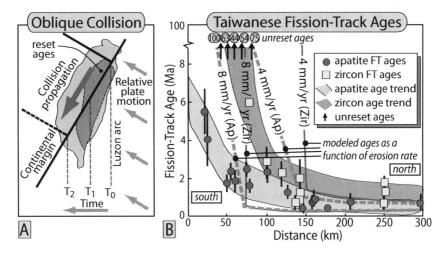

Fig. 10.23   Exhumational steady state and Taiwanese fission-track ages.
A. Schematic tectonic framework of Taiwan. Oblique collision of the Luzon arc causes a southward migration of the deformation front over time (T_0–T_1–T_2). Reset fission-track ages (darker shading) delimit regions where more than 4 km of erosion has occurred. B. Fission-track ages projected on a northeast–southwest transect show two nested zones: reset zircon fission-track ages with their higher closure temperature (~230°C) lie farther from the leading edge of deformation than do reset apatite ages (closure temperature ~100–120°C). Modeled distribution of ages for different exhumation rates (dashed lines) are consistent with sustained erosion of at least 4 mm/yr within the reset zone. Overall, these data are interpreted to indicate an exhumational steady state in a northwest–southeast cross-section. The steady-state region is expanding toward the southwest as the collision progresses. Modified after Willett *et al.* (2003) and Fuller *et al.* (2006).

then compared with the sample's depositional age to define the lag time, which is then plotted versus time. Finally, changes in lag times during the depositional record are used to assess (i) when erosion was most rapid in the past (the shortest lag times) and (ii) whether intervals of uniform lag times exist, because such intervals would be consistent with an exhumational steady state (Bernet *et al.*, 2001). This detrital approach has the advantage of allowing us to go back in time to see how cooling-age distributions have changed during growth or decay of an orogen. One limitation of this approach is that we never know precisely where the detrital grains were derived. Consequently, we cannot assess whether the spatial distribution of cooling ages was constant, as the formal definition of exhumational steady state requires.

In a *topographic steady state*, average topographic characteristics, such as mean elevation and relief, may vary spatially, but are independent of time, such that the same topographic pattern would be predicted to persist

through time (Figs 1.8 and 10.1). Perhaps surprisingly, in an orogen in which rock is being advected laterally, a balance between rock uplift rates and erosion rates at a point may be insufficient to define a topographic steady state. In particular, an imbalance can be created due to lateral advection of sloping topography that causes an increase or decrease in surface height at a point that is independent of rock uplift (u), subsidence ($-u$), erosion (e), or deposition ($-e$), such that:

$$\mathrm{d}h/\mathrm{d}t = u - e - v\,\mathrm{d}h/\mathrm{d}x \qquad (10.3)$$

where v equals the velocity of lateral advection in the plane of a given cross-section, $\mathrm{d}h/\mathrm{d}t$ equals the change in height with time t, and $\mathrm{d}h/\mathrm{d}x$ equals the average slope of the topography that is being laterally advected (Fig. 10.25). Hence, in order to sustain a steady-state topography, if the surface slope trends downward with respect to the advection direction, then not only must erosion balance rock uplift, but it must also balance $v\,\mathrm{d}h/\mathrm{d}x$.

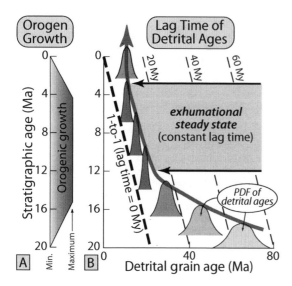

Fig. 10.24   Exhumational steady state and lag times of detrital mineral populations.
A. Model for orogenic growth in which accelerating and decelerating growth bracket an interval of steadiness between 15 and 4 Ma. B. Probability distributions of detrital cooling ages versus time. The lag time for the youngest population of ages within a given horizon is expected to get younger during the growth state, older during decay, and remain constant if an exhumational steady state is attained. Here the lag time is ~10 Myr during the steady-state phase and would represent average cooling rates of 10–20°C/Myr for apatite and zircon fission-track ages, respectively. Modified after Burbank *et al.* (2007).

Morphologies of steady state

In terms of steady-state topography, we might expect, for example, that areas with persistently faster rock uplift rates would have persistently steeper slopes, higher maximum elevation, and greater relief. More significantly, we would require that the spatial pattern of topography remain steady through time. Such temporal steadiness is nearly impossible to assess at geological time scales, because we lack accurate means of reconstructing past topography. Instead, we can sometimes use an ergodic substitution in which we know that a range is propagating at a particular rate in a certain direction and perhaps know it has been propagating for a given length of time. If we see a trend in topographic characteristics that increases or decreases from a site where

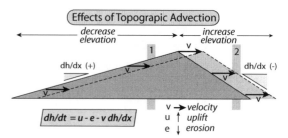

Fig. 10.25   Effects of topographic advection on steady state.
Lateral advection of sloping topography either adds to or subtracts from existing topography. In the configuration shown here, advection causes a decrease in elevation on the left side and an increase in elevation on the right side. To maintain a steady-state topography, erosion would need to be more rapid on the right (site 2) than on the left (site 1).

nascent deformation has just begun to a site where deformation has been ongoing for some time, we can deduce that a steady state does not exist: topography is instead changing with time. If, as we encounter older and older topography, however, we find more uniform topographic characteristics, this uniformity, especially in the face of propagating deformation, may indicate a topographic steady state. For example, in the normal-faulted Lemhi Range of Idaho, initially elevation, relief, and steepness all steadily increase with distance from the fault tip (Fig. 10.21C and D). Still farther from the tip (>15 km), however, these indices become independent of distance, thereby suggesting that a topographic steady state has been attained (Densmore *et al.*, 2005, 2007). Moreover, the rather abrupt change at ~15 km suggests that the surface processes are responding to some thresholds, such that each increment of rock uplift is efficiently counter-balanced by an increase in erosion rate in order to sustain nearly constant steepness, relief, and elevation (Fig. 10.26).

A similar situation may also occur in contractional ranges. For example, a series of relief transects of fission-track ages in the eastern Kyrgyz Range (Sobel *et al.*, 2006) serves to define an eastward propagation of a thrust-faulted range since middle Miocene time (Fig. 10.27). On a traverse toward the fault tip, catchment sizes, mean elevation, mean slope, and maximum

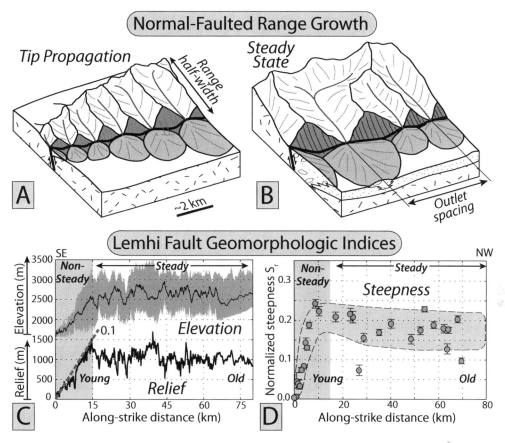

Fig. 10.26   Growth of normal-faulted ranges and their topographic indices.
A. Model for topographic evolution in a normal-faulted range adjacent to a propagating fault tip with increases in relief, catchment size, mean and maximum elevation, facet and fan size, and range half-width scaled by distance.
B. Model for steady-state topography in a normal-faulted range in which systematic topographic change is absent.
C,D. Elevation, relief, and normalized steepness for the normal-faulted Lemhi Range, Idaho, where transient topography yields to steady-state topography at ~15 km from the fault tip (0 km). Modified after Densmore *et al.* (2005, 2007).

elevation all decrease. This zone of changing topography is part of a surface uplift zone where rock uplift is outpacing erosion. Farther from the range tip, these indices even out, suggesting that a topographic steady state is attained in this bedrock range after 5–7 Myr of deformation.

In the Kyrgyz and Lemhi Ranges, along-strike changes in morphologic parameters clearly separate parts of a range where the topography is growing (pre-steady state) from parts with generally constant morphology that could be classified as steady state. Sometimes, the presence of distinctive morphologic forms can also serve to identify a range that is clearly not in steady state. For example, the preservation of

erosion surfaces, especially broad planation surfaces that have been carved across resistant bedrock, is incompatible with a steady state (Fig. 10.15). Such planation surfaces are preserved in the Otago fold-and-thrust belt (Fig. 10.18), the Tien Shan of Kyrgyzstan (Burbank *et al.*, 1999), and the northern Tibetan Plateau, sometimes at elevations exceeding 4–5 km. Generally, the presence at high altitudes of low-relief surfaces, thick fills, or readily erodible sedimentary strata is unexpected in steady-state topography.

In areas where steady state appears to prevail and well-defined spatial variations in rock uplift are apparent, at least two end-member

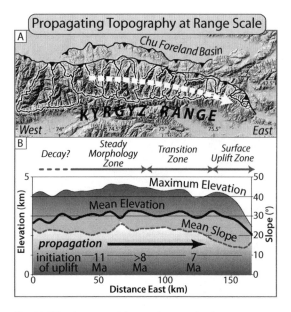

Fig. 10.27 Topographic steady state in the thrust-bounded Kyrgyz Range.
A. Map view of catchment boundaries and major faults on the northern and eastern flank of the Kyrgyz Range in northern Kyrgyzstan. Note the decrease in catchment size in the propagation direction (toward the east).
B. Along-strike changes in slope, mean elevation, and maximum elevation in the western Kyrgyz Range. The timing of the initiation of uplift is derived from apatite fission-track ages in relief transects (Bullen *et al.*, 2003; Sobel *et al.*, 2006). The eastern tip of the range displays increases in slope, mean elevation, and peak heights, whereas these characteristics remain steady in the central range, where a topographic steady state is deduced. Modified after Sobel *et al.* (2006).

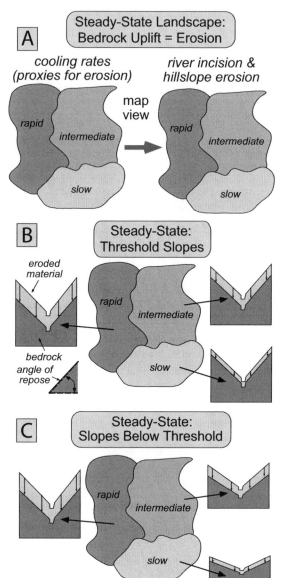

Fig. 10.28 Different ways to attain a topographic steady state.
A. If variations in cooling rates or rock uplift correlate spatially with variations in denudation or incision, a steady-state topography is commonly inferred. B. If slopes everywhere are at threshold angles for stability, then hillslope angles will be independent of the erosion rate (assuming uniform rock strength), and denudation will be solely a function of the rate of local base-level lowering. C. If slopes are proportional to the denudation rate, then spatial variations in slope would correlate with variations in rock uplift.

topographic signatures can be envisioned to define the relationship between slopes, rock uplift, and denudation rates (Fig. 10.28). In the first scenario, slopes vary in some proportion to the rock uplift rate, under the assumption that the flux of material from hillslopes can be summarized as a diffusive process that is strongly related to slope angle (Penck, 1953). Alternatively, if most of the hillslope sediment flux is delivered by landslides, then most slopes would lie close to the threshold angle for failure (Fig. 10.28) and, hence, would have similar slopes even among areas with differing rock uplift rates. In both cases, the rate of river incision could balance the rock uplift rate. Some studies seem to bridge between these end-members (Fig. 7.3B and C): at lower erosion rates, slope is linearly proportional to the erosion rate, whereas at higher erosion rates, slope is largely independent of the rate, as would be expected if threshold slopes were prevalent.

To assess whether a landscape more closely mimics one end-member or the other, it is necessary to document the spatial variation in rock uplift, denudation, and slope angles. A good case study emerges from the northwest Himalaya in the vicinity of Nanga Parbat, where some of the greatest relief on Earth is found: across a distance of about 20 km, over 7 km of relief occurs between the Indus River and the summit of Nanga Parbat. Thermochronological studies in the vicinity of Nanga Parbat (Zeitler, 1985) document greater than 10-fold differences in bedrock cooling rates and inferred long-term denudation rates. These rates appear to vary systematically across the region (Fig. 10.29). As described previously (Fig. 7.14), short-term bedrock incision rates by rivers show a similar spatial pattern of denudation (Burbank et al., 1996b). Based on a 90-m DEM, average hillslope angles can be defined (300 × 300 m moving windows were used here) and then compared between areas of very rapid versus slower denudation. Histograms of the slope distributions (Fig. 10.30A) show a remarkable similarity among these mountainous regions even though they differ by factors of 3–5 in erosion rates (Burbank et al., 1996b). The mean slope angle for each of these areas is $32° ± 2°$, an angle similar to the angle of repose in dry non-cohesive materials. In the context of the fractured bedrock and the very large hillslope lengths that characterize this region, these slopes can be interpreted as threshold slopes in which the slope angle is generally independent of denudation rate (Fig. 10.28).

Climate and tectonics

From a geophysical perspective, we might expect that enhanced rates of erosion that result from climate changes could change the patterns and rates of deformation. Such climate–tectonic linkages could take many forms. One such proposal was that of Molnar and England (1990)

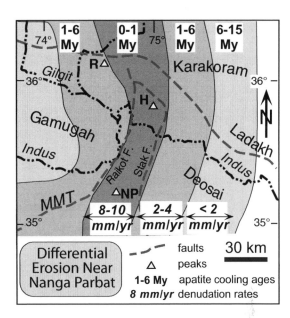

Fig. 10.29   Nested cooling ages in the northwest Himalaya near Nanga Parbat and the Indus River, northern Pakistan.
Based on numerous apatite fission-track ages (Zeitler, 1985), specific ranges of ages delineate a nested suite of cooling zones. The north–south zone centered on Nanga Parbat (NP) and Haramosh (H) cooled below approximately 110°C within the last 1 Myr. Ages get progressively older farther from this central zone of young ages. Spatial variability in long-term cooling rates is approximately matched by spatially variable incision rates (given in mm/yr) derived from dating of strath terraces (see Fig. 7.14). Each cooling and incision rate varies by about 10-fold. Peaks over 7500 m in the area include Nanga Parbat, Haramosh, and Rakaposhi (R). MMT: Main Mantle Thrust (equivalent to the Indo-Asian suture). Modified after Burbank et al. (1996b).

in which they suggested that accelerated late Quaternary erosion in mountain belts would have generated a geophysical response whereby mass removal due to erosion would impel isostatic uplift of the eroded regions (Fig. 1.6). Whereas the Molnar and England (1990) model dealt primarily with the motion of rocks in a vertical domain, some more recent climate–tectonic models have also emphasized widening or narrowing of self-similar orogens in response to decelerating or accelerating erosion, respectively (Meade and Conrad, 2008; Stolar et al., 2007; Tomkin and Roe, 2007; Whipple and Meade, 2006) (Fig. 10.2). In most cases, the

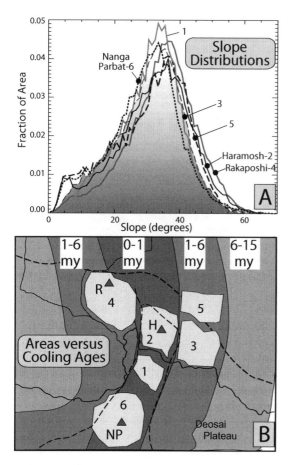

Fig. 10.30 Slope distributions from zones with contrasting erosion rates in the northwest Himalaya near Nanga Parbat.
A. Slopes are calculated as best-fit planes to a 4×4 matrix in this 3-arcsecond (about 90 m) DEM. Note that, even though these span areas with ten-fold contrasts in denudation rates, there are few significant differences among their slope distributions. B. Areal boundaries used for slope distributions. These regions are designed to sample regions with different denudation and rock uplift rates (see Fig. 10.29 for rate estimates). NP: Nanga Parbat; H: Haramosh; R: Rakaposhi. Modified after Burbank *et al.* (1996b).

changing erosion rates are hypothesized to be linked to variations in climate through time, such as due to rainfall, discharge, or glacial sliding velocities. Although the intuitive nature of these models is appealing, few studies have clearly demonstrated the proposed linkage of changes in patterns of rock uplift or orogen width to changes in erosion rates.

Similarly, given the lateral changes in rainfall rates that typify orographic precipitation in many orogens, several numerical models predict that deformation will be enhanced where rainfall is high (Fig. 1.9) (e.g., Willett, 1999; Beaumont *et al.*, 2001). Again, a logical link exists between high rainfall, high discharge, high stream power, and high erosion rates. Quite a few studies have, in fact, shown spatial correlations between zones of young cooling ages (hence, rapid erosion) with regions of heavier precipitation (Grujic *et al.*, 2006; Reiners *et al.*, 2003; Thiede *et al.*, 2004; Willett, 1999). But, do such spatial correlations demonstrate cause-and-effect linkages, or do the different time frames over which each of the measurements are integrated dictate that no meaningful correlation can be made between them? Overall, we need to critically assess what data would allow us to test more reliably whether the observed correlations actually represent causal links.

Most of the studies cited above are "snapshots" that compare current distributions of cooling ages, for example, with current patterns of rainfall. Instead, we might be more convinced by a history that compares cooling and erosion over several million years. Such a comparison would assume more importance if a change could be demonstrated in one of the variables and a synchronous change can be shown to occur in a variable considered to represent a response to that controlling variable. We can imagine many such pairings: more intense rainfall with more rapid erosion; more rapidly flowing glaciers with faster bedrock erosion; development of a rain shadow with a decrease in erosion; or a decrease in precipitation with a widening of the zone of active orogenesis. An apparent response to an observed change is more persuasive than a simple correlation among factors that are presumed to be linked.

Have convincing correlated changes of this type been observed? Unfortunately, at present, few persuasive examples exist (Whipple, 2009). One might wonder why such a paucity persists for an intellectual problem that fascinates many people. Possible explanations are numerous. Natural data are generally noisy, and we are looking for perhaps

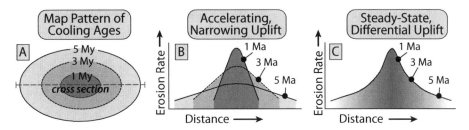

Fig. 10.31   Contrasting interpretations of cooling-age distributions.
A. Concentric zones of cooling ages indicate that erosion is most rapid in the center. Horizontal line shows cross-section location of parts B and C. B,C. Map pattern of ages could be produced by (B) an acceleration of erosion rates into an increasingly narrow zone over the past 5 Myr or by (C) a spatially focused pattern of erosion that persists for millions of years.

subtle effects. We almost always have to rely on proxy data: cooling ages, rather than actual erosion rates; isotopes, rather than elevation; sediments from non-specific sources, rather than *in situ* bedrock; etc. Each type of proxy data separates us farther from the actual process of interest and requires a suite of assumptions to interpret. Because mountain ranges erode, we simply cannot go back to the bedrock to determine the distribution of cooling ages at 5 or 10 Ma. Instead, we have to examine the detrital record found in sediments of those ages in nearby basins. But then, we usually cannot be certain from where those sediments were actually derived, and we are still forced to reconstruct a scenario to explain their characteristics.

Another challenge when we examine a "snapshot" of current data is to try to determine the time frame for which it is relevant. Consider, for example, a map pattern showing a bull's eye with progressively younger cooling ages toward the center (Fig. 10.31). One interpretation of these data would be that a significant acceleration of erosion has occurred in the central part of the mapped area, whereas a rather different interpretation would argue that a gradient of erosion from rapid in the center to slower on the margins has persisted for an unknown length of time, but for at least a few million years. Although data to distinguish between these alternatives may be difficult to obtain, the implications of these different interpretations are profound: accelerating versus steady erosion; and a spatial and temporal evolution of erosion versus a spatial and temporal persistence of erosion (Fig. 10.31).

Orographic rainfall and topography

As previously described, when moisture-laden winds hit mountains, air is forced to rise and, therefore, to cool and ultimately to condense and precipitate the moisture within it. Hence, loci of high rainfall develop on the upwind side of a range, whereas rain shadows commonly develop on the downwind side. The patterns of such orographic rainfall can vary depending on the size of the mountain range (Can air flow around it, rather than over it?) and wind speed, as well as the height and shape of the range. Such interactions have been studied both from a theoretical vantage point (Galewsky, 2009; Roe, 2005; Smith and Barstad, 2004), as well as from an observational perspective, relying on data either from networks of weather stations (e.g., Bookhagen and Strecker, 2008; Burbank *et al.*, 2003), or from remote sensing of rainfall and snowfall (Bookhagen and Burbank, 2010). Whereas weather stations can provide higher accuracy and precision at a point, space-based measurements can provide both spatially broader and more uniform records. With improved remote sensing, measurement of both snowfall and the snowmelt contribution to runoff is becoming more tractable. Similarly, these new high-quality data enable the relationship of precipitation to topography to be examined in more detail (Bookhagen and Burbank, 2010).

Every summer, for example, we expect that monsoon rains will soak the southern flank of the Himalaya. But, is the distribution and relative intensity of monsoon rainfall predictable? To answer this question, we can now exploit a

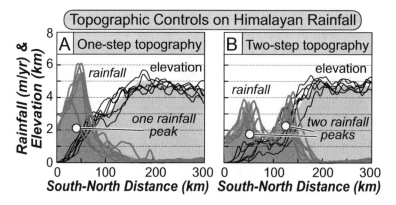

Fig. 10.32 **Topographic controls on monsoonal rainfall in the Himalaya.**
Topographic and rainfall data are compared from swaths that are 300 km long and 50 km wide, with the long axis oriented perpendicular to the strike of the range. Data for either rainfall or elevation are averaged across each 50-km-wide swath. A. Data from six swaths in which topography rises steadily from the lowlands to the high peaks. A single peak of high rainfall occurs near the front of this topographic ramp. B. Data from six swaths in which topography rises in an initial low step (1–2 km high) followed by a larger step (2–4 km high) farther into the range. Monsoon rainfall produces a peak associated with each topographic step. Note that the total amount of rainfall in each swath is about equal, but its spatial distribution depends strongly on topography. Modified after Bookhagen and Burbank (2010).

decade of remotely sensed rainfall throughout the Himalaya. For example, along the southern Himalaya, monsoon rainfall commonly displays either a single, range-parallel band of high rainfall or two such bands (Bookhagen and Burbank, 2006) (Plate 10). Analysis of the topography associated with these single or dual bands of rainfall reveals two clear end-members (Fig. 10.32) that relate broad topographic characteristics to rainfall patterns. In particular, where the range-front topography rises as a steadily climbing ramp, a single peak of high rainfall exists. In contrast, where the Himalayan front is defined by two topographic steps (generally separated by a broad expanse of Lesser Himalaya of nearly uniform altitude), two bands of high rainfall prevail.

Contrary to some expectations, zones of high rainfall commonly do not spatially coincide with the highest topography. Instead, these zones are typically offset upwind of a range crest by several to many kilometers, particularly for the unusually high topography of the Andes and Himalaya. In fact, given the typical moisture and wind conditions, the peaks of high rainfall can be confined to rather discrete altitude ranges that may lie several kilometers below the height of the summits that define the range crest (Fig. 10.33).

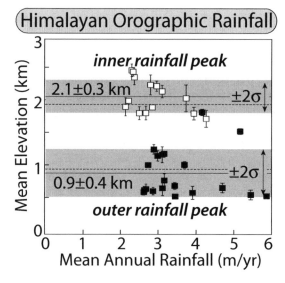

Fig. 10.33 **Elevation control on peaks of Himalayan monsoon rainfall.**
Along the southern Himalayan topographic front, peaks of high rainfall typically lie within discrete altitudinal bands. Commonly, these peaks are offset many kilometers upwind of the highest topography and lie several kilometers upwind of the topographic steps that drive the rise, cooling, and condensation of moist air masses that produce these peaks (see Fig. 10.32). Modified after Bookhagen and Burbank (2006).

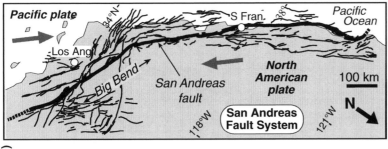

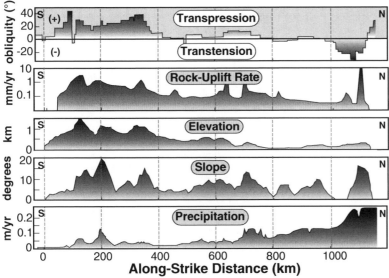

Fig. 10.34 Topography, dynamics, and rainfall along the San Andreas Fault system.

Along ~1000 km of the San Andreas system in California, obliquity of the fault trend with respect to directions of plate motion defines transpressional and transtensional regimes. Other data are averaged from an 80-km-wide corridor centered on the fault. Rock-uplift rates are derived from thermochronological data and an estimated surface-uplift rate. These rates generally correlate with stronger transpression and higher elevation. Modified after Spotila *et al.* (2007).

Latitudinal gradients in climate and tectonics

Elongate, north–south oriented ranges can span latitudinal bands with contrasting climate regimes. Prevailing winds can change from easterly to westerly flows, the wet side of a range may switch from east to west, and local climates can vary from desert to rainforest conditions. The Cordillera of the Americas provides the best modern examples of ranges that span broad latitudinal zones. Hence, it is reasonable to ask which aspects, if any, of tectonic, topographic, and climatic trends correlate along these ranges.

One might expect that along strike-slip faults, correlations between climate and tectonics would be weak. In fact, along the San Andreas Fault in California, tectonics appears more important than climate in driving topographic change (Fig. 10.34). Data averaged along the

length of the fault show that the higher rock uplift rates, steeper slope angles, and higher elevation tend to correlate with higher rates of transpression, whereas rainfall appears largely inversely correlated with elevation (Spotila *et al.*, 2007) – the opposite of the common linkage of orographic rainfall to topography. In general, climate appears largely decoupled from local topography and tectonics along this very active strike-slip system.

The Andes traverse 60° of latitude, which includes several climate zones dominated by either easterlies or westerlies. The western flank of the Andes encompasses both tropical rainforests and hyperarid deserts. Despite the overall continuity of subduction beneath the western Andes, we might expect major, along-strike differences in sediment fluxes, erosion rates, and perhaps tectonics as a result of this climate zonation. A recent synthesis of Andean

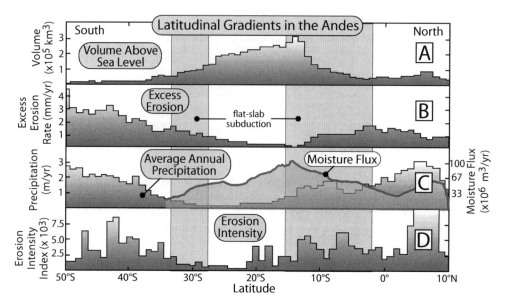

Fig. 10.35 Latitudinal gradients in Andean topography, rainfall, and erosion metrics.
A. Volume of topography above sea level in 1° latitudinal bins. B. Excess erosion (see text for explanation).
C. Average annual precipitation and integrated moisture flux – the latter from Bookhagen and Strecker (2008).
These cannot be directly compared because the moisture flux must be divided by area to give a rainfall rate.
D. Erosion intensity index derived from local slope times discharge, calculated as the product of the upstream
drainage area and its average rainfall. Modified after Montgomery *et al.* (2001).

topography and climate (Montgomery *et al.*, 2001) suggests that latitudinal variability is correlative among several facets of the Andes (Fig. 10.35). According to this analysis, the volume of the Andes above sea level is anticorrelated with annual rainfall: high preserved volumes of rock occur where rainfall is low. Assuming that convergence has been steady over the length of the subduction zone, a comparable amount of rock is assumed to have been added per increment of time along each increment of the Andes. This along-strike similarity underpins the calculation of an excess erosion rate that is a function of the volume deficit within a given 1° bin when compared with the bin with the largest volume (and presumably the least erosion). The excess erosion broadly correlates with rainfall (Fig. 10.35). An erosion intensity index was derived as a rough approximation of stream power based on the average rainfall, upstream drainage area, and average slope. This index also broadly correlates with excess erosion and anticorrelates with volume above sea level. Montgomery *et al.*'s

(2001) analysis also suggested that differences in zonal precipitation (easterlies versus westerlies) shifted drainage divides away from moisture sources and produced large-scale topographic asymmetries across the Andes.

One potential drawback of Montgomery *et al.*'s (2001) analysis is its reliance on local rain gauges to define precipitation variations. Where few stations are present, data must be interpolated across long distances. When these data are compared with spatially continuous, but remotely sensed, rainfall data (Bookhagen and Strecker, 2008), the integrated moisture flux into the Andes differs significantly from the pattern of average annual precipitation (Fig. 10.35). Whereas part of the difference is because the flux needs to be divided by area to give a rainfall estimate, part is due to very different estimates of the actual amount of rainfall. Such differences caution against over-reliance on data from widely scattered weather stations, but also indicate the need for remotely sensed data to be well calibrated against measured climate variables.

Given the latitudinal climatic gradients along the length of the Andes, it is not surprising that the amount of sediment delivered to the offshore trench varies significantly along strike. In general, where arid climates prevail, the trench is sediment starved, whereas in regions with more erosive climates, the trench may hold sediment more than 2 km thick. Because sediments contain abundant fluids and because fluids cannot support shear stresses, interplate stresses within subduction zones are proposed to be low where sediment thicknesses are high, and vice versa (Lamb and Davis, 2003; Melnick and Echtler, 2006). In addition, thick blankets of sediment may mantle irregular sea-floor topography and, thereby, reduce the stresses that the uneven topography would produce in the absence of sediments (Fisher *et al.*, 1998). Along the length of the Andes, the maximum range height and the depth of the sea floor in the offshore trench are correlated (Fig. 10.36A). In general, trench depth is a function of the thickness of sediment fill, such that a shallower trench is associated with thicker fill.

If the lithosphere of the high Andes is weak, as several tectonic characteristics suggest (Lamb and Davis, 2003), then the potential energy difference between the range crest and the trench (the buoyancy stress contrast) must be partially supported by lateral push from the subducting plate. Calculations of the shear stresses on the plate interface that would be required to balance the buoyancy stress contrasts (Fig. 10.36B) suggest that, where these rise above 35 MPa and 100 MPa, respectively, the trench fill is thin (<500 m) or absent. If this model is correct, then the aridification of the western Andes has important consequences for the growth of the range. When such aridification occurred is poorly documented, but it may be associated with the mid-Miocene growth of the Antarctic ice sheets and the strengthening of the cool Peru–Chile Current along South America's western coast (Lamb and Davis, 2003).

The role of sediment in the southern Andean trench has also been invoked to explain a major change from forearc extension and subsidence to forearc uplift, inversion, and shortening (Melnick and Echtler, 2006). Here, offshore of

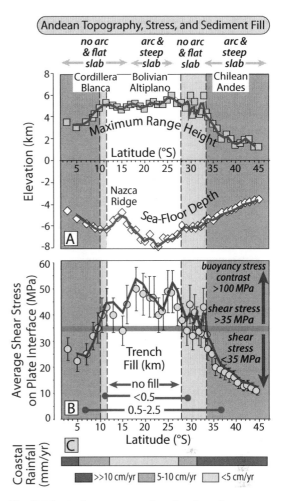

Fig. 10.36   Andean topography, plate-interface stresses, trench fill, and climate.
A. Maximum range height and sea-floor depth are well correlated along the length of the Andes. B. Average shear stress on the plate interface required to balance the buoyancy stresses due to the contrast in potential energy between the High Andes and the trench. Calculated shear stresses >35 MPa and buoyancy stress contrasts >100 MPa occur where trench fill is thin or absent. The mean is corrected for the relative plate convergence direction. C. Latitudinal zonations of coastal rainfall. Low rainfall generally coincides with thin trench fill. Modified after Lamb and Davis (2003).

the Patagonian Andes, extensive glaciation during the past 5 Myr has dumped a large quantity of sediment into the trench, where seismic data reveal an increase in sediment thickness from <1 km to >2 km since 5 Ma. Many of these water-rich sediments have either been accreted to the forearc, causing a reduction in taper of the Andean wedge, or been subducted, where their fluid-rich character would be expected to reduce interplate stresses. The net result has been uplift of the forearc basin by ~1.5 km during the past 3 Myr. Interestingly, the thick trench fill only occurs south of the topographic high (the Juan Fernández Ridge), which acts as a barrier to northward sediment transport. North of this barrier, sediment thicknesses in the trench are less than 0.5 km, subsidence is ongoing, and extension is active (Melnick and Echtler, 2006). This contrast in tectonic behavior as an apparent function of sediment thickness suggests that climatically driven changes in sediment supply can strongly impact the dynamics of forearcs and coastal regions.

The glacial buzzsaw

The "glacial buzzsaw" hypothesis proposes that glacial erosion tends to bevel off topography near the snowline (Fig. 10.37), such that trends in topography will mimic snowline trends (Brozović et al., 1997). Because the flux of glacial ice is generally highest near the glacial equilibrium line (the long-term snowline), and because glacial erosion rates are thought to scale with ice flow rates (Humphrey et al., 1986), enhanced erosion in the vicinity of the equilibrium line (MacGregor et al., 2000) could be expected to lower landscapes toward the snowline and flatten the topography in this zone. If correct, this scenario leads to the surprising conclusion that climatic gradients can play a more important role in controlling steady-state topography at the regional scale than do variations in tectonic forcing (Brozović et al., 1997).

Support for the glacial buzzsaw hypothesis is found in both the rapidly deforming northwest Himalaya (Fig. 10.38) and the slowly deforming Cascades of Washington (Mitchell and Montgomery, 2006). In both areas, the mean topography and snowlines follow similar topographic gradients that are independent of known tectonic gradients. Comparisons of mean elevations (Fig. 10.38) with the cooling and/or denudation data for the northwest Himalaya (Fig. 10.29) clearly show that zones of rapid rock uplift do not

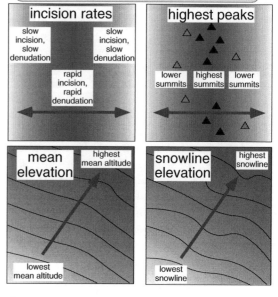

Fig. 10.37   Glacial buzzsaw: elevation follows snowline gradients.
Generalized relationships in landscapes where tectonics and topography are decoupled, and climate and topography are coupled. (Top) Highest incision and denudation rates correspond with the zone of the highest peaks. In this cartoon, the tectonic gradient is oriented east–west. (Bottom) The gradient of mean elevation trends at a high angle to the tectonic and/or erosion gradients, but is parallel to the snowline gradient.

correspond with zones of high mean elevation, although these rapidly uplifting zones do contain many of the highest summits. The observation that the regional topographic trend lies nearly orthogonal to the zonation of tectonic rates in the northwest Himalaya forces one to conclude that the mean topography is quite independent of variations in tectonic forcing at this scale (Fig. 10.37). Although modern Himalayan snowlines sit about 700–1000 m above the height of the mean topography (Brozović et al., 1997), the snowline and the mean topography during the last glaciation would have nearly coincided, thereby lending further support to the basic buzzsaw hypothesis.

Detailed studies of ranges that encompass contrasting rates of rock uplift, however, do suggest that topography is not completely independent of tectonics (Brocklehurst and Whipple, 2007). The most obvious effects in zones of

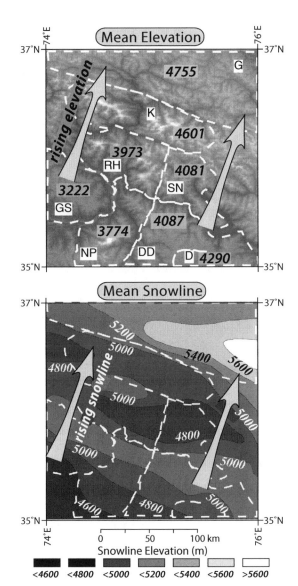

Fig. 10.38   Mean elevations and mean snowlines in the northwest Himalaya.

(Top) Mean elevation (meters) of subregions in the northwest Himalaya. D: Deosai Plateau; DD: dissected Deosai; NP: Nanga Parbat; RH: Rakaposhi–Haramosh; SN: Skardu north; K: Karakoram; G: Ghujerab. (Bottom) Elevation of modern snowline in the same region as shown above. Note that snowlines also rise toward the northwest and generally lie ~700–1000 m above the mean elevations. The topographic and snowline gradients trend at nearly right angles to the tectonic gradients in this region (Fig. 10.28). Modified after Brozović *et al.* (1997).

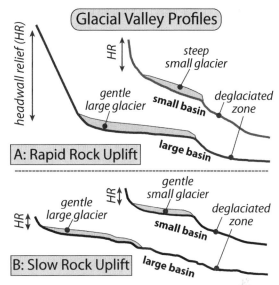

Fig. 10.39   Glaciated topography as a function of rock-uplift rate.

Glacial valley profiles resulting from average Quaternary conditions: ~400 m of snowline lowering. A. Glacial valley profiles under rapid rock-uplift regimes. Both headwall relief and the gradient of small glaciers increase, whereas large glaciers retain their low gradients. B. Under slow rock-uplift regimes, both large and small glaciers have gentle gradients and headwall relief is modest. Modified after Brocklehurst and Whipple (2007).

rapid rock uplift include significant increases in the height of cirque headwalls, a modest rise in mean elevation (due largely to the headwall lengthening), and a steepening of moderate- to small-sized glaciers (Fig. 10.39).

Even in rapid uplift zones, large glaciers retain low gradients, suggesting that they erode effectively irrespective of rock-uplift rate. At least two factors could serve to enhance glacial erosion in regions of rapid rock uplift. First, these areas often contain very high peaks, even if the mean elevation is not particularly high. Such peaks have been termed "topographic lightning rods" because of their role in causing more precipitation on their flanks (Brozović *et al.*, 1997). Greater ice fluxes should lead to more rapid erosion. Second, although one's eyes are attracted to the high summits, only a few percent of the total topography will typically occupy the upper 10–20% of the range in elevation. For example, in the Himalaya, our

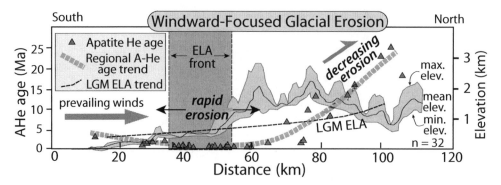

Fig. 10.40   Enhanced glacial erosion on the upwind flank of the St. Elias orogen, southeast Alaska.
Young (U–Th)/He ages in a zone between 35 and 60 km on the upwind flank of the St. Elias orogen imply rapid erosion that coincides spatially with the position of the glacial-age equilibrium line. The observation that the deformation front has propagated outward 10–15 km during the past 800 kyr appears inconsistent with the contention that enhanced glacial erosion caused a contraction of the zone of deformation, as might be expected from theory. Modified from Berger and Spotila (2008).

eyes are attracted to the 7- to 8-km high peaks, but only 2–4% of the topography lies above 6 km. Most of the land surface is concentrated near the snowline. As a result, if rock uplift drives surface uplift, more of the land surface is pushed into the accumulation zone for the glaciers. This enhanced accumulation zone, too, will enhance the ice flux and, hence, glacial erosion (Brozović *et al.*, 1997). This response is an interesting example of a negative feedback in which an increase in mean elevation creates the conditions that should lead to a lowering of the mean elevation through enhanced erosion.

In the St. Elias orogen of southern Alaska (Fig. 10.40), a zone of rapid erosion has been delineated based on apatite (U–Th)/He cooling ages (Berger *et al.*, 2008; Berger and Spotila, 2008). This zone correlates spatially with the position of the glacial equilibrium line during the Last Glacial Maximum and is interpreted as a response to repetitive snowline lowering during Middle to Late Pleistocene time. As a consequence, enhanced glacial erosion has been invoked as positive feedback that locally accelerates the rate of rock uplift. Moreover, this region has been cited as an example of an orogen where a climatically modulated increase in erosion caused a reduction in the cross-sectional area of the orogen (Fig. 10.2) (Berger *et al.*, 2008; Whipple, 2009). Such a reduction implies that the leading

edge of deformation retracts during the interval of enhanced erosion (Davis *et al.*, 1983). Although the leading edge of deformation can be difficult to define in the past, thickening and thinning of rather well-dated growth strata in the offshore seismic stratigraphy from this region (Bruns and Schwab, 1983; Meigs *et al.*, 2008) actually suggest that thrust sheets have been steadily propagating outward and the orogenic belt has been widening, rather than contracting as speculated, during this Middle to Late Pleistocene interval of rapid glacial erosion (Meigs, 2010). This apparent contradiction of a modern paradigm provides a cautionary note against attempts to generalize both orogen behavior and climate–tectonic interactions on the basis of modern snapshots of cooling ages and snowlines in the absence of careful attention to spatial and temporal reconstructions of the structural history of an orogen.

One persistent, but unresolved, problem relates to the rate at which glaciers can actually erode bedrock. Is it as fast as or faster than rivers? The fact that moving ice covers the bed of glaciers makes it difficult to observe and calibrate erosion processes at the glacial bed. A qualitative answer, however, can be derived simply by examining the topography of glaciated and non-glaciated regions. At least two observations suggest that glaciers erode bedrock at least as fast as rivers do. First, when side-by-side

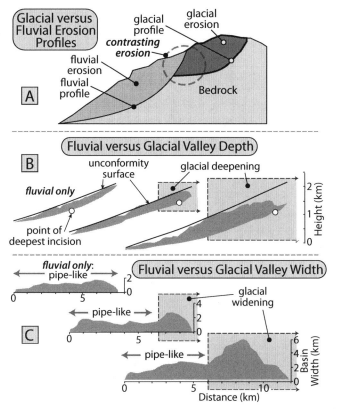

Fig. 10.41   Fluvial versus glacial erosion.
A. Cartoon of generalized long profiles for valleys with and without former glaciers in their headwaters. The glaciated valley is typified by a break in slope to a much gentler gradient in its upper reaches. This low-gradient zone delineates the contribution from glacial erosion that is absent in a catchment that lacks former glaciers. B, C. Examples of fluvial versus glacio-fluvial valleys from the Kyrgyz Range, where an unconformity surface serves as a geomorphic marker against which to calibrate erosion. The point of deepest incision moves up-valley and is significantly deeper for the more glaciated catchments. The fluvial parts of the valleys are pipe-like: uniform width along their length. Such uniformity is not uncommon when parallel drainages develop on steep slopes. Large-scale widening of the valleys occurs within their glacial portions. Modified after Oskin and Burbank (2005).

catchments are compared, one that was glaciated in its upper reaches and one that was not, the shape of the fluvial parts of the long profile of each valley will match, but the upper part of the catchment that once held a glacier will show a significant drop in gradient in comparison to the equivalent part of the non-glaciated valley (Brocklehurst and Whipple, 2002). The gradient change represents a divot of bedrock that was removed by the glacier, but whose equivalent was not eroded by the river system in the adjacent valley (Fig. 10.41A). Geomorphic studies of glaciated valleys for which a reference frame exists, such as a pre-existing erosion surface (Oskin and Burbank, 2005; Small and Anderson, 1998), show very efficient headwall retreat and valley widening by glaciers in comparison to nearby fluvial valleys (Fig. 10.41B and C).

The second observation that supports the greater or equal celerity of glacial erosion compared to rivers is the fact that, in many rapidly eroding ranges, even though glaciers are at the smallest size that they have been for perhaps nearly a million years, they rest in the bottom of deep, steep-walled, high-relief, and typically low-gradient valleys that could only have been carved by ice in the past, given the currently retracted size of the glaciers. Compared to nearby river valleys, the previously glaciated valleys almost always sit lower in the landscape, a testament to the efficiency of glacial erosion in the past.

Whereas glaciers may erode as fast as or faster than rivers, these rates are only true for temperate or warm-based glaciers that are sliding on their beds because they are at the pressure-melting point. Cold-based glaciers deform internally, but do not slide. As a consequence, they do not erode the bedrock beneath them, and no glacial buzzsaw exists where the glaciers are frozen to their beds.

Steady-state topography assumes a different form and time scale in rapidly uplifting ranges that are sufficiently high to contain cold-based glaciers (Fig. 10.42). For example, peaks mantled by cold-based glaciers do not erode. Instead,

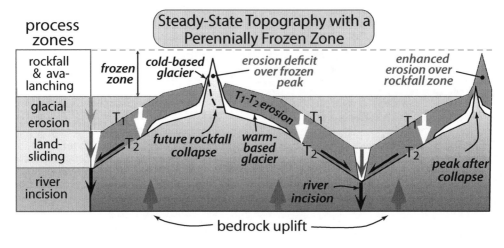

Fig. 10.42 Oscillating steady state in the presence of cold-based glaciers.
Process zones (left side) of river incision, landsliding, and glacier erosion are the same for ranges with and without glaciers that are frozen to their beds. Where cold-based glaciers exist, glacial erosion drops to zero. Bedrock erosion in the frozen zone occurs primarily due to episodic large rockfalls.

they grow upwards at the same rate as rock uplift occurs beneath them. Farther down the slopes of the mountains, the climate may be sufficiently warm to permit warm-based glaciers to exist. As described above, such glaciers are likely to be capable of eroding as rapidly as the rocks are uplifting. The topographic consequence of the dichotomy between rapid erosion that is focused low on the slopes versus no erosion on the summits is similar to extruding a steeple that keeps growing taller. Eventually, the summit steeple on a mountain will exceed the strength of the rocks beneath it and collapse in a large rockfall. The time between successive summit collapses provides a rough time scale for the fitful oscillations around steady state of a range with frozen summits (Fig. 10.42).

Mismatches between tectonics, topography, and climate

Although many studies have shown spatial correlations between zones of high modern precipitation and zones of young cooling ages and, hence, inferred rapid, long-term erosion (e.g., Reiners *et al.*, 2003; Thiede *et al.*, 2004), sometimes we can learn more by studying mismatches between climate and various proxies

for past erosion. One such example comes from the central Himalaya, where a pronounced gradient in monsoonal rainfall appears decoupled from the long-term erosion rates (Burbank *et al.*, 2003). Whereas a 10-fold decrease in rainfall occurs across the Himalaya, apatite fission-track ages (Blythe *et al.*, 2007) appear quite uniformly young across the entire monsoon gradient (Fig. 10.43). This long-term gradient also contrasts with a four-fold northward decrease in modern erosion rates as deduced from river sediment loads (Gabet *et al.*, 2008).

This apparent mismatch of long-term rates with modern rainfall forces one to ask why erosion would be just as fast in a rather dry area (<40 cm/year of rain) as in a very wet area (>4 m/year). Let us accept that the data were correctly measured, so that these differences are real. Is it possible that the weather data, which were measured over six years, are not valid for the long term? Certainly, the magnitude of rainfall changes from year to year (see inset, Fig. 10.44), but the physics of the way storms interact with topography is constant, so the basic gradient in average rainfall should be persistent through time.

Rapid erosion tends to be event driven, especially in non-glaciated landscapes. Hence, we might ask whether the gradient in mean

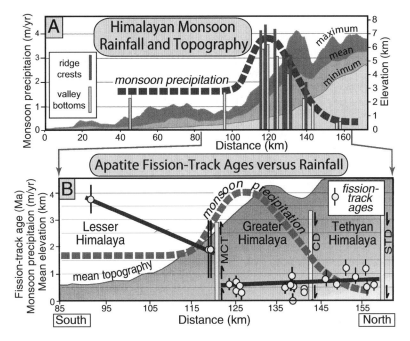

Fig. 10.43 Modern climate, topography, and cooling ages in the central Himalaya.
Data from the Marsyandi catchment in central Nepal. A. Weather station data define a 10-fold gradient in monsoon rainfall, with the maximum rainfall occurring ~15 km from the highest topography and on its upwind flank. Apatite fission-track cooling ages (closure temperature ~ 140°C for very rapid cooling) display a large offset across the Main Central Thrust (MCT), but no significant gradient to the north, where ages adjacent to the South Tibetan Detachment (STD) are about as young as those adjacent to the MCT. CD: Chame detachment. Modified after Burbank *et al.* (2003) and Blythe *et al.* (2007).

monsoonal rainfall is typical of spatial gradients in individual storms. Analysis of storm data shows that a difference does exist: large storms penetrate farther into the range and deliver proportionately greater amounts of rainfall (Craddock *et al.*, 2007). As a result, the north–south gradient in rainfall delivered by large storms is only four-fold across the range, rather than the 10-fold annual gradient. But, even a four-fold gradient is much greater than the gradient in cooling ages.

Most of the Marsyandi cooling ages, especially in the north, come from valley bottoms. Are these ages representative of mean bedrock erosion rates? Young fission-track ages like these tend to have large (≥50%) uncertainties that could mask spatial gradients. Valley bottoms are sites where hydrothermal fluids are concentrated, and a higher flux should create steeper geothermal gradients (Derry *et al.*, 2009). The presence of hot springs along the Marsyandi and the theory

behind fluxes of hot water suggest perhaps that the ages are too young, although no clear reason exists to suspect that the northern ages are more perturbed by hydrothermal fluxes than are the southern ages. Moreover, relief transects of ages that rise far above the valley bottom in the southern and mid parts of this study area (Blythe *et al.*, 2007; Whipp *et al.*, 2007) suggest that long-term erosion rates could be as fast or faster in regions with less annual rainfall and smaller storm rainfall. Additional relief transects in the arid north, preferably using thermochronometers with higher resolution, are needed to confirm the proposed climate-erosion decoupling.

How might topography change in order to maintain a persistent erosion rate, despite a decrease in rainfall? Given that erosion rates in this area far exceed documented rates of soil production, most of the hillslope erosion is driven by landsliding. Such slides occur when

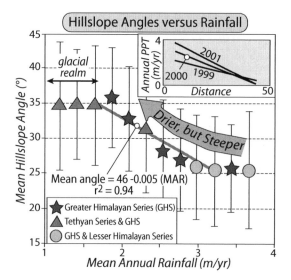

Fig. 10.44   Correlations of mean annual rainfall with hillslope angles in the central Himalaya.
As rainfall decreases from 3.3 to 1.5 m/yr, mean hillslope angles increase about 10°. These changes appear independent of rock type, but the rainfall–hillslope correlation does not persist through the glacial realm. Inset shows rainfall gradients over three years. Data from the Marsyandi catchment in central Nepal. Modified after Gabet *et al.* (2004a).

pore pressures on steep slopes are driven above a local threshold for stability by storms (Gabet *et al.*, 2004b). So, as the climate becomes increasingly dry, slopes might get progressively steeper, such that the critical pore pressure could be reached with lesser rainfall amounts. An increase in average slope angles in the Marsyandi occurs concurrently with the decrease in rainfall (Gabet *et al.*, 2004a) (Fig. 10.44).

A (perhaps) final possibility is that glacial erosion in the dry realm during colder climates compensates for slower erosion there during interglacial climates, such as today's. Ice-age glacial cover was much greater north of the Himalayan crest than south of it. Whereas most of the southern Himalayan flank is dominated by fluvial erosion and hillslope processes even during glacial times, expanded glaciers and periglacial weathering in the north during glacial times may compensate for the reduced erosion that occurs when climate is warmer and glaciers are retracted, as is the case today. This

scenario suggests that erosion rates experience much greater oscillations in the northern, drier, glacial areas than in the southern, wetter, fluvial regions, but that the integrated rates over multiple climate cycles may be comparable.

Certainly more research is needed to document how erosion rates change during glacial times and whether different segments of an orogen erode at different rates under glacial versus non-glacial climates. Concurrently, we need to understand better how snowline gradients change in different climate regimes, how storm tracks vary and how moisture enters into ranges as climate changes, what geomorphic thresholds modulate erosion fluxes, and how to best interpret cooling ages in terms of erosion rates as topography evolves.

Dynamic topography

Dynamic topography represents topographic anomalies that result from subsurface loads or lithospheric stresses that raise or depress topographic surfaces with respect to the geoid. Commonly, these are long-wavelength anomalies that may span hundreds of kilometers, and their effects are usually subtle when viewed at decadal scales or short wavelengths. Overall, dynamic topography produces some spectacular geomorphic effects that were not well recognized until recently.

Geophysical studies in the southern Sierra Nevada of California have identified a "drip" of high-density, mantle lithosphere that has detached from beneath the Sierras and is sinking into the asthenosphere (Zandt, 2003). The downward pull of this excess mass is causing subsidence of the overlying Central Valley (Fig. 10.45) (Ducea and Saleeby, 1998; Saleeby and Foster, 2004). As sediments continue to aggrade in the Central Valley, they onlap the subsiding bedrock and create a classic "drowned" topography, although, in this case, the drowning is caused by sediment aggradation, rather than by relative sea-level rise. The zone of onlap corresponds spatially with the approximate extent of the "drip" and includes topographic anomalies, such as Tule Lake (Fig. 10.45A), which lies in a closed depression in what was originally a longitudinally drained valley that fed San Francisco

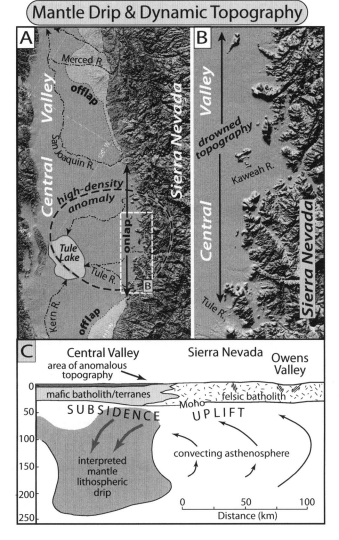

Fig. 10.45 Dynamic topography associated with a mantle drip in the southern Sierra Nevada, California.
A. Topographic overview of the Central Valley and adjacent Sierra. Tule Lake lies in a closed depression that is generated by subsidence in response to a lithospheric drip. Sediment onlap dominates above the drip, whereas offlap predominates elsewhere. B. Drowned topography along the Sierra foothills where steep "islands" of bedrock rise abruptly above the aggrading basin floor. C. Schematic cross-section of subsidence in the Central Valley and uplift of the Sierra that is driven by the mantle drip. Modified after Saleeby and Foster (2004).

Bay. This closed depression clearly indicates that, today, the pace of subsidence above the drip is outstripping the rate of sediment delivery from the Sierra. Both north and south of the anomaly, more typical offlap of Holocene sediment prevails, as is expected given the general late Quaternary decrease in sediment flux from the Sierra.

At the scale of tectonic plates, subduction of dense lithosphere tends to flex plates downward, whereas upwelling and outward push at spreading ridges tends to flex them upwards. The geoid, in contrast, should be high-standing above the denser lithospheric bodies associated with subduction zones. A spectacular example of flexure at the plate scale is found in the coastal record of Australia (Sandiford and Quigley, 2009) (Fig. 10.46). The Indo-Australian Plate is bounded on the north and east largely by subduction zones and on its south and west by mid-ocean ridges. The down-to-the-northeast pull and up-to-the-southwest push of dynamic lithospheric stresses exerted by these boundaries are interpreted to cause the plate to tilt down to the north-northeast. The topographic consequence of this tilt is that the southern part of Australia is emerging, whereas the northern part is sinking. The geomorphic manifestation of this long-term tilt is that the 15-Ma Miocene

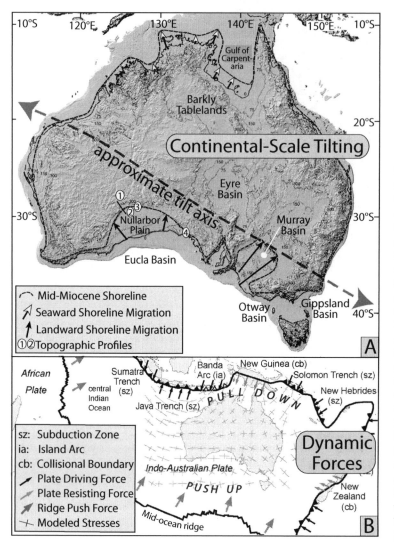

Fig. 10.46 Dynamic topography and continental-scale tilting of the Australian Plate.
A. Asymmetric mid-Miocene shoreline records provide clear evidence for differential tilting of the Australian continent around a southeast-trending axis. The shoreline is displaced up to 300 km inland and 250 m above sea level in southern Australia, but lies up to 200 km offshore in northern Australia. Spectacular geomorphic examples of displaced Neogene shorelines are from the northern edge of the Nullarbor Plain, where numbers delineate the topographic profiles of Fig. 10.47. B. Geodynamic forces on the Indo-Australian Plate margins. Primarily subduction and/or collision boundaries on the north and east "pull" the plate downward. Spreading ridges on the south and west "push" the plate upward. Together, these forces drive a down-to-the-north tilt of the plate.

shoreline in southern Australia sits as much as 300 km inland, whereas it lies up to 200 km offshore in northern Australia. These offset shorelines record ~300 m of north-down tilt at a rate of ~20 m/Myr (Sandiford and Quigley, 2009). The observation that last interglacial shorelines appear differentially offset across the continent by about 2 m (Murray-Wallace and Belperio, 1991) suggests that the tilting today is being sustained at the average Neogene rate.

The broad Nullarbor Plain and Murray Basin in southern Australia (Fig. 10.46) provide some of the clearest evidence for abandoned shore-lines and for progressive long-wavelength tilting (Sandiford, 2007). Remarkably flat shoreline platforms that are 25–40 km wide, yet contain only 5–15 m of relief, delineate the Eocene and mid-Miocene shorelines (Fig. 10.47A). The mid-Miocene shoreline marking the northern limit of the Nullarbor Plain trends inland to the north-northwest for over 700 km and rises from about 50 m to over 250 m above sea level across that distance (Fig. 10.47B). With respect to the recon-structed tilt axis, this transect lies farther away from the axis as it moves inland (Fig. 10.46), such that the higher mid-Miocene shoreline that

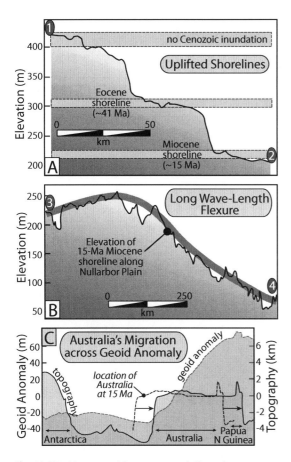

Fig. 10.47 Topographic transects delineating deformed shorelines and the geoidal anomaly associated with the Antarctic–Australian region. A. Topographic profile of uplifted Cenozoic shorelines along transect 1–2 (Fig. 10.46). Two broad, low-relief shoreline platforms are clearly delineated. B. Transect 3–4 (Fig. 10.46) along the mid-Miocene shoreline. The shoreline's greater elevation in the western (left) part of this 700-km-long transect is due to its position farther south of the tilt axis. C. Northward migration of the Australian continent causes it to traverse the geoidal anomaly associated with the subduction zones along the plate's northern margin and results in marine inundation of northern Australia.

prevails farther from the coast is consistent with the overall continental-scale tilting.

Some of the inundation of northern Australia can be explained by its steady motion towards a geoidal high over the past 15 Myr (Fig. 10.47C). This shift could explain about 40 m of the flooding of northern Australia, while no significant change of geoidal height is predicted for southern Australia during this interval (Sandiford, 2007). Overall, changes in the geoid that occur as a subduction zone is approached will be superimposed on any large-scale tilting that is concurrently driven by the dynamic stresses and resultant topography associated with subduction zones and spreading ridges. The late Cenozoic tilting of Australia provides an illustrative example of continental-scale dynamic topography for which the geomorphic record provides some of the most compelling evidence.

Sustained, large-scale continental collisions typically produce significantly thickened crust. A combination of conductive and radioactive heating of the thickened crust can induce partial melting and, therefore, significant weakening of the lower part of the crust. As discussed earlier (Fig. 1.9), such a condition has been deduced for much of the Tibetan Plateau (Beaumont et al., 2001; Nelson et al., 1996).

What might be the geomorphic and tectonic manifestations of a thick crust with a weak lower layer? First, given Tibet's great height (average elevation of 5 km; Fielding et al., 1994) with respect to surrounding lowlands, a strong potential energy gradient coincides with the plateau's margin (Hodges, 2000). In combination with a warm, ductile lower crust, this energy gradient might be expected to drive outward crustal flow, especially of the lower crust, towards the lowlands (Fig. 1.9). Such flow could create a topographically descending ramp off the plateau margin. Second, if the crust in the lowlands varies in strength, the ease with which lower crustal flow would occur might be expected to vary as a function of that strength: stronger rocks would impede flow (Clark et al., 2005a). Such impedance could cause relative uplift on the edge of the plateau and a steeper topographic gradient at the plateau's margin (Fig. 10.48). Third, the wavelength and magnitude of any crustal deflection created in response to a discrete structure, such as a normal fault, provides an indication on the depth of compensation: shorter wavelengths reflect shallower compensation,

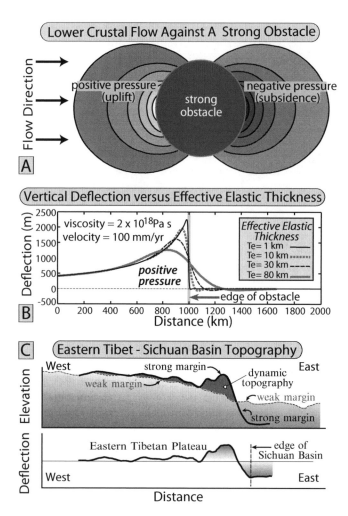

Fig. 10.48   Impact of a strong crustal obstacle on lower crustal flow at a plateau margin.

A. Strong obstacle causes positive pressure on the upstream side of the flow and negative pressure on the downstream side, particularly in comparison to adjacent regions that lack strong obstacles. B. Upward (downward) deflection is predicted on the upstream (downstream) side as a response to the pressure gradient. The magnitude and wavelength of deflection depends on the effective elastic thickness (T_e). C. Example of topographic and deflection profiles across an inferred strong versus weak (or normal) margin in eastern Tibet. Both the topography and deflection mimic that predicted for a strong obstacle to flow. The inferred "weak" margin reveals a steadily descending ramp.

and vice versa (Fig. 10.49). Fourth, where surface erosion at the plateau margin is intense, the ductile, lower crustal channel could be drawn toward the surface (Fig. 1.9C) and active faults would likely bound such a channel (Beaumont *et al.*, 2001). Conversely, with less intense erosion, surface faulting might be significantly reduced, and the ductile lower crust could remain deeply buried, despite a strong topographic gradient across the plateau's margin.

Many of these characteristics are observed along the eastern margin of the Tibetan Plateau (Clark and Royden, 2000; Royden *et al.*, 1997). From the southeastern corner of Tibet, a huge, low-gradient ramp extends south and east to the lowlands (Fig. 10.50). North of the Sichuan

Basin, a second ramp extends towards the northeast. These two ramps are interpreted as zones in which lower crustal flow is channeled outward from the Tibetan Plateau. In between these ramps, the Sichuan Basin appears to be underlain by old, stiff crust that resists the interpreted outward flow of lower crust from beneath Tibet (Clark *et al.*, 2005a). As suggested by topographic profiles (Fig. 10.48C), the plateau surface appears deflected upwards adjacent to this more rigid Sichuan crust. In comparison to the Himalayan front, erosion is generally slower in eastern Tibet, and active surface faulting is considerably more subdued, although certainly is not absent – as demonstrated by the 2008 M=7.9 Wenchuan earthquake (Kirby *et al.*, 2008). The inferred pattern of outward crustal

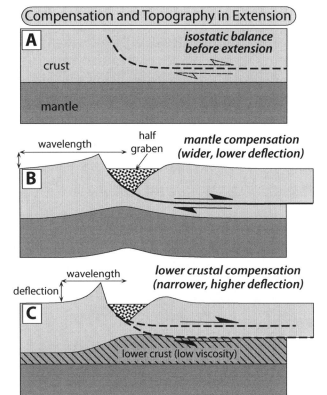

Compensation and Topography in Extension

A

crust

isostatic balance before extension

mantle

B

wavelength   half graben

mantle compensation (wider, lower deflection)

C

deflection   wavelength

lower crustal compensation (narrower, higher deflection)

lower crust (low viscosity)

Fig. 10.49 Models for compensation and topography in extensional regimes.
A. The "negative" load of a half-graben fill induces regionally compensated footwall uplift. For the same amount of extension, compensation in the mantle yields a broader, but lower, footwall uplift (B) than does lower-crustal compensation (C). Modified after Masek *et al.* (1994a).

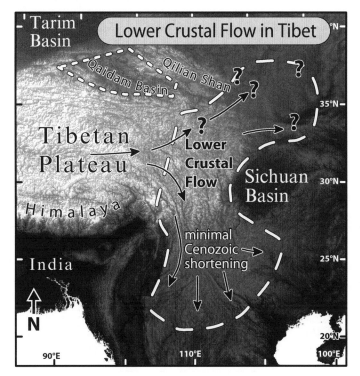

Lower Crustal Flow in Tibet

Tarim Basin

Qaidam Basin

Qilian Shan

Tibetan Plateau

Lower Crustal Flow

Sichuan Basin

Himalaya

minimal Cenozoic shortening

India

N

90°E   110°E   100°E

35°N
30°N
25°N
20°N

Fig. 10.50 Proposed lower crustal flow in eastern Tibet.
The topography of the Tibetan Plateau reveals a broad, gentle ramp to the southeast, steep topographic gradients adjacent to the Sichuan Basin, and less well-defined ramp on the north side of the Sichuan Basin. Stiff, old crust underlying the Sichuan Basin is interpreted to impede outward flow of ductile lower crust from beneath Tibet. Courtesy of Richard Lease.

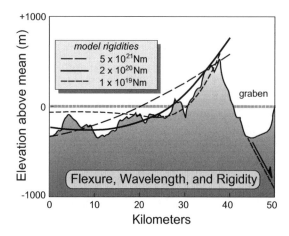

Fig. 10.51   Tibetan topography and flexure across a rift-flank uplift.
Predicted deformation is depicted for three different rigidities. Only the general trend of the topography, not the small-scale wiggles, should be compared to the curves. The mismatches between the curves and the actual topographic trends provide an estimate of the precision attainable with this approach. For these data, rigidities in the range $(2–50) \times 10^{20}$ N m provide the best match. Modified after Masek *et al.* (1994a).

flow in southeast Tibet is broadly consistent with both the regional geodetic patterns (Fig. 5.15A) and the history of regional shear inferred for the Three Rivers region (Fig. 10.20).

Although the presence across much of Tibet of a widespread, ductile lower crust remains debated, Masek *et al.* (1994a) showed 15 years ago that the deflections associated with many Tibetan grabens was most consistent with compensation associated with a weak lower crust (Fig. 10.51). The deflections associated with normal faults are largely isostatic in nature and occur because the air and sediment that fill the grabens act as a negative load (in comparison to the original, unfaulted crust). Owing to the long length scale associated with the flexural rigidity of the lithosphere, the isostatic response to this negative load is more regional than the load itself and induces uplift of the flanks of the graben (Masek *et al.*, 1994a; Small and Anderson, 1995; Weissel and Karner, 1989). Even though the Tibetan crust is 60–70 km thick in the vicinity of several of these grabens, the flexural wavelength is consistent with an effective elastic thickness of <10 km and implies that

the lower crust, rather than the upper mantle, is the zone where isostatic compensation occurs.

Shaping landscapes during orogenic growth

Outward flow of the lower crust is one of the many potential modes of orogenic growth (Fig. 10.3). Wherever erosion is both localized and intense, some crustal response is expected to the focused removal of mass. Typically, the climate–erosion link is conceptualized in the context of heavy rainfall, such as the monsoon rains of the Himalaya. But, the key is truly erosion, irrespective of the presence of a specific climatic driver. Large rivers that run through arid landscapes can potentially erode just as fast as if they were flowing through a rainforest. The key role played by the river is in its efficient removal of any combination of bedrock from beneath it and of detritus along its flanks.

The two largest river catchments in the Himalaya are those of the Tsangpo and the Indus (Fig. 10.52A). These rivers each flow parallel to the range for more than 1000 km in southern Tibet before turning south and exiting to the foreland near the eastern or western edge of the indentor formed by the Indian craton. As these rivers leave the plateau and head toward the lowlands, they enter deep, steep gorges that flow along the flank of perhaps the most rapidly uplifting massifs in the Himalaya (Zeitler *et al.*, 2001b). The foreland-veering rivers and uplifting massifs occur at what have been termed "indentor corners," and the uplifting massifs of Nanga Parbat in the west and Namche Barwa in the east have been referred to as "tectonic aneurysms" – a term meant to invoke accelerated deformation due to weakening of the confining stresses (Zeitler *et al.*, 2001a). In each case, the weakening is accomplished by the rapid incision of large rivers in the topographic "gap" adjacent to the massif (Fig. 10.52B). This incision increases the near-surface thermal gradient, weakens the crust, and creates a positive feedback that enhances the development of a crustal-scale shear zone (Koons and Kirby, 2007). Whereas the climate is moderately wet where Tsangpo flows past Namche Barwa, it is considerably drier than

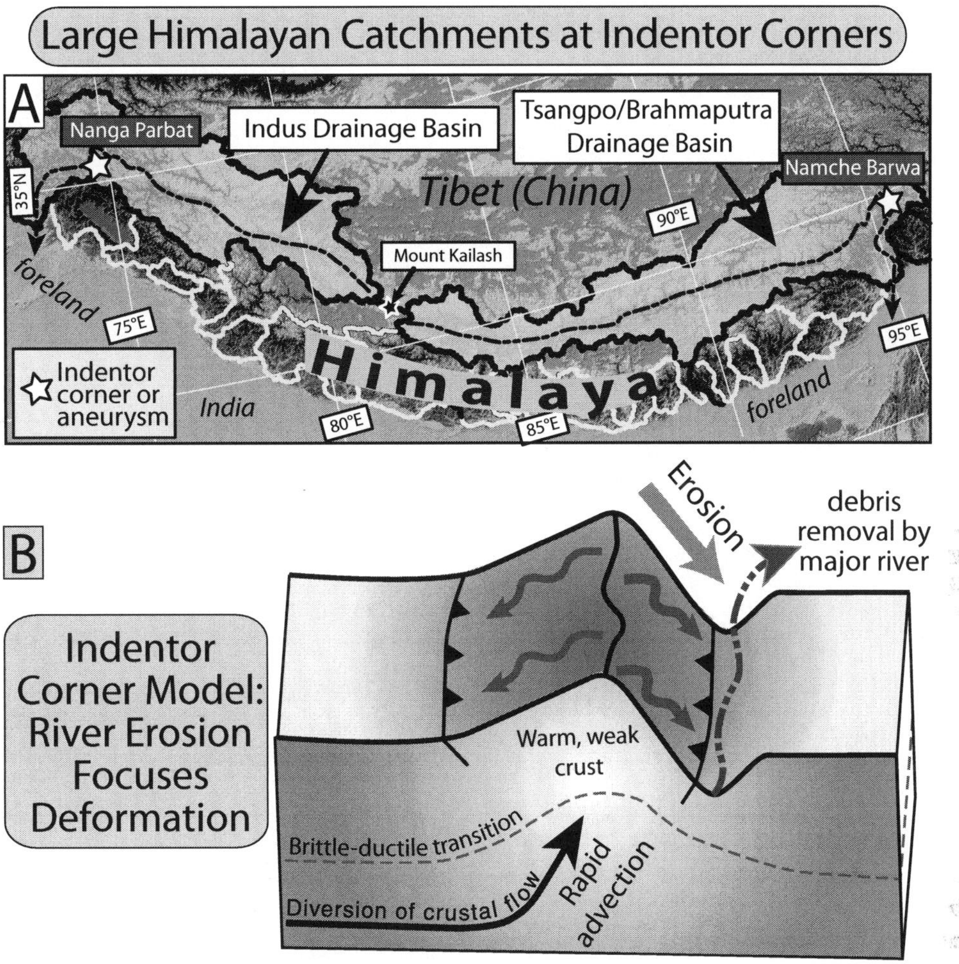

Fig. 10.52 Large Himalayan catchments and indentor corners.
A. The elongate Indus and Tsangpo catchments in southern Tibet turn southward toward the foreland near the edge of the Indian Plate indentor. Two rapidly uplifting and eroding massifs (Nanga Parbat, Namche Barwa) are localized at these indentor corners. Map courtesy of Bodo Bookhagen. B. Indentor corner or "tectonic aneurysm" model. Efficient export of rapidly eroded debris by large rivers, plus bedrock incision by the same rivers, helps to create weak, warm crust by incising crust in the topographic trough adjacent to an uplift. The weakened crust localizes shear and creates a positive feedback that helps focus deformation. Modified from Zeitler *et al.* (2001a,b).

along much of the southern Himalayan front. Yet, the erosion is particularly intense in this massif (Stewart *et al.*, 2008). The desert varnish that coats the Indus gorge adjacent to Nanga Parbat on the western side of the orogen attests to the locally arid climate. Nonetheless, the discharge from this large river effectively removes the erosional detritus generated by rapid rock uplift, while the river's bedrock incision augments the growth of the shear zone bounding the massif (Koons *et al.*, 2002). In the absence of such large

orogenic rivers, the magnitude of exhumation, the rates of rock uplift, and the localization of crustal strain at the edges of the Indian indentor would all be much reduced.

Whereas large rivers that drain orogenic uplands can focus erosion and rock uplift, glaciers can play an important role in transforming a plateau margin from a slowly eroding, low-relief landscape to one that is rapidly eroding and exhibits high relief. Recent studies of hundreds of glaciers on the margins and within the Tibetan Plateau

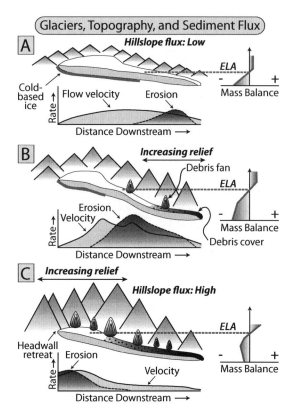

Glaciers, Topography, and Sediment Flux

Fig. 10.53　Glacial–topographic interactions.
Glacial velocity and debris cover are generalized from remote sensing of glaciers on the Tibetan Plateau. Note that the local climate is sufficiently cold that high-altitude glaciers are cold based. A. Initial glaciers are frozen to their beds in upper reaches and embedded in low-relief landscapes. Erosion is focused near the toe, and hillslope debris production is minimal. B. Erosion beneath the lower glacier steepens the glacier, increases topographic relief adjacent to the toe, and increases the hillslope sediment flux and resultant supraglacial debris in this reach. The cold-based region of the glacier shrinks and the locus of erosion and velocity migrate up-glacier. More of the glacier lies below the climatic equilibrium-line altitude (ELA) but melting is greatly reduced beneath the supraglacial debris. C. In a mature stage, glacial erosion is focused in the headwall area. Relief is high, especially in upper reaches. Extensive supraglacial debris is driven by widespread, high hillslope fluxes. Most of the glacier lies below the ELA. Modified after Scherler *et al.* (2011).

(Scherler *et al.*, 2011) suggest linkages among supraglacial debris cover, ice-flow dynamics, glacial steepness and erosion, and landscape characteristics. Glaciers in low-relief, high-altitude

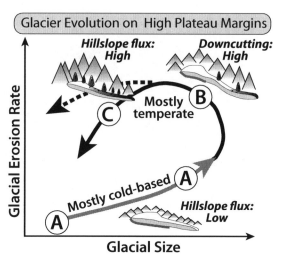

Glacier Evolution on High Plateau Margins

Fig. 10.54　Conceptual model for glacial-landscape evolution on a plateau margin.
High-altitude, low-relief, and slowly eroding initial glaciers (A) are mostly cold based. Glacial downcutting (B) steepens glaciers, reduces the cold-based region's size, increases relief, and increases sediment fluxes. Late-stage glacial profiles (C) tend to be gentler and warm based. As the glacial landscape comes toward an equilibrium, the relief, sediment flux, and subglacial erosion rate depend on the rock-uplift rate. Modified after Scherler *et al.* (2011).

areas are largely nourished by direct snowfall, have little debris cover, and show a symmetrical distribution of ice flow around the equilibrium line (Fig. 10.53A). The upper part of such glaciers may be frozen to their beds (cold-based), such that erosion (most of which is subglacial) is focused towards the glacier's toe. This erosion steepens the glacier, gradually decreases the cold-based region of the glacier, and increases both the relief of adjacent peaks and the hillslope debris flux from them, especially in the lower reaches of the glacier. Meanwhile, the locus of erosion and the zones of highest ice velocity shift progressively upstream, thereby driving retreat of headwalls and a decrease in both the mean elevation and the cold-based fraction of the glacier. Overall, these changes in erosion, relief, and hillslope sediments suggest a progressive landscape evolution (Scherler *et al.*, 2011) from low- to high-relief landscape and an increase (at least initially) in the glacial erosion rates through time as the glacier steepens, erodes more effectively, and transforms from cold-based to temperate (Fig. 10.54). The final form of both

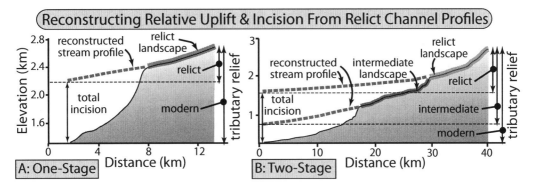

Fig. 10.55   Relict channels, knickpoints, and evidence for pulsed uplift.
Relict landscapes lie above knickpoints in these topographic profiles from the Sierra Nevada, California. Projections of the relict profiles based on the channel concavity serve to define the channel incision, as well as the relief in the relict landscapes. A. Single knickpoint and one inferred phase of rock uplift and incision. B. Two knickpoints and two pulses of inferred rock uplift. Modified after Clark *et al.* (2005b).

the glaciers and the landscape likely depends on the underlying rate of rock uplift. High rates will encourage high relief, rapid erosion both beneath and adjacent to the glacier, and abundant down-glacier transport of supraglacial debris, whereas low rates will lead to lower relief, lower glacial erosion rates, and smaller supraglacial loads (Fig. 10.54).

The existence of high mountain ranges or plateaus prompts questions about how and when they attained their current height. Did they rise steadily over time, pop up in a single interval of rapid uplift, or experience several punctuated periods of rapid rock uplift? Both thermochronology (Reiners, 2005) and river profiles (Harkins *et al.*, 2007) can be used to reconstruct intervals of relative rock uplift or base-level lowering. When used in combination, these approaches sometimes allow both the magnitude and timing of events of rock uplift to be deduced.

One such study in the Sierra Nevada of California uses relict landscapes, modern river profiles, and (U–Th)/He cooling ages to delineate a pulsed uplift history (Clark *et al.*, 2005b). Whereas the cooling ages have specific relevance directly to the Sierra, the techniques used to indentify relict landscapes and pulsed uplift could be widely applicable. Topographic profiles across the Sierran landscape reveal multiple knickpoints that separate regions of

contrasting steepness (Fig. 10.55). Clearly, the presence of knickpoints that are not lithologically controlled indicates that this landscape is in a transient state. When plotted in log area–log slope space (Figs 8.4 and 8.8), data from reaches above the knickpoints define the concavities of relict channels. When this concavity is used to project the relict channel profile laterally to a position above a current local base level (as defined by a tributary–main stem junction or a basin floor), the difference in elevation between the projected profile and the current elevation of the local base level provides an estimate both of the magnitude of incision since that reconstructed profile was abandoned and of the paleorelief of the relict channel (Fig. 10.54A). Where more than one knickpoint occurs along a profile, several stages of incision can be identified (Fig. 10.54B). The presence of relict landscapes and multiple knickpoints in several different locations supports interpretations of pulsed uplift (Clark *et al.*, 2005b). In the Sierra, the most recent pulse appears to be younger than ~3.5 Ma and is consistent with the removal of a dense, lithospheric root from beneath the range (Fig. 10.45) (Saleeby and Foster, 2004). An older interval of rock uplift is interpreted to have begun sometime after 32 Ma, but currently awaits better temporal constraints (Clark *et al.*, 2005b). Despite the somewhat weak time constraints, the geomorphic analyses provide a robust argument for pulsed, rather than steady, rock uplift during the Cenozoic. Similar approaches may provide insights on numerous other ranges for which their topographic development and rock uplift history are poorly known.

Summary

Interpretations of landscapes that have evolved over millions of years provide a significant challenge to tectonic geomorphology owing to the likely superimposition of numerous climatic, tectonic, and geomorphic events over that time period. The explosion of thermochronological, isotopic, geomorphic, and topographic techniques, however, has provided new tools for investigating landscapes at even long time scales. Any pristine feature, such as a fault scarp, in a long-lived landscape relates only to the most recent phases of its evolution. Such features, however, can provide a useful template and should provoke these questions: If the processes that created this feature were repeated innumerable times, would it generate this large-scale landscape? If not, why not? Because plate motions tend to be rather steady over millions of years, the ultimate driving forces and input of energy into a deforming region may remain approximately steady through time. On the one hand, this steadiness could lead to predictable patterns of deformation that simply become structurally amplified through time and are only obscured through erosional modification. On the other hand, as rocks deform and rotate, initial geometries may be overprinted by later ones, and simple correlation of structures with the plate motions may become increasingly obscured. When interpreting terrains with prolonged and complex histories, the challenge becomes one of discerning the signature of earlier tectonic events in the now-degraded geomorphological record, and of utilizing structural, stratigraphic, and chronological data to reconstruct successions of past deformational and erosional episodes.

In recent years, several new tools or data sets have assisted studies of long-term tectonic landscapes. New dating techniques permit far broader and more reliable dating of events and processes within landscapes than was previously possible. Digital topography allows rapid characterization of the land surface. We have seen that the regional overviews that can be readily attained with DEMs provide a key basis for comparing different areas and for quantifying some of their geomorphically and geophysically important characteristics.

Although the crust is heterogeneous and can deform in some unexpected ways, we find some predictable patterns of deformation and landscape evolution. The sinuosity of mountain fronts has been shown to be a good indicator of fault activity. Drainage systems not only help delineate present-day structural geometries, but their persistence as geomorphic entities means that they also record aspects of earlier history. Moreover, we recognize that they can record major events of river capture, pulsed uplift of mountain ranges, and large scale strain that deforms catchments.

We have seen that the concept of a dynamic equilibrium or long-term steady state in which rock uplift is balanced by erosion within a deforming landscape is easy to articulate, but challenging to test. It is likely rare that such equilibrium exists at human time scales. The problems lie in part in the mismatch between the time and length scales over which the available geological tools operate. When one compares the periodicities of climate change (thousands of years), the effect that climate exerts on rates of erosion, and the shifting loci of deformation within an orogen, it seems apparent that the most useful definition of a steady state is one that spans sufficient time (>100 kyr) to average among the fluctuating forcing functions of landscape evolution.

In addition to these temporal challenges, plentiful challenges are related to the spatial distribution of data in long-term landscapes. For a given mountain range, direct and unambiguous geomorphic or structural measures of rock uplift and of erosion rates are typically available from only a few, if any, sites, and it is difficult to know whether these data reliably represent the rates in the remainder of the mountain range. The interpretation of raw data also often relies on a model of how the system has worked through time. For example, interpretation of radio-metrically determined cooling rates in terms of erosional histories commonly relies on assumptions about the thermal field through which the rock cooled, which itself is dictated by the spatial and temporal distribution of erosion rates. And, again, such interpretations are typically based on a few scattered dates. New ways to integrate data from across ranges and to assess the patterns of long-term rock uplift and erosion are needed. Fortun-ately, many new approaches are now routinely incorporated into tectonic-geomorphic analyses.

Numerical modeling is another approach to understanding how landscapes may evolve. Considerable progress has been made in recent decades in quantifying the ways in which fluvial, hillslope, and glacial processes relate to erosion, sediment transport, water flux, slope gradients, and rock strength. These data underpin new generations of numerical models of surface processes. Similarly, improved geodetic, seismic, and structural data aid in constraining new numerical models for the geophysical processes involved in building mountains: plastic defor-mation in collisional orogens; crustal deformation through the seismic cycle on individual faults; fault propagation and linkage; and broad-scale isostatic compensation associated with evolving topographic loads. In the following chapter, the fundamentals of modeling landscape evolution are described, and examples of the integration of surface and tectonic processes in landscape evolution are examined at a range of spatial and temporal scales.

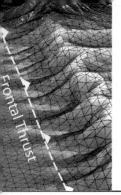

Numerical modeling of landscape evolution

11

We have demonstrated throughout this book that the processes involved in generating landscapes – both those geophysical processes that generate rock uplift and those surface processes that attack and rearrange the rock – are complex in both time and space. The processes interact with one another to generate interesting and complicated feedbacks in the system. Within the last two decades, Earth scientists have increasingly turned to the use of numerical models as a means of exploring the richness of these linkages. As computers have increased in both speed and memory size, these models have become more capable of treating the real-world complexity by discretizing space and time into smaller and smaller bites. These models have allowed scientists to understand in a quantitative sense the linkages among the processes. They have provided a set of visual images of how a mountain range or hillslope or channel ought to evolve under a prescribed set of rules. They have forced us to document process rates in the landscape that must be embedded in the models, and to face difficult questions such as: "How did a particular landscape begin?" Finally, and perhaps most importantly, by embedding our knowledge of surface and tectonic processes in a logical (rigid, unforgiving) framework, these models allow us to investigate where our understanding of these processes is flawed or weak, toward which new research efforts should be focused.

In this chapter, we explore the elements of such models and provide examples from models that are designed to operate at different time and length scales. Some, for instance, are designed to illustrate or mimic the evolution of individual fault scarps, whereas others are aimed at mountain range or even orogen scales. Each will, therefore, require a different set of assumptions that make the problem tractable, given the computer resources at the time the model was generated. In all cases, the goals of such modeling exercises are to enhance our ability to derive insights into the operation of the many processes that conspire to generate tectonic landscapes.

We have seen throughout this book that a landscape owes its shape to the combination of tectonic and climatic forces. Rock is moved with respect to the geoid, either upward or downward, by tectonic processes, some of which result in smooth displacement (creep), and others in discrete displacement (seismic). This displacement of rock changes the local elevation of the Earth's surface. Such displacement affects geomorphic rates to the extent that geomorphic processes are altitude dependent, directly in some instances (e.g., the dominant process switches to glacial erosion for elevations above the snowline) and indirectly in others (e.g., as streams seek a base level, most often sea level, that is independent of the local rock mass). In addition, gradients in the vertical deformation

Tectonic Geomorphology, Second Edition. Douglas W. Burbank and Robert S. Anderson.
© 2012 Douglas W. Burbank and Robert S. Anderson. Published 2012 by Blackwell Publishing Ltd.

from place to place result in tilting of the Earth's surface. Because geomorphic processes are largely driven by gravity and because the component of gravity operating parallel to the surface is dictated by the slope of the surface, process rates will be affected by displacement gradients. Of course, as we have already discussed, in particular instances, direct connections between geomorphic and tectonic processes exist, including, for example, seismically induced mass wasting. Because tectonic and geomorphic processes are so tightly coupled, and their results so tightly intertwined in the landscape, one must often resort to numerical modeling of landscapes in order to extract the tectonic signal. In fact, if the tectonic signal is known, such tectonically active landscapes can be "inverted" to obtain the geomorphic process rates.

We will discuss in this chapter a wide variety of models of landscape evolution that have been developed largely within the last couple of decades. These models will serve to illustrate the couplings within the landscape system and to demonstrate both how tectonic and geomorphic systems conspire to generate particular landforms and how documentation of the shapes of these landforms can be used to infer tectonic process rates. We hope to impress upon the reader that the generation of numerical landscape evolution models and the collection of relevant field data are intimately coupled intellectual exercises.

Approaches

The approach one takes to the incorporation of both tectonic and geomorphic processes in a landscape evolution model depends on the questions being asked of the landscape. Is the goal one of deducing the paleoseismic record, or of deducing long-term changes in slip rates on a fault? Is the exercise a generic one, in which a class of landforms is being addressed, or is it site-specific, in which the attributes of a particular site are being used as a means of assessing the "fit" of a particular model.

We will focus on numerical models that rely on solving equations on a grid of points. Although such models are more common in the computer age, it is well worth noting that there are still solutions to differential equations for the physics (and sometimes chemistry) involved. In certain restricted but important cases, these equations can be solved analytically (with paper, pencil, and brain), which allows very rapid assessment of the dependence of the model solution on one or another process or process rate. These analytic solutions also form important tests of numerical codes. Unfortunately, the real world is complex in that (i) several processes are acting simultaneously, (ii) some of these processes are nonlinear, (iii) the tectonic and climatic forcing of the system is non-uniform in both space and time, and (iv) the geometric boundaries of the features are complex. In general, these complexities preclude analytical solutions to the problems and require that we turn to numerical models.

Numerical models come in several flavors: finite difference, finite element, and boundary element. Finite difference models operate on a discretized space, and solve for the change in some property of each cell in the space (e.g., its elevation) by approximating the differential equation at finite (as opposed to infinitesimal) temporal steps. For example, the differential equation

$$\frac{\partial z}{\partial t} = aA \frac{\partial z}{\partial x}$$

could be written as

$$\Delta z_i = aA \frac{z_i - z_{i-1}}{\mathrm{d}x} \mathrm{d}t$$

where $\mathrm{d}x$ is the node spacing in the x direction, $\mathrm{d}t$ is the time step, and i is the index of the node. The change in the elevation of node i is calculated, and then the new elevation Z_i is obtained by summing the old elevations with the changes in elevation.

Boundary element and finite element methods differ in essence by the characterization of the boundaries and of the region of interest (see Crouch and Starfield, 1983). Mathematically, this translates into finite element solutions that are exact at the nodes but approximate elsewhere, and into boundary element methods that are approximate at the nodes but exact elsewhere. Each method has its pros and cons. Boundary element methods are relatively simple and are able to handle multiple discontinuities (each properly influencing the other). Finite element methods cannot handle discontinuities (although faults can be simulated by so-called slippery nodes or "shear zones"), but they are more amenable to more complex rheological and thermal states. (One might argue, however, that once a problem gets that complicated, the numerical technique loses it pedagogic value, and the number of model parameters becomes large enough that it rapidly becomes very specialized.)

The choice of model scale in a numerical code – how much that model can resolve in both time and space – depends on the goals of the modeling exercise. One must decide whether the feature being studied is simple enough to allow characterization and, hence, modeling in only one dimension – i.e., $z(x)$, where the elevation z is the dependent variable being assessed on a one-dimensional grid of points in x – or whether the model requires two dimensions – i.e., $z(x,y)$. We note as an aside that, although the landscape being explored in these two-dimensional models is in fact a three-dimensional object, the model is still strictly speaking two-dimensional (some call it a two-dimensional planform), because the vertical dimension is a dependent variable.

The choice of modeling strategy is not simple. To understand the choices, one must have in mind a target feature and know to what degree this feature is describable in one or two dimensions. Is the scarp being assessed essentially a linear feature whose profile is everywhere the same? Is the tectonic deformation profile symmetrical in some way (radially or cylindrically)? Modeling strategies generally consist of setting up the problem, embedding in the code the differential equations for the various processes to be modeled (all of which will have free parameters, such as the diffusivity or the rate of regolith generation or the fault slip rate), and finally sweeping through a set of model runs to explore the dependence of the final results on (i) initial conditions, (ii) boundary conditions, and (iii) various model parameters that set the relative importance of one or another process. The complexity and computational requirements increase many-fold as one moves to higher dimensions. Such demands, therefore, limit the degree to which a particular parameter can be explored, because any exploration requires many model runs.

For example, we can clearly discriminate between models whose goal is to explore the degradation of a single fault scarp, from those in which an entire mountain range is to be addressed. The length scales for the former might be 1–100 m, whereas those of the latter might be 10–100 km. Obviously, the details of the single fault being addressed can be treated in the smaller fault model, whereas decisions must be made about how to treat the reality of numerous faults and their geometrical complexity when modeling at the mountain range scale. Given the computational limitations, this restriction might require that the fault scarp model be capable of resolving meters, whereas the mountain range will be resolved at 100 m. Although this distinction seems at first to be a simple scaling problem, it becomes clear quickly that the lower resolution limits the detail with which certain processes can be treated, and it requires that other processes either be ignored altogether, or more likely be parameterized in some manner in the model. This scaling issue forms a large part of the art of modeling and of the challenge faced by the modeling community.

The building blocks

We first illustrate the components of a landscape evolution model, all of which must be linked in the final model. Please note that this recitation is by no means an exhaustive catalog of such model components! Numerical models discretize space and time into small increments, dx and dy for two-dimensional space, dx alone for one-dimensional space, and dt for time. The numerical landscape (in either finite element or finite difference cases) lives within this discretized space as a set of nodes whose attributes include the elevation of the point, the thickness of the crust at that point, the amount of soil or regolith or channel sediment at that point, and so on. One must prescribe rules that allow these attributes to change in a physically reasonable way. In general, these rules simply represent physical processes, the rate at which they operate being dependent on other attributes of the point of concern, such as the distance from a fault, which might dictate the local tectonic uplift rate, the local slope, the local rainfall, and so on. These rules are generally mathematical representations of differential equations, the solutions to which require both initial and boundary conditions. The differential equations represent abstracts of the physics of the problem, including the conservation of mass and of momentum, equations for the response of crustal materials to stress fields, rate of weathering of bedrock, and so on. We have also seen in previous chapters the importance of being able to define how a landscape begins (as a sea cliff with a known geometry; as a river with a given profile). In the model world, the effect of the starting geometry translates into the need to specify *initial conditions*. In addition, we have to worry about the edges of the model, that is, its *boundary conditions*. The numerical model ends abruptly, but the real space it is meant to represent is connected to the rest of the world through exchange of materials and forces. Models, therefore, differ not only in how the processes are mimicked, but also in how these boundary and initial conditions are set up. If faults are present within the model space and form internal boundaries to blocks within the volume, boundary conditions must be applied to each of the faults.

The rules

It is inevitable that any model is a simplification or idealization of the real world. The "rules" discussed here are the rules used within the model and are not necessarily the rules by which the real world operates. The art of modeling lies in choosing a set of model rules that are appropriate for the given problem and that capture the essence of the problem. These model rules evolve as our knowledge of the physical processes evolves. It is here that the links must be forged tightly with both the tectonics and the geomorphic communities. Because the role of the pattern of precipitation on the landscape appears to matter significantly, one must also establish links to the atmospheric sciences community. Such linkage has indeed been a thrust of the community in the last decade, with the development of models of orographic precipitation (e.g., Roe *et al.*, 2003; Roe, 2005) and of observational capabilities that allow us to document the complex pattern of precipitation as the atmosphere interacts with the topography (e.g., Bookhagen and Burbank, 2006, 2010) (Fig. 10.32).

Tectonics

The sophistication of tectonic models varies widely. The most commonly employed model for tectonic deformation associated with discrete faults within some material is based upon the expected elastic deformation associated with a dislocation across which no shear stresses are transmitted. In all such models, the fault plane or planes must be defined (Fig. 11.1). The full location requires one point on the fault, the fault strike and dip, and the vertical and horizontal extent of the fault plane. The slip on the fault must also be either determined or specified. In boundary element models, the slip on any dislocation within the volume can be calculated in several ways. The boundary conditions on the faults within the space can be either displacements (same as specifying the slip on the faults) or stresses (stress drop can be specified; zero shear stress can be specified). These latter modes can be driven by specifying remote stresses on the volume or by specifying the strain within the volume. For example, a right-lateral shear strain within the volume can be imposed by dictating that the east edge of the block has moved to the south relative to the west edge of the block. If the slip is instead specified, it can be defined in two equivalent ways (Fig. 11.1): either by specifying the fault rake and the total slip, or by

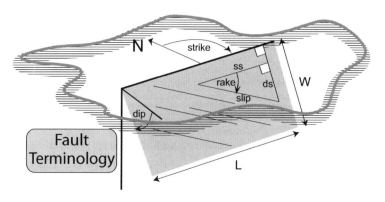

Fig. 11.1 Definition sketch for the elements of a dipping fault that must be prescribed in models of landscape evolution with specific faults.
The spatial orientation of the fault is dictated by the strike and dip of the fault, and the size of the fault is dictated by the length L (horizontal) and height (also sometimes called width) W (down-dip) of the fault. The slip on the fault may be prescribed by specifying either the dip-slip (ds) and strike-slip (ss) components, or the total slip and the rake of the slip event. Parallel lines scribed on the fault surface are meant to indicate slickenlines associated with slip in one event.

specifying the dip-slip and the strike-slip components of displacement. In general, the slip on the fault is assumed to be either uniform or variable from point to point on the fault plane.

Given these specifications of the fault, the boundary conditions on the fault, and the edges of the volume, one then calculates the expected material dislocations (and, if desired, the strains and stresses, e.g., Fig. 4.6) at any other point within the material. Because most measurements of displacement are made on the surface of the Earth, it is usually the deformation field (horizontal and vertical) at the surface of the Earth that is solved for and reported by the code. Given that elastic problems are linear, one may superpose the solutions for any number of discrete dislocations (faults) within the material. For example, fault bends and step-overs are dealt with by breaking the fault into several discrete segments, each with its own deformation pattern. One may also calculate the displacement field associated with a complex pattern of slip on a single planar fault by breaking up the fault into a series of smaller planes. The simplest pattern of slip is, of course, uniform. More complex patterns can be constructed that vary either smoothly, e.g., in an elliptical pattern (Crider and Pollard, 1998; Willemse et al., 1996), or more irregularly. The deformation field associated

with a single fault and driven by an imposed displacement of the edges of the volume are expected to differ for both normal and reverse faults and for faults that do and that do not cut the surface (Fig. 11.2).

Models differ in how they deal with the material at depths greater than those expected to behave in an elastic manner. The simplest assume that the entire Earth behaves as an elastic half-space, but more realistic models employ a region at moderate depths that relaxes viscously (e.g., King et al., 1988).

One of the more widely available boundary element codes (3DDEF; Gomberg, 1993; Gomberg and Ellis, 1993, 1994) is now being used to assess the cross-talk between faults in complicated real-world settings. Given the ability to calculate the changes in the state of stress within the material imposed by the slip on one fault, one may calculate how one rupture loads up other nearby faults, placing them either closer to or further from failure, as we have discussed elsewhere in this book (e.g., Fig. 4.6) – and see also Stein and Ekstrom (1992), Simpson et al. (1994), Simpson and Reasenberg (1994), Li et al. (1998), Harris and Day (1993), and Harris et al. (1995).

Those models that operate at the orogen scale cannot possibly take into account all the

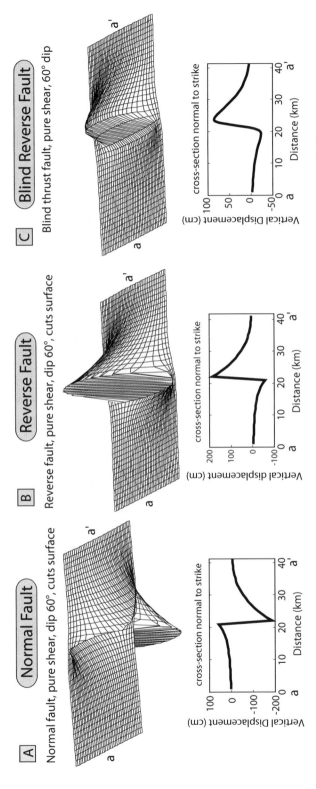

Fig. 11.2 Three-dimensional deformation models of three cases: A. Normal fault, B. Reverse fault, and C. Blind reverse fault.
Dislocations have the same scales and orientations (60° dip, same strike) in each model, although the top of the dipping plane is buried in the blind fault case, C. Each is driven by far-field pure shear stresses, rather than by a prescribed slip on the fault. At the base of each mesh plot is a cross-section of the vertical displacement field. Note the similarity of the normal and reverse cases, the principal difference being in the asymmetry: in the normal fault case, the maximum displacement is downward, on the hanging-wall block; whereas in the reverse case, it is upward, again on the hanging-wall block. Note that the vertical exaggeration is more than 100-fold.

discontinuities in the crustal material. The strategy here is instead to capture the large-scale forcing within the lithospheric mass as it responds to lithospheric-scale tectonic forcing, e.g., continental collision rates. One strategy is kinematic: a vertical and/or horizontal deformation pattern is simply assumed. Other models that more properly capture the dynamics, although still in their infancy, will ultimately be more useful in the exploration of the nature of the feedbacks in these systems. The dynamic models differ in the assumptions made about the rheology of the lithosphere at long time and length scales. For example, in some models, the rheology is assumed to be captured by a non-linear flow law in which the flow parameter corresponding to the effective viscosity of the material is temperature dependent, such that the shear strain rate changes as a power of the local shear stress. This sort of model, therefore, requires proper modeling of the thermal evolution of the lithosphere as well. Whereas such models are quite complex, they can embody important feedbacks between the tectonic forcing and the geomorphic response.

Flexure

Over long time scales, deep Earth materials behave as a fluid. All fluids flow in response to pressure gradients, moving from sites of high pressure to sites of low pressure. Because the pressure at depth depends on the density and height of the column of rock above it, the tectonic and geomorphic modification of the topography represents a rearrangement of surface loads, which can force deep Earth materials to flow. This flow in turn modifies the surface topography at long wavelengths and on time scales of many thousands of years, and it operates in such a fashion as to reduce the lateral pressure gradients at depth. This process is one of *isostatic adjustment*, the static case being one of *isostatic equilibrium* (*isostatic* meaning "equal pressure"). That the near-surface rock is too strong to deform in a ductile way over these same time scales requires that it respond instead as a broad, flexing elastic plate. The wavelength of the response of the surface

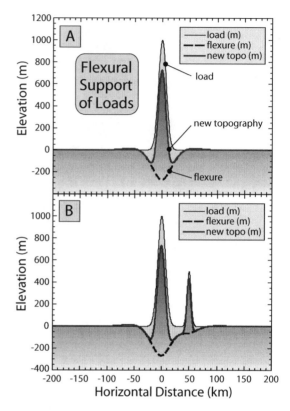

Fig. 11.3 Flexural response to a distributed set of line loads.
Load shape in A is a single linear mountain range with a Gaussian cross-section 1 km tall. Load in B has in addition a second crest centered at 50 km to the right of the first, with 500-m amplitude. Final topography is the sum of the load and the flexure from each individual load. Effective elastic thickness of 10 km is used in both cases.

to this motion at depth, called the *flexural wavelength*, is dictated by the average rigidity of the near-surface materials. The flexural response (Fig. 11.3) is calculated as if it were the bending of a beam of uniform thickness. This thickness is fictitious in that one could not drill to this depth and find either a material or a chemical discontinuity; rather, this represents the thickness of a beam that reflects the average or integrated strength of the near-surface rock – see the full discussion in Watts' (2001) wonderful book on isostasy and flexure.

The complexity of the flexural component of the models depends in part on the symmetry of the problem. For simple one-dimensional cases, one may easily turn to analytic solutions

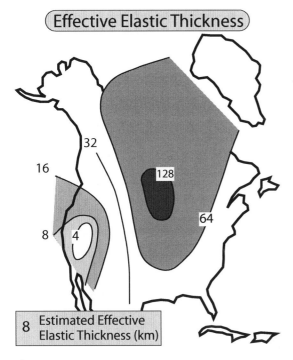

**Fig. 11.4   Effective elastic thicknesses across
North America.**
Estimated distribution of effective elastic thicknesses
derived from analysis of gravity and topographic data.
Modified from Bechtel *et al.* (1990).

reviewed in Turcotte and Schubert (1982). For
two-dimensionally distributed loads, one must
turn to the solutions for a point load or disk
load (see, e.g., Lambeck, 1988; Watts, 2001).
Although these, too, are analytic solutions, the
flexural response to complex loads, i.e., those
that vary spatially in such a way that they cannot
be simplified into a simple symmetrical pattern,
is calculated by turning to a numerical code that
sums the vertical displacement associated with
each of many point or line loads. We show a
simple case employing the superposition of
many line loads meant to represent one or two
mountain crests (Fig. 11.3). Finally, one must
know something about the expected flexural
rigidity in a particular landscape. For example, a
map of the broad-scale features of the effective
thickness in North America, as estimated from
analysis of the gravity and topographic fields,
reveals both the strong rigid lithosphere of the

shield and the distinctive lower effective thicknesses in the western United States (Fig. 11.4).

Surface processes

To first order, landscapes subjected simply to
geomorphic processes tend to degrade through
time, high places losing material, low places
gaining. The specific processes and their rates
vary from climate to climate, and from one
lithology to the next. This general behavior can
be captured mathematically with the diffusion
equation, in which changes in the elevation of a
point depend entirely on the local curvature.
In other words, in one dimension,

$$\frac{\partial z}{\partial t} = \kappa \frac{\partial^2 z}{\partial x^2} \tag{11.1}$$

where x is horizontal distance, z is elevation, and
κ is a landscape diffusion coefficient. The simplicity of the diffusion equation is appealing, although
it is not without drawbacks. In particular, simple
diffusion-based models of surface processes generally ignore the need to produce regolith before
it can be transported (a problem only if the feature involves bedrock to begin with), and they
cannot account for channels in a landscape. In
addition, if we wish to explore the dependence
of the surface processes on climatic factors such
as precipitation or temperature, which change in
both space and time, we need to prescribe geomorphic rules that explicitly acknowledge these
dependences. The development of such rules is
evolving rapidly, as both field instrumentation
and dating methods allow documentation of these
spatial and temporal variations in process rates
(see, e.g., Dietrich *et al.*, 2003; Roering, 2008).

Surface processes can be broken down loosely
into those that produce mobile material, or
regolith, from intact bedrock (usually by the processes of *weathering*), and those that move such
material about on the landscape (by transport
processes). At any point where the rate at which
the mobile material is stripped away outpaces
the rate of generation of mobile material, the
landscape will be bare bedrock. We refer to such
landscapes or points on the landscape as being
weathering limited; the pace of weathering limits

the pace of landscape change. In such instances, one cannot simply operate on the numerical landscape with a transport rule – which might lead to diffusive behavior – but must as well embed in the model a rule, or mathematical representation, for the generation of regolith (see the next section; Anderson and Humphrey, 1989).

In such circumstances, one therefore needs a rule for what sets the rate at which regolith is produced. Once produced from bedrock, particles are moved down slopes, first on hillslopes and then in channels. One must, therefore, determine where the channels lie within a landscape, how effective they are in transporting sediment, and, if bare bedrock is exposed in the channel, how fast the bedrock is being eroded.

Regolith production

Our understanding of the long-term controls on the conversion of bedrock to regolith is incomplete. In fact, this lack of understanding has led to the establishment of a number of experimental sites in both the United States and Europe, dubbed Critical Zone Observatories, in which both the processes and rates of regolith production, and the ecological and hydrologic services that they perform, are targets of investigation (Anderson et al., 2008). To first order, we would expect that, the wetter the climate, the faster will be the weathering of rock. Gilbert (1877) hypothesized that, within any particular landscape (he was then considering arid landscapes of the Henry Mountains in Utah), the regolith production rates should be a strong function of the regolith thickness as well. Bare bedrock can effectively shed water, thereby protecting itself from chemical attack, whereas even a thin regolith cover should allow significant chemical attack of the underlying rock. Beyond a given thickness, however, the bedrock lies at such a great depth beneath the surface that wetting events may not penetrate, and the conversion rate might be expected to diminish. A mathematical statement of this conceptual picture has been incorporated in a number of numerical landscape evolution models within the last few decades. Although appealing, this conceptual picture has been difficult to document in the real

world, because the weathering rates are far too low to measure on human time scales. Hope has emerged, however, with the aid of cosmogenic radionuclides in constraining the constants in at least a few geological and climatic settings – see, e.g., Pavich and Hack (1985), Pavich (1986), and McKean et al. (1993), who use garden-variety ^{10}Be, and Heimsath et al. (1997, 1999, 2000) and Small et al. (1999), who use ^{10}Be and ^{26}Al produced in situ; see also Chapter 3 on dating for a discussion of these methods.

Hillslopes

Material is moved down hillslopes by a number of processes, whose types and rates are modulated by the local climate and slope materials. All hillslope transport processes are dictated to some degree by the local slope angle. The simplest model rule is therefore something like:

$$Q = -k \frac{\partial z}{\partial x} \tag{11.2}$$

where $\partial z/\partial x$ is the local gradient of the topography, or the slope, in the x direction. The constant k reflects the long-term efficiency of sediment motion, which might be expected to be a function of the climate. A major goal of modern geomorphology is to explore the dependence of this hillslope transport efficiency on aspects of the climate and on the material properties of the regolith (e.g., Dietrich et al., 2003; Anderson, 2002). This arena has seen much recent progress, with considerable attention to specific transport processes. Gabet (2000) and Heimsath et al. (2002) have proposed rules that acknowledge the roles of biological transport actors; Anderson (2002) has proposed a plausible rule for transport by frost creep. Roering (2008) has summarized several potential rules and has explored the impact of the choice of rule on the look of the resulting modeled topography. Several workers (e.g., Anderson and Humphrey, 1989; Roering et al., 2001; Roering, 2008) have modified this rule (Eqn 11.2) in order to mimic regolith landslides by enhancing the transport nonlinearly as a threshold slope angle is approached (Fig. 11.7B).

This type of rule is not capable of treating such potentially important processes as bedrock-involved landsliding. Unfortunately, in some rapidly eroding landscapes (Burbank *et al.*, 1996b), this process is the dominant means of pulling rock from the adjoining hillslopes toward the bounding channels. Algorithms that attempt to capture this process come in several flavors. The simplest is one in which there is assumed to be a threshold bedrock slope angle above which a slope cannot be maintained; any material found above an envelope that points in a "V" upward from the channel at the threshold slope angle is, therefore, shaved off and delivered instantly to the channel. Another algorithm essentially modifies the hillslope efficiency, k, such that it increases rapidly as an angle of repose is approached from lower angles (Anderson and Humphrey, 1989; Howard, 1994, 1997; see discussion in Roering, 2008). Yet other approaches explicitly employ failure thresholds that entail both local slope and height and incorporate an assumed rock-mass strength (Densmore *et al.*, 1998; Schmidt and Montgomery, 1995, 1996). In these latter cases, the failed masses must then be distributed in the landscape in a realistic manner, capturing the material properties of the masses subsequent to failure.

Channel initiation

Most models of landscape evolution embed some rule for where in the landscape the channels initiate. As shown convincingly in map and field analysis by Montgomery and Dietrich (1988, 1989, 1992), channels begin wherever a stream-power threshold has been exceeded. This threshold is thought to reflect the material strength at the channel head and both the water discharge and resultant shear stresses at that point. This initial incision point is usually captured in the numerical models by embedding a threshold product of the local slope, S, with the local drainage area, A, thereby producing a slope–area product in which upstream area is a proxy for discharge. Nodes at which this product is greater than some constant are considered channels, whereas all others remain hillslopes. The sensitivity of the landscape character to the choice of the channel initiation threshold or to the balance between advective and diffusive processes is explored in detail through numerical models by Whipple and Tucker (1999) and Perron *et al.* (2008).

Bedrock incision

Our knowledge of the complex process by which rivers etch into bedrock has only recently been addressed by field and modeling studies. The physical processes include abrasion by particles entrained in the flow, plucking of blocks from the bed, and cavitation (Hancock *et al.*, 1998). Most model rules that purport to capture the essence of this suite of processes use a stream-power argument (Howard and Kerby, 1983; Rosenbloom and Anderson, 1994; Seidl and Dietrich, 1992) – see also the review in Whipple (2004), and chapter 13 in Anderson and Anderson (2010). This simplification is often abstracted further to a rule in which the rate of erosion is dependent on the local channel slope, S, and the drainage area at that point, A, i.e., the slope–area product:

$$\frac{\partial z_c}{\partial t} = dSA \tag{11.3}$$

where d is a coefficient of erosion. This equation is undoubtedly only a crude approximation of bedrock channel behavior. For instance, the coefficient d collapses any number of sins, including the resistance of the rock, the effective discharge of the river, the runoff efficiency, and so on, into a single model parameter, as discussed for example in Hancock *et al.* (1998).

Alluvial transport

Many transport rules have been used in geomorphic models. We can only touch on a couple of classes here. Most model rules embed the observation that sediment transport increases with the boundary shear stress of the flow, or the discharge of the water, above some entrainment threshold. The simplest such statement is the DuBoys equation, relating sediment flux per unit width of flow, Q_s, to the local boundary shear stress, τ_b:

$$Q_s = b(\tau_b - \tau_c) \tag{11.4}$$

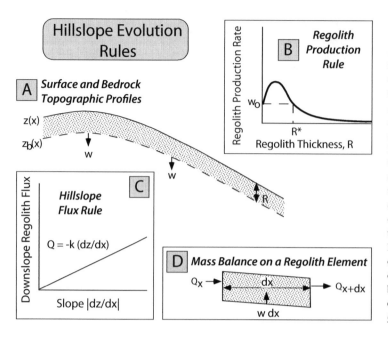

Fig. 11.5   Schematic illustration of the components of a hillslope evolution model that might be used to assess scarp evolution.
A. Topographic, $z(x)$, and bedrock, $z_b(x)$, profiles, with in general a non-uniform regolith thickness, $R(x)$.
B. Regolith production rule, in which the rate, w, is here taken to be dependent only on the local regolith thickness. C. Hillslope regolith specific discharge rule, one in which the discharge is linearly proportional to the local slope of the landscape.
D. Mass balance on a regolith element depends on the rate of rock-to-regolith conversion, w, which adds mass to the base of the element, and on the downslope sediment flux both in to, Q_x, and out of, Q_{x+dx}, the element.

where τ_c is the shear stress necessary to entrain the sediment. Because the shear stress ($\tau_b = \rho g H S$) requires a knowledge of the flow depth, H, most model strategies rewrite this rule in terms of the water discharge, Q, which is a quantity that is more directly tied to the climatic forcing of the landscape. This assumption results in a yet simpler statement:

$$Q_s = cQ \qquad (11.5)$$

Examples

Given the set of rules stated above, a wide array of models can be generated that differ in their initial and boundary conditions, and in the pattern of uplift and subsidence imposed by tectonic processes. We start with simple fault-scarp models in the next section and then move to larger features.

Scarp degradation modeling

In many instances, none of the absolute dating methods discussed in Chapter 3 can be applied to a particular landform, be it a fault scarp, a lake shoreline, or a marine terrace. The reason is

most often a lack of datable material. In these circumstances, researchers have turned to theoretical models of how a scarp is expected evolve in the face of the surface processes active on it. The idea is simple: because, in most cases, scarp shape evolves monotonically from sharp-edged toward smoother forms (Nature abhors sharp corners), a knowledge of the initial shape of the scarp and documentation of the detailed shape of the scarp, through surveying in the field, may be used to estimate the age of the feature.

Theory

Starting with the extensive work of Culling in the 1960s (Culling, 1960, 1963, 1965), most models are based on the diffusion equation. A useful introduction can also be found in Carson and Kirkby (1972). In direct analogy with the well-understood thermal problem (Carslaw and Jaeger, 1986), the relevant equations derive from two statements: (i) mass is conserved, and (ii) the flux of mass is proportional to the local topographic slope (Fig. 11.5). The first statement may be written, for the one-dimensional case, as

$$\frac{\partial z}{\partial t} = -\frac{1}{\rho_b} \frac{\partial Q_x}{\partial x} \qquad (11.6)$$

where Q_x is the discharge of mass (formally, the specific discharge, or mass per unit contour length per unit time) in the x direction, through whatever processes are operating, ρ_b is the bulk density of the material being transported, and z is the elevation of the surface. The surface will decline in elevation (erosion) if the gradient in mass discharge is positive (more mass leaves the element than arrives), and it will increase in elevation (deposition) if the gradient in discharge is negative (more mass arrives than leaves). This statement is very general and broadly applicable; it holds for hillslopes, river beds, and sea floors alike. What changes from one system to the next are the surface process and how that process is governed by measurable quantities in the landscape, such as local slope, the distance from a ridge crest, the distance downstream, and so on. What is needed to close the system is a process rule that describes what controls the hillslope fluxes, Q_x. In the simplest case on a single hillslope, of which a scarp is an example, the discharge of mass is taken to be simply proportional to the local slope. Stated mathematically,

$$Q_x = -k\frac{\partial z}{\partial x} \qquad (11.7)$$

where k is a proportionality constant reflecting the efficiency of the process. Note that the negative sign reflects the fact that mass is being fluxed in the downslope direction. The local slope in two dimensions is $\partial z/\partial x$, whereas in two horizontal dimensions x and y the slope is ∇z. This expression is a subset of a large suite of possible models for the flux of mass on a hillslope, in which a general dependence exists on x, possibly to some power, and in which slope may come in to some power (e.g., Carson and Kirkby, 1972).

In general, then, the rule may be stated as:

$$Q_x = -kx^m\left(\frac{\partial z}{\partial x}\right)^n \qquad (11.8)$$

Under conditions in which the transport efficiency, k, is uniform with x, and the discharge of sediment is dictated linearly by the local slope, i.e., $m=1$ and $n=0$, the combination

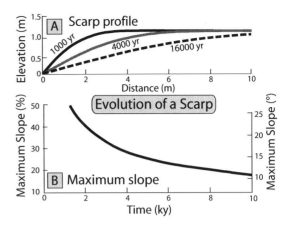

Fig. 11.6   The evolution of a scarp profile by diffusion.
A. The top half of a scarp profile at several times after initiation of a 2-m vertical step in the surface.
B. Evolution of the maximum slope from 1 to 10 kyr. Note that the maximum slope falls rapidly at first, and more gently later, a change reflecting the dependence of diffusive processes on the square root of time. Modified after Hanks *et al.* (1984).

of the mass flux and mass conservation equations results in the following diffusion equation:

$$\frac{\partial z}{\partial t} = \kappa\frac{\partial^2 z}{\partial x^2} \qquad (11.9)$$

The rate of change of the elevation of the surface depends solely on the local curvature (or rate of elevation change) of the surface. Here, the diffusion coefficient, often called the landscape diffusivity, κ, reflects both the bulk density of the material being transported and the efficiency of the transport process: $\kappa = k/\rho_b$. Diffusivity always has units of L^2/T. Note that this formulation is easily extended to both x and y directions, although we will focus on simple landforms described by a single profile, $z(x)$.

Solution of the diffusion equation, a second-order partial differential equation, requires specification of initial conditions. In addition, estimation of the age of a feature requires knowledge of the diffusion constant. In analogy with the problem of thermal evolution (see Turcotte and Schubert, 2002) within a slab with an initial step in temperature, scarp evolution may be represented as an error function (Fig. 11.6).

One key feature of this solution is that the slope at the mid-point of the scarp, which is often the steepest slope on the feature, decays as the square root of time, making the measurement of this slope in the field a useful target for constraining the scarp age (Hanks et al., 1984). The simplicity of this approach is indeed appealing, although one must be very aware of the geomorphic and seismic setting (Was the scarp formed in one event or is it a composite feature built through multiple events?) to apply this technique properly for the dating of scarp-like features.

Applications

Early application of diffusive analysis to tectonically active landscapes included the treatment of wave-cut scarps in the Lake Bonneville basin, Utah, whose age had been determined independently (by ^{14}C), and which were cut by recent faulting or warped by isostatic deformation. Knowledge of the present profiles, an estimate of the initial profile shape by appeal to modern wave-cut features, and knowledge of the age led to constraints on the diffusivity, κ, of 5–100 m^2/kyr (or 0.005–0.1 m^2/yr). Based on these estimates of diffusivity, ages of scarps of unknown age in arid landscapes nearby were subsequently estimated (Bucknam and Anderson, 1979; Colman and Watson, 1983; Mayer, 1986; Nash, 1984, 1986; Wallace, 1978).

Using another feature of the solutions to the diffusion equation, Avouac et al. (1993) have surveyed numerous profiles across scarps separating terrace risers in arid Tibet (Fig. 2.15) to deduce the ages of the terraces. They then plot the pattern of the derivative of the topography (in other words, the slope) versus distance. This pattern should be a broadening Gaussian through time if diffusion controlled the evolution of the terrace riser (Avouac, 1993; Avouac et al., 1993). Avouac et al. use the best fits to the breadth of this distribution to constrain what they call the scarp degradation (m^2), which is the product of the diffusivity and time since the scarp step was generated. If the diffusivity is uniform from scarp to scarp, then the scarp degradation should be a good surrogate for scarp age.

More recently, Clarke and Burbank (2010b) have addressed the potential role of climate in modification of scarp profiles. Using examples from New Zealand and Idaho, they attempted to model measured profiles on slopes facing either the equator or the poles. They deduced a non-linear hillslope transport function and found that the hillslopes in Idaho displayed an effective diffusivity that was roughly twice that of their New Zealand counterparts, and that slopes facing the poles were consistently steeper than those facing the equator (Fig. 11.7). Given that the climate, scarp ages, and sediment character are quite similar between these two sites, the differences in diffusivity seem likely to reflect vegetation and moisture differences, as well as perhaps the roles of freeze–thaw cycles and burrowing mammals: the latter drive diffusive transport in some Northern Hemisphere sites (Gabet, 2000), but were largely absent in New Zealand until very recently.

Complexities

Geomorphologists have recognized several pitfalls that face the modeling of scarp evolution with the simple diffusion equation. In the development outlined in the previous section, and in the applications cited, several assumptions were made, the following among them: (i) The proper initial condition is that left by whatever erosional or tectonic process caused the topographic step in the first place. (ii) The diffusion coefficient is spatially uniform along the profile, which allowed pulling the k out of the derivative when the two equations were combined. (iii) The diffusion coefficient is not dependent upon the aspect (facing direction) of the slope, i.e., it would be the same from profile to profile within the same landscape. (iv) The simple linear relation of flux to local slope is valid, i.e., it is not nonlinear and is not related to other factors, such as distance from the top of the slope. (v) Weathering of the material to be transported may be ignored. We summarize subsequent work addressing some of these complexities in the following paragraphs.

Initial conditions

It was early recognized, in dealing with fault (and other) scarps in arid regions, that another process

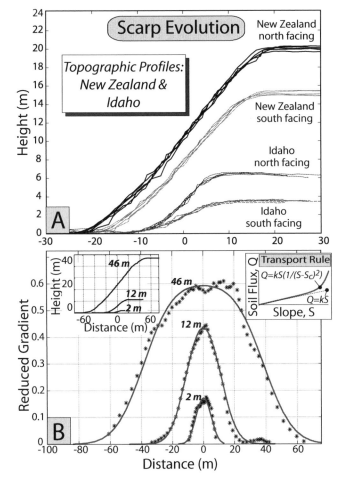

Fig. 11.7 Scarp degradation in contrasting climates.
A. Topographic profiles of 15- to 20-kyr-old fluvial scarps in New Zealand and in Idaho showing rounding of corners and relatively straight riser slopes. Equator-facing scarps at both sites degrade faster than pole-facing scarps. B. Modeled slopes for three New Zealand profiles, shown in the left inset. Right inset shows hillslope transport rule, in which the linear dependence of hillslope transport on slope at low slopes is modified by more efficient transport as slopes approach a threshold slope. Modified after Roering *et al.* (2001). The effective diffusivity of the landscape in New Zealand is shown to be roughly half that in Idaho, such that scarps degrade about twice as fast in Idaho. Modified after Clarke and Burbank (2010b).

not well described by diffusion was operating in the early stages of scarp decay. Rapid failure of the scarp by slumping and talus formation leads to angle-of-repose slopes typically within a matter of decades to centuries – times much shorter than the presumed ages of the scarps (Pierce and Colman, 1986). This angle-of-repose slope is, therefore, likely the more appropriate initial condition for diffusion-based modeling of the scarp. Estimates of scarp age are made knowing that a slight lag is associated with the development of the angle-of-repose slope. Alternatively, one may postulate a process rule that mimics the generation of the angle-of-repose slope and start the calculation from the initial oversteepened step. Examples of this treatment include that in Anderson and Humphrey (1989), Howard (1997),

and Howard *et al.* (1994), where the downslope sediment flux departs from a linear relation with slope angle, increasing rapidly as the slope steepens toward the angle of repose.

Non-uniform diffusivity

Late glacial fluvial terrace scarps in central Idaho (Pierce and Colman, 1986) and New Zealand (Clarke and Burbank, 2010b) display a dependence of the diffusion coefficient on the orientation of the slope (slope aspect; Fig. 11.7). These studies document a several-fold difference between north- and south-facing slopes, which they attribute largely to the dependence of vegetation density on solar radiation and moisture availability. Both studies also note a dependence on height of the scarp, tall scarps having higher effective diffusion coefficients, and Pierce and Colman (1986) attribute this greater diffusivity to the operation of slopewash processes that are dependent on slope length – in other words, the equation describing hillslope discharge is no longer simply dependent on slope, but is dependent on position on the slope, x.

Weathering-limited situations

In landforms that are not entirely composed of loose, transportable materials, one must take into account the transformation of the cohesive material (or bedrock) into material capable of being transported by the available surface processes. This requirement was recognized early by Gilbert (1909), in his classic word picture of how convex hilltops operate. More recently, Anderson and Humphrey (1989) have modeled the evolution of scarps and moraines in the face of weathering and linear diffusion. In such cases, one must posit a second rule that represents this transformation of rock into transportable material – call it the weathering rate, W. Anderson and Humphrey (1989) followed the proposal of Ahnert (1970), who was following Gilbert's (1877) word picture, in which the rate of transformation depends solely on the thickness of the regolith (defined as the layer of transportable material). Although other rules could be entertained, Anderson and Humphrey (1989) use a simple exponential, in which the maximum rate of regolith production was achieved on bare bedrock (no regolith) and decayed exponentially with increasing regolith thickness:

$$\frac{\partial z_b}{\partial t} = W = -W_0\, e^{-R/R^*} \qquad (11.10)$$

where z_b is the local elevation of the regolith–bedrock interface, R is the regolith thickness, W_0 is the rate of bare bedrock weathering, and R^* is a regolith thickness at which the rate of regolith production is $1/e$ (or roughly 1/3) of that on bare bedrock (recall that e is roughly 2.72). The rule set must now include both the rate of lowering of the bedrock interface and the rate of change of the regolith due to surface processes, which is controlled by the diffusion equation (see Fig. 11.5). The relevant diffusion equation is now

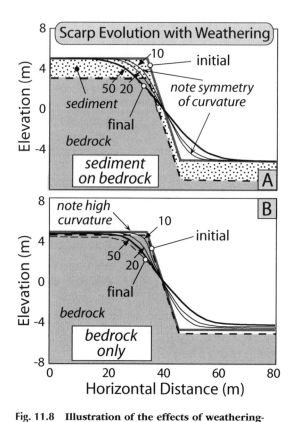

Fig. 11.8   Illustration of the effects of weathering-limited case on the evolution of scarp profiles. Bedrock interface, early (thick dashed line) and at termination of the simulation at 50 ka (thin dashed line); and surface topography at 0 (initial), 10, 20, and 50 ka (full lines). A. Pure diffusion case is allowed by dictating that the initial bedrock interface be deeply buried, such that the bedrock is only etched into late in the simulation. B. Shallow initial bedrock interface (0.2 m) restricts rate of rounding of the scarp crest, whereas surface processes still effectively smear the growing colluvial wedge. This shallow bedrock breaks the symmetry of the pure diffusion case, the top of the scarp showing consistently higher curvature than the base. Modified after Anderson and Humphrey (1989).

$$\frac{\partial R}{\partial t} = \kappa\, \frac{\partial^2 z}{\partial x^2} \qquad (11.11)$$

and the surface elevation at any point in time is $z = z_b + R$. With these modifications, Anderson and Humphrey (1989) show that the symmetry for a scarp's crest and toe that one expects from pure diffusion in the absence of weathering is broken (Fig. 11.8). The base of the slope rounds as

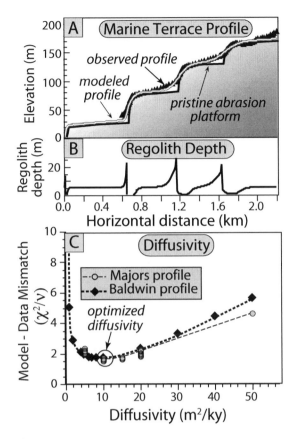

Fig. 11.9 Santa Cruz marine terrace profiles.
A. The observed stair-stepped topography of marine terraces in Santa Cruz, California, are assumed to be underlain by a wedge of sediment that is a combination of original marine terrace deposits (here assumed to be uniformly 6 m thick upon retreat of the sea) and colluvium derived from the decay of the sea cliff backing any particular terrace. These wedges overlie "pristine abrasion platforms" that are inserted into the model at the distances required by the locations of observed paleo-sea cliffs and at the times prescribed by the assumed ages of the sea's abandonment of the inner edge of each terrace. B. The regolith depth as a function of position. Note the spike associated with the base of the cliff and the taper that grows through time. Regolith is absent on the bedrock-dominated edge of the first cliff, but has a finite but low depth on the upper edges of older cliffs. The terrace surfaces of old or narrow platforms dip significantly more steeply toward the sea than do younger and broader surfaces, suggesting that caution is warranted in interpreting dips of these surfaces as tilt of the original platforms. C. Sensitivity analyses of the diffusion parameters controlling the shapes of the model hillslope profiles. Tests used the reduced χ^2 statistic relating model calculations to real

sediment accumulates from upslope. In contrast, rounding of the top of the slope requires weathering of the rock to produce sediment that can then be passed downslope. If weathering is slow, the rounding of the crest is slower than the rounding of the base, and the profile becomes asymmetric. Especially at early times, the sharp curvature at the hillcrest (or scarp crest) results in rapid removal of the existing regolith; thereafter, the rate at which this portion of the slope declines is dictated by the weathering rate on bare bedrock, i.e., it is *weathering limited*. All the while, the material transported downslope forms the smoothed colluvial apron at the base of the slope.

Given that this time lag in producing transportable material could potentially explain the asymmetry of the marine terrace risers studied by Hanks *et al.* (1984), Rosenbloom and Anderson (1994) incorporated such a rule set into their models of hillslope evolution and demonstrated that considerably better fits to the profiles could be achieved when both weathering and diffusive surface processes were treated explicitly (Fig. 11.9). It was not necessary to appeal to surface processes that resulted in nonlinear diffusion.

In an application to glacial moraines, which have been used in eastern California to constrain both the ages of the glaciations and slip rates on the Sierran frontal fault (Bursik and Gillespie, 1993), Hallet and Putkonen (1994) have shown that the interpretation of cosmogenic radionuclide (CRN) results from boulders on moraine surfaces is not straightforward. They argue that use of a numerical model of moraine evolution that employs a slightly modified form of the diffusion equation described previously can fit the moraine topographic profiles. By tracking, in addition, the build-up of cosmogenic radionuclides in boulders that are slowly exhumed from within the moraine during its shape evolution (see Chapter 3), they show that: (i) on old moraines, the boulders commonly used to date them should yield only a minimum age for the moraine; and (ii) the mean

data. For both surveyed profiles, the minimum χ^2 corresponds to $\kappa = 10\,\mathrm{m^2/kyr}$. Adapted from Rosenbloom and Anderson (1994).

CRN age of boulders sampled from a moraine surface can in fact decrease with the true age of the moraine. This example highlights the need to take into account the dynamic nature of the geomorphic surfaces on which CRN techniques are currently being employed. Such integration requires the marriage of cosmogenic and numerical landscape evolution modeling techniques.

Conclusions from modeling at the hillslope scale

Proper use of this sort of modeling to estimate the ages of scarps and similar topographic features requires all of the following: (i) knowledge of initial conditions; (ii) awareness of talus-like processes that violate the linear diffusion process rule in the early stages of topographic evolution; (iii) treatment of out-of-plane processes that might remove material from the cross-section, thereby violating the one-dimensional conservation of mass equation; (iv) explicit treatment of the generation of transportable particles if cohesive materials or bedrock are involved in the original scarp; (v) awareness of a possible dependence of the diffusion coefficient on slope aspect, through a dependence on process efficiency and possible process type with orientation; and (vi) estimation of the diffusion coefficient by appeal to nearby or climatically relevant estimates from other well-dated sites in similar materials.

Given these complexities, it is perhaps best not to rely on a single slope within the scarp (e.g., mid-point slope) to infer age, but rather to construct a forward numerical model of the entire scarp that explicitly addresses the issues outlined previously. Modeled scarp profiles can then be assessed against field-collected profile data using goodness-of-fit estimates for a range of possible models, in which the full range of possible weathering rates and diffusion coefficients may be explored.

Marine terrace generation

Whereas several models exist of the decaying sea cliffs that bound a set of marine terraces, far fewer studies have targeted the generation of the terrace platforms in the first place. Here we summarize a study that addressed the issue of both generation and demise of these useful geomorphic features (Anderson et al., 1999). To capture the evolution of a marine terraced landscape in a numerical model, one must include uplift of the landmass, oscillation of sea level, and erosion of the coastline. Anderson et al. (1999) treated this system of processes while focusing on parameters relevant to the suite of five marine terraces at Santa Cruz, California. In a two-dimensional shore-normal model (Fig. 11.10), they employed a long-term sea-level curve based upon the $\delta^{18}O$ curve (Fig. 2.3), and rock uplift prescribed to occur uniformly with shore-normal distance. The erosion rate was specified to occur at a rate that was damped by a crude approximation of wave interaction with the offshore bathymetry, i.e., a broad, shallow shelf would dissipate more wave energy than a steep, narrow shelf. Combination of these processes leads to a set of marine terraces, the number of preserved terraces being governed largely by the rate of rock uplift, given that the sea-level history to which all coastlines have been subjected has been essentially the same.

The numerical model reproduces the means by which older terraces generated during a lower sea-level highstand can be cut away by subsequent, higher sea-level maxima. This selective preservation is analogous to the moraine survival problem beautifully explained by Gibbons et al. (1984), in which younger, longer glaciers can obliterate the moraines left by earlier, shorter glaciers, resulting in a small number of recorded advances (see Box 2.3). Whereas the algorithm for the erosion of sea floor is crude, the model nonetheless raises the need for knowledge of these submarine processes. In particular, the shape (width, convexity, etc.) of the continental shelf left in the wake of repeated beveling of the margin by the sea requires greater scrutiny (Adams et al., 2005). Fortunately, the increasing availability of high-resolution imaging of the sea floor provides a basis for tackling this problem.

One might also ask why most active tectonic coastlines display only a small suite of terraces – two, three or five, but rarely more. Why is this? If the rock uplift has been ongoing for millions of

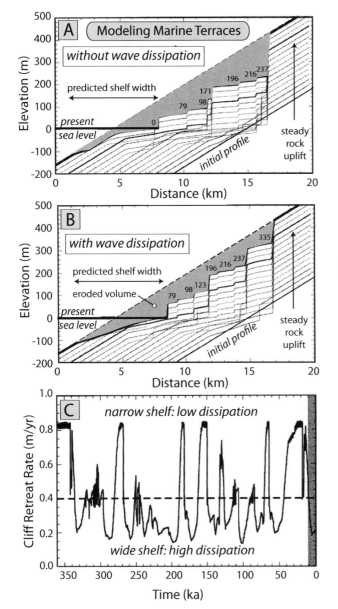

Fig. 11.10   Evolution of marine terraced profile.

A,B. Modeled evolution of marine terraced profile on an uplifting coastline, and C, the modeled cliff recession-rate history. Sea cliff retreat is shown for 350 kyr models, given uniform and steady 1 mm/yr of uplift of the rock mass, and sea-level history derived from the marine oxygen isotope curve (Fig. 2.3). Eroded rock is shown in gray. Models differ only in that in A wave energy is assumed not to be dissipated by offshore bathymetry, such that cliff retreat rates are steady (dashed line in C), whereas in B wave dissipation is allowed, being strongest on wide nearshore platforms. This dissipation results in the variations in the cliff retreat rates shown in the solid line in C. The final profiles differ in the width of the final offshore platforms, being narrower in the case allowing wave dissipation, and in the total amount of erosion, being greater without wave dissipation. In addition, the final sequence of recorded marine terraces differs. With wave dissipation allowed, more terraces are recorded, e.g., the 123 ka notch. Such small terraces are erased in the case of more efficient cliff retreat when dissipation is ignored. Modified after Anderson *et al.* (1999).

years (in some cases), why do we not see many tens of terraces, even accounting for the terrace survival problem? This problem has been approached as well in simulations both of the channels that cut through the terrace sequence and of the decay of the sea cliffs once abandoned by relative sea-level drop (Anderson *et al.*, 1999). Whereas cliff decay is a slow, essentially diffusive,

process, it appears that the fluvial incision of the terraced landscape, as well as the hillslope processes that are slaved to these channels (Fig. 11.11B), are more efficient at cutting away terraces. This degradation can be seen easily in a DEM of the Santa Cruz terraces (Fig. 11.11A) as the terrace edges become less distinct and their flats less extensive. As the rivers ramify inland,

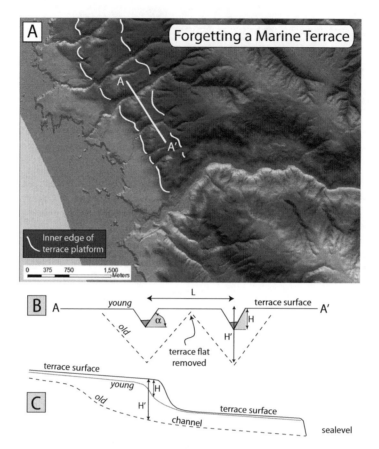

Fig. 11.11   How marine terraces are forgotten in the landscape.
With increasing age, marine terraces are progressively removed or forgotten from the landscape. A. DEM of a portion of the terraced landscape in Santa Cruz, California, showing progressive loss of well-preserved terraces with elevation. Two of the more prominent, younger terraces are delineated by white lines at approximately the back edge of each terrace. B. Cartoon of a cross-section across two channels and intervening terrace tread (e.g., along the line AA′ in the DEM), at early (solid) and late (dashed) times. C. Cartoon of a long profile of terrace and incising channel, the latter shown at early (solid) and later (dashed) times. Deepening of the stream channel below the terrace surfaces, from H to H′, incites broadening of the adjacent valley walls, which in turn bite into the remnants of the flat marine terrace treads. The resulting time scale for removal of the terrace flat is dependent on the distance between adjacent channels, *L*, and is inversely dependent upon the rock uplift rate, which in turn scales the stream incision rate. Modified after Anderson *et al.* (1999) and Anderson and Anderson (2010).

the tributaries etch rapidly into the terraced flats. In this landscape, terraces older than the fifth terrace are lost from the record. In this fashion, the number of identifiable terraces and the pristinity of the terrace features – as reflected in the distinctness of the paleo-sea cliff edges and in the fraction of the original terrace flat that remains – are governed by the efficiency of fluvial and hillslope processes as the landscape responds to relative base-level fall.

Modeling river incision

Models of river incision have evolved over the last decade to include how rivers should respond to patterns of rock uplift or base-level fall (e.g., Whipple and Tucker, 1999), how this response is modified by orographic precipitation patterns (see review in Roe, 2005), and how knickpoints propagate in fluvial networks. Whipple (2004) has nicely summarized bedrock rivers, the processes responsible for their erosion, and some modeling approaches that have been taken to address these. We advertise only a few such studies here.

Fluvial strath terraces

As discussed in Chapter 2, whereas marine terraces are distinct elevational datums, strath terraces cut by rivers record the vertical and horizontal position of a river. The height of a dated strath above the modern river has been used to constrain the mean vertical incision rate of the river. But the specific processes responsible for the formation and abandonment of a strath or a suite of straths have been less frequently addressed. The conceptual model for this sequence is set out in Fig. 2.12 and is discussed further in Chapter 7 (see Fig. 7.13). Strath terraces are formed by alternation of vertical incision of a river into bedrock to abandon one strath, and lateral planation of the river to widen the next strath. In the "sediment loading model," it is crucial to be able to model the river's response to temporal variations of the sediment supply. Using the set of 15 distinct terraces along the present Wind River in Wyoming as their inspiration, Hancock and Anderson (2002) attempted to capture the essence of the strath problem in a numerical model. Acknowledging that this system has glacial headwaters, they modulated both sediment and water delivery to the channel at its upstream end. Both vertical incision and horizontal incision were allowed. Crucially, when the alluvial cover of the channel is greater than a prescribed scour depth threshold, the river cannot access the bedrock beneath it and vertical incision is disallowed. On the other hand, bedrock is still available to incise laterally, and the river can widen its floodplain, etching a strath. If, for some reason, the alluvial cover thins to allow vertical incision into bedrock, the strath is abandoned. The model results are summarized in Fig. 11.12.

They found that the stronger control on formation and abandonment of straths was modulation of sediment supply in the headwaters. In glacial settings, sediment supply is likely to vary many-fold as the footprint and erosivity of the

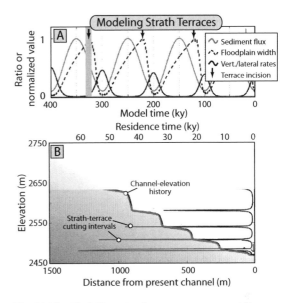

Fig. 11.12   Modeling strath terrace sequences.
A. History of normalized sediment inputs (solid gray line), normalized valley-floor width (i.e., floodplain width) at river level (dashed line), and the ratio of vertical to lateral erosion rates (solid black line) during a simulation with a 10-fold variation in sediment supply at a 100-kyr periodicity. Inputs and valley-floor widths are normalized by dividing the value at each time by the maximum value during each simulation. The valley floor, i.e., the "floodplain," widens most significantly when vertical erosion rates are low, producing a ratio of vertical to lateral erosion that is small or zero. Floodplains are abandoned to form terraces when vertical erosion rates and the ratio of vertical to lateral erosion rates increase, leading to renewed downcutting and narrowing of the active valley. Terraces are formed at the transition (arrows) from valley-floor widening to narrowing that represents channel incision. Terrace formation does not occur at the times of either maximum inputs or minimum inputs, but instead significantly lags the timing of maximum sediment supply. In the si mulations, all terraces generated are strath terraces. B. The topographic profile (thick gray line) and channel residence time as a function of elevation during downcutting in this simulation (thin black line). Terraces are related to periods of lateral planation during long channel residence within a narrow elevation range. Residence time at individual terrace levels reaches up to many tens of thousands of years, indicating that the river spent most of the simulation forming terraces. Modified after Hancock and Anderson (2002) and Anderson and Anderson (2010).

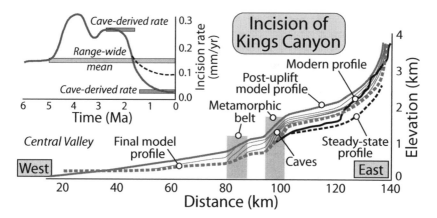

Fig. 11.13   Rejuvenation of Kings River, Sierra Nevada.
Response of South Fork Kings River to late Cenozoic tectonic and climatic events. Example of stream-power-based numerical simulation. Steady river profile (dashed) with steps corresponding to quartzite in two metamorphic belts is subjected to ~1.5 km of crestal uplift, causing a tilt, down to the west. Model profiles at 1-Myr intervals (thin gray lines) show that, over the next 9 Myr, a wave of rapid incision begins at the hinge line and propagates up-river. (Inset) The 6-Myr incision history at the cave site where cave dates have been constrained from [10]Be burial age methods. A wave of rapid incision passes between about 5 and 2 Ma, followed by return to low pre-uplift rates (dashed curve after 2 Ma). Further reduction in late Quaternary rates (solid curve after 2 Ma) reflects sediment mantling of bed associated with large glaciers in headwaters. Final modeled river profile (gray dashed line) fits modern profile (black) to just upstream of cave site, above which glacial erosion, not represented in the river incision rule, has dominated the past few million years. Modified after Stock *et al.* (2004).

glacier varies strongly over Quaternary times. But several other key points were illustrated by the modeling exercise. First, straths can be abandoned progressively, resulting in a spread in the abandonment age of a surface. This spread reflects the wave of sediment thinning in the aftermath of sediment supply reduction in the headwaters. Second, the mean rate of incision one deduces from dividing the height of a terrace by the age of the terrace can vary greatly (see Box 7.2). Given when we live, in the midst of an interglacial, soon after the abandonment of the Last Glacial Maximum (LGM) strath, the rate we are likely to deduce from the LGM terrace is probably higher than that we would deduce from older straths. Whereas some workers have interpreted such age structure to signify acceleration of rock uplift, it is not necessarily so. Despite being driven by a perfectly sinusoidal sediment and water supply (Fig. 11.12A), the model results indicate that the formation and abandonment of straths is not a steady process. Rather, it is very spotty in time, with long periods of lateral incision, punctuated by short periods of vertical incision.

Interpretation of strath records for any rock uplift signal must, therefore, be done with full attention to the punctuated nature of incision.

Caves as straths internal to a mountain

In a twist to the strath story, several researchers have demonstrated that caves in the walls of gorges can serve the same role as strath terraces in allowing documentation of incision rates (e.g., Granger *et al.*, 1997, 2001). Stock *et al.* (2004) employ multiple caves in the walls of King's Canyon in California to document a record of incision history that displays a striking reduction of incision rate toward the present (Fig. 3.21): rates between 3 and 1 Ma are 0.15–0.25 mm/yr, whereas those from 1 Ma to present are an order of magnitude slower (Fig. 11.13). A numerical model of river incision was employed to test whether this reduction was consistent with tilting of the range deduced from other geomorphic and stratigraphic evidence. We have discussed the underlying cause of this late Cenozoic tilting in Chapter 10 (Fig. 10.45). Stock *et al.* (2004) used a stream-power-based incision rule and appealed to a model of tilting of the range in which a dense keel was allowed to disconnect from the base of the crust. The flexural–isostatic rock uplift pattern resulting from this removal, and flexural rigidity that decayed strongly toward the Basin and Range, produced a strong asymmetry in rock uplift, tilting the range toward the west. The river responded to this tilting by initiating a knickpoint at the western edge of the range. This knickpoint subsequently propagated eastward.

Stock *et al.* (2004) argue that the rapid incision in the 3–1 Ma interval reflects the translation of the steep knickpoint past the site of the caves, with low rates of incision left in the wake of the knickpoint. Note that, in the absence of several dated caves, one could not tell this story: one long-term average rate derived from a single dated terrace or cave is insufficient to discern changes in incision through time. This wave of rapid incision associated with the passage of the knickpoint up the main stem of Kings Canyon in turn sends knickpoints up its tributaries, which show strong convexities in their profiles. Importantly, a large fraction of the core of the range is effectively "ignorant" of these events, because the hillslopes on the interfluves in the large regions between major drainages have yet to be influenced by the drop in base level of these major streams or their tributaries (Stock *et al.*, 2005; Clark *et al.*, 2005b).

Sets of knickpoints

Given the importance of fluvial incision in landscape evolution, and the fact that, in some cases, the fluvial response to change in base level is through propagation of knickpoints, significant attention has focused on natural experiments that display numerous knickpoints. In both New Zealand's Waipaoa catchment (Crosby and Whipple, 2006; Crosby *et al.*, 2007) and in Colorado's Roan Plateau (Berlin and Anderson, 2007), tens of knickpoints can be attributed to a single change in the base level of the system. The Waipaoa case we have already discussed in Chapter 8 (see Figs 8.7 and 8.11). In the case of the Roan Plateau, incision of the adjacent Colorado River has cast off a wave of incision

into flat-lying Cenozoic stratigraphy to the north. The present locations of 37 box-headed canyons, some with spectacular waterfalls, represent the target for models of how the fluvial system has operated (Fig. 11.14). Berlin and Anderson (2007) proceeded by crafting a generic incision model in which horizontal incision rate was governed by the drainage area upstream of the knickpoint, taken to some power:

$$c = \frac{\mathrm{d}x}{\mathrm{d}t} = aA^p \qquad (11.12)$$

where c is the celerity or wave speed of the knickpoint in the upstream (x) direction, A is the upstream drainage area, and a and p are empirically determined constants. The system is initiated by a base-level drop on the Colorado River at a time loosely constrained by geological observations on the southern edge of the plateau. The single knickpoint works its way upstream to the first tributary junction, at which point it bifurcates into two knickpoints. Each of these knickpoints will proceed more slowly than the first, because the drainage areas of the tributaries are smaller than the trunk below the junction. This process of bifurcation, repeated at each tributary junction, therefore leads to stepwise slowing of the growing set of knickpoints (Fig. 8.10A). Any particular model run is assessed by the misfit between the final locations of model knickpoints and actual canyon heads. The model that best explains the present pattern of canyon heads (Fig. 11.14) is one in which the power $p=0.5$ (as found in the Waipaoa case), meaning that the speed of the knickpoint goes as the square root of drainage area.

As discussed in Chapter 9, a similar situation exists along the Yellow River (Fig. 9.19), in which the propagation of knickpoints through the Yellow River and its tributaries has served to extract a large volume of sediment from a series of basins in the last half-million years (Harkins *et al.*, 2007).

Mountain range-scale models

At the scale of mountain ranges, numerical models have evolved significantly over the last two decades. Tucker and Hancock (2010) provide

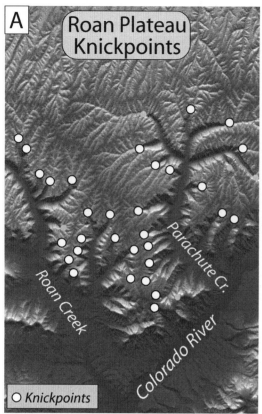

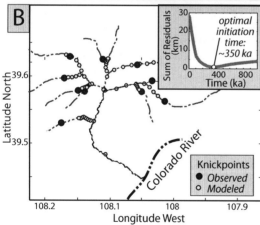

Fig. 11.14 Knickpoints on the Roan Plateau.
A. Map of Parachute and Roan Creeks, Roan Plateau, Colorado, showing modern locations of knickpoints.
B. Results of 350-kyr celerity model of knickpoint migration in Parachute Creek, showing present positions of knickpoints (black dots) and locations of model knickpoints at 20 time increments (open circles). Most knickpoints are well fit with this simple model. (Inset) Goodness of fit measured by sum of misfits of knickpoint

an excellent review of the literature of landscape evolution models, including discussion of the history of modeling, the model ingredients (or rules), and a list of recently developed models. Early attempts involved one-dimensional models of the evolution of simple ranges bounded by a single fault. These studies included a uniform-slip dislocation model of a single planar fault, diffusion of resulting topography, and flexural compensation for the rearrangement of rock mass. In a pair of prescient papers (King *et al.*, 1988; Stein *et al.*, 1988), these authors both developed the fundamental principles behind such linked models (Fig. 4.30), and applied them to specific settings in the American West. They explored both normal- and thrust-faulted ranges. Importantly, they found that the flexural rigidity of the crust played importantly into setting the width of the zone over which the sediments eroded from the rising mountain mass were spread (Fig. 11.15). For example, they argue for a very low rigidity in the Basin and Range province (e.g., effective elastic thicknesses of only a few kilometers – see Fig. 11.4) in order to match the very narrow depositional basins near the active faults. The simplicity of the models, being one-dimensional with only three major components, allowed efficient exploration of the roles of each of the processes in the problem. Note, however, that the entirety of the geomorphic suite of processes is boiled down into a single diffusion coefficient. No channels exist. No weathering exists. None of the complex feedbacks between these processes are allowed. Nonetheless, much insight was gained through this suite of models. Was it necessary to incorporate these other processes? Was it worth the effort of moving to two dimensions, as is necessary if attempting to distinguish channels and hillslopes? At the same time, models along these lines were also being constructed of thrust-faulted settings, with more attention being paid to the stratigraphic package

locations, as a function of simulation time. Best fit is 350 kyr for the given choice of $k=5\times10^{-10}\,\mathrm{m^{-1}yr^{-1}}$, and a critical upstream area of $0.5\,\mathrm{km^2}$. Whereas the value of k directly scales with the assumed time of initiation of the base-level fall event, the value of p is robust. Modified after Berlin and Anderson (2007).

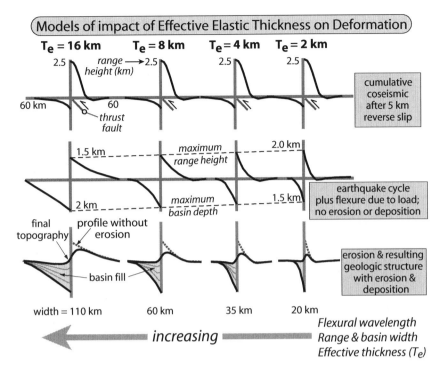

Fig. 11.15   Matrix of mountain range-scale models in which the effective elastic thickness (T_e) is varied.
(Top row) The pattern of vertical displacements associated with a prescribed total slip of 5 km on the thrust fault is shown. All patterns are the same and show 2.5 km of maximum rock uplift on the hanging-wall block, and <1 km of maximum subsidence on the footwall block. Note the clear coseismic asymmetry in which hanging-wall uplift far exceeds footwall subsidence. (Middle row) The pattern is now allowed to flex in response to relaxation at depth: the resulting sum of dislocation plus isostatically driven flexure is shown. The difference in the plots reflects the decline in the assumed effective elastic thickness from 16 km to 2 km from left to right. The width scale of the resulting mountain range–basin pair declines dramatically from 100 to 20 km. (Bottom row) A crude redistribution of mass by erosion of the mountain crest and deposition in the adjacent basin. The scale and the asymmetry of the geological structure vary significantly with the flexural rigidity. Modified after King *et al.* (1988).

generated in the bounding basin (e.g., Flemings and Jordan, 1989, 1990). Again, no attention was paid to the details of the processes generating the debris, nor of the delivery mechanisms of that debris to the basin edge. The focus was on the details of the stratigraphic package, and how these reflected either episodes of rapid motion along the fault or climate variations.

The first numerical model both to move to a two-dimensional planform and to incorporate channels as critical elements in the evolving landscape was that of Koons (1989), who focused on the Southern Alps of New Zealand. Koons incorporated a simple, geometrically reasonable, and geologically defensible rock uplift pattern. He inserted a set of streams that

lived in specific places within this uplift pattern, that did not move, and that draped smoothly and logarithmically from a channel head near the uplift maximum to the base level of the sea, and he dictated that the surface processes that operated upon the interfluves between these stream channels could be approximated by diffusion that was allowed to vary spatially. In particular, in an effort to mimic the strong orographic forcing caused by the fact that the Southern Alps are embedded in the westerlies, he tied the landscape diffusivity to the rainfall rate. As shown in Fig. 11.16, the essence of the ridge topography is simulated well using this simple set of model rules. Initially, the ridgeline is highest near the coast on an interfluve that is

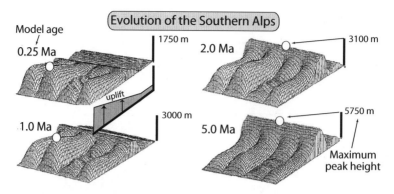

Fig. 11.16   First-generation planview landscape evolution model incorporating channels and non-uniform precipitation and rock uplift patterns, Southern Alps.
Modeled temporal landscape evolution of the Southern Alps, New Zealand. Channels are dictated to have a particular shape draping between the mountain crest and the sea. Diffusion modulates the hillslopes, and both differential rock uplift and rainfall patterns are incorporated. Circles represent the highest topographic point in each simulation. The complex ridge pattern reflects the essence of that observed in the Southern Alps of New Zealand. Modified after Koons (1989).

wider than the others in the simulation. This position reflects both the asymmetric rock uplift pattern and the fact that the steady-state relief depends on the square of the interfluve width (Koons, 1989). The simplicity of this model, which is as simple as a two-dimensional planform model can get, allows rapid exploration of the various controls on the topography. Koons did this exploration very effectively by varying the spatial pattern of the uplift pattern and of the rainfall pattern (hence diffusivity).

One of the drawbacks to Koon's (1989) model is that the channels are not interactive: their profiles are dictated, rather than reflecting the real dynamics of the channel system. Nonetheless, just as much was learned from the Stein *et al.* (1988) models in one dimension, much is learned from Koons' (1989) two-dimensional approach. With two-dimensional models, the modeling community took a strong step toward realizing the relative importance of one or another process, and in particular alerted the geomorphic community to the importance of bedrock channel incision.

In the 1990s, several attempts were made to incorporate more interactive channels and hillslopes into the mountain range evolution models. This generation of models required integration of rules for the sediment transport down channels and for the incision of bedrock by channels. Anderson (1994) addressed the problem of topographic evolution in a region of 30×100 km near a restraining bend in a major strike-slip fault (the San Andreas Fault). The rock-uplift pattern was simply dictated at the outset, as a two-dimensional Gaussian rock uplift pattern whose scale was set by the requirement for conservation of crustal volume arriving at the bend. Crust was then translated through this pattern, such that any parcel of crust experienced a time-varying history of uplift dictated by what portion of the uplift pattern it intersected. The coarse, 1-km resolution of the model required significant abstraction of the hillslope-scale processes. Even small-scale channels with drainage areas of less than $1 \, km^2$ were subsumed in the numerical pixels. The remaining major channels were inserted in the calculation space, were tied to the bedrock as it slid horizontally relative to the fixed fault bend, and were allowed to evolve as bedrock channels with their drainage area acting as a proxy for local stream power. No attempt was made to incorporate orographic effects: rainfall was taken to be uniform. Arguing that landslides were the dominant delivery mechanism for debris to the channels, Anderson (1994) simply set the hillslope angles in the landscape and

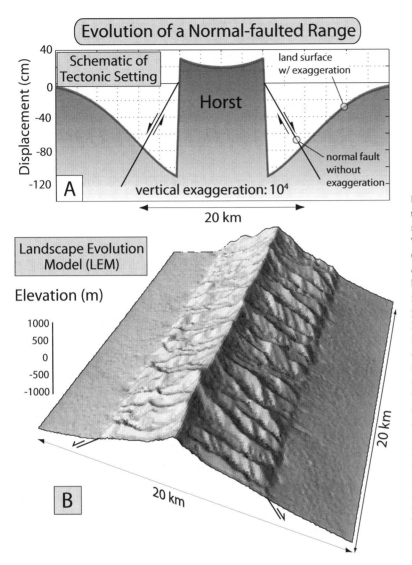

Fig. 11.17 Numerical model of topographic evolution of a normal-fault-bounded range. The numerical model ZSCAPE (Densmore *et al.*, 1998; Ellis *et al.*, 1999) of a normal-fault-bounded range in which erosion is driven by a suite of surface processes, including bedrock landsliding. Resolution of the model = 100 m. A. Cross-section of the tectonic forcing of the system with two opposing normal faults driving both vertical and horizontal deformation fields. Dislocations dip 60°. B. Resulting topography after one million years. Precipitation pattern is uniform. Channels are self-formed, and incise at rates driven by local stream power. Landsliding dominates the hillslope evolution. Triangular facets form on both faulted mountain fronts. Adapted from Densmore *et al.* (1998).

allowed the hillslopes to come down at the rate dictated by the incision rate of the local channel. Flexure was taken into account in two dimensions, allowing both the topographic envelope and the crustal response to erosional unloading to be followed through time.

Densmore *et al.* (1998) incorporated more realistic landsliding rules in their model, dubbed ZSCAPE (Fig. 11.17). The tectonic component of this model is based on the Gomberg and Ellis (1993, 1994) and Gomberg (1993) 3DDEF boundary element code for tracking the displacement field to be expected from a set of prescribed dislocations in an elastic medium. The setting addressed by Densmore *et al.* (1998) and by Ellis *et al.* (1999) is the Basin and Range province of western North America. High-angle normal faults are prescribed. The surface processes for which model components exist in the code include regolith production, diffusion of regolith, landsliding, fluvial sediment transport, and fluvial bedrock incision. The code is flexible enough to deal with spatially non-uniform precipitation. The 100-m pixels resolve many-fold more details than Anderson's (1994) model. Initial conditions of a flat or a slightly tilted topography were explored. Many of the salient characteristics of such mountain ranges are captured well in the model, including both triangular facets at the mountain front and benches in spur ridges. Exploration of the relative importance of various surface processes and comparisons with actual topography demonstrated that bedrock-involved landsliding was likely the dominant process in this setting.

Orogen-scale models

In a series of models that extended the early work and ideas of Koons (1989) on the Southern Alps, Beaumont and coworkers (e.g., Beaumont *et al.*, 1992, 1996; Willett *et al.*, 1993) have added interactive channels and an explicit treatment of both the geometry and the rheology of the crust in sometimes quite complicated collisional settings. The crust is allowed to deform by two mechanisms: shallow crust behaves as a cohesionless frictional–plastic (Coulomb) material with a specified yield strength; at depth, the deformation mechanism alters to one of thermally activated power-law creep. The switch-over (loosely speaking, the brittle–ductile transition) between these two mechanisms is determined dynamically within the model and depends on the thermal and mineralogical structure. In the surface-process components of these models, the code is simplified considerably by treating both hillslope and channel processes in any particular node (see Kooi and Beaumont, 1994, 1996).

In an important contribution, Kooi and Beaumont (1996) have explored more generic mountain range evolution (Fig. 11.18). They illustrate that the older conceptual models of many major geomorphologists as diverse as King, Gilbert, and Davis (Fig. 1.2), each of whom developed their conceptual models with particular real-world landscapes in mind, can be illustrated with such a generic model by paying attention to one or another phase of the development, or by setting the tectonic and surface-process rates differently.

Considerable attention has been devoted to the two-sided orogen (Koons, 1990; Willett *et al.*, 1993; Willett, 1999). Much of this work followed

on the notion that a strong coupling could exist between (i) the surface processes that actively rearrange the load on the crust and (ii) the deformation field at depth. This coupling is exciting in that it allows the geomorphology and those who study it into the tectonic game. This proposed coupling is illustrated in the work of Willett (1999) (Figs 10.1 and 10.22), who has constructed a model of collisional orogens with examples from Taiwan and the Himalaya in mind. These models examine cross-sections oriented normal to the orogenic collision, and the rheology of the crustal materials involved in the collision is assessed explicitly (Fig. 11.19). The surface-process component of this model is simple, as it must be one dimension. Reflecting the assumption that the dominant surface process is bedrock fluvial incision and that the hillslopes will essentially come along for the ride, Willett uses an incision rule based on stream power. The spatial distribution of power varies as a function of the precipitation falling upstream of the node being assessed. Orographic forcing of the precipitation is allowed. Importantly, these models illustrate the strong coupling of the spatial pattern of erosion rates, themselves governed by patterns of precipitation, with the deep crustal strain-rate field. This class of models quantifies the conceptual ideas of Koons (1989) that show the potential influence of the atmospheric conditions (primarily the wind direction) on the asymmetry of the strain field in collisional settings (Fig. 11.19). That this degree of coupling can be simulated in apparently realistic models has acted as a catalyst for a wide range of earth scientists to collect the sort of field information (topography, meteorology, exhumation patterns) that will act both as tests of the present models, and as constraints on further models at the mountain range scale. For example, recent work in New Guinea reveals that this coupling has been operating to localize the strain in the collisional orogen in Iryan Jaya (e.g., Weiland and Cloos, 1996).

In their review of geomorphic modeling, Tucker and Hancock (2010) illustrate the utility of the recent version of the CHILD (Channel–Hillslope Integrated Landscape Development) model (Tucker *et al.*, 2001). In this and several of the other existing landscape evolution models,

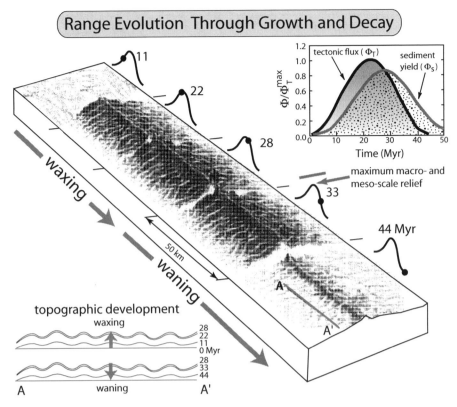

Fig. 11.18 Landscape model incorporating a triangular uplift pattern and a self-forming channel pattern, showing several snapshots of the landscape as it grows and then decays in relief.
Triangular uplift occurs with the same across-strike pattern, but at different rates in each of the five ages (11, 22, 28, 33, and 44 Myr). The total sediment flux leaving the model landscape lags the rate at which mass is being added by tectonic forcing, as shown in the inset (top right). A topographic profile (A–A′) across the flank of the range depicts the waxing and waning of relief (bottom left). Modified after Kooi and Beaumont (1996).

the early square mesh has been replaced with a Voronoi mesh (an irregular triangular mesh), which allows more realistic treatment of river geometries. In Fig. 11.20, we reproduce an example of the oceanic edge of a landscape model that is meant to illustrate the differences between subsiding and uplifting blocks. Modern landscape evolution models are increasingly capable of capturing the essence of many landscape elements, including those that cross the boundary into the oceanic realm.

In a final example illustrating use of the CHILD model, we examine a numerical prediction of the landscape in the frontal zone of Himalayan deformation in western Nepal (Fig. 11.21 and Plate 9). Here the slip rates on the underlying Main Frontal Thrust fault are estimated to be ~20 mm/yr

(Fig. 8.21). The Siwalik Hills form above the southward (leftward) verging Main Frontal Thrust at the southern margin of the Himalaya, seen in the background. The Karnali River flows along the base of the range's backlimb. The mountain-range model has been run long enough to attain a steady state in which the form of the topography is no longer changing. The asymmetry of the real topography, and the crenulation of the range crest are well captured in this model.

Atmospheric interactions

Whereas we all know that the atmosphere delivers most of the "events" that govern the patterns and the rates of geomorphic processes and that these patterns reflect the tangling of the

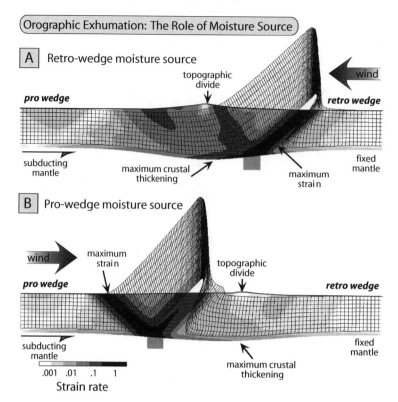

Fig. 11.19 Cross-sectional model of crust, erosion, and topography in a two-sided collisional orogen.
The two panels display contrasting topography and exhumation patterns associated with how moisture enters the orogen based on wind direction: in the same direction as the subducting slab or opposing it. The initial mesh, attached to material particles, was rectilinear, and has been deformed significantly. Substrate detachment point is given by the gray rectangular block. Crust deforms according to a temperature-dependent viscous constitutive law below the Coulomb yield stress. The final topography is shown as the solid bold line in each model. The portion of the rock mass within the "Lagrangian tracking mesh" above this line has been exhumed and has been lost to erosion. Note that, despite identical convergence tectonics, the zones of strain and patterns of erosion are strikingly different based on whether storms enter the orogen from the left or right sides. Modified after Willett (1999).

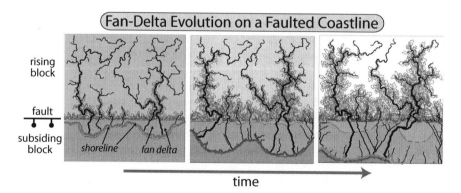

Fig. 11.20 Numerical model of fan-delta evolution.
Three frames from a numerical model of landscape evolution (CHILD; Tucker *et al.*, 2001), showing the development of topography and fan-delta complexes in response to differential vertical motion on adjacent, normal-fault-bounded blocks; the upstream block is rising. The shoreline position is shown by the heavy gray line; lighter colors indicate higher elevations. Modified after Tucker and Hancock (2010).

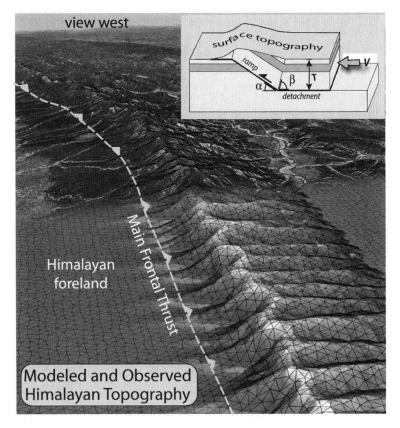

Fig. 11.21 CHILD model of a thrust-generated mountain range.
Satellite imagery draped over digital topography (top part of image) of the Siwalik Hills, western Nepal, compared with a numerical landscape evolution model (bottom part) of a mountain range formed in the hanging wall of an emergent thrust fault. One can see the Voronoi mesh employed in the CHILD model. Scale varies in this perspective. In this region, the Siwalik Hills and the model mountain range are both ~13 km wide and ~1300 m in total relief. The model topography shown forms above a planar fault ramp dipping 30° to the right (see inset), a 5-km-thick hanging wall, a 20 mm/yr fault-slip rate, and a stream-power coefficient, K, equal to $2 \times 10^{-5}\,\mathrm{yr}^{-1}$. Exponents m and n in the stream-power equation are 0.5 and 1, respectively. Himalayan images from Google Earth (copyrights 2005 Google, 2005 EarthSat, and 2005 DigitalGlobe). Figure courtesy of Scott Miller. [A color version of this appears as Plate 9.]

atmosphere with topography, only recently have we developed models to acknowledge these interactions in a quantitative fashion. The most important of these interactions is what we broadly term "orographic effects" or "orography," the role of mountains in governing the pattern of precipitation. The last decade has seen advances in our ability both to observe meteorological events and to model them. We begin with the observations.

Weather stations are simply too sparse to document precipitation patterns at the spatial density we require. Happily, observation platforms in the form of satellites have come to the rescue

in documenting precipitation patterns. In particular, one can extract from the TRMM (Tropical Rainfall Measurement Mission) satellite data stream spectacular products (some with spatial resolution of ~5 km) that reveal both spatial and temporal patterns of precipitation in mid- to low-latitude regions of the world. For example, Bookhagen and Burbank (2006, 2010) have shown that the precipitation pattern in the Himalaya displays orographic effects (Fig. 10.32) that help to explain the water discharge patterns (hydrographs) of the major rivers draining the range (Fig. 11.22 and Plate 10).

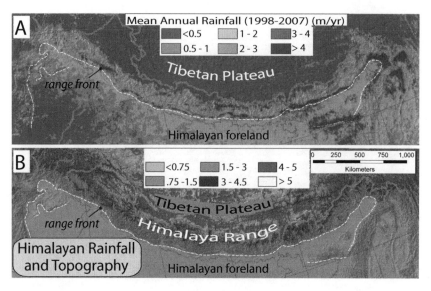

Fig. 11.22   Orographic precipitation and topography in the Himalaya.
A. Decadal record of remotely sensed rainfall from the TRMM satellite with ~5-km spatial resolution. Note nearly continuous band of high rainfall along the range front (white dashed line) and second, discontinuous band lying on the southern flank of the High Himalaya. B. Topography of the northern foreland, the Himalaya, and the southern Tibetan Plateau. See swaths in Fig. 10.32 for relationship of topography to orographic rainfall. Modified after Bookhagen and Burbank (2010). [A color version of this appears as Plate 10.]

Models of orographic precipitation have been around for many decades. Ron Smith (1979) nicely summarized the state of knowledge in the late 1970s and has continued to participate in advancing our knowledge of these important interactions. In the early 2000s, Roe developed simpler versions of these models of atmosphere–topographic interactions that were designed to be coupled to landscape evolution models. In Roe's models (Fig. 11.23), one specifies the wind direction and speed, the water content of the atmosphere, and its temperature. The air mass then encounters a landmass and is made to rise at a rate dictated by the air speed and the slope of the landscape. Here, the relevant slope is some envelope of the topography, although later models demonstrate the role of smaller-scale steering of the air masses by local valleys (Anders *et al.*, 2008). The water in the air mass condenses to form "hydrometeors," and falls at a rate governed by the settling speed of either the raindrops or the snowflakes. Were it not for the finite travel distances of these hydrometeors from their site of formation due to wind velocity, the precipitation

pattern in these models would simply reflect the slope of the landmass. Roe and Baker (2006) detail the pattern of precipitation that takes into account the trajectories of the hydrometeors. The effect is to shift the pattern of precipitation downwind of that associated with the slope of the topography. This shift allows some snow or rain to waft over the crest of a mountain range to fall on the dry side, where the slopes alone would suggest no precipitation should fall (Fig. 11.23).

Roe *et al.* (2002) applied this simple model to the evolution of a river profile in an uplifting rock mass. Using a simple stream-power rule for river incision, in which incision rate is a function of the product of water discharge and slope of the river, requires that one calculate the spatial pattern of water discharge. Although many early models of river incision simply took the water discharge to be a product of drainage area and some given precipitation (in fact, most models use drainage area as a proxy for discharge), a more proper formulation for the discharge at a point on a river is to integrate the product of the precipitation with the area at all elevations from the crest of

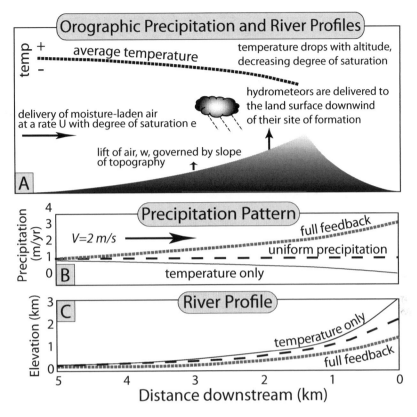

Fig. 11.23 One-dimensional models of orographic precipitation.
Changes in bedrock steady-state stream profiles caused by orographic feedbacks. A. Competing effects of slope and elevation on precipitation. Slope of the topography forces rise of the air mass, promoting precipitation. The temperature drop with elevation results in reduction of the degree of saturation. Precipitation falls to the surface downwind of where it condenses. B. Resulting change in precipitation pattern from the far-field rate for temperature effect only, and for both temperature and slope effects: the latter dominates. Uniform precipitation case shown for comparison (dashed). C. Resulting steady-state channel profiles for temperature-only and full-feedback model. Rock uplift rates are spatially steady and are equal for each simulation. The local channel slope needed to accomplish incision of rising rock mass is everywhere lower in the full model, resulting in a gentler profile and lower total relief in the channel network. Modified after Roe *et al.* (2002).

the range to a site on the river whose incision rate we wish to know. Roe *et al.* (2002) show that this effect results in significantly different slopes of the rivers draining a range whose precipitation pattern is strongly controlled by orographic effects (Fig. 11.23). With the increasing availability of high-resolution spatial records of rainfall variation over the past decade (see Bookhagen and Burbank, 2010), more realistic calculations of upstream discharge should become more routine.

Anders *et al.* (2008) take this analysis further and demonstrate the role of the phase of the precipitation in governing the erosional pattern in

the landscape. As snow falls at roughly 10-fold slower rates than rain, the snow wafts farther downwind than rain from its site of nucleation, allowing snowfall to decouple from the local slopes that generate the lift in the air mass; and it falls more diffusely on higher elevations (see Foster *et al.*, 2010). They quantify this effect using a "delay time" that captures the time it takes for the precipitation to fall from its site of nucleation. That this delay time ought to correlate with the mean annual temperature is corroborated in measurements (Fig. 11.24A). In models of landscape evolution in which the phase of the

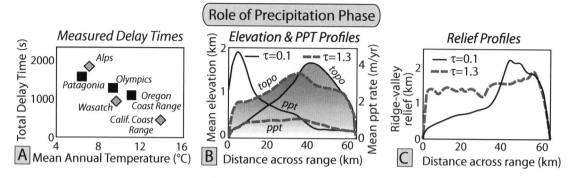

Fig. 11.24   Role of precipitation phase.
A. Measured delay times of precipitation in ranges dominated by snow (low mean annual temperatures) and rainfall (higher temperatures). B,C. Results of coupled models of landscape evolution and orographic precipitation with a preferred wind direction (from left). Delay times (τ) reflect the phase of precipitation (solid lines = short delay = rain; dashed lines = long delay times = snow). B. Mean elevation and precipitation rate profiles across the range. C. Ridge-valley relief across the range. Modified after Anders *et al.* (2008).

precipitation is explicitly taken into account, they show that the feedbacks are indeed strong. The closely coupled system associated with rainfall (short delay times) generates strong asymmetry in precipitation, elevation, and ridge-valley relief, whereas snowfall-dominated ranges (long delay times) show smaller variation in elevation, precipitation, and relief (Fig. 11.24B and C).

But, topography is not one-dimensional: mountain ranges do not wrap around the globe. Galewsky (2009) has elegantly dealt with the broad-scale two-dimensionality of mountain ranges. His models predict that air masses can dodge being lifted by a range if the range is short enough to go around, and that a range can "block" the air mass, causing it to rise well before the slope of the range would otherwise dictate (Fig. 11.25). These effects give rise to precipitation patterns that differ significantly from those predicted in simple one-dimensional models.

In the numerical analysis of the interactions of air masses with topography, a diagnostic relationship is Nh/U, in which N is a measure of atmospheric stability, h is the maximum relief of the range, and U is the wind speed prior to encountering the range (Galewsky, 2009). High wind speeds and low relief ($Nh/U \ll 0$) permit storms to flow directly over mountain ranges without deflection (Fig. 11.25A). Such conditions are typical of one-dimensional orographic precipitation models, which predict precipitation

maxima on the windward slopes of the range (e.g., Fig. 11.23). In contrast, low wind speeds and high relief ($Nh/U \gg 1$) tend to block flow over a range, deflect moisture around its ends, create a zone of relatively stagnant air on their windward flanks, and cause precipitation well upwind of the range (Fig. 11.25B). Another key predictive parameter in these models is the symmetry of the range, β, which represents the ratio of the length to the width of a range. For conical ranges, $\beta = 1$, whereas for elongate ranges, $\beta \gg 1$. Conical ranges deflect winds and moisture much less than do elongate ranges (Fig. 11.25C). Finally, rotation due to Coriolis effects can enhance range-parallel flow and introduce greater asymmetries in precipitation patterns (Fig. 11.25D). Galewsky's (2009) exploration of configurations of different individual and paired ranges provides predictions of wind and moisture patterns that show striking variation dependent on range height, spacing, and sequencing (high range upwind of a low range, or vice versa). Such models provide a template for considering precipitation in growing ranges and in active orogenic belts where successions of elongate ranges grow above laterally extensive faults.

Finally, we illustrate here perhaps the newest tool to be employed in addressing the interaction of the atmosphere and topography (Fig. 11.26 and Plate 11). We are beginning to utilize weather forecasting models to predict

Atmospheric Interaction with Two-dimensional Mountain Ranges

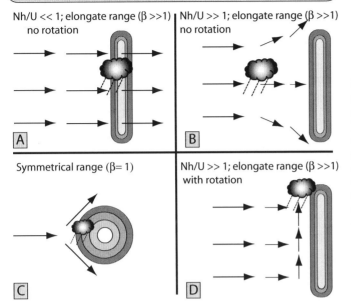

Fig. 11.25   Interaction of incoming moisture with a two-dimensional mountain range.

Diagrams depicting the roles of the non-dimensional number Nh/U, range symmetry β, and rotation of winds due to Coriolis effects on the pattern of precipitation associated with a mountain range. Changes in the ratio (Nh/U) of relief (h) to wind speed (U) are reflected in the degree of blocking of the flow (A versus B). Changes in range symmetry (β) govern the likelihood of flow splitting (A versus C). Coriolis-induced rotation (B versus D) moves the maximum of precipitation off the axis of topographic symmetry. Modified after Galewsky (2009).

Weather Forecasting Models, WRF

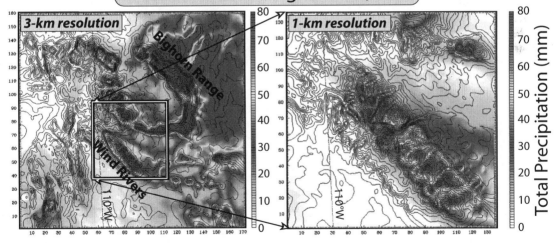

Fig. 11.26   Use of research-grade Weather Research and Forecasting (WRF) models.

A WRF model run at 3 km (left) and at 1 km (right) resolution during a storm event on 4–7 May 2008. Contours represent total precipitation (mm) since the beginning of simulation. Shades (colors) depict precipitation up to hour 72 (3 km run) and hour 60 (1 km run). Note the significant difference in precipitation pattern between the 3 km simulation and the higher-resolution 1 km run. In addition, the temperature at the time of precipitation is very near 0°C in the Bighorn Range, while it is −10°C in the Wind River Range, thereby suggesting a rain–snow mix in the Bighorns and dry snow in the Wind Rivers. Figures courtesy of Gary Clow, US Geological Survey. [A color version of this figure appears as Plate 11.]

precipitation patterns. The most recently developed of these is WRF, the Community Weather Research and Forecasting (WRF) model. WRF models formally ingest NCEP's (National Centers for Environmental Prediction) reanalyzed weather data, and are capable of downscaling these results to 1-km resolution. Given that 1 km is indeed the scale of much valley and ridge topography, we now have a tool to model the precipitation patterns on digital landscapes at the proper spatial and temporal resolution to resolve weather events.

Clever application of these weather models will be needed when coupling them to landscape evolution models operating on long time scales. The spatial scales now match quite well for many applications (data, for example, are at sub-hillslope scales for mountain ranges), but the time scales are seriously mismatched. The WRF models, for example, require model updates at time scales of seconds in order to honor the atmospheric physics. Landscapes evolve on time scales of thousands to millions of years. Development of robust methodologies to bridge these time scales represents a significant challenge to the tectonic-geomorphic community in the coming years.

Glacial models

In the first edition of this book, we reported the lack of attention to the role of glaciers in landscape evolution models. Glaciers are important agents of landscape modification in high alpine settings: exactly the settings to which the tectonic geomorphologist is drawn. The situation has improved significantly in the decade since that writing. In order to erode the landscape, a glacier must slide against its bed and, thereby, abrade or quarry the underlying rock (Iverson, 2002; Hallet, 1979, 1996). Recent attention to the sliding of glaciers has been motivated both by its geomorphic import and its role in delivering ice rapidly to the oceans at the fringe of the Greenland Ice Sheet. Although the sliding process is quite complicated, in that it is intimately intertwined with the seasonally

evolving subglacial hydrologic system, models of glaciers have parameterized sliding in several ways. The problem is analogous to the climate versus weather problem that we discussed above in that the processes that appear to govern sliding operate on short (sub-seasonal) time scales, whereas we wish to address problems of landscape change that occur on time scales of tens to hundreds of thousands of years.

One-dimensional models of long-valley profile evolution were initiated by Oerlemans (1984) and were recast by MacGregor et al. (2000). The more recent models (MacGregor et al., 2009) have incorporated at least a crude hydrologic sub-model that allows the water system to evolve seasonally. These models demonstrate the essence of glacial erosion: glaciated valleys flatten down-valley of the long-term equilibrium-line altitude (ELA) and steepen in their headwalls relative to their fluvial initial profiles. This profile change simply reflects the pattern of ice discharge in glacial systems (Anderson et al., 2006): it increases from zero at the headwall to a maximum at the ELA and then declines to zero at the terminus. Note that this discharge pattern contrasts with fluvial systems in which the typical water discharge increases monotonically to the final sink of a lake or ocean. Water cannot leave the river system except through groundwater loss or evaporation (which are usually minor), whereas ice can leave the glacial system by melting.

Two-dimensional models of alpine glacial systems are now becoming more common and can be tested against moraine patterns that record the occupation of the landscape by ice at the Last Glacial Maximum (Kessler et al., 2006). These two-dimensional models are beginning to be used to address the role of glaciers in modifying two-dimensional landscapes (e.g., Tomkin, 2007). Kessler et al. (2008) demonstrate how the topographic steering of ice through gaps in rift-related escarpments serves to erode these gaps, defeat drainage divides, and promote the growth of major fjord systems that subsequently deliver much of the ice from continental-scale ice sheets to the ocean.

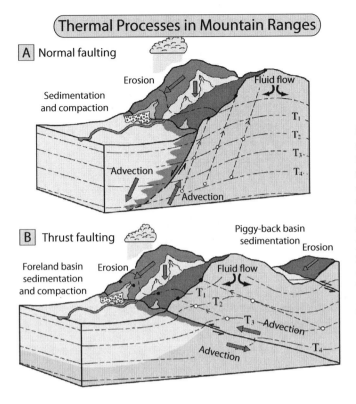

Thermal Processes in Mountain Ranges

A Normal faulting

Erosion

Fluid flow

Sedimentation and compaction

T_1
T_2
T_3
T_4

Advection

Advection

B Thrust faulting

Piggy-back basin sedimentation

Erosion

Foreland basin sedimentation and compaction

Erosion

Fluid flow

T_1
T_2
T_3 Advection
T_4

Advection

Fig. 11.27 Processes governing thermal evolution in faulted mountain ranges. In both normal-faulted cases (A) and thrust-faulted cases (B), heat moves by both conduction and advection. The advection occurs both because the rock and groundwater are in motion and because the surface itself is evolving due to geomorphic processes. Isotherms are warped by these processes, and they intersect the evolving surface at different elevations. The data that we use to interpret long-term rates of fault motion in these settings are the times at which a rock sampled at the surface crossed its particular closure temperatures (T_1, T_2, T_3). Careful interpretation of these ages requires models of the evolving thermal structure of the mountain mass. Modified after Ehlers (2005).

Modeling the thermal field

One of the most common applications of landscape evolution models has been in the interpretation of thermochronometry data. The thermochronometers being used have been pushed to lower and lower closure temperatures, so that they constrain mean exhumation rates of less and less rock. The closure temperatures of the systems have become low enough that the rocks are recording temperatures that occur close enough to the Earth's surface for the isotherms to be modified significantly by the presence of the surface. This warping of the thermal structure must be modeled in order to interpret the depth at which the rocks start their thermal clocks (Fig. 7.20). In such cases, we may not assume that isotherms are horizontal, planar surfaces unaffected by topography (Fig. 11.27 and Plate 4).

Early models solved for the expected thermal structure beneath a uniformly eroding landscape under steady conditions: a topographic form is assigned, and it steadily erodes, retaining its shape. This uniformity is a good place to start, of course, although, as we shall see, this set of assumptions does not allow assessment of several important scenarios. At the surface, the temperatures are controlled by the atmospheric processes that are commonly summarized with a lapse rate, declining with elevation at a rate of, say, 6°C/km. Because this gradient is much smaller than that within the Earth (nominally 25°C/km), the thermal structure is nearly uniform at the surface, so that the topographic surface is almost an isotherm. At great depths, the isotherms are flat: the Earth at these depths knows nothing of thermal variations at the surface. Between the topographic surface and these depths, the isotherms must warp from something closely mimicking the surface to something that is flat. The isotherms are compressed beneath valleys, spread apart beneath ridgetops, and deformed near faults that juxtapose rocks with contrasting thermal states. The spatial pattern of closure ages for particular thermochronometers is predictable (Fig. 7.20), and

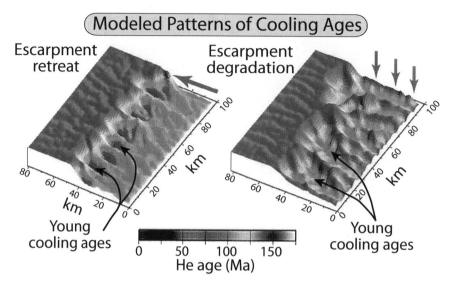

Fig. 11.28   Modeling the generic thermal evolution of an eroding escarpment.
A. Modeled topographic evolution of an escarpment in which geomorphic rules lead to escarpment retreat, and associated predictions of the field of He ages. B. Modeled topographic evolution of an escarpment in which geomorphic rules lead to escarpment degradation, and associated predictions of the field of He ages. Note that the patterns of cooling ages are significantly different when retreat, rather than downwearing, dominates. Modified from Braun and van der Beek (2004). [A color version of this appears as Plate 12.]

the differences in closure ages for two or more chronometers in a single sample are predictable.

At a minimum, even these simple steady models must, therefore, address the following list of thermal aspects of the problem: thermal conduction and potential spatial variations in the thermal diffusivity of the rocks; the role of radiogenic heat production in the rocks, which again can be spatially variable (for example, a silicic pluton embedded in sedimentary rocks); and surface temperature boundary conditions, which will vary with elevation.

But many geological and geomorphic situations of interest in mountainous landscapes are not steady, and the advection of heat can be very complex (see Fig. 11.27). In these cases, we must abandon simple models for the more complex. In models capturing these cases, the thermal algorithms (and the physics they are meant to portray) are the same, but are played out on a two-dimensional grid. One such model is called Pecube (Braun, 2003, 2005). The complexity lies largely in driving the spatial distribution of surface elevation change – in effect, one must

marry a landscape evolution model with a thermal model, and deal with the complexity in both systems. Imagine a model in which the model landscape evolves over millions of years, the final landscape perfectly representing the present-day topography. One could then chose any pixel from the model and assess the thermal history of the rock that emerges on the surface at that point. This history could then be compared with thermochronometric data, and the misfit between the data and the model predictions can be calculated. Effectively, each model is then given a score. After many models are run, one chooses the one yielding the best score (lowest misfit between model and data).

In Fig. 11.28 and Plate 12, we illustrate the application of this strategy to the evolution of a major escarpment (Braun and van der Beek, 2004). Here the question is whether the escarpment evolves by backwearing, which is equivalent to horizontal motion of the scarp, by downwearing, or by some combination of the two. Models of generic escarpments that have evolved with these two end-member styles, governed by differing suites of geomorphic rules, are shown. The land-surface model employed here is CASCADE (Braun and Sambridge, 1997), whereas the thermal model is Pecube. At every pixel, the time since passage of the rock now at the surface through a particular temperature, here the closure temperature for the apatite (U–Th)/He system, is shown. The spatial patterns of the calculated He ages can then be compared with data. The models shown are two of many models run in this investigation. Braun (2005) discusses several strategies for how one might search more efficiently through the many possible models, rather than simply sweeping through the full ranges of the many parameters in such models.

A major challenge in the application of such a strategy to real landscapes, rather than generic ones, is that one must choose both a proper initial condition for the model (an initial landscape some millions of years ago) and a suite of geomorphic processes and rates that will cause the landscape to evolve to the present one. You can imagine that many such scenarios exist. No simple "inverse problem" treatment of the geomorphic system is feasible. Several strategies have emerged to deal with this issue. One is to run zillions of forward models in which initial conditions are chosen and process rates are prescribed, and then model results are assessed by how well they reproduce simultaneously the thermochronometric data and the topography. A second strategy, which can work in certain special circumstances, is to assert that the ancient topography looked like a faint echo of the present and subsequently "morphs" into the present topography. This strategy assures that every run of the model will result in the modern topography, meaning that many fewer models must be run. But, the special circumstances are rather restrictive.

This strategy works in landscapes in which much of the present topography is a relict, frozen-in version of the past topography. In Fig. 11.29 and Plate 13, we show results from Schildgen *et al.*'s (2009) attempt to model the incised edge of the Andean Altiplano into which river canyons have been deeply etched, while much of the remaining landscape has been geomorphically inactive, such that persistent older surfaces are common. As an initial condition, the model strategy takes a topography that is tacked to the present topography of these remnant surfaces and that smoothly drapes across the present river canyons. The difference between this and the present topography is, therefore, mostly the canyon. The model then simply morphs between this initial landscape and the modern: 10% of the canyon relief is in place by a certain age, 30% by some other age, and so on. The thermal history of a rock that emerges at the surface at any sample site can then be assessed. Such models also require assumptions about the boundary conditions at the base of the model: in Schildgen's work, she assumed a particular basal temperature and gradient in that temperature associated with position of the section with respect to the subducting slab. Comparing the modeled apatite He ages with those measured in numerous samples in vertical transects through the canyon walls suggests that the canyon was cut in the interval 9–5 Ma (Fig. 11.29B).

Dynamic topography

At the broadest continental scale, the elevation of the Earth's surface can be significantly affected by stresses imposed on the lithosphere by motion of the mantle in what has been dubbed "dynamic topography." The passage of lithospheric slabs beneath a site, the dropping of drips of anomalously heavy crust, the impact of mantle plumes with the base of the lithosphere, all these can raise or lower the Earth's surface by on the order of 1 km vertically over length scales of thousands of kilometers. It has, for example, been argued that the interior of North America was pulled downward while the Farallon slab was subducting beneath it and has bobbed upward subsequent to the passage of that slab more deeply into the mantle. The geological evidence for this lies largely in the relative sea-level pattern recorded in the Cretaceous seaway (marine) sediments, and the present 1-km elevation of a large fraction of that old seaway – see, e.g., Mitrovica *et al.* (1989) and Forte *et al.* (2007) for discussion of the resulting state of stress in the continental

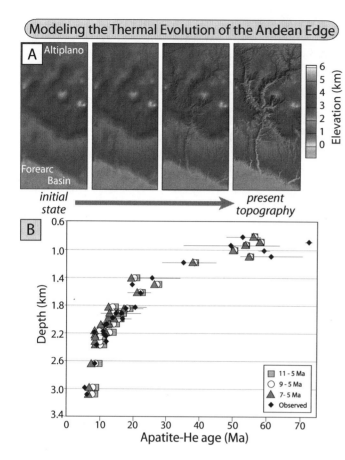

Fig. 11.29 Modeling the thermal evolution of the edge of the Andes.
A. Models of the topography through time, taking the initial state (left) as a splined fit across the modern canyon. Second frame shows 10% of the modern canyon in place; third frame 30%, and fourth frame 100%.
B. Measured apatite He ages at sample locations (black diamonds) and modeled ages for various assumed times and durations of canyon insertion into the landscape. The best-fit time for canyon cutting (circles) is 9–5 Ma. Modified after Schildgen *et al.* (2009). [A color version of this appears as Plate 13.]

interior. Gurnis (2001) has argued that such processes are responsible for the marine sedimentation in interior Australia (Figs 10.46 and 10.47): even though Australia has seen no collisional tectonics for a long time, it has nonetheless overridden the sites of several sinking lithospheric slabs. Modeling of such effects requires modeling mantle flowfields, which in turn requires knowledge of the mantle viscosity structure (e.g., Mitrovica and Forte, 2004).

Another region that has been addressed using dynamic mantle flow models is the Colorado Plateau, the enigmatic region in western North America that has been surprisingly immune to active tectonism in the Cenozoic. Recent work of Flowers *et al.* (2008) suggests a complex pattern of sedimentation and subsequent exhumation in the last 100 Ma. Their story may be at least in part corroborated by models of mantle dynamics (Liu and Gurnis, 2010), as

discussed in commentary to their model (Flowers, 2010). Using a strategy of "reverse tectonics," Liu and Gurnis (2010) roll the clock backward, and scroll the Farallon slab from its present location, as deduced from tomographic images of the mantle beneath North America, back to where it is likely to have been several million years ago. They can, therefore, predict the spatial–temporal pattern of large-scale uplift and subsidence of the overlying North American lithosphere. Such calculations suggest that the Colorado Plateau region should have experienced broad pre-Laramide subsidence and Laramide (80–50 Ma) uplift, with amplitudes of a little more than 1 km (Fig. 11.30). These large-scale models, and those in which smaller-scale convection associated with transitions between the Colorado Plateau and adjacent extending regions are captured (Moucha *et al.*, 2008; van Wijk *et al.*, 2010), may go a long way toward

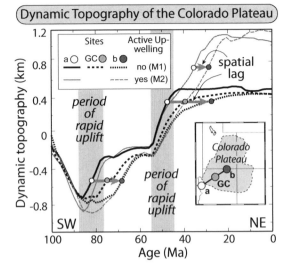

Fig. 11.30   Modeling the uplift history of the Colorado Plateau.
Results of models of the dynamic topography associated with passage of the Farallon slab beneath western North America. The topographic effects of active mantle upwelling are incorporated only in the second (M2). Histories of three sites (shown in inset map centered on Colorado Plateau) reveal early subsidence of 600–800 m, followed by two-stage uplift, with periods of maximum rate of uplift shaded. Notably, these models predict little change in elevation over the past 20 Myr. Lags between sites (heavy arrows) indicate a northeast-migrating wave of Cretaceous to Cenozoic subsidence and uplift in response to mantle slabs and suggest tilting of the surface through time. Modified after Liu and Gurnis (2010).

explaining the complex temporal–spatial elevation history across the Colorado Plateau (Flowers, 2010). Together, these studies both advertise the role of mantle flow models in addressing elevation histories of broad regions, and call for more corroborating evidence in the form of paleoelevation histories in such enigmatic regions.

Present limits and the next generation of models

The progress made by the community in the last decade has been tremendous. Many of the deficiencies identified in the past have been addressed. Coastal, river, and hillslope evolution

components have developed significantly to acknowledge the operation of specific processes in the face of more complete distributions of atmospheric and oceanic events. Glacial models have evolved to fill in the gap at high elevations accessed by the highest of our mountain ranges. Importantly, new models are beginning to couple landscape evolution models both upward and downward as they incorporate both atmospheric and mantle feedbacks.

As we have tried to illustrate, considerably more cross-over now exists between relevant research fields: Atmospheric scientists have become more interested in the rich interactions between topography and the atmosphere. Thermochronologists are now teaming up with landscape modelers and geophysicists specializing in thermal problems to develop models that predict the thermal evolution of rocks that emerge at the surface. Geodynamicists with models of mantle flow are now calculating the impacts of such flowfields on the Earth's surface elevations. The resulting models are now capable of predicting the spatial and temporal patterns of a variety of measurable quantities. Old data sets can now be evaluated – see, e.g., Braun's (2005) re-evaluation of House et al.'s (1998) He age pattern in the Sierras. Just as importantly, however, such models can be and should be used to inform better strategies for collection of new data in the field. Given the high expense of both collecting samples and making the measurements of, say, [10]Be or (U–Th)/He, it behooves the tectonic geomorphologist to make use of these models so that the samples that are collected can exert maximum leverage on our knowledge.

Significant challenges remain in developing strategies to handle the huge discrepancies between (i) time scales and length scales at which real geomorphic processes act and (ii) those at which landscapes evolve significantly enough to alter either atmospheric or geodynamic processes. Whereas great excitement understandably surrounds applications of weather-scale models in landscape evolution, at present these models cannot encompass enough geological time to fill out the full probability of storms that the climate might produce. To overcome such limitations, clever new strategies must be developed that both honor everyday atmospheric physics, while acknowledging the importance of extreme events in geomorphology that reflect the nonlinearity of many geomorphic processes.

We remain challenged by having to choose proper initial conditions in our models: these can play an important role in setting the final look of the landscape. Their significance is reflected by the fact that the most robust feature of a landscape, and hence the most difficult to change, is the planview shape of the drainage pattern: streams most efficiently cut downward, and, unless strongly forced, do not wander significantly once established in a bedrock channel. Unfortunately, it is most difficult to establish with confidence through geological evidence what the channel pattern might have been at the inception of an episode of mountain building. Despite such challenges, the process of stream capture, as well as the distortion of channel patterns through crustal deformation of the rock mass in which they are embedded, need to be addressed in future models. These processes may have played key roles in the evolution of portions of several great collisional ranges.

Finally, the tectonic-geomorphic community faces the challenge of acknowledging and somehow addressing the role of rock type in governing the pace of erosion. More specifically, we must learn what it is about a rock that matters in establishing its susceptibility to erosion by specific processes. The community wallows about in a world of "k"s – call it "the k-problem" – and in general simply chooses a "k" or set of "k"s that yield the pattern found in nature: the stream profile, the hillslope profile, the glacial valley profile, and so on. It is indeed time to attack this issue directly. In tectonically active settings, faults juxtapose rocks of very different character. The tectonic activity itself can modify the rock by causing it to crack as it is made to go around fault bends (Molnar et al., 2007). It should not be surprising to the geomorphologist with any knowledge of specific processes responsible for lowering either river or glacial beds that the availability of cracks for plucking or quarrying processes to exploit is crucial in

determining the rate at which this process can remove rock – see Duhnforth et al. (2010) for the case of the Yosemite landscape. Similarly, whether rocks arrive at the surface pre-fractured due to tectonics or are fractured by top-down geomorphic processes appears to modulate the size and depth of landslides (Clarke and Burbank, 2010a, 2011). Given the importance of rock type and of the roles of faults in both juxtaposing rocks of differing types and generating cracks in rock masses, the quantification of these effects and their incorporation into landscape evolution models is a challenge that should be faced head-on.

These challenges and the general importance of modeling are now acknowledged by the full community. A recent manifestation of this is considerable investment in a Community Surface Dynamics Modeling System (CSDMS) (Anderson et al., 2004). This effort is designed to provide a clearing house for existing models, to promote efficient linkages between these models, and to make data sets relevant to such models readily available.

Summary

We have briefly explored a variety of modeling techniques and results that cover a range of spatial and temporal scales and that allow treatment of landscape elements from river profiles to fault scarps to mountain ranges. Both the tectonic forcing of the system and the suite of geomorphic processes acting to modify the tectonically generated landscape must somehow be incorporated into the model. One must chose a modeling strategy that suits the problem. This choice includes decisions about whether to work in one dimension or two, as well as the style of model (analytical or numerical; finite difference, finite element or boundary element). One must comb the literature to assemble any existing data against which the model can be tested, and can then utilize initial model results to guide the efficient collection of new data that will maximally discriminate among the possible landscape evolution scenarios.

We reiterate that the purpose of modeling in this field is most often the development of insight into the complexity of the various processes interacting to produce a tectonic landscape. We always make simplifications in these efforts. The art lies in the decisions about what degree of detail must be preserved in the model in order to remain faithful enough to the real world to answer the specific questions being asked of it. No model of a scarp will treat every sand grain in the scarp. No model of a mountain range will treat every little bend in the faults bounding it. Finally, the development of models of landscape evolution must go hand in hand with the generation of data sets that are capable of constraining these models. Considerable excitement lies in the interplay among these efforts, as new models force the generation of new data, and new data force the incorporation of new elements in the models. In the last decade, this interplay has been exemplified in the development of both new thermochronometers and the required marriage of models for evolution of both the surface and the subsurface thermal fields in the interpretation of these data. Just as improved global tomography has improved our ability to model and assess dynamic topography, so have global precipitation data sets enabled prediction and testing of interactions of orographic rainfall with geomorphic processes. As our ability to quantify the magnitude and timing of geomorphic and tectonic processes continues to expand, new models, both simple and complex, will provide a critical integrative framework to improve our interpretations and understanding of how the Earth deforms and how tectonic landscapes respond to, record, and modulate that deformation.

References

Abbott, L.D., Silver, E.A., Anderson, R.S., Smith, R., Ingle, J.C., Kling, S.A., Haig, D., Small, E., Galewsky, J., and Sliter, W. (1997) Measurement of tectonic surface uplift rate in a young collisional mountain belt. *Nature*, **385**, 501–507.

Abdrakhmatov, K.Y., Aldazhanov, S.A., Hager, B.H., Hamburger, M.W., Herring, T.A., Kalabaev, K.B., Makarov, V.I., Molnar, P., Panasyuk, S.V., Prilepin, M.T., Reilinger, R.E., Sadybakasov, I.S., Souter, B.J., Trapeznikov, Y.A., Tsurkov, V.Y., and Zubovich, A.V. (1996) Relatively recent construction of the Tien Shan inferred from GPS measurements of present-day crustal deformation rates. *Nature*, **384**, 450–453.

Adams, J. (1990) Paleoseismicity of the Cascadia subduction zone: evidence from turbidites off the Oregon–Washington margin. *Tectonics*, **9**, 569–583.

Adams, K.D., Wesnousky, S.G., and Bills, B.G. (1999) Isostatic rebound, active faulting, and potential geomorphic effects in the Lake Lahontan basin, Nevada and California. *Geological Society of America Bulletin*, **111**, 1739–1756.

Adams, P.N., Anderson, R.S., and Revenaugh, J. (2002) Microseismic measurement of wave-energy delivery to a rocky coast. *Geology*, **30**, 895–898.

Adams, P.N., Storlazzi, C.D., and Anderson, R.S. (2005) Nearshore wave-induced cyclical flexing of sea cliffs. *Journal of Geophysical Research*, **110**, F02002, doi:10.1029/2004JF000217.

Ahnert, F. (1970) Functional relationships between denudation, relief, and uplift in large mid-latitude basins. *American Journal of Science*, **268**, 243–263.

Aitken, M.J. (1985) *Thermoluminescence Dating*. Academic Press, London.

Aitken, M.J. (1998) *An Introduction to Optical Dating: The Dating of Quaternary Sediments by the Use of Photon-Stimulated Luminescence*. Oxford University Press, New York.

Aki, K. (1984) Asperities, barriers, characteristic earthquakes and strong motion prediction. *Journal of Geophysical Research*, **89**, 5867–5872.

Alexander, J., Bridge, J.S., Leeder, M.R., Collier, R.E.L., and Gawthorpe, R.L. (1994) Holocene meander-belt evolution in an active extensional basin, southwestern Montana. *Journal of Sedimentary Research*, **B64**, 542–559.

Allen, C.R. (1986) Seismological and paleoseismological techniques of research in active tectonics. In *Active Tectonics*, pp. 148–154. National Academy Press, Washington, DC.

Allen, P.A. (2008) From landscapes into geological history. *Nature*, **451**, 274–276.

Allmendinger, R.W. (1998) Inverse and forward numerical modeling of trishear fault-propagation folds. *Tectonics*, **17**, 640–656.

Amidon, W.H., Burbank, D.W., and Gehrels, G.E. (2005) U–Pb zircon ages as a sediment mixing tracer in the Nepal Himalaya. *Earth and Planetary Science Letters*, **235**, 244–260.

Amos, C.B., and Burbank, D. (2007) Channel width response to differential uplift. *Journal of Geophysical Research*, **112**, F02010, doi:10.1029/2006JF000672.

Amos, C.B., Burbank, D., Nobes, D., and Read, S.A.L. (2007) Geomorphic constraints on listric thrust faulting: implications for active deformation in the Mackenzie Basin, South Island, New Zealand. *Journal of Geophysical Research*, **112**, B03S11, doi:10.1029/2006JB004291.

Amos, C.B., Burbank, D.W., and Read, S.A.L. (2010) Along-strike growth of the Ostler fault, New Zealand, and consequences for drainage deflection above a non-propagating thrust. *Tectonics*, **29**, TC4021, doi:10.1029/2009TC002613.

Anders, A.M., Roe, G.H., Montgomery, D.R., and Hallet, B. (2008) Influence of precipitation phase on the form of mountain ranges. *Geology*, **36**, 479–482.

Anders, M.H., and Schlische, R.W. (1994) Overlapping faults, intrabasin highs, and the growth of normal faults. *Journal of Geology*, **102**, 165–180.

Andersen, M.B., Stirling, C.H., Potter, E.-K., Halliday, A.N., Blake, S.G., McCulloch, M.T., Ayling, B.F., and O'Leary, M.J. (2010) The timing of sea-level high-stands during Marine Isotope Stages 7.5 and 9: constraints from the uranium-series dating of fossil corals from Henderson Island. *Geochimica et Cosmochimica Acta*, **74**, 3598–3620.

Anderson, D.W., and Rymer, M.J. (1983) *Tectonics and Sedimentation Along Faults of the San Andreas System*. Pacific Section, Society of Economic Paleontologists and Mineralogists, Los Angeles.

Anderson, R.S. (1994) Evolution of the Santa Cruz Mountains, California, through tectonic growth and geomorphic decay. *Journal of Geophysical Research*, **99**, 20161–20179.

Anderson, R.S. (1998) Near-surface thermal profiles in alpine bedrock: implications for the frost-weathering of rock. *Arctic and Alpine Research*, **30**, 362–372.

Anderson, R.S. (2002) Modeling of tor-dotted crests, bedrock edges and parabolic profiles of the high alpine surfaces of the Wind River Range, Wyoming. *Geomorphology*, **46**, 35–58.

Anderson, R.S., and Anderson, S.P. (2010) *Geomorphology: The Mechanics and Chemistry of Landscapes*. Cambridge University Press, Cambridge.

Anderson, R.S., and Humphrey, N.F. (1989) Interaction of weathering and transport processes in the evolution of arid landscapes. In Cross, T.A. (ed.), *Quantitative Dynamic Stratigraphy*, pp. 349–361. Prentice-Hall, Englewood Cliffs, NJ.

Anderson, R.S., and Menking, K.M. (1994) The Santa Cruz marine terraces: evidence for two coseismic uplift mechanisms. *Geological Society of America Bulletin*, **106**, 649–664.

Anderson, R.S., Repka, J.L., and Dick, G.S. (1996) Explicit treatment of inheritance in dating depositional surfaces using in situ [10]Be and [26]Al. *Geology*, **24**, 47–51.

Anderson, R.S., Densmore, A.L., and Ellis, M.A. (1999) The generation and degradation of marine terraces. *Basin Research*, **11**, 7–20.

Anderson, R.S., Dietrich, W.E., Furbish, D., Hanes, D., Howard, A., Paola, C., Pelletier, J., Slingerland, R., Stallard, R., Syvitski, J.P.M., Vörösmarty, C., and

Wiberg, P. (2004) *Community Surface Dynamics Modeling System (CSDMS) Science Plan. A Report to the National Science Foundation*. National Center for Earth-surface Dynamics (NCED), University of Minnesota.

Anderson, R.S., Molnar, P., and Kessler, M.A. (2006) Features of glacial valley profiles simply explained. *Journal of Geophysical Research*, **111**, F01004, doi:10.1029/2005JF000344.

Anderson, S.P., and Anderson, R.S. (1990) Debris-flow benches; dune-contact deposits record paleo-sand dune positions in North Panamint Valley, Inyo County, California. *Geology*, **18**, 524–527.

Anderson, S.P., Bales, R.C., and Duffy, C.J. (2008) Critical Zone Observatories: building a network to advance interdisciplinary study of Earth surface processes. *Mineralogical Magazine*, **72**(1), 7–10.

Attal, M., Tucker, G.E., Whittaker, A.C., Cowie, P.A., and Roberts, G.P. (2008) Modeling fluvial incision and transient landscape evolution: influence of dynamic channel adjustment. *Journal of Geophysical Research*, **113**, F03013, doi:10.1029/2007JF000893.

Atwater, B.F. (1992) Geologic evidence for earthquakes during the past 200 years along the Copalis River, southern coastal Washington. *Journal of Geophysical Research*, **97**, 1901–1919.

Atwater, B.F., and Yamaguchi, D.K. (1991) Sudden, probably coseismic submergence of Holocene trees and grass in coastal Washington State. *Geology*, **19**, 706–709.

Atwater, B.F., Stuiver, M., and Yamaguchi, D.K. (1991) Radiocarbon test of earthquake magnitude at the Cascadia subduction zone. *Nature*, **353**, 156–158.

Atwater, B.F., Nelson, A.R., Clague, J.J., Carver, G.A., Yamaguchi, D.K., Bobrowsky, P.T., Bourgeois, J., Darienzo, M.E., Grant, W.C., Hemphill-Haley, E., Kelsey, H.M., Jacoby, G.C., Nishenko, S.P., Palmer, S.P., Peterson, C.D., and Reinhart, M.A. (1995) Summary of coastal geologic evidence for past great earthquakes at the Cascadia subduction zone. *Earthquake Spectra*, **11**, 1–18.

Avouac, J.-P. (1993) Analysis of scarp profiles: evaluation of errors in morphologic dating. *Journal of Geophysical Research*, **98**, 6745–6754.

Avouac, J.-P. (2003) Mountain building, erosion, and the seismic cycle in the Nepal Himalaya. *Advances in Geophysics*, **46**, 1–80.

Avouac, J.-P., and Peltzer, G. (1993) Active tectonics in southern Xinjiang, China: analysis of terrace riser and normal fault scarp degradation along the Hotan–Qira fault system. *Journal of Geophysical Research*, **98**, 21773–21807.

Avouac, J.-P., Tapponnier, P., Bai, M., You, H., and Wang, G. (1993) Active thrusting and folding along the northern Tien Shan and late Cenozoic rotation of the Tarim relative to Dzungaria and Kazakhstan. *Journal of Geophysical Research*, **98**, 6755–6804.

Axelrod, D.I. (1957) Late Tertiary floras and the Sierra Nevada uplift. *Geological Society of America Bulletin*, **68**, 19–46.

Bada, J.L., Luyendyk, B.P., and Maynard, J.B. (1970) Marine sediments; dating by the racemization of amino acids. *Science*, **170**, 730–732.

Bailey, R.A., Dalrymple, G.B., and Lanphere, M.A. (1976) Volcanism, structure, and geochronology of Long Valley caldera, Mono Country, California. *Journal of Geophysical Research*, **81**, 725–744.

Baker, D.M., Lillie, R.J., Yeats, R.S., Johnson, G.D., Yousuf, M., and Zamin, A.S.H. (1988) Development of the Himalayan thrust zone: Salt Range, Pakistan. *Geology*, **16**, 3–7.

Bakun, W.H., and McEvilly, T.V. (1984) Recurrence models and Parkfield, California, earthquakes. *Journal of Geophysical Research*, **89**, 3051–3058.

Balco, G., Stone, J.O.H., and Jennings, C. (2005) Dating Plio-Pleistocene glacial sediments using the cosmic-ray-produced radionuclides ^{10}Be and ^{26}Al. *American Journal of Science*, **305**, 1–41.

Balco, G., Stone, J.O., Lifton, N.A., and Dunai, T.J. (2008) A complete and easily accessible means of calculating surface exposure ages or erosion rates from ^{10}Be and ^{26}Al measurements. *Quaternary Geochronology*, **3**, 174–195.

Bard, E., Arnold, M., Hamelin, B., Tisnerat-Laborde, N., and Cabioch, G. (1998) Radiocarbon calibration by means of mass spectrometric ^{230}Th/^{234}U and ^{14}C ages of corals; an updated database including samples from Barbados, Mururoa and Tahiti. *Radiocarbon*, **40**, 1085–1092.

Bard, E., Hamelin, B., Fairbanks, R.G., and Zindler, A. (1990) Calibration of the ^{14}C timescale over the past 30,000 years using mass spectrometer U-Th ages from Barbados corals. *Nature*, **345**, 405.

Batt, G.E., Brandon, M.T., Farley, K.A., and Roden-Tice, M. (2001) Tectonic synthesis of the Olympic Mountains segment of the Cascadia wedge, using 2-D thermal and kinematic modeling of isotopic ages. *Journal of Geophysical Research*, **106**, 26731–26746.

Bawden, G.W., Thatcher, W., Stein, R.S., Hudnut, K.W., and Peltzer, G. (2001) Tectonic contraction across Los Angeles after removal of groundwater pumping effects. *Nature*, **412**, 812–815, doi:10.1038/35090558.

Beanland, S., and Clark, M.M. (1994) The Owens Valley fault zone, eastern California, and surface faulting associated with the 1872 earthquake. *U.S. Geological Survey Bulletin*, **B1982**, 29p.

Beaumont, C., Fullsack, P., and Hamilton, J. (1992) Erosional control of active compressional orogens. In McClay, K.R. (ed.), *Thrust Tectonics*, pp. 1–18. Chapman and Hall, London.

Beaumont, C., Ellis, S., Hamilton, J., and Fullsack, P. (1996) Mechanical model for subduction–collision tectonics of Alpine-type compressional orogens. *Geology*, **24**, 675–678.

Beaumont, C., Jamieson, R.A., Nguyen, M.H., and Lee, B. (2001) Himalayan tectonics explained by extrusion of a low-viscosity crustal channel coupled to focused surface denudation. *Nature*, **414**, 738–742.

Beaumont, C., Jamieson, R.A., Nguyen, M.H., and Medvedev, S. (2004) Crustal channel flows: 1. Numerical models with applications to the tectonics of the Himalayan–Tibetan orogen. *Journal of Geophysical Research*, **109**, B06406, doi:10.1029/2003JB002809.

Bechtel, T.D., Forsyth, D.W., Sharpton, V.L., and Grieve, R.A.F. (1990) Variations in effective elastic thickness of the North American lithosphere. *Nature*, **343**, 636–638.

Behr, W.M., Rood, D.H., Fletcher, K.E., Guzman, N., Finkel, R., Hanks, T.C., Hudnut, K.W., Kendrick, K.J., Platt, J.P., Sharp, W.D., Weldon, R.J., and Yule, J.D. (2010) Uncertainties in slip-rate estimates for the Mission Creek strand of the southern San Andreas fault at Biskra Palms Oasis, southern California. *Geological Society of America Bulletin*, **122**, 1360–1377.

Bendick, R., and Flesch, L. (2007) Reconciling lithospheric deformation and lower crustal flow beneath central Tibet. *Geology*, **35**, 895–898.

Benedict, J.B. (1967) Recent glacial history of an alpine area in the Colorado Front Range, USA; Part 1, Establishing a lichen-growth curve. *Journal of Glaciology*, **6**, 817–832.

Bennett, E.R., Youngson, J.H., Jackson, J.A., Norris, R.J., Raisbeck, G.M., Yiou, F., and Fielding, E. (2000) Growth of South Rough Ridge, Central Otago, New Zealand: using *in situ* cosmogenic isotopes and geomorphology to study an active, blind reverse fault. *Journal of Geophysical Research*, **110**, B02404, doi:10.1029/2004JB003184.

Bennett, R.A., Friedrich, A.M., and Furlong, K.P. (2004) Codependent histories of the San Andreas and San Jacinto fault zones from inversion of fault displacement rates. *Geology*, **32**, 961–964.

Benson, L.V., Currey, D.R., Dorn, R.I., Lajoie, K.R., Oviatt, C.G., Robinson, S.W., Smith, G.I., and Stine, S. (1990) Chronology of expansion and contraction of four Great Basin lakes during the past 35,000 years. *Palaeogeography, Palaeoclimatology, Palaeoecology*, **78**, 241–286.

Berger, A.L., and Spotila, J.A. (2008) Denudation and deformation in a glaciated orogenic wedge: the St. Elias orogen, Alaska. *Geology*, **36**, 523–526.

Berger, A.L., Gulick, S.P.S., Spotila, J.A., Upton, P., Jaeger, J.M., Chapman, J.B., Worthington, L.A., Pavlis, T.L., Ridgway, K.D., Willems, B.A., and McAleer, R.J. (2008) Quaternary tectonic response to intensified glacial erosion in an orogenic wedge. *Nature Geoscience*, **1**, 793–799.

Berger, G.W. (1988) Dating Quaternary events by luminescence. In Easterbrook, D.J. (ed.), *Dating Quaternary Sediments*, Special Paper 227, pp. 13–50. Geological Society of America, Boulder, CO.

Berlin, M.M., and Anderson, R.S. (2007) Modeling of knickpoint retreat on the Roan Plateau, western Colorado. *Journal of Geophysical Research*, **112**, F03S06, doi:10.1029/2006JF000553.

Bernet, M., Zattin, M., Garver, J.I., Brandon, M.T., and Vance, J.A. (2001) Steady-state exhumation of the European Alps. *Geology*, **29**, 35–38.

Bernet, M., Brandon, M.T., Garver, J.I., and Molitor, B.R. (2004) Fundamentals of detrital zircon fission-track analysis for provenance and exhumation studies with examples from the European Alps. In Bernet, M., and Spiegel, C. (eds), *Detrital Thermochronology – Provenance Analysis, Exhumation, and Landscape Evolution of Mountain Belts*, Special Paper 378, pp. 25–36. Geological Society of America, Boulder, CO.

Beroza, G.C., and Jordan, T.H. (1990) Searching for slow and silent earthquakes using free oscillations. *Journal of Geophysical Research*, **95**, 2485–2510.

Beschel, R.E. (1961) Dating rock surfaces by lichen growth and its application to glaciology and physiography (lichenometry). *Geology of the Arctic*, **2**, 1044–1062.

Bettinelli, P., Avouac, J.-P., Flouzat, M., Jouanne, F., Bollinger, L., Willis, P., and Chitrakar, G. (2006) Plate motion of India and interseismic strain in the Nepal Himalaya from GPS and DORIS measurements. *Journal of Geodesy*, **80**, 567–589.

Bettinelli, P., Avouac, J.-P., Flouzat, M., Bollinger, L., Ramillien, G., Rajaure, S., and Sapkota, S. (2008) Seasonal variations of seismicity and geodetic strain in the Himalaya induced by surface hydrology. *Earth and Planetary Science Letters*, **266**, 332–344.

Bevis, M., Alsdorf, D., Kendrick, E., Fortes, L.P., Forsberg, B., Smalley, R., Jr., and Becker, J. (2005) Seasonal fluctuations in the mass of the Amazon River system and Earth's elastic response.

Geophysical Research Letters, **32**, L16308, doi:10.1029/2005GL023491.

Biddle, K.T., and Christie-Blick, N. (1985) Strike-slip deformation, basin formation, and sedimentation. In Biddle, K.T., and Christie-Blick, N. (eds), *Strike-Slip Deformation, Basin Formation, and Sedimentation*, Special Publication 37, pp. 375–385. Society of Economic Paleontologists and Mineralogists, Tulsa, OK.

Bierman, P.R. (1994) Using *in situ* produced cosmogenic isotopes to estimate rates of landscape evolution; a review from the geomorphic perspective. *Journal of Geophysical Research*, **99**, 13885–13896.

Bierman, P.R. (2007) Cosmogenic glacial dating, 20 years and counting. *Geology*, **35**, 575–576.

Bierman, P.R., and Gillespie, A.R. (1991) Range fires; a significant factor in exposure–age determination and geomorphic surface evolution. *Geology*, **19**, 641–644.

Bigi, A., Hasbargen, L.E., Montanari, A., and Paola, C. (2006) Knickpoints and hillslope failures: interactions in a steady-state experimental landscape. In Willett, S.D., Hovius, N., Brandon, M.T., and Fisher, D.M. (eds), *Tectonics, Climate and Landscape Evolution*, Special Paper 398, pp. 295–308. Geological Society of America, Boulder, CO.

Bilham, R., Larson, K., Freymuller, J., *et al.* (1997) GPS measurements of present-day convergence across the Nepal Himalaya. *Nature*, **386**, 61–64.

Bilham, R., Gaur, V.K., and Molnar, P. (2001) Himalayan seismic hazard. *Science*, **293**, 1442–1444.

Birkeland, P.W. (1990) Soil-geomorphic research; a selective overview. *Geomorphology*, **3**, 207–224.

Blanpied, M.L., Lockner, D.A., and Byerlee, J.D. (1991) Fault stability inferred from granite sliding experiments at hydrothermal conditions. *Geophysical Research Letters*, **18**, 609–612.

Bloom, A.L., Broecker, W.S., Chappell, J.M., Matthews, R.K., and Mesolella, K.J. (1974) Quaternary sea level fluctuations on a tectonic coast: new ^{230}Th/^{234}U dates from the Huon Peninsula, New Guinea. *Quaternary Research*, **4**, 185–205.

Blum, J.D., Gazis, C.A., Jacobson, A.D., and Chamberlain, C.P. (1998) Carbonate versus silicate weathering in the Raikhot watershed within the High Himalayan Crystalline Series. *Geology*, **26**, 411–414.

Blythe, A.E., Burbank, D.W., Carter, A., Schmidt, K., and Putkonen, J. (2007) Plio-Quaternary exhumation history of the central Nepalese Himalaya: 1. Apatite and zircon fission track and apatite [U–Th]/He analyses. *Tectonics*, **26**, TC3002, doi:10.1029/2006TC001990.

Bollinger, L., Henry, P., and Avouac, J.P. (2006) Mountain building in the Nepal Himalaya: thermal and kinematic model. *Earth and Planetary Science Letters*, **244**, 58–71.

Bookhagen, B., and Burbank, D. (2006) Topography, relief, and TRMM-derived rainfall variations along the Himalaya. *Geophysical Research Letters*, **33**, L08405, doi:10.1029/2006GL026037.

Bookhagen, B., and Burbank, D.W. (2010) Toward a complete Himalayan hydrological budget: spatiotemporal distribution of snowmelt and rainfall and their impact on river discharge. *Journal of Geophysical Research*, **115**, F03019, doi:10.1029/2009JF001426.

Bookhagen, B., and Strecker, M.R. (2008) Orographic barriers, high-resolution TRMM rainfall, and relief variations along the eastern Andes. *Geophysical Research Letters*, **35**, L06403, doi:10.1029/2007GL032011.

Bookhagen, B., Echtler, H.P., Melnick, D., Strecker, M.R., and Spencer, J.Q.G. (2006) Using uplifted Holocene beach berms for paleoseismic analysis on the Santa María Island, south-central Chile. *Geophysical Research Letters*, **33**, L15302, doi:10.1029/2006GL026734.

Bourgois, J., Bigot-Cormier, F., Bourles, D., Braucher, R., Dauteuil, O., Witt, C., and Michaud, F. (2007) Tectonic record of strain buildup and abrupt coseismic stress release across the northwestern Peru coastal plain, shelf, and continental slope during the past 200 kyr. *Journal of Geophysical Research*, **112**, B04104, doi:10.1029/2006JB004491.

Brandon, M.T. (2002) Decomposition of mixed grain-age distributions using BINOMFIT. *On Track*, **24**, 13–18.

Braun, J. (2002) Quantifying the effect of recent relief changes on age–elevation relationships. *Earth and Planetary Science Letters*, **200**, 331–343.

Braun, J. (2003) Pecube: a new finite-element code to solve the 3D heat transport equation including the effects of a time-varying, finite amplitude surface topography. *Computers and Geosciences*, **29**, 787–794.

Braun, J. (2005) Quantitative constraints on the rate of landform evolution derived from low-temperature thermochronology. *Reviews of Mineralogy and Geochemistry*, **58**, 351–374.

Braun, J., and Sambridge, M. (1997) Modelling landscape evolution on geological time scales: a new method based on irregular spatial discretization. *Basin Research*, **9**, 27–52.

Braun, J., and van der Beek, P. (2004) Evolution of passive margin escarpments: what can we learn from low-temperature thermochronology? *Journal of Geophysical Research*, **109**, F04009, doi:10.1029/2004JF000147.

Braun, J., van der Beek, P., and Batt, G. (2006) *Quantitative Thermochronology: Numerical Methods for the Interpretation of Thermochronological Data*. Cambridge University Press, Cambridge.

Brewer, I.D., Burbank, D.W., and Hodges, K.V. (2003) Modelling detrital cooling-age populations: insights from two Himalayan catchments. *Basin Research*, **15**, 305–320.

Brewer, I.D., Burbank, D.W., and Hodges, K.V. (2005) Downstream development of a detrital cooling-age signal: insights from ^{40}Ar/^{39}Ar muscovite thermochronology in the Nepalese Himalaya. In Willett, S.D., Hovius, N., Brandon, M.T., and Fisher, D. (eds), *Tectonics, Climate, and Landscape Evolution*, Special Paper 398, pp. 321–338. Geological Society of America, Boulder, CO, doi:10.1130/2006.2398(20).

Briggs, R.W., Sieh, K., Meltzner, A.J., Natawidjaja, D., Galetzka, J., Suwargadi, B., Hsu, Y.-J., Simons, M., Hananto, N., Suprihanto, I., Prayudi, D., Avouac, J.-P., Prawirodirdjo, L., and Bock, Y. (2006) Deformation and slip along the Sunda Megathrust in the great 2005 Nias–Simeulue earthquake. *Science*, **311**, 1897–1901.

Briggs, R.W., Sieh, K., Amidon, W.H., Galetzka, J., Prayudi, D., Suprihanto, I., Sastra, N., Suwargadi, B., Natawidjaja, D., and Farr, T.G. (2008) Persistent elastic behavior above a megathrust rupture patch: Nias island, West Sumatra. *Journal of Geophysical Research*, **113**, B12406, doi:10.1029/2008JB005684.

Brocklehurst, S.H., and Whipple, K.X. (2002) Glacial erosion and relief production in the eastern Sierra Nevada, California. *Geomorphology*, **42**, 1–24.

Brocklehurst, S.H., and Whipple, K.X. (2007) Response of glacial landscapes to spatial variations in rock uplift rate. *Journal of Geophysical Research*, **112**, F02035, doi:10.1029/2006JF000667.

Brown, E.T., Stallard, R.F., Larsen, M.C., Raisbeck, G.M., and Yiou, F. (1995) Denudation rates determined from the accumulation of in situ-produced ^{10}Be in the Luquillo experimental forest, Puerto Rico. *Earth and Planetary Science Letters*, **129**, 193–202.

Brown, K.M., Tryon, M.D., DeShon, H.R., Dorman, L.M., and Schwartz, S.Y. (2005) Correlated transient fluid pulsing and seismic tremor in the Costa Rica subduction zone. *Earth and Planetary Science Letters*, **238**, 189–203.

Brozović, N., Burbank, D.W., Fielding, E., and Meigs, A.J. (1995) The spatial and temporal topographic evolution of Wheeler Ridge California: new insights from digital elevation data. *Geological Society of America Abstracts with Programs*, **27**, 396.

Brozović, N., Burbank, D.W., and Meigs, A.J. (1997) Climatic limits on landscape development in the northwestern Himalaya. *Science*, **276**, 571–574.

Brune, J.N. (1996) Precariously balanced rocks and ground-motion maps for southern California. *Bulletin of the Seismological Society of America*, **86**, 43–54.

Brune, J.N. (2002) Precarious-rock constraints on ground motion from historic and recent earthquakes in southern California. *Bulletin of the Seismological Society of America*, **92**, 2602–2611.

Brune, J.N., Anooshehpoor, A., Purvance, M.D., and Brune, R.J. (2006) Band of precariously balanced rocks between the Elsinore and San Jacinto, California, fault zones: constraints on ground motion for large earthquakes. *Geology*, **34**, 137–140.

Bruns, T.R., and Schwab, W.C. (1983) Structure maps and seismic stratigraphy of the Yakataga segment of the continental margin, northern Gulf of Alaska. Miscellaneous Field Studies Map, MF-1424. United States Geological Survey, Washington, DC.

Bucknam, R.C., and Anderson, R.E. (1979) Estimation of fault-scarp ages from a scarp-height–slope-angle relationship. *Geology*, **7**, 11–14.

Buddemeier, R.W., and Taylor, F.W. (2000) Sclerochronology. In Noller, J.S., Sowers, J.M., and Lettis, W.R. (eds), *Quaternary Geochronology: Methods and Applications*, pp. 25–40. American Geophysical Union, Washington, DC.

Buech, F., Davies, T.R., and Pettinga, J.R. (2010) The Little Red Hill seismic experimental study: topographic effects on ground motion at a bedrock-dominated mountain edifice. *Bulletin of the Seismological Society of America*, **100**, 2219–2229.

Bull, W.B. (1964) *Geomorphology of Segmented Alluvial Fans in Western Fresno County, California*, Professional Paper 352-E, pp. 89–129. United States Geological Survey, Washington, DC.

Bull, W.B. (1991) *Geomorphic Responses to Climatic Change*. Oxford University Press, London.

Bull, W.B. (1996) Dating San Andreas fault earthquakes with lichenometry. *Geology*, **24**, 111–114.

Bull, W.B., and Brandon, M.T. (1998) Lichen dating of earthquake-generated regional rockfall events, Southern Alps, New Zealand. *Geological Society of America Bulletin*, **110**, 60–84.

Bull, W.B., and Cooper, A.F. (1986) Uplifted marine terraces along the Alpine fault, New Zealand. *Science*, **234**, 1225–1228.

Bull, W.B., and McFadden, L.D. (1977) Tectonic geomorphology north and south of the Garlock Fault, California. In Doehring, D.O. (ed.), *Geomorphology in Arid Regions*, pp. 115–138. State University of New York at Binghamton, Binghamton, NY.

Bull, W.B., King, J., Kong, F., Moutoux, T., and Phillips, W.M. (1994) Lichen dating of coseismic landslide hazards in alpine mountains. *Geomorphology*, **10**, 253–264.

Bullen, M.E., Burbank, D.W., Abdrakhmatov, K.Y., and Garver, J. (2001) Late Cenozoic tectonic evolution of the northwestern Tien Shan: constraints from magnetostratigraphy, detrital fission track, and basin analysis. *Geological Society of America Bulletin*, **113**, 1544–1559.

Bullen, M.E., Burbank, D.W., and Garver, J.I. (2003) Building the northern Tien Shan: integrated thermal, structural, and topographic constraints. *Journal of Geology*, **111**, 149–165.

Burbank, D.W. (1983) Multiple episodes of catastrophic flooding in the Peshawar basin during the past 700,000 years. *Geological Bulletin of the University of Peshawar*, **16**, 4349.

Burbank, D.W. (1992) Causes of recent Himalayan uplift deduced from deposited patterns in the Ganges basin. *Nature*, **357**, 680–682.

Burbank, D.W., and Beck, R.A. (1991) Rapid, long-term rates of denudation. *Geology*, **19**, 1169–1172.

Burbank, D.W., and Raynolds, R.G.H. (1984) Sequential late Cenozoic structural disruption of the northern Himalayan foredeep. *Nature*, **311**, 114–118.

Burbank, D.W., and Tahirkheli, R.A.K. (1985) Magnetostratigraphy, fission-track dating, and stratigraphic evolution of the Peshawar intermontane basin, northern Pakistan. *Geological Society of America Bulletin*, **96**, 530–552.

Burbank, D.W., and Vergés, J. (1994) Reconstruction of topography and related depositional systems during active thrusting. *Journal of Geophysical Research*, **99**, 20281–20297.

Burbank, D.W., Beck, R.A., and Mulder, T. (1996a) The Himalayan foreland. In An, Y., and Harrison, T.M. (eds), *Asian Tectonics*, pp. 149–188. Cambridge University Press, Cambridge.

Burbank, D.W., Leland, J., Fielding, E., Anderson, R.S., Brozović, N., Reid, M.R., and Duncan, C. (1996b) Bedrock incision, rock uplift and threshold hillslopes in the northwestern Himalayas. *Nature*, **379**, 505–510.

Burbank, D.W., Meigs, A., and Brozović, N. (1996c) Interactions of growing folds and coeval depositional systems. *Basin Research*, **8**, 199–223.

Burbank, D.W., McLean, J.K., Bullen, M.E., Abdrakh-matov, K.Y., and Miller, M.M. (1999) Partitioning of intermontane basins by thrust-related folding, Tien Shan, Kyrgyzstan. *Basin Research*, **11**, 75–92.

Burbank, D., Hudnut, K., Ryerson, R., Rubin, C., Schwartz, D., Wernicke, B., and Wesnousky, S. (2002) *The Plate Boundary Observatory: Results of First Workshop on Geological Research*, http://www.unavco.org/research_science/publications/proposals/pbo/pbo.html.

Burbank, D.W., Blythe, A.E., Putkonen, J., Pratt-Sitaula, B., Gabet, E., Oskin, M., Barros, A., and Ojha, T.P. (2003) Decoupling of erosion and precipitation in the Himalayas. *Nature*, **426**, 652–655.

Burbank, D.W., Brewer, I.D., Sobel, E.R., and Bullen, M.E. (2007) Single-crystal dating and the detrital record of orogenesis. In Nichols, G., Williams, E., and Paola, C. (eds), *Sedimentary Processes, Environments and Basins: A Tribute to Peter Friend*, pp. 253–281. Blackwell Publishing, Malden, MA.

Burchfiel, B.D., Chen, Z., Hodges, K.V., Liu, Y., Royden, L.H., Deng, C., and Xu, J. (1992) *The South Tibetan Detachment System, Himalayan Orogen: Extension Contemporaneous with and Parallel to Shortening in a Collisional Mountain Belt*, Special Paper 269. Geological Society of America, Boulder, CO.

Bürgmann, R., Rosen, P.A., and Fielding, E.J. (2000) Synthetic aperture radar interferometry to measure earth's surface topography and its deformation. *Annual Review of Earth and Planetary Sciences*, **28**, 169–209.

Bürgmann, R., Hilley, G., Ferretti, A., and Novali, F. (2006) Resolving vertical tectonics in the San Francisco Bay area from permanent scatterer InSAR and GPS analysis. *Geology*, **34**, 221–224.

Bursik, M.I., and Gillespie, A.R. (1993) Late Pleistocene glaciation of Mono Basin, California. *Quaternary Research*, **39**, 24–35.

Byrd, J.O., Smith, R.B., and Geissman, J.W. (1994) The Teton Fault, Wyoming: topographic signature, neo-tectonics, and mechanisms of deformation. *Journal of Geophysical Research*, **99**, 20095–20122.

Campbell, R., McCarroll, D., Loader, N.J., Grudd, H., Robertson, I., and Jalkanen, R. (2007) Blue intensity in *Pinus sylvestris* tree-rings: developing a new paleoclimate proxy. *The Holocene*, **17**, 821–828.

Carslaw, H.S., and Jaeger, J.C. (1986) *Conduction of Heat in Solids*. Oxford University Press, New York.

Carson, M.A., and Kirkby, M.J. (1972) *Hillslope Form and Process*. Cambridge University Press, London.

Carter, W., Shrestha, R., and Slatton, K. (2007) Geodetic laser scanning. *Physics Today*, **60**, 41–48.

Cartwright, J.A., Trudgill, B.D., and Mansfield, C.S. (1995) Fault growth by segment linkage: an explanation for scatter in maximum displacement and trace length data from the Canyonlands grabens of SE Utah. *Journal of Structural Geology*, **17**, 1319–1326.

Carver, G.A., Jayko, A.S., Valentine, D.W., and Li, W.H. (1994) Coastal uplift associated with the 1992 Cape Mendocino earthquake, Northern California. *Geology*, **22**, 195–198.

Cassel, E.J., Graham, S.A., and Chamberlain, C.P. (2009) Cenozoic tectonic and topographic evolution of the northern Sierra Nevada, California, through stable isotope paleoaltimetry in volcanic glass. *Geology*, **37**, 547–550.

Castle, R.O., Estrem, J.E., and Savage, J.C. (1984) Uplift across the Long Valley caldera, California. *Journal of Geophysical Research*, **89**, 11507–11516.

Cattin, R., and Avouac, J.-P. (2000) Modeling mountain building and the seismic cycle in the Himalaya of Nepal. *Journal of Geophysical Research*, **105**, 13389–13407.

Cavalié, O., Lasserre, C., Doin, M.P., Peltzer, G., Sun, J., Xu, X., and Shen, Z.K. (2008) Measurement of inter-seismic strain across the Haiyuan fault (Gansu, China), by InSAR. *Earth and Planetary Science Letters*, **275**, 246–257.

Cerling, T.E., and Craig, H. (1994) Geomorphology and in-situ cosmogenic isotopes. *Annual Review of Earth and Planetary Sciences*, **22**, 273–317.

Chamberlain, C.P., and Poage, M.A. (2000) Reconstructing the paleotopography of mountain belts from the isotopic composition of authigenic minerals. *Geology*, **28**, 115–118.

Chamberlain, C.P., Poage, M.A., Craw, D., and Reynolds, R.C. (1999) Topographic development of the Southern Alps recorded by the isotopic composition of authigenic clay minerals, South Island, New Zealand. *Chemical Geology*, **155**, 279–294.

Chappell, J. (1974) Geology of coral terraces, Huon Peninsula, New Guinea: a study of Quaternary tectonic movements and sea-level changes. *Geological Society of America Bulletin*, **85**, 553–570.

Chappell, J., Omura, A., Esat, T., McCulloch, M., Pandolfi, J., Ota, Y., and Pillans, B. (1996) Reconciliation of late Quaternary sea levels derived from coral terraces at Huon Peninsula with deep sea oxygen isotope records. *Earth and Planetary Science Letters*, **141**, 227–236.

Chediya, O.K. (1986) *Morphostructure and Neo-tectonics of the Tien Shan*. Academia Nauk Kyrgyz CCP, Frunze (Bishkek), Kyrgyzstan.

Chen, J., Chen, Y., Ding, G., Wang, Z., Tian, Q., Yin, G., Shan, X., and Wang, Z. (2004) Surficial slip distribution and segmentation of the 426 km long surface rupture of the 14 November 2001 Ms 8.1 earthquake on the East Kunlun Fault, northern Tibetan Plateau, China. *Seismology and Geology*, **26**, 378–392 (in Chinese with English abstract).

Chen, Y.-G., Lai, K.-Y., Lee, Y.-H., Suppe, J., Chen, W.-S., Lin, Y.-N.N., Want, Y., Hung, J.-H., and Kuo, Y.-T. (2007) Coseismic fold scarps and their kinematic behavior in the 1999 Chi-Chi earthquake Taiwan. *Journal of Geophysical Research*, **112**, B03S02, doi:10.1029/2006JB004388.

Cheng, H., Edwards, R.L., Hoff, J., Gallup, C.D., Richards, D.A., and Asmerom, Y. (2000) The half-life of uranium-234 and thorium-230. *Chemical Geology*, **169**, 17–33.

Chlieh, M., Avouac, J.-P., Hjorleifsdottir, V., Song, T.-R.A., Ji, C., Sieh, K., Sladen, A., Hebert, H., Prawirodirdjo, L., Bock, Y., and Galetzka, J. (2007) Coseismic slip and afterslip of the great Mw 9.15 Sumatra–Andaman earthquake of 2004. *Bulletin of the Seismological Society of America*, **97**(Suppl), S152–S173.

Christie-Blick, N., and Biddle, K.T. (1985) Deformation and basin formation along strike-slip faults. In Biddle, K.T., and Christie-Blick, N. (eds), *Strike-Slip Deformation, Basin Formation, and Sedimentation*, Special Publication 37, pp. 1–34. Society of Economic Paleontologists and Mineralogists, Tulsa, OK.

Church, M., and Slaymaker, O. (1989) Disequilibrium of Holocene sediment yield in glaciated British Columbia. *Nature*, **337**, 452–454.

Clark, D.H., Bierman, P.R., and Larsen, P. (1995) Improving *in situ* cosmogenic chronometers. *Quaternary Research*, **44**, 367–377.

Clark, J.A., Farrell, W.E., and Peltier, W.R. (1978) Global changes in postglacial sea level: a numerical calculation. *Quaternary Research*, **9**, 265–287.

Clark, M.K., and Royden, L.H. (2000) Topographic ooze: building the eastern margin of Tibet by lower crustal flow. *Geology*, **28**, 703–706.

Clark, M.K., Schoenbohm, L.M., Royden, L.H., Whipple, K.X., Burchfiel, B.C., Zhang, X., Tang, W., Wang, E., and Chen, L. (2004) Surface uplift, tectonics, and erosion of eastern Tibet from large-scale drainage patterns. *Tectonics*, **23**, TC1006, doi:10.1029/2002TC001402.

Clark, M.K., Bush, J.W.M., and Royden, L.H. (2005a) Dynamic topography produced by lower crustal flow against rheological strength heterogeneities bordering the Tibetan Plateau. *Geophysical Journal International*, **162**, 575–590, doi:10.1111/j.1365-246X.2005.02580.x.

Clark, M.K., Maheo, G., Saleeby, J., and Farley, K.A. (2005b) The non-equilibrium landscape of the southern Sierra Nevada, California. *GSA Today*, **15**, 4–10.

Clarke, B.A., and Burbank, D.W. (2010a) Bedrock fracturing, threshold hillslopes, and limits to the magnitude of bedrock landslides. *Earth and Planetary Science Letters*, **297**, 577–586.

Clarke, B.A., and Burbank, D.W. (2010b) Evaluation hillslope diffusion and terrace riser degradation in New Zealand anf Idaho. *Journal of Geophysical Research*, **115**, F02013, doi:02010.01029/02009JF 001279.

Clarke, B.A., and Burbank, D.W. (2011) Quantifying bedrock fracture densities in the shallow subsurface and the implications for bedrock landslides and erodability. *Journal of Geophysical Research*, submitted.

Clift, P.D., and Blusztajn, J. (2005) Reorganization of the western Himalayan river system after five million years ago. *Nature*, **438**, 1001–1003.

Colman, S.M. (1986) Levels of time information in weathering measurements, with examples from weathering rinds on volcanic clasts in the Western United States. In Colman, S.M., and Dethier, D.P. (eds), *Rates of Chemical Weathering of Rocks and Minerals*, pp. 379–393. Academic Press, Orlando, FL.

Colman, S.M., and Dethier, D.P. (eds) (1986) *Rates of Chemical Weathering of Rocks and Minerals*. Academic Press, Orlando, FL.

Colman, S.M., and Pierce, K.L. (1992) Varied records of early Wisconsinan alpine glaciation in the Western United States derived from weathering-rind thickness. In Clark, P.U., and Lea, P.D. (eds), *The Last Interglacial–Glacial Transition in North America*, Special Paper 270, pp. 269–278. Geological Society of America, Boulder, CO.

Colman, S.M., and Watson, K. (1983) Ages estimated from a diffusion equation model for scarp degradation. *Science*, **221**, 263–265.

Cotton, C.A. (1922) *Geomorphology of New Zealand*, Part I, *Systematic*. New Zealand Board of Science and Art, Wellington, New Zealand.

Cowgill, E. (2007) Impact of riser reconstructions on estimation of secular variation in rates of strike-slip faulting: revisiting the Cherchen River site along the Altyn Tagh Fault, NW China. *Earth and Planetary Science Letters*, **254**, 239–255.

Cowie, P.A., and Scholz, C.H. (1992) Displacement–length scaling relationship for faults: data synthesis and discussion. *Journal of Structural Geology*, **14**, 1149–1156.

Cowie, P.A., Gupta, S., and Dawers, N.H. (2000) Implications of fault array evolution for synrift depocentre development; insights from a numerical fault growth model. *Basin Research*, **12**, 241–261.

Cowie, P.A., Underhill, J.R., Behn, M.D., Lin, J., and Gill, C.E. (2005) Spatio-temporal evolution of strain accumulation derived from multi-scale observations of Late Jurassic rifting in the northern North Sea: a critical test of models for lithospheric extension. *Earth and Planetary Science Letters*, **234**, 401–419.

Cox, A., Doell, R.R., and Dalrymple, G.B. (1964) Reversals of the earth's magnetic field. *Science*, **144**, 1537–1543.

Cox, R.T. (1994) Analysis of drainage-basin symmetry as a rapid technique to identify areas of possible Quaternary tilt-block tectonics: an example from the Mississippi embayment. *Geological Society of America Bulletin*, **104**, 571–581.

Craddock, W.H., Burbank, D.W., Bookhagen, B., and Gabet, E.J. (2007) Bedrock channel geometry along an orographic precipitation gradient in the upper Marsyandi River valley in central Nepal. *Journal of Geophysical Research*, **112**, F03007, doi:10.1029/2006JF000589.

Craddock, W.H., Kirby, E., Harkins, N.W., Zhang, H., Shi, X., and Liu, J. (2010) Rapid fluvial incision along the Yellow River during headward basin integration. *Nature Geoscience*, **3**, 209–213.

Craw, D., Koons, P.O., Winslow, D., Chamberlain, C.P., and Zeitler, P. (1994) Boiling fluids in a region of rapid uplift, Nanga Parbat massif, Pakistan. *Earth and Planetary Science Letters*, **128**, 169–182.

Creer, K.M. (1962) The dispersion of the geomagnetic field due to secular variation and its determination for remote times from paleomagnetic data. *Journal of Geophysical Research*, **67**, 3461–3476.

Creer, K.M. (1967) Application of rock magnetism to investigations of the secular variation during geological time. In *Magnetism and the Cosmos*, NATO Advanced Study Institute on Planetary and Stellar Magnetism, University of Newcastle upon Tyne, 1965, pp. 45–59. Oliver and Boyd, Edinburgh.

Crespi, J.M., Chan, Y.-C., and Swaim, M.S. (1996) Synorogenic extension and exhumation of the Taiwan hinterland. *Geology*, **24**, 247–250.

Crider, J.G., and Pollard, D.D. (1998) Fault linkage: three-dimensional mechanical interaction between echelon normal faults. *Journal of Geophysical Research*, **103**, 24373–24391.

Crook, R., Jr. (1986) Relative dating of Quaternary deposits based on P-wave velocities in weathered granitic clasts. *Quaternary Research*, **25**, 281–292.

Crook, R., Jr., and Gillespie, A.R. (1986) Weathering rates in granitic boulders measured by P-wave speeds. In Colman, S.M., and Dethier, D.P. (eds), *Rates of Chemical Weathering of Rocks and Minerals*, pp. 395–417. Academic Press, Orlando, FL.

Crosby, B.T., and Whipple, K.X. (2006) Knickpoint initiation and distribution within fluvial networks: 236 waterfalls in the Waipaoa River, North Island, New Zealand. *Geomorphology*, **82**, 16–38, doi:10.1016/j.geomorph.2005.08.023.

Crosby, B.T., Whipple, K.X., Gasparini, N.M., and Wobus, C.W. (2007) Formation of fluvial hanging valleys: theory and simulation. *Journal of Geophysical Research*, **112**, F03S10, doi:10.1029/2006JF000566.

Crouch, S.L., and Starfield, A.M. (1983) *Boundary Element Methods in Solid Mechanics: with Applications in Rock Mechanics and Geological Engineering*. Allen & Unwin, London.

Crowley, J.L., Schoene, B., and Bowring, S.A. (2007) U–Pb dating of zircon in the Bishop Tuff at the millennial scale. *Geology*, **35**, 1123–1126.

Culling, W.E.H. (1960) Analytical theory of erosion. *Journal of Geology*, **68**, 336–344.

Culling, W.E.H. (1963) Soil creep and the development of hillside slopes. *Journal of Geology*, **71**, 127–161.

Culling, W.E.H. (1965) Theory of erosion on soil-covered slopes. *Journal of Geology*, **73**, 230–254.

Culmann, C. (1875) *Die Graphische Statik*. Meyer and Zeller, Zurich.

Dadson, S.J., Hovius, N., Chen, H., Dade, W.B., Hsieh, M.-L., Willett, S.D., Hu, J.-C., Horng, M.-J., Chen, M.-C., Stark, C.P., Lague, D., and Lin, J.-C. (2003) Links between erosion, runoff variability and seismicity in the Taiwan orogen. *Nature*, **426**, 648–651.

Dadson, S.J., Hovius, N., Chen, H., Dade, W.B., Lin, J.-C., Hsu, M.-L., Lin, C.-W., Horng, M.-J., Chen, T.-C., Milliman, J., and Stark, C.P. (2004) Earthquake-triggered increase in sediment delivery from an active mountain belt. *Geology*, **32**, 733–736.

Dahlen, F.A., and Suppe, J. (1988) Mechanics, growth, and erosion of mountain belts. In Clark, S.P.J., Burchfiel, B.C., and Suppe, J. (eds), *Processes in Continental Lithospheric Deformation*, pp. 161–178. Geological Society of America, Denver, CO.

Davis, D., Suppe, J., and Dahlen, F.A. (1983) Mechanics of fold-and-thrust belts and accretionary wedges. *Journal of Geophysical Research*, **88**, 1153–1172.

Davis, K., Burbank, D.W., Fisher, D., Wallace, S., and Nobes, D. (2005) Thrust-fault growth and segment linkage in the active Ostler fault zone, New Zealand. *Journal of Structural Geology*, **27**, 1528–1546.

Davis, W.M. (1899) The geographical cycle. *Geographical Journal*, **14**, 481–504.

Dawers, N.H., and Anders, M.H. (1995) Displacement–length scaling and fault linkage. *Journal of Structural Geology*, **17**, 607–614.

Dawers, N.H., Anders, M.H., and Scholz, C.H. (1993) Growth of normal faults: displacement–length scaling. *Geology*, **21**, 1107–1110.

Delouis, B., Giardini, D., Lundgren, P., and Salichon, J. (2002) Joint inversion of InSAR, GPS, teleseismic, and strong-motion data for the spatial and temporal distribution of earthquake slip: application to the 1999 Izmit mainshock. *Bulletin of the Seismological Society of America*, **92**, 278–299.

DeMets, C., and Dixon, T.H. (1999) New kinematic models for Pacific–North America motion from 3 Ma to present. I: Evidence for steady motion and biases in the NUVEL-1A model. *Geophysical Research Letters*, **26**, 1921–1924.

DeMets, C., Gordon, R.G., Argus, D.F., and Stein, S. (1990) Current plate motions. *Geophysical Journal International*, **101**, 425–478.

DeMets, C., Gordon, R.G., and Stein, D.F. (1994) Effect of recent revisions to the geomagnetic reversal time scale on estimates of current plate motions. *Geophysical Research Letters*, **21**, 2191–2194.

Densmore, A.L., and Hovius, N. (2000) Topographic fingerprints of bedrock landslides. *Geology*, **28**, 371–374.

Densmore, A.L., Ellis, M.A., and Anderson, R.S. (1998) Landsliding and the evolution of normal-fault-bounded mountains. *Journal of Geophysical Research*, **103**, 15203–15219.

Densmore, A.L., Dawers, N.H., Gupta, S., and Guidon, R. (2005) What sets topographic relief in extensional footwalls? *Geology*, **33**, 453–456.

Densmore, A.L., Gupta, S., Allen, P.A., and Dawers, N.H. (2007) Transient landscapes at fault tips. *Journal of Geophysical Research*, **112**, F03S08, doi:10.1029/2006JF000560.

Denton, G.H., and Karlen, W. (1973) Lichenometry: its application to Holocene moraine studies in southern Alaska and Swedish Lapland. *Arctic and Alpine Research*, **5**, 347–372.

Derry, L.A., Evans, M.J., Darling, R., and France-Lanord, C. (2009) Hydrothermal heat flow near the Main Central Thrust, central Nepal Himalaya. *Earth and Planetary Science Letters*, **286**, 101–109.

Dietrich, W.E., Bellugi, D., Heimsath, A.M., Roering, J.J., Sklar, L., and Stock, J.D. (2003) Geomorphic transport laws for predicting the form and evolution of landscapes. In Wilcock, P., and Iverson, R. (eds), *Prediction in Geomorphology*, AGU Geophysical Monograph Series, 135, pp. 103–132. American Geophysical Union, Washington, DC.

Dixon, T.H. (1991) An introduction to the global positioning systems and some geological applications. *Reviews of Geophysics*, **29**, 249–276.

Dixon, T., Norabuena, E., and Hotaling, L. (2003) Paleoseismology and Global Positioning System; earthquake-cycle effects and geodetic versus geologic fault slip rates in the Eastern California shear zone. *Geology*, **31**, 55–58.

Dolan, J.F., Christofferson, S.A., and Shaw, J.H. (2003) Recognition of paleoearthquakes on the Puente Hills blind thrust fault, California. *Science*, **300**, 115–118, doi:10.1126/science.1080593.

Dolan, J.F., Bowman, D.D., and Sammis, C.G. (2007) Long-range and long-term fault interactions in Southern California. *Geology*, **35**, 855–858.

Dong, S.P., Han, Z.J., and An, Y.F. (2008) Surface deformation at the epicenter of the May 12, 2008 Wenchuan M8 earthquake, at Yingxiu town of Sichuan Province, China. *Science in China Series E: Technological Sciences*, **51**, 151–163.

Dorsey, R.J. (2002) Stratigraphic record of Pleistocene initiation and slip on the Coyote Creek fault, Lower Coyote Creek, southern California. In Barth, A. (ed.), *Contributions to Crustal Evolution of the South-western United States*, Special Paper 365, pp. 251–269. Geological Society of America, Boulder, CO.

Douglas, B.C. (1991) Global sea level rise. *Journal of Geophysical Research*, **96**, 6981–6992.

Dowdeswell, J.A., Unwin, B., Nuttall, A.M., and Wingham, D.J. (1999) Velocity structure, flow instability and mass flux on a large Arctic ice cap from satellite radar interferometry. *Earth and Planetary Science Letters*, **167**, 131–140.

Dragert, H., Wang, K., and James, T.S. (2001) A silent slip event on the deeper Cascadia subduction interface. *Science*, **292**, 1525–1528.

Ducea, M.N., and Saleeby, J.B. (1998) A case for delamination of the deep batholithic crust beneath the Sierra Nevada. *International Geology Review*, **40**, 78–93.

Dühnforth, M., Anderson, R.S., Ward, D., and Stock, G.M. (2010) Bedrock fracture control of glacial erosion processes and rates. *Geology*, **38**, 423–426.

Duller, G.A.T. (1996) Recent developments in luminescence dating of Quaternary sediments. *Progress in Physical Geography*, **20**, 127–145.

Duvall, A., Kirby, E., and Burbank, D. (2004) Tectonic and lithologic controls on bedrock channel profiles and processes in coastal California. *Journal of Geophysical Research*, **109**, F03002, doi:10.1029/2003JF000086.

Edgar, D.E. (1973) *Geomorphic and hydraulic properties of laboratory rivers*. M.Sc. thesis, Colorado State University, Fort Collins, CO.

Edmond, J.M. (1992) Himalayan tectonics, weathering processes, and the strontium isotope record in marine limestones. *Science*, **258**, 1594–1597.

Edwards, R.L., Taylor, F.W., and Wasserburg, G.J. (1988) Dating earthquakes with high-precision thorium-230 ages of very young corals. *Earth and Planetary Science Letters*, **90**, 379–381.

Edwards, R.L., Beck, J.W., Burr, G.S., Donahue, D.J., Chappell, J.M., Bloom, A.L., Druffel, E.R.M., and Taylor, F.W. (1993) A large drop in atmospheric $^{14}C/^{12}C$ and reduced melting in the Younger Dryas, documented with ^{230}Th ages of corals. *Science*, **260**, 962–968.

Edwards, R.L., Cheng, H., Murrell, M.T., and Goldstein, S.J. (1997) Proactinium-231 dating of carbonates by thermal ionization mass spectrometry: implications for Quaternary climate change. *Science*, **276**, 782–786.

Ehlers, T.A. (2005) Crustal thermal processes and the interpretation of thermochronometer data. *Reviews of Mineralogy and Geochemistry*, **58**, 315–350.

Ehlers, T.A., and Farley, K.A. (2003) Apatite (U–Th)/He thermochronometry: methods and applications to problems in tectonics and surface processes. *Earth and Planetary Science Letters*, **206**, 1–14.

Ehlers, T.A., and Poulsen, C.J. (2009) Influence of Andean uplift on climate and paleoaltimetry estimates. *Earth and Planetary Science Letters*, **281**, 238–248.

Ehlers, T.A., Farley, K.A., Rusmore, M.E., and Woodsworth, G.J. (2006) Apatite (U–Th)/He signal of large-magnitude accelerated glacial erosion, southwest British Columbia. *Geology*, **34**, 765–768.

Eiler, J.M. (2007) "Clumped-isotope" geochemistry – the study of naturally-occurring, multiply-substituted isotopologues. *Earth and Planetary Science Letters*, **262**, 309–327.

Ellis, M.A., and Densmore, A.L. (2006) First-order topography over blind thrusts. In Willett, S.D., Hovius, N., Brandon, M.T., and Fisher, D.M. (eds), *Tectonics, Climate and Landscape Evolution*, Special Paper 398, pp. 251–266. Geological Society of America, Boulder, CO.

Ellis, M.A., Densmore, A.L., and Anderson, R.S. (1999) Evolution of mountainous topography in the Basin and Range Province. *Basin Research*, **11**, 21–42.

England, P., and McKenzie, D. (1982) A thin viscous sheet model for continental deformation. *Geophysical Journal International*, **70**, 295–321.

England, P., and Molnar, P. (1990) Surface uplift, uplift of rocks, and exhumation of rocks. *Geology*, **18**, 1173–1177.

Erslev, E.A. (1991) Trishear fault-propagation folding. *Geology*, **19**, 617–620.

Fairbanks, R.G. (1989) A 17,000-year glaci-eustatic sea-level record: influence of glacial melting rates on the Younger Dryas event and deep-ocean circulation. *Nature*, **342**, 637–642.

Fang, X., Yan, M., Van der Voo, R., Rea, D.K., Song, C., Pares, J.M., Gao, J., Nie, J., and Dai, S. (2005) Late Cenozoic deformation and uplift of the NE Tibetan Plateau: Evidence from high-resolution magneto-stratigraphy of the Guide Basin, Qinghai Province, China. *Geological Society of America Bulletin*, **117**, 1208–1225.

Farr, T.G., Rosen, P.A., Caro, E., Crippen, R., Duren, R., Hensley, S., Kobrick, M., Paller, M., Rodriguez, E., Roth, L., Seal, D., Shaffer, S., Shimada, J., Umland, J., Werner, M., Oskin, M., Burbank, D., and Alsdorf, D. (2007) The Shuttle Radar Topography Mission. *Reviews of Geophysics*, **45**, RG2004, doi:10.1029/2005RG000183.

Faulkner, D.R., Lewis, A.C., and Rutter, E.H. (2003) On the internal structure and mechanics of large strike-slip fault zones: field observations of the Carboneras fault in southeastern Spain. *Tectonophysics*, **367**, 235–251.

Ferretti, A., Prati, C., and Rocca, F. (2001) Permanent scatterers in SAR interferometry. *IEEE Transactions on Geoscience and Remote Sensing*, **39**, 8–20.

Fielding, E.J., Isacks, B.L., Barazangi, M., and Duncan, C. (1994) How flat is Tibet? *Geology*, **22**, 163–167.

Finnegan, N.J., Hallet, B., Montgomery, D.R., Zeitler, P.K., Stone, J.O., Anders, A.M., and Liu, Y. (2008) Coupling of rock uplift and river incision in the Namche Barwa–Gyala Peri massif, Tibet. *Geological Society of America Bulletin*, **120**, 142–155.

Fisher, D.M., Gardner, T.W., Marshall, J.S., Sak, P.B., and Protti, M. (1998) Effect of subducting sea-floor roughness on fore-arc kinematics, Pacific coast, Costa Rica. *Geology*, **26**, 467–470.

Fitch, T.J., and Scholz, C.H. (1971) Mechanism of underthrusting in southwest Japan: a model of convergent plate interactions. *Journal of Geophysical Research*, **76**, 7260–7292.

Fitzgerald, P.G., Sorkhabi, R.B., Redfield, T.F., and Stump, E. (1995) Uplift and denudation of the central Alaska Range: a case study in the use of apatite fission track thermochronology to determine absolute uplift parameters. *Journal of Geophysical Research*, **100**, 20175–20191.

Fleming, A., Summerfield, M., Stone, J., Fifield, L., and Cresswell, R. (1999) Denudation rates for the southern Drakensberg Escarpment, SE Africa, derived from in-situ-produced cosmogenic ^{36}Cl; initial results. *Journal of the Geological Society of London*, **156**, 209–212.

Fleming, K., Johnston, P., Zwartz, D., Yokoyama, Y., Lambeck, K., and Chappell, J. (1998) Refining the eustatic sea-level curve since the last glacial maximum using far- and intermediate-field sites. *Earth and Planetary Science Letters*, **163**, 327–342.

Flemings, P.B., and Jordan, T.E. (1989) A synthetic stratigraphic model of foreland basin development. *Journal of Geophysical Research*, **94**, 3851–3866.

Flemings, P.B., and Jordan, T.E. (1990) Stratigraphic modeling of foreland basins: interpreting thrust deformation and lithospheric rheology. *Geology*, **18**, 430–434.

Flowers, R.M. (2010) The enigmatic rise of the Colorado Plateau. *Geology*, **38**, 671–672.

Flowers, R.M., Wernicke, B.P., and Farley, K.A. (2008) Unroofing, incision, and uplift history of the southwestern Colorado Plateau from apatite (U–Th)/He thermochronometry. *Geological Society of America Bulletin*, **120**, 571–587.

Fort, M. (1987) Sporadic morphogenesis in a continental subduction setting: an example from the Annapurna Range, Nepal Himalaya. *Zeitschrift für Geomorphologie*, **63**, 9–36.

Forte, A.M., Mitrovica, J.X., Moucha, R., Simmons, N.A., and Grand, S.P. (2007) Descent of the ancient Farallon slab drives localized mantle flow below the New Madrid seismic zone. *Geophysical Research Letters*, **34**, L04308, doi:10.1029/2006GL027895.

Foster, D., Brocklehurst, S.H., and Gawthorpe, R.L. (2010) Glacial–topographic interactions in the Teton Range, Wyoming. *Journal of Geophysical Research*, **115**, F01007, doi:10.1029/2008JF001135.

Frankel, K.L., Brantley, K., Dolan, J., Finkel, R., Klinger, R., Knott, J., Machette, M., Owen, L., Phillips, F., Slate, J., and Wernicke, B. (2007) Cosmogenic ^{10}Be and ^{36}Cl geochronology of offset alluvial fans along the northern Death Valley fault zone: implications for transient strain in the eastern California shear zone. *Journal of Geophysical Research*, **112**, B06407, doi:10.1029/2006JB004350.

Freymueller, J., Cohen, S., and Fletcher, H. (2000) Spatial variations in present-day deformation, Kenai Peninsula, Alaska, and their implications. *Journal of Geophysical Research*, **105**, 8079–8101.

Friedrich, A.M., Wernicke, B., Niemi, N.A., Bennett, R.A., and Davis, J.L. (2003) Comparison of geodetic and geologic data from the Wasatch region, Utah, and implications for the spectral character of Earth deformation at periods of 10 to 10 million years. *Journal of Geophysical Research*, **108**, 2199, doi:10.1029/2001JB000682.

Fritts, H.C. (1976) *Tree Rings and Climate*. Academic Press, New York.

Fruneau, B., Achache, J., and Delacourt, C. (1996) Observation and modelling of the Saint-Etienne-de-Tinee landslide using SAR interferometry. *Tectonophysics*, **265**, 181–190.

Fu, B., Awata, Y., Du, J., and He, W. (2005) Late Quaternary systematic stream offsets caused by repeated large seismic events along the Kunlun fault, northern Tibet. *Geomorphology*, **71**, 278–292.

Fuller, C.W., Willett, S.D., Hovius, N., and Slingerland, R. (2003) Erosion rates for Taiwan mountain basins: new determinations from suspended sediment records and a stochastic model of their temporal variation. *Journal of Geology*, **111**, 71–87.

Fuller, C.W., Willett, S.D., Fisher, D., and Lu, C.Y. (2006) A thermomechanical wedge model of Taiwan constrained by fission-track thermochronometry. *Tectonophysics*, **425**, 1–24.

Fuller, T.K., Perg, L.A., Willenbring, J.K., and Lepper, K. (2009) Field evidence for climate-driven changes in sediment supply leading to strath terrace formation. *Geology*, **37**, 467–470.

Gabet, E.J. (2000) Gopher bioturbation: field evidence for non-linear hillslope diffusion. *Earth Surface Processes and Landforms*, **25**, 1419–1428.

Gabet, E., Pratt-Sitaula, B., and Burbank, D.W. (2004a) Climatic controls on hillslope angle and relief in the Himalayas. *Geology*, **32**, 629–632.

Gabet, E.J., Burbank, D.W., Putkonen, J.K., Pratt-Sitaula, B.A., and Ojha, T. (2004b) Rainfall thresholds for landsliding in the Himalayas of Nepal. *Geomorphology*, **63**, 131–143.

Gabet, E.J., Burbank, D.W., Pratt-Sitaula, B., and Putkonen, J. (2008) Modern erosion rates in the High Himalayas of Nepal. *Earth and Planetary Science Letters*, **267**, 482–494.

Galewsky, J. (2009) Rain shadow development during the growth of mountain ranges: an atmospheric dynamics perspective, *Journal of Geophysical Research*, **114**, F01018, doi:10.1029/2008JF001085.

Galewsky, J., Silver, E.A., Gallup, C.D., Edwards, R.L., and Potts, D.C. (1996) Foredeep tectonics and carbonate platform dynamics in the Huon Gulf, Papua New Guinea. *Geology*, **24**, 819–822.

Gardner, T.W. (1983) Experimental study of knickpoint migration and longitudinal profile evolution in cohesive homogeneous material. *Geological Society of America Bulletin*, **94**, 664–672.

Garzione, C.N., Dettman, D.L., Quade, J., DeCelles, P.G., and Butler, R.F. (2000) High times on the Tibetan Plateau: paleoelevation of the Thakkhola graben, Nepal. *Geology*, **28**, 339–342.

Gawthorpe, R.L., and Hurst, J.M. (1993) Transfer zones in extensional basins: their structural style and influence on drainage development and stratigraphy.

Journal of the Geological Society of London, **150**, 1137–1152.

Ghose, S., Mellors, R.J., Korjenkov, A.M., Hamburger, M.W., Pavlis, T.L., Pavlis, G.L., Omuraliev, M., Mamyrov, E., and Muraliev, A.R. (1997) The M_s=7.3 1992 Suusamyr, Kyrgyzstan, earthquake in the Tien Shan: 2. Aftershock focal mechanisms and surface deformation. *Bulletin of the Seismological Society of America*, **87**, 23–38.

Ghose, S., Hamburger, M.W., and Virieux, J. (1998) Three-dimensional velocity structure and earthquake locations beneath the northern Tien Shan of Kyrgyzstan, central Asia. *Journal of Geophysical Research*, **103**, 2725–2748.

Ghosh, P., Garzione, C., and Eiler, J.M. (2006) Rapid uplift of the Altiplano revealed through ^{13}C–^{18}O bonds in paleosol carbonates. *Science*, **311**, 511–515, doi:10.1126/science.1119365.

Gibbons, A.B., Megeath, J.D., and Pierce, K.L. (1984) Probability of moraine survival in a succession of glacial advances. *Geology*, **12**, 327–330.

Gilbert, G.K. (1877) *Report on the Geology of the Henry Mountains [Utah]*. Publication of the Powell Survey.

Gilbert, G.K. (1879) *Geology of the Henry Mountains (Utah)*. United States Government Printing Office, Washington, DC.

Gilbert, G.K. (1890) *Lake Bonneville*, Monographs of the United State Geological Survey. Government Printing Office, Washington, DC.

Gilbert, G.K. (1909) The convexity of hilltops. *Journal of Geology*, **17**, 344–350.

Gillespie, A.R. (1982) *Quaternary Glaciation and Tectonism in the Southeastern Sierra Nevada, Inyo County, California*. California Institute of Technology, Pasadena, CA.

Gillespie, A., and Molnar, P. (1995) Asynchronous maximum advances of mountain and continental glaciers. *Reviews of Geophysics*, **33**, 311–364.

Goldfinger, C., Nelson, C.H., and Johnson, J.E. (2003) Holocene earthquake records from the Cascadia subduction zone and northern San Andreas fault based on precise dating of offshore turbidites. *Annual Review of Earth and Planetary Sciences*, **31**, 555–577.

Goldfinger, C., Morey, A.E., Nelson, C.H., Gutierrez-Pastor, J., Johnson, J.E., Karabanov, E., Chaytor, J., and Eriksson, A. (2007) Rupture lengths and temporal history of significant earthquakes on the offshore and north coast segments of the Northern San Andreas Fault based on turbidite stratigraphy. *Earth and Planetary Science Letters*, **254**, 9–27.

Gomberg, J.S. (1993) Tectonic deformation in the New Madrid seismic zone; inferences from map view and cross-sectional boundary element models. *Journal of Geophysical Research*, **98**, 6639–6664.

Gomberg, J.S., and Ellis, M. (1993) *3D-DEF; a user's manual (a three-dimensional, boundary element modeling program)*. USGS Open File reports, OF 93-0547. United States Geological Survey, Washington, DC.

Gomberg, J.S., and Ellis, M. (1994) Topography and tectonics of the central New Madrid seismic zone – results of numerical experiments using a three-dimensional boundary element program. *Journal of Geophysical Research*, **99**, 20299–20310.

Goodbred, S.L.J., and Kuehl, S.A. (1999) Holocene and modern sediment budgets for the Ganges–Brahmaputra river system: evidence for highstand dispersal to flood-plain, shelf, and deep-sea depocenters. *Geology*, **27**, 559–562.

Goode, J.K., and Burbank, D.W. (2009) Numerical study of degradation of fluvial hanging valleys due to climate change. *Journal of Geophysical Research*, **114**, F01017, doi:10.1029/2007JF000965.

Goode, J.K., and Burbank, D.W. (2011) The temporal evolution of minor channels on growing folds and its bearing on fold kinematics. *Journal of Geophysical Research*, **116**, B04407, doi:10.1029/2010JB 007617.

Gosse, J.C., and Phillips, F.M. (2001) Terrestrial *in situ* cosmogenic nuclides: theory and application. *Quaternary Science Reviews*, **20**, 1475–1560.

Granger, D.E., and Muzikar, P. (2001) Dating sediment burial with cosmogenic nuclides: theory, techniques, and limitations. *Earth and Planetary Science Letters*, **188**, 269–281.

Granger, D.E., and Smith, A.L. (2000) Dating buried sediments using radioactive decay and muogenic production of ^{26}Al and ^{10}Be. *Nuclear Instruments and Methods in Physics Research B: Beam Interactions with Materials*, **172**, 822–826.

Granger, D.E., Kirchner, J.W., and Finkel, R. (1996) Spatially averaged long-term erosion rates measured from in situ-produced cosmogenic nuclides in alluvial sediment. *Journal of Geology*, **104**, 249–257.

Granger, D.E., Kirchner, J.W., and Finkel, R.C. (1997) Quaternary downcutting rate of the New River, Virginia, measured from differential decay of cosmogenic ^{26}Al and ^{10}Be in cave-deposited alluvium. *Geology*, **25**, 107–110.

Granger, D.E., Fabel, D., and Palmer, A.N. (2001) Pliocene–Pleistocene incision of the Green River, Kentucky, determined from radioactive decay of cosmogenic ^{26}Al and ^{10}Be in Mammoth Cave sediments. *Geological Society of America Bulletin*, **113**, 825–836.

Grapes, R., and Wellman, H. (1988) *The Wairarapa Fault*. Victoria University of Wellington, Wellington, New Zealand.

Grapes, R.H., and Wellman, H.W. (1993) *Field Guide to the Wharekauhau Thrust (Palliser Bay) and Wairarapa Fault (Pigeon Bush)*, Miscellaneous Publication, 79B, pp. 27–44. Geological Society of New Zealand, Wellington, New Zealand.

Grauch, V.J.S. (2001) High-resolution aeromagnetic data, a new tool for mapping intrabasinal faults: example from the Albuquerque basin, New Mexico. *Geology*, **29**, 367–370.

Gregory, K. (1994) Palaeoclimate and palaeoelevation of the 35 Ma Florrisant flora, Front Range, Colorado. *Palaeoclimates*, **1**, 23–57.

Gregory, K.M., and Chase, C.G. (1992) Tectonic significance of paleobotanically estimated climate and altitude of the late Eocene erosion surface, Colorado. *Geology*, **20**, 581–585.

Grujic, D., Coutand, I., Bookhagen, B., Bonnet, S., Blythe, A., and Duncan, C. (2006) Climatic forcing of erosion, landscape, and tectonics in the Bhutan Himalayas. *Geology*, **34**, 801–804.

Gupta, S., Cowie, P.A., Dawars, N.H., and Underhill, J.R. (1998) A mechanism to explain rift-basin subsidence and stratigraphic patterns through fault-array evolution. *Geology*, **26**, 595–598.

Gurnis, M. (2001) Sculpting the Earth from inside out. *Scientific American*, **284**(March), 40–47. [Also updated and reprinted in: (2005) *Our Ever Changing Earth*, Scientific American Special Edition, pp. 56–63.]

Hack, J.T. (1973) Stream-profile analysis and stream-gradient indices. *United States Geological Survey Journal of Research*, **1**, 421–429.

Hack, J.T. (1975) Dynamic equilibrium and landscape evolution. In Melhorn, W.N., and Flemal, R.C. (eds), *Theories of Landform Evolution*, pp. 87–102. Allen and Unwin, Boston, MA.

Hacker, B.R. (2007) Ascent of the ultrahigh-pressure Western Gneiss Region, Norway. In Cloos, M., Carlson, W.D., Gilbert, M.C., Liou, J.G., and Sorensen, S.S. (eds), *Convergent Margin Terranes and Associated Regions: A Tribute to W.G. Ernst*, Special Paper 419, pp. 171–184. Geological Society of America, Boulder, CO.

Hales, T.C., and Roering, J.J. (2007) Climatic controls on frost cracking and implications for the evolution of bedrock landscapes. *Journal of Geophysical Research*, **112**, F02033, doi:10.1029/2006JF000616.

Hallet, B. (1979) A theoretical model of glacial abrasion. *Journal of Glaciology*, **23**, 39–50.

Hallet, B. (1996) Glacial quarrying: a simple theoretical model. *Annals of Glaciology*, **22**, 1–8.

Hallet, B., and Molnar, P. (2001) Distorted drainage basins as markers of crustal strain east of the Himalaya. *Journal of Geophysical Research*, **106**, 13697–13709.

Hallet, B., and Putkonen, J. (1994) Surface dating of dynamic landforms: young boulders on aging moraines. *Science*, **265**, 937–940.

Hallet, B., Walder, J.S., and Stubbs, C.W. (1991) Weathering by segregation ice growth in microcracks at sustained sub-zero temperatures: verification from an experimental study using acoustic emissions. *Permafrost and Periglacial Processes*, **2**, 283–300

Hallet, B., Hunter, L., and Bogen, J. (1996) Rates of erosion and sediment evacuation by glaciers: a review of field data and their implications. *Global and Planetary Change*, **12**, 213–235.

Hampel, A., and Hetzel, R. (2006) Response of normal faults to glacial–interglacial fluctuations of ice and water masses on Earth's surface. *Journal of Geophysical Research*, **111**, B06406, doi:10.1029/2005JB00412.

Hampel, A., Hetzel, R., and Densmore, A.L. (2007) Postglacial slip-rate increase on the Teton normal fault, northern Basin and Range Province, caused by melting of the Yellowstone ice cap and deglaciation of the Teton Range? *Geology*, **35**, 1107–1110.

Hancock, G.S., and Anderson, R.S. (2002) Numerical modeling of fluvial strath-terrace formation in response to oscillating climate. *Geological Society of America Bulletin*, **114**, 1131–1142.

Hancock, G.S., Anderson, R.S., and Whipple, K.X. (1998) Beyond power: bedrock river incision process and form. In Tinkler, K.J., and Wohl, E.E. (eds), *Rivers Over Rock: Fluvial Processes in Bedrock Channels*, Geophysical Monograph Series, 107, pp. 35–60. American Geophysical Union, Washington, DC.

Hanks, T.C., Bucknam, R.C., Lajoie, K.R., and Wallace, R.E. (1984) Modification of wave-cut and faulting controlled landforms. *Journal of Geophysical Research*, **89**, 5771–5790.

Harden, J.W. (1982) A quantitative index of soil development from field descriptions; examples from a chronosequence in central California. *Geoderma*, **28**, 1–28.

Hardy, S., and Ford, M. (1997) Numerical modeling of trishear fault propagation folding. *Tectonics*, **16**, 841–854.

Hardy, S., and Poblet, J. (1994) Geometric and numerical model of progressive limb rotation in detachment folds. *Geology*, **22**, 371–374.

Harkins, N., and Kirby, E. (2008) Fluvial terrace riser degradation and determination of slip rates on strike-slip faults: an example from the Kunlun fault, China. *Geophysical Research Letters*, **35**, L05406, doi:10.1029/2007GL033073.

Harkins, N., Kirby, E., Heimsath, A., Robinson, R., and Reiser, U. (2007) Transient fluvial incision in the headwaters of the Yellow River, northeastern Tibet, China. *Journal of Geophysical Research*, **112**, F03S04, doi:10.1029/2006JF000570.

Harp, E.L., and Jibson, R.W. (1996) Landslides triggered by the 1994 Northridge, California, earthquake. *Bulletin of the Seismological Society of America*, **86**, S319–S332.

Harris, R.A., and Day, S.M. (1993) Dynamics of fault interaction; parallel strike-slip faults. *Journal of Geophysical Research*, **98**, 4461–4472.

Harris, R.A., Simpson, R.W., and Reasenberg, P.A. (1995) Influence of static stress changes on earthquake locations in southern California. *Nature*, **375**, 221–224.

Hartshorn, K., Hovius, N., Dade, W.B., and Slingerland, R.L. (2002) Climate-driven bedrock incision in an active mountain belt. *Science*, **297**, 2036–2038.

Hasbargen, L., and Paola, C. (2000) Landscape instability in an experimental drainage basin. *Geology*, **28**, 1067–1070.

Hearty, P.J., and Miller, G.H. (1987) Global trends in isoleucine epimerization. Data from the circum-Atlantic, the Mediterranean, and the South Pacific. *Geological Society of America Abstracts with Programs*, **19**, 698.

Heimsath, A.M. (1999) *The soil production function*, Ph.D. thesis. University of California.

Heimsath, A.M., Dietrich, W.E., Nishiizumi, K., and Finkel, R.C. (1997) The soil production function and landscape equilibrium. *Nature*, **388**, 358–361.

Heimsath, A.M., Dietrich, W.E., Nishiizumi, K., and Finkel, R.C. (1999) Cosmogenic nuclides, topography, and the spatial variation of soil depth. *Geomorphology*, **27**(1–2), 151–172.

Heimsath, A.M., Chappell, J., Dietrich, W.E., Nishiizumi, K., and Finkel, R.C. (2000) Soil production on a retreating escarpment in southeastern Australia. *Geology*, **28**, 788–790.

Heimsath, A.M., Chappell, J., Spooner, N.A., and Questiaux, D.G. (2002) Creeping soil. *Geology*, **30**, 111–114.

Heimsath, A.M., Furbish, D.J., and Dietrich, W.E. (2005) The illusion of diffusion: field evidence for depth-dependent sediment transport. *Geology*, **33**, 949–952.

Heki, K., Miyazaki, S., and Tsuji, H. (1997) Silent fault slip following an interplate thrust earthquake at the Japan Trench. *Nature*, **386**, 595–598.

Heller, P.L., Angevine, C.L., Winslow, N.S., and Paola, C. (1988) Two-phase stratigraphic model of foreland basin development. *Geology*, **16**, 501–504.

Hemphill-Haley, E. (1995) Diatom evidence for earthquake-induced subsidence and tsunami 300 yr ago in southern coastal Washington. *Geological Society of America Bulletin*, **107**, 367–378.

Herman, F., Braun, J., and Dunlap, W.J. (2007) Tectonomorphic scenarios in the Southern Alps of New Zealand. *Journal of Geophysical Research*, **112**, B04201, doi:10.1029/2004JB003472.

Herman, F., Rhodes, E.J., Braun, J., and Heiniger, L. (2010) Uniform erosion rates and relief amplitude during glacial cycles in the Southern Alps of New Zealand, as revealed from OSL-thermochronology. *Earth and Planetary Science Letters*, **297**, 183–189.

Herring, T.A., *et al.* (1986) Geodesy by radio interferometry: evidence for contemporary plate motion. *Journal of Geophysical Research*, **91**, 8341–8347.

Hetzel, R., and Hampel, A. (2005) Slip rate variations on normal faults during glacial–interglacial changes in surface loads. *Nature*, **435**, 81–84.

Hetzel, R., and Hampel, A. (2006) Long-term rates of faulting derived from cosmogenic nuclides and short-term variations caused by glacial–interglacial volume changes of glaciers and lakes. *International Journal of Modern Physics B*, **20**, 261–276.

Hilley, G.E., and Arrowsmith, J.R. (2008) Geomorphic response to uplift along the Dragon's Back pressure ridge, Carrizo Plain, California. *Geology*, **36**, 367–370.

Hilley, G.E., Blisniuk, P.M., and Strecker, M.R. (2005) Mechanics and erosion of basement-cored uplift provinces. *Journal of Geophysical Research*, **110**, B12409, doi:10.1029/2005JB003704.

Hodges, K.V. (2000) Tectonics of the Himalaya and southern Tibet from two perspectives. *Geological Society of America Bulletin*, **112**, 324–350.

House, M.A., Wernicke, B.P., and Farley, K.A. (1998) Dating topography of the Sierra Nevada, California, using apatite (U–Th)/He ages. *Nature*, **396**, 66–69.

Hovius, N. (1996) Regular spacing of drainage outlets from linear mountain belts. *Basin Research*, **8**, 29–44.

Hovius, N., Stark, C.P., and Allen, P.A. (1997) Sediment flux from a mountain belt derived by landslide mapping. *Geology*, **25**, 231–234.

Howard, A.D. (1994) A detachment-limited model of drainage basin evolution. *Water Resources Research*, **30**, 2261–2285.

Howard, A.D. (1997) Badland morphology and evolution: interpretation using a simulation model. *Earth Surface Processes and Landforms*, **22**, 211–227.

Howard, A.D., and Kerby, G. (1983) Channel changes in badlands. *Geological Society of America Bulletin*, **94**, 739–752.

Howard, A.D., Dietrich, W.E., and Seidl, M.A. (1994) Modeling fluvial erosion on regional to continental scales. *Journal of Geophysical Research*, **99**, 13971–13986.

Huber, N.K. (1981) *Amount and Timing of Late Cenozoic Uplift and Tilt of the Central Sierra Nevada, California: Evidence from the Upper San Joaquin River Basin*, Professional Paper 1197. United States Geological Survey, Washington, DC.

Hubert-Ferrari, A., Suppe, J., Gonzalez-Mieres, R., and Wang, X. (2007) Mechanisms of active folding of the landscape (southern Tian Shan, China). *Journal of Geophysical Research*, **112**, B03S09, doi:10.1029/2006JB004362.

Hudnut, K.W., Borsa, A., Glennie, C., and Minster, J.B. (2002) High-resolution topography along surface rupture of the 16 October 1999 Hector Mine, California, earthquake (Mw 7.1) from airborne laser swath mapping. *Bulletin of the Seismological Society of America*, **92**, 1570–1576.

Hughen, K., Lehman, S., Southon, J., Overpeck, J., Marchal, O., Herring, C., and Turnbull, J. (2004) ^{14}C activity and global carbon cycle changes over the past 50,000 years. *Science*, **303**, 202–207.

Humphrey, N.F., and Heller, P.L. (1995) Natural oscillations in coupled geomorphic systems: an alternative origin for cyclic sedimentation. *Geology*, **23**, 499–502.

Humphrey, N.F., and Konrad, S.K. (2000) River incision or diversion in response to bedrock uplift. *Geology*, **28**, 43–46.

Humphrey, N., Raymond, C., and Harrison, W. (1986) Discharges of turbid water during mini-surges of Variegated Glacier, Alaska, USA. *Journal of Glaciology*, **32**, 195–207.

Hurtrez, J.-E., Lucazeau, F., Lavé, J., and Avouac, J.-P. (1999) Investigation of the relationships between basin morphology, tectonic uplift, and denudation from the study of an active fold belt in the Siwalik Hills, central Nepal. *Journal of Geophysical Research*, **104**, 12779–12796.

Hyndman, R.D., and Wang, K. (1995) The rupture zone of Cascadia great earthquakes from current deformation and the thermal regime. *Journal of Geophysical Research*, **100**, 22133–22154.

Innes, J.L. (1984) Lichenometric dating of moraine ridges in Northern Norway; some problems of application. *Geografiska Annaler, Series A: Physical Geography*, **66**, 341–352.

Innes, J.L. (1985) Lichenometry. *Progress in Physical Geography*, **9**, 187–254.

Iverson, N.R. (2002) Processes of glacial erosion. In Menzies, J. (ed.), *Modern and Past Glacial Environments*, pp. 131–145. Elsevier, New York.

Jackson, J., Norris, R., and Youngson, J. (1996) The structural evolution of active fault and fold systems in central Otago, New Zealand: evidence revealed by drainage patterns. *Journal of Structural Geology*, **18**, 217–234.

Jackson, M., and Bilham, R. (1994a) Constraints on Himalayan deformation inferred from vertical velocity fields in Nepal and Tibet. *Journal of Geophysical Research*, **99**, 13897–13912.

Jackson, M., and Bilham, R. (1994b) 1991–1992 GPS measurements across the Nepal Himalaya. *Geophysical Research Letters*, **21**, 1169–1172.

Jackson, M., Barrientos, S., Bilham, R., Kyestha, D., and Shrestha, B. (1992) Uplift in the Nepal Himalaya revealed by spirit leveling. *Geophysical Research Letters*, **19**, 1539–1542.

Jacoby, G.C., Jr., Sheppard, P.R., and Sieh, K.E. (1988) Irregular recurrence of large earthquakes along the San Andreas fault: evidence from trees. *Science*, **241**, 196–199.

Jerolmack, D.J., and Mohrig, D. (2007) Conditions for branching in depositional rivers. *Geology*, **35**, 463–466.

Johanson, I.A., Fielding, E.J., Rolandone, F., and Burgmann, R. (2006) Coseismic and postseismic slip of the 2004 Parkfield earthquake from space-geodetic data. *Bulletin of the Seismological Society of America*, **96**, S269–S282.

Jordan, T.E., Flemings, P.B., and Beers, J.A. (1988) Dating thrust-fault activity by use of foreland-basin strata. In Kleinspehn, K.L., and Paola, C. (eds), *New Perspectives in Basin Analysis*, pp. 307–330. Springer, New York.

Kamb, B., and Engelhardt, H. (1987) Waves of accelerated motion in a glacier approaching surge – the mini-surges of Variegated Glacier, Alaska, USA. *Journal of Glaciology*, **33**, 27–46.

Kanamori, H., and Kikuchi, M. (1993) The 1992 Nicaragua earthquake; a slow earthquake associated with subducted sediments. *Seismological Society of America, 88th Annual Meeting*, vol. 64, pp. 13–14.

Kaufman, D.S., and Miller, G.H. (1992) Overview of amino acid geochronology. *Comparative Biochemistry and Physiology*, **102B**, 199–204.

Kaufman, D.S., Miller, G.H., and Andrews, J.T. (1992) Amino acid composition as a taxonomic tool for molluscan fossils; an example from Pliocene–Pleistocene Arctic marine deposits. *Geochimica et Cosmochimica Acta*, **56**, 2445–2453.

Keller, E.A., Bonkowski, M.S., Korsch, R.J., and Shlemon, R.J. (1982) Tectonic geomorphology of the San Andreas fault zone in the southern Indio Hills, Coachella Valley, California. *Geological Society of America Bulletin*, **93**, 46–56.

Keller, E.A., Zepeda, R.L., Rockwell, T.K., Ku, T.L., and Dinklage, W.S. (1998) Active tectonics at Wheeler Ridge, southern San Joaquin Valley, California. *Geological Society of America Bulletin*, **110**, 298–310.

Keller, E.A., Gurrola, L., and Tierney, T.E. (1999) Geomorphic criteria to determine direction of lateral propagation of reverse faulting and folding. *Geology*, **27**, 515–518.

Kelsey, H.M., and Bockheim, J.G. (1994) Coastal landscape evolution as a function of eustasy and surface uplift rate, southern Cascadia margin, USA. *Geological Society of America Bulletin*, **106**, 840–854.

Kelsey, H.M., Sherrod, B.L., Nelson, A.R., and Brocher, T.M. (2008) Earthquakes generated from bedding plane-parallel reverse faults above an active wedge thrust, Seattle fault zone. *Geological Society of America Bulletin*, **120**, 1581–1597.

Kessler, M.A., Anderson, R.S., and Stock, G.M. (2006) Modeling topographic and climatic control of east–west asymmetry in Sierra Nevada glacier length during the Last Glacial Maximum. *Journal of Geophysical Research – Earth Surface*, **111**, F02002, doi:10.1029/2005JF000365.

Kessler, M.A., Anderson, R.S., and Briner, J.P. (2008) The insertion of fjords into continental margins. *Nature Geoscience*, **1**, 365–369, doi:10.1038/ngeo201.

King, G., and Ellis, M. (1990) The origin of large local uplift in extensional regions. *Nature*, **348**, 689–693.

King, G., and Stein, R. (1983) Surface folding, river terrace deformation rate and earthquake repeat time in a reverse faulting environment: the Coalinga, California earthquake of May, 1983. In *The 1983 Coalinga, California, Earthquake*, Special Publication 66, pp. 261–274. California Division of Mines and Geology, Sacramento, CA.

King, G.C.P., Stein, R.S., and Rundle, J.B. (1988) The growth of geological structure by repeated earthquakes 1. Conceptual framework. *Journal of Geophysical Research*, **93**, 13307–13318.

Kirby, E., and Whipple, K. (2001) Quantifying differential rock-uplift rates via stream profile analysis. *Geology*, **29**, 415–418.

Kirby, E., Johnson, C., Furlong, K.P., and Heimsath, A. (2007) Transient channel incision along Bolinas Ridge, California: evidence for differential rock uplift adjacent to the San Andreas fault. *Journal of Geophysical Research*, **112**, F03S07, doi:10.1029/2006JF000559.

Kirby, E., Whipple, K., and Harkins, N. (2008) Topography reveals seismic hazard. *Nature Geoscience*, **1**, 485–487.

Kirchner, J.W., Finkel, R.C., Riebe, C.S., Granger, D.E., Clayton, J.L., King, J.G., and Megahan, W.F. (2001) Mountain erosion over 10 yr, 10 k.y., and 10 m.y. time scales. *Geology*, **29**, 591–594.

Klinger, Y., Etchebes, M., Tapponnier, P., and Narteau, C. (2011) Characteristic slip for 5 great earthquakes along the Fuyun fault in China. *Nature Geoscience*, **4**, doi:10.1038/NGE01158.

Kooi, H., and Beaumont, C. (1994) Escarpment evolution on high-elevation rifted margins: insights derived from a surface processes model that combines diffusion, advection, and reaction. *Journal of Geophysical Research*, **99**, 12191–12209.

Kooi, H., and Beaumont, C. (1996) Large-scale geomorphology: classical concepts reconciled and integrated with contemporary ideas via a surface processes model. *Journal of Geophysical Research*, **101**, 3361–3386.

Koons, P.O. (1989) The topographic evolution of collisional mountain belts: a numerical look at the Southern Alps, New Zealand. *American Journal of Science*, **289**, 1041–1069.

Koons, P.O. (1990) The two-sided orogen: collision and erosion from the sand box to the Southern Alps. *Geology*, **18**, 679–682.

Koons, P.O. (1995) Modeling the topographic evolution of collisional belts. *Annual Review of Earth and Planetary Sciences*, **23**, 375–408.

Koons, P.O., and Kirby, E. (2007) Topography, denudation, and deformation: the role of surface processes in fault evolution. In Handy, M.R., and Hirth, G. (eds), *Tectonic Faults: Agents of Change on a Dynamic Earth*, pp. 205–230. MIT Press, Cambridge, MA.

Koons, P.O., Zeitler, P.K., Chamberlain, C.P., Craw, D., and Meltzer, A.S. (2002) Mechanical links between erosion and metamorphism in Nanga Parbat, Pakistan Himalaya. *American Journal of Science*, **302**, 749–773.

Koppes, M.L., and Hallet, B. (2006) Erosion rates during rapid deglaciation in Icy Bay, Alaska. *Journal of Geophysical Research*, **111**, F02023, doi:10.1029/2005JF000349.

Korup, O. (2006) Rock-slope failure and the river long profile. *Geology*, **34**, 45–48.

Ku, T.-L. (1976) The uranium-series methods of age determination. *Annual Review of Earth and Planetary Sciences*, **4**, 347–379.

Ku, T.L., Bull, W.B., Freeman, S.T., and Knauss, K.G. (1979) Th^{230}–U^{234} dating of pedogenic carbonates in gravelly desert soils of Vidal Valley, southeastern California. *Geological Society of America Bulletin*, **90**, 1063–1073.

Lajoie, K.R. (1986) Coastal tectonics. In *Active Tectonics*, pp. 95–124. National Academy Press, Washington, DC.

Lal, D. (1988) *In situ* produced cosmogenic isotopes in terrestrial rocks and some applications to geochronology. *Annual Review of Earth and Planetary Sciences*, **16**, 355–388.

Lal, D. (1991) Cosmic ray labeling of erosion surfaces: *in situ* nuclide production rates and erosion. *Earth and Planetary Science Letters*, **104**, 424–439.

Lamb, M.P., Dietrich, W.E., and Sklar, L.S. (2008) A model for fluvial bedrock incision by impacting suspended and bed load sediment. *Journal of Geophysical Research*, **113**, F03025, doi:10.1029/2007JF000915.

Lamb, S., and Davis, P. (2003) Cenozoic climate change as a possible cause for the rise of the Andes. *Nature*, **425**, 792–797.

Lambeck, K. (1988) *Geophysical Geodesy, the Slow Deformations of the Earth*. Clarendon Press, Oxford.

Lambeck, K., and Chappell, J. (2001) Sea level change through the last glacial cycle. *Science*, **292**, 679–685.

Lambeck, K., Yokoyama, Y., and Purcell, T. (2002) Into and out of the Last Glacial Maximum: sea-level change during Oxygen Isotope Stages 3 and 2. *Quaternary Science Reviews*, **21**, 343–360.

Langbein, J., Dzurisin, D., Marshall, G., Stein, R., and Rundle, J. (1995) Shallow and peripheral volcanic sources of inflation revealed by modeling two-color geodimeter and leveling data from Long Valley caldera, California, 1988–1992. *Journal of Geophysical Research*, **100**, 12487–12495.

Larsen, S., and Reilinger, R. (1992) Global positioning system measurements of deformations associated with the 1987 Superstition Hills earthquake: evidence for conjugate faulting. *Journal of Geophysical Research*, **97**, 4885–4902.

Lavé, J., and Avouac, J.P. (2000) Active folding of fluvial terraces across the Siwalik Hills, Himalayas of central Nepal. *Journal of Geophysical Research*, **105**, 5735–5770.

Lavé, J., and Avouac, J.P. (2001) Fluvial incision and tectonic uplift across the Himalaya of central Nepal. *Journal of Geophysical Research*, **106**, 26561–26591.

Lavé, J., and Burbank, D.W. (2004) Denudation processes and rates in the Transverse Ranges, southern California: erosional response of a transitional landscape to external and anthropogenic forcing. *Journal of Geophysical Research*, **109**, F01006, doi:10.1029/2003JF000023.

Lavé, J., Yule, D., Sapkota, S., Basant, K., Madden, C., Attal., M., and Pandey, R. (2005) Evidence for a great Medieval earthquake (~1100 AD) in the Central Himalayas, Nepal. *Science*, **307**, 1302–1305.

Leeder, M.R., and Jackson, J.A. (1993) The interaction between normal faulting and drainage in active extensional basins, with examples from the western United States and central Greece. *Basin Research*, **5**, 79–102.

Leeder, M.R., Harris, T., and Kirkby, M.J. (1998) Sediment supply and climate change; implications for basin stratigraphy. *Basin Research*, **10**, 7–18.

Lensen, G.J. (1964) The general case of progressive fault displacement of flights of degradational terraces. *New Zealand Journal of Geology and Geophysics*, **7**, 864–870.

Leprince, S., Barbot, S., Ayoub, F., and Avouac, J.-P. (2007) Automatic and precise orthorectification, coregistration, and subpixel correlation of satellite images, application to ground deformation measurements. *IEEE Transactions in Geoscience and Remote Sensing*, **45**, 1529–1558.

Li, Y., Schweig, E.S., Tuttle, M.P., and Ellis, M.A. (1998) Evidence for large prehistoric earthquakes in the northern New Madrid seismic zone, central United States. *Seismological Research Letters*, **69**, 270–276.

Libby, W.F. (1955) *Radiocarbon Dating*. University of Chicago Press, Chicago.

Lindvall, S.C., Rockwell, T.K., and Hudnut, K.W. (1989) Evidence for prehistoric earthquakes on the Superstition Hills fault from offset geomorphic features. *Bulletin of the Seismological Society of America*, **79**, 342–361.

Lisiecki, L.E., and Raymo, M.E. (2005) A Pliocene–Pleistocene stack of 57 globally distributed benthic D^{18}O records. *Paleoceanography*, **20**, PA1003, doi:10.1029/2004PA001071.

Lisowski, M., Savage, J.C., and Prescott, W.H. (1991) The velocity field along the San Andreas Fault in central and southern California. *Journal of Geophysical Research*, **96**, 8369–8389.

Little, T.A., Grapes, R., and Berger, G.W. (1998) Late Quaternary strike slip on the eastern part of the Awatere Fault, South Island, New Zealand. *Geological Society of America Bulletin*, **110**, 127–148.

Liu, J., Klinger, Y., Sieh, K., and Rubin, C. (2004) Six similar sequential ruptures of the San Andreas fault, Carrizo Plain, California. *Geology*, **32**, 649–652.

Liu, L., and Gurnis, M. (2010) Dynamic subsidence and uplift of the Colorado Plateau. *Geology*, **38**, 663–666, doi:10.1130/G30368.1.

Locke, W.W., III, Andrews, J.T., and Webber, P.J. (1979) *A Manual for Lichenometry*, Technical Bulletin 26. British Geomorphological Research Group, London.

Loso, M.G., and Doak, D.F. (2005) The biology behind lichenometric dating curves. *Oecologia*, **147**, 223–229, doi:10.1007/s00442-005-0265-3.

Lowry, A.R. (2006) Resonant slow fault slip in subduction zones forced by climate load stress. *Nature*, **442**, 802–805.

Lu, H., Burbank, D.W., and Li, Y. (2010) Alluvial sequence in the north piedmont of the Chinese Tian Shan over the past 550 kyr and its relationship to climate change. *Palaeogeography, Palaeoclimatology, Palaeoecology*, **285**, 343–353.

Lund, S.P. (1996) A comparison of Holocene paleomagnetic secular variation records from North America. *Journal of Geophysical Research*, **101**, 8007–8024.

MacGregor, K.C., Anderson, R.S., Anderson, S.P., and Waddington, E.D. (2000) Numerical simulations of longitudinal profile evolution of glacial valleys. *Geology*, **28**, 1031–1034.

MacGregor, K.R., Anderson, R.S., and Waddington, E.D. (2009) Numerical modeling of glacial erosion and headwall processes in alpine valleys. *Geomorphology*, **103**, 189–204.

Machette, M.N., Personius, S.F., and Nelson, A.R. (1992a) Paleoseismicity of the Wasatch Fault zone: a summary of recent investigation, interpretations, and conclusions. In Gori, P.L. (ed.), *Assessment of Regional Earthquake Hazards and Risk Along the Wasatch Front, Utah*, Professional Paper 1500, pp. A1–A30. United States Geological Survey, Washington, DC.

Machette, M.N., Personius, S.F., and Nelson, A.R. (1992b) The Wasatch Fault zone, USA. *Annales Tectonicae*, **6**, 5–39.

Mahsas, A., Lammali, K., Yelles, K., Calais, E., Freed, A.M., and Briole, P. (2008) Shallow afterslip following the 2003 May 21, Mw=6.9 Boumerdes earthquake, Algeria. *Geophysical Journal International*, **172**, 155–166.

Mancktelow, N.S., and Grasemann, B. (1997) Time-dependent effects of heat advection and topography on cooling histories during erosion. *Tectonophysics*, **270**, 167–195.

Manighetti, I., King, G.C.P., Gaudemer, Y., Scholz, C., and Doubre, C. (2001) Slip accumulation and lateral propagation of active normal faults in Afar. *Journal of Geophysical Research*, **106**, 13667–13696.

Marshall, J.S., and Anderson, R.S. (1995) Quaternary uplift and seismic cycle deformation, Peninsula de Nicoya, Costa Rica. *Geological Society of America Bulletin*, **107**, 463–473.

Masek, J.G., Isacks, B.L., Fielding, E.J., and Browaeys, J. (1994a) Rift-flank uplift in Tibet: evidence for crustal asthenosphere. *Tectonics*, **13**, 659–667.

Masek, J.G., Isacks, B.L., Gubbels, T.L., and Fielding, E.J. (1994b) Erosion and tectonics at the margins of continental plateaus. *Journal of Geophysical Research*, **B99**, 12941–13956.

Massonnet, D., Rossi, M., Carmona, C., Adragna, F., Peltzer, G., Feigl, K., and Rabaute, T. (1993) The displacement of the Landers earthquake mapped by radar interferometry. *Nature*, **364**, 138–142.

Massonnet, D., Feigl, K., Rossi, M., and Adragna, F. (1994) Radar interferometric mapping of deformation in the year after the Landers earthquake. *Nature*, **369**, 227–230.

Matmon, A., Simhai, O., Amit, R., Haviv, I., Porat, N., McDonald, E., Benedetti, L., and Finkel, R. (2009) Desert pavement-coated surfaces in extreme deserts present the longest-lived landforms on Earth. *Geological Society of America Bulletin*, **121**, 688–697.

Mayer, L. (1986) Tectonic geomorphology of escarpments and mountain fronts. In *Active Tectonics*, pp. 125–135. National Academy Press, Washington, DC.

McElwain, J.C. (2004) Climate-independent paleoaltimetry using stomatal density in fossil leaves as a proxy for CO_2 partial pressure. *Geology*, **32**, 1017–1020.

McFadden, L.D., Tinsley, J.C., and Bull, W.B. (1982) Late Quaternary pedogenesis and alluvial chronologies of the Los Angeles basin and San Gabriel Mountain areas, southern California. In Tinsley, J.C., Matti, J.C., and McFadden, L.D. (eds), *Late Quaternary Pedogenesis and Alluvial Chronologies of the Los Angeles Basin and San Gabriel Mountain Areas, Southern California, and Holocene Faulting and Alluvial Stratigraphy Within the Cucamonga Fault Zone*, Field Trip 12, pp. 1–13. Cordilleran Section of the Geological Society of America.

McGill, S.F., and Rubin, C. (1999) Surficial slip distribution on the central Emerson Fault during the 28 June (1992) Landers earthquake, California. *Journal of Geophysical Research*, **104**, 4811–4833.

McGill, S.F., and Sieh, K. (1991) Surficial offsets on the central and eastern Garlock fault associated with prehistoric earthquakes. *Journal of Geophysical Research*, **96**, 21597–21621.

McKean, J.A., Dietrich, W.E., Finkel, R.C., Southon, J.R., and Caffee, M.W. (1993) Quantification of soil production and downslope creep rates from cosmogenic ^{10}Be accumulations on a hillslope profile. *Geology*, **21**, 343–346.

McSaveney, M.J., Graham, I.J., Begg, J.G., Beu, A.G., Hull, A.G., Kim, K., and Zondervan, A. (2006) Late Holocene uplift of beach ridges at Turakirae Head, south Wellington coast, New Zealand. *New Zealand Journal of Geology and Geophysics*, **49**, 337–358.

Meade, B.J. (2007) Present-day kinematics at the India–Asia collision zone. *Geology*, **35**, 81–84.

Meade, B.J., and Conrad, C.P. (2008) Andean growth and the deceleration of South American subduction: time evolution of a coupled orogen–subduction system. *Earth and Planetary Science Letters*, **275**, 93–101.

Medwedeff, D.A. (1992) Geometry and kinematics of an active, laterally propagating wedge thrust, Wheeler Ridge, California. In Mitra, S., and Fisher, G.W. (eds), *Structural Geology of Fold and Thrust Belts*, pp. 3–28. Johns Hopkins University Press, Baltimore, MD.

Meghraoui, M., Jaegy, R., Lammali, K., and Albarede, F. (1988a) Late Holocene earthquake sequences on the El Asnam (Algeria) thrust fault. *Earth and Planetary Science Letters*, **90**, 187–203.

Meghraoui, M., Philip, H., Albarede, F., and Cisternas, A. (1988b) Trench investigations through the trace of the 1980 El Asnam thrust fault: evidence for paleoseismicity. *Bulletin of the Seismological Society of America*, **78**, 979–999.

Meigs, A. (2010) Orogenic widening in the Quaternary argues against climate control of deformation in the Chugach/St. Elias orogenic belt, southern Alaska. *Geological Society of America Abstracts with Programs*, **41**, 305.

Meigs, A., Johnston, S., Garver, J., and Spotila, J. (2008) Crustal-scale structural architecture, shortening, and exhumation of an active, eroding orogenic wedge (Chugach/St Elias Range, southern Alaska). *Tectonics*, **27**, TC4003, doi:10.1029/2007TC002168.

Melnick, D., and Echtler, H.P. (2006) Inversion of forearc basins in south-central Chile caused by rapid glacial age trench fill. *Geology*, **34**, 709–712.

Merritts, D., and Bull, W.B. (1989) Interpreting Quaternary uplift rates at the Mendocino triple junction, northern California, from uplifted marine terraces. *Geology*, **17**, 1020–1024.

Merritts, D., and Hesterberg, T. (1994) Stream networks and long-term surface uplift in the New Madrid seismic zone. *Science*, **265**, 1081–1084.

Merritts, D.J., Vincent, K.R., and Wohl, E.E. (1994) Long river profiles, tectonism, and eustasy: a guide to interpreting fluvial terraces. *Journal of Geophysical Research*, **99**, 14031–14050.

Métivier, F., Gaudemer, Y., Tapponier, P., and Klein, M. (1999) Mass accumulation rates in Asia during the Cenozoic. *Geophysical Journal International*, **137**, 280–318.

Meunier, P., Hovius, N., and Haines, J.A. (2008) Topographic site effects and the location of earthquake induced landslides. *Earth and Planetary Science Letters*, **275**, 221–232.

Meynadier, L., Valet, J.-P., Weeks, R., Shackleton, N.J., and Hagee, V.L. (1992) Relative geomagnetic intensity of the field during the last 140ka. *Earth and Planetary Science Letters*, **114**, 39–57.

Miller, G.H., and Brigham-Grette, J. (1989) Amino acid geochronology; resolution and precision in carbonate fossils. *Quaternary International*, **1**, 111–128.

Miller, G.H., Magee, J.W., Johnson, B.J., Fogel, M., Spooner, N.A., McCulloch, M.T., and Ayliffe, L.K. (1999) Pleistocene extinction of *Genyornis newtoni*: human impact on Australian megafauna. *Science*, **283**, 205–208.

Miller, G.H., Fogel, M.L., Magee, J.W., Gagan, M.K., Clarke, S., and Johnson, B.J. (2005) Ecosystem collapse in Pleistocene Australia and a human role in megafaunal extinction. *Science*, **309**, 287–290.

Miller, S.A. (1996) Fluid-mediated influence of adjacent thrusting on the seismic cycle at Parkfield. *Nature*, **382**, 799–802.

Miller, S.R., and Slingerland, R.L. (2006) Topographic advection on fault-bend folds: inheritance of valley positions and the formation of wind gaps. *Geology*, **34**, 769–772.

Miller, S.R., Slingerland, R.L., and Kirby, E. (2007) Characteristics of steady state fluvial topography above fault-bend folds. *Journal of Geophysical Research*, **112**, F04004, doi:10.1029/2007JF000772.

Milliman, J.D., and Meade, R.H. (1983) World-wide delivery of river sediment to the oceans. *Journal of Geology*, **91**, 1–21.

Milliman, J.D., and Syvitski, J.P.M. (1992) Geomorphic/tectonic control of sediment discharge to the ocean: the importance of small mountainous rivers. *Journal of Geology*, **100**, 525–544.

Milne, G.A., and Mitrovica, J.X. (2008) Searching for eustasy in deglacial sea-level histories. *Quaternary Science Reviews*, **27**, 2292–2302.

Minster, J.B., and Jordan, T.H. (1987) Vector constraints on western U.S. deformation from space geodesy, neotectonics, and plate motions. *Journal of Geophysical Research*, **92**, 4798–4804.

Minster, J.B., Jordan, T.H., Hager, B.H., Agnew, D.C., and Royden, L.H. (1990) Implications of precise positioning. In Rundle, J.B. (ed.), *Geodesy in the Year 2000*, pp. 23–45. National Academy Press, Washington, DC.

Mitchell, S.G., and Montgomery, D.R. (2006) Influence of a glacial buzzsaw on the height and morphology of the Cascade Range in central Washington State, USA. *Quaternary Research*, **65**, 96–107.

Mitrovica, J.X., and Forte, A.M. (2004) A new inference of mantle viscosity based upon joint inversion of convection and glacial isostatic adjustment data. *Earth and Planetary Science Letters*, **225**, 177–189.

Mitrovica, J.X., Beaumont, C., and Jarvis, G.T. (1989) Tilting of continental interiors by the dynamical effects of subduction. *Tectonics*, **8**, 1079–1094.

Mohr, J.J., Reeh, N., and Madsen, S.N. (1998) Three-dimensional glacial flow and surface elevation measured with radar interferometry. *Nature*, **391**, 273–276.

Molnar, P., and England, P. (1990) Late Cenozoic uplift of mountain ranges and global climatic change: chicken or egg? *Nature*, **346**, 29–34.

Molnar, P., and Gipson, J.M. (1994) Very long baseline interferometry and active rotations of crustal blocks in the western Transverse Ranges, California. *Geological Society of America Bulletin*, **106**, 595–606.

Molnar, P., England, P., and Martinod, J. (1993) Mantle dynamics, uplift of the Tibetan Plateau, and the Indian monsoon. *Reviews of Geophysics*, **31**, 357–396.

Molnar, P., Brown, E.T., Burchfiel, B.C., Deng, Q., Feng, X., Li, J., Raisbeck, G.M., Shi, J., Wu, Z., Yiou, F., and You, H. (1994) Quaternary climate change and the formation of river terraces across growing anticlines on the north flank of the Tien Shan, China. *Journal of Geology*, **102**, 583–602.

Molnar, P., Anderson, R.S., and Anderson, S.P. (2007) Tectonics, fracturing of rock, and erosion. *Journal of Geophysical Research*, **112**, F03014, doi:10.1029/2005JF000433.

Montgomery, D.R., and Brandon, M.T. (2002) Topographic controls on erosion rates in tectonically active mountain ranges. *Earth and Planetary Science Letters*, **201**, 481–489.

Montgomery, D.R., and Dietrich, W.E. (1988) Where do channels begin? *Nature*, **336**, 232–234.

Montgomery, D.R., and Dietrich, W.E. (1989) Source areas, drainage density, and channel initiation. *Water Resources Research*, **25**, 1907–1918.

Montgomery, D.R., and Dietrich, W.E. (1992) Channel initiation and the problem of landscape scale. *Science*, **255**, 826–830.

Montgomery, D.R., Balco, G., and Willett, S.D. (2001) Climate, tectonics, and the morphology of the Andes. *Geology*, **29**, 579–582.

Morley, C.K. (1989) Extension, detachments, and sedimentation in continental rifts (with particular reference to east Africa). *Tectonics*, **8**, 1175–1192.

Morris, J.D. (1991) Applications of ^{10}Be to problems in the earth sciences. *Annual Review of Earth and Planetary Sciences*, **19**, 313–350.

Moucha, R., Forte, A.M., Rowley, D.B., Mitrovica, J.X., Simmons, N.A., and Grand, S.P. (2008) Mantle convection and the recent evolution of the Colorado Plateau and the Rio Grande rift valley. *Geology*, **36**, 439–442.

Mueller, K., and Talling, P. (1997) Geomorphic evidence for tear faults accommodating lateral propagation of an active fault-bend fold, Wheeler Ridge, California. *Journal of Structural Geology*, **19**, 397–412.

Muhs, D.R. (1992) The last interglacial–glacial transition in North America; evidence from uranium-series dating of coastal deposits. In Clark, P.U., and Lea, P.D. (eds), *The Last Interglacial–Glacial Transition in North America*, Special Paper 270, pp. 31–51. Geological Society of America, Boulder, CO.

Muhs, D.R., Kennedy, G.L., and Rockwell, T.K. (1994) Uranium-series ages of marine terrace corals from the Pacific coast of North America and implications for last-interglacial sea level history. *Quaternary Research*, **42**, 72–87.

Mulch, A., Graham, S.A., and Chamberlain, C.P. (2006) Hydrogen isotopes in Eocene river gravels and paleoelevation of the Sierra Nevada. *Science*, **313**, 87–89.

Murphy, M.A., An, Y., Harrison, T.M., Dürr, S.B., Chen, Z., Ryerson, F.J., Kidd, W.S.F., Wang, X., and Zhou, X. (1997) Did the Indo-Asian collision alone create the Tibetan plateau? *Geology*, **25**, 719–722.

Murray-Wallace, C.V., and Belperio, A.P. (1991) The last interglacial shoreline in Australia – a review. *Quaternary Science Reviews*, **10**, 441–461.

Nash, D.B. (1984) Morphologic dating of fluvial terrace scarps and fault scarps near West Yellowstone, Montana. *Geological Society of America Bulletin*, **95**, 1413–1424.

Nash, D.B. (1986) Morphologic dating and modeling degradation of fault scarps. In *Active Tectonics*, pp. 181–194. National Academy Press, Washington, DC.

Natawidjaja, D.H., Sieh, K., Ward, S.N., Cheng, H., Edwards, R.L., Galetzka, J., and Suwargadi, B.W.

(2004) Paleogeodetic records of seismic and aseismic subduction from central Sumatran microatolls, Indonesia. *Journal of Geophysical Research*, **109**, B04306, doi:10.1029/2003JB002398.

Natawidjaja, D.H., Sieh, K., Galetzka, J., Suwargadi, B.W., Cheng, H., Edwards, R.L., and Chlieh, M. (2007) Interseismic deformation above the Sunda Megathrust recorded in coral microatolls of the Mentawai islands, West Sumatra. *Journal of Geophysical Research*, **112**, B02404, doi:10.1029/2006JB004450.

Nelson, A.R., Johnson, S.Y., Kelsey, H.M., Wells, R.E., Sherrod, B.L., Pezzopane, S.K., Bradley, L., and Koehler, R.D. (2003) Late Holocene earthquakes on the Toe Jam Hill fault, Seattle fault zone, Washington. *Geological Society of America Bulletin*, **115**, 1388–1403.

Nelson, K.D., Zhao, W., Brown, L.D., Kuo, J., Che, J., Liu, X., Klemperer, S.L., Makovsky, Y., Meissner, R., Mechie, J., Kind, R., Wenzel, F., Ni, J., Nabelek, J., Chen, L., Tan, H., Wei, W., Jones, A.G., Booker, J., Unsworth, M., Kidd, W.S.F., Hauck, M., Alsdorf, D., Ross, A., Cogan, M., Wu, C., Sandvol, E.A., and Edwards, M.A. (1996) Partially molten middle crust beneath southern Tibet; synthesis of Project INDEPTH results. *Science*, **274**, 1684–1688.

Nicol, A., Walsh, J., Berryman, K., and Villamor, P. (2006) Interdependence of fault displacement rates and paleoearthquakes in an active rift. *Geology*, **34**, 865–868.

Niemi, N.A., Oskin, M., Burbank, D.W., Heimsath, A.M., and Gabet, E.J. (2005) Effects of bedrock landslides on cosmogenically determined erosion rates. *Earth and Planetary Science Letters*, **237**, 480–498.

Nishiizumi, K., Winterer, E.L., Kohl, C.P., Klein, J., Middleton, R., Lal, D., and Arnold, J.R. (1989) Cosmic ray production rates of [10]Be and [26]Al in quartz from glacially polished rocks. *Journal of Geophysical Research*, **94**, 17907–17915.

Nishiizumi, K., Kohl, C.P., Arnold, J.R., Klein, J., Fink, D., and Middleton, R. (1991) Cosmic ray produced [10]Be and [26]Al in Antarctic rocks: exposure and erosion history. *Earth and Planetary Science Letters*, **104**, 440–454.

Noller, J.S., Sowers, J.M., and Lettis, W.R. (2000) *Quaternary Geochronology: Methods and Applications*, AGU Reference Shelf 4. American Geophysical Union, Washington, DC.

Oberlander, T.M. (1985) Origin of drainage transverse to structures in orogens. In Morisawa, M., and Hack, J.T. (eds), *Tectonic Geomorphology: Proceedings, 15th Annual Binghamton Geomorphology Symposium*, pp. 155–182. Allen and Unwin, Boston, MA.

Oerlemans, J. (1984) Numerical experiments on large-scale glacial erosion. *Zeitschrift für Gletscherkunde und Glazialgeologie*, **20**, 107–126.

Ohmori, H. (1992) Morphological characteristics of the scar created by large-scale rapid mass movement. *Japanese Geomorphological Union Transactions*, **13**, 185–202.

Olsson, I.U. (1968) Modern aspects of radiocarbon dating. *Earth Science Reviews*, **4**, 203–218.

Oskin, M., and Burbank, D.W. (2005) Alpine landscape evolution dominated by cirque retreat. *Geology*, **33**, 933–936.

Oskin, M., and Burbank, D.W. (2007) Transient landscape evolution of basement-cored uplifts: example of the Kyrgyz Range, Tian Shan. *Journal of Geophysical Research*, **112**, F03S03, doi:10.1029/2006JF000563.

Ouchi, S. (1985) Response of alluvial rivers to slow active tectonic movement. *Geological Society of America Bulletin*, **96**, 504–515.

Ouimet, W.B., Whipple, K.X., Royden, L.H., Sun, Z., and Chen, Z. (2007) The influence of large landslides on river incision in a transient landscape: eastern margin of the Tibetan Plateau (Sichuan, China). *Geological Society of America Bulletin*, **119**, 1462–1476.

Owen, L., Caffee, M.W., Finkel, R., and Seong, Y.B. (2008) Quaternary glaciation of the Himalayan–Tibetan orogen. *Journal of Quaternary Science*, **23**, 513–531.

Pan, B., Burbank, D., Wang, Y., Wu, G., Li, J., and Guan, Q. (2003) A 900 k.y. record of strath terrace formation during glacial–interglacial transitions in northwest China. *Geology*, **31**, 957–960.

Paola, C., Heller, P.L., and Angevine, C.L. (1992) Large-scale dynamics of grain-size variation in alluvial basins, 1: Theory. *Basin Research*, **4**, 73–90.

Parker, R.A. (1977) *Experimental study of basin evolution and its hydrologic implications*, Ph.D. thesis, Colorado State University, Fort Collins, CO.

Pavich, M. (1986) Processes and rates of saprolite production and erosion on a foliated granitic rock of the Virginia Piedmont. In Colman, S., and Dethier, D. (eds), *Rates of Chemical Weathering of Rocks and Minerals*, pp. 552–590. Academic Press, Orlando, FL.

Pavich, M.J., and Hack, J.T. (1985) Appalachian Piedmont morphogenesis: weathering, erosion, and Cenozoic uplift. In Morisawa, M., and Hack, J.T. (eds), *Tectonic Geomorphology: Proceedings of the 15th Annual Binghamton Geomorphology Symposium*, pp. 299–319. Unwin Hyman, Boston, MA.

Pavich, M.J., Brown, L., Klein, J., and Middleton, R. (1984) ^{10}Be accumulation in a soil chronosequence. *Earth and Planetary Science Letters*, **68**, 198–204.

Pazzaglia, F.J., and Brandon, M.T. (2001) A fluvial record of long-term steady-state uplift and erosion across the Cascadia forearc high, western Washington State. *American Journal of Science*, **301**, 385–431.

Peakall, J., Ashworth, P.J., and Best, J.L. (2007) Meander-bend evolution, alluvial architecture, and the role of cohesion in sinuous river channels: a flume study. *Journal of Sedimentary Research*, **77**, 197–212.

Peltzer, G., Crampé, F., Hensley, S., and Rosen, P. (2001) Transient strain accumulation and fault interaction in the Eastern California Shear Zone. *Geology*, **29**, 975–978.

Penck, A., and Brückner, E. (1909) *Die Alpen im Eiszeitalter*. Tauchnitz, Leipzig.

Penck, W. (1953) *Morphological Analysis of Landforms*. St. Martin's Press, New York.

Perg, L.A., Anderson, R.S., and Finkel, R.C. (2001) Use of a new ^{10}Be and ^{26}Al inventory method to date marine terraces, Santa Cruz, California, USA. *Geology*, **29**, 879–882.

Perron, J.T., Dietrich, W.E., and Kirchner, J.W. (2008) Controls on the spacing of first-order valleys. *Journal of Geophysical Research*, **113**, F04016, doi:10.1029/2007JF000977.

Petit, C., Gunnell, Y., Gonga-Saholiariliva, N., Meyer, B., and Seguinot, J. (2009a) Faceted spurs at normal fault scarps: insights from numerical modeling. *Journal of Geophysical Research*, **114**, B05403, doi:10.1029/2008JB005955.

Petit, C., Meyer, B., Gunnell, Y., Jolivet, M., San'kov, V., Strak, V., and Gonga-Saholiariliva, N. (2009b) Height of faceted spurs, a proxy for determining long-term throw rates on normal faults: evidence from the North Baikal Rift System, Siberia. *Tectonics*, **28**, TC6010, doi:10.1029/2009TC002555.

Philip, H., and Meghraoui, M. (1983) Structural analysis and interpretation of the surface deformations of the El Asnam earthquake of October 10, 1980. *Tectonics*, **2**, 17–49.

Phillips, F.M., Leavy, B.D., Jannik, N.O., Elmore, D., and Kubik, P.W. (1986) The accumulation of cosmogenic chlorine-36 in rocks; a method for surface exposure dating. *Science*, **231**, 41–43.

Phillips, F.M., Zreda, M.G., Gosse, J.C., Klein, J., Evenson, E.B., Hall, R.D., Chadwick, O.A., and Sharma, P. (1997) Cosmogenic ^{36}Cl and ^{10}Be ages of Quaternary glacial and fluvial deposits of the Wind River Range, Wyoming. *Geological Society of America Bulletin*, **109**, 1453–1463.

Pierce, K.L., and Colman, S.M. (1986) Effect of height and orientation (microclimate) on geomorphic degradation rates and processes, late-glacial terrace scarps in central Idaho. *Geological Society of America Bulletin*, **97**, 869–885.

Pierce, K.L., Obradovich, J.D., and Friedman, I. (1976) Obsidian hydration dating and correlation of Bull Lake and Pinedale glaciations near West Yellowstone, Montana. *Geological Society of America Bulletin*, **87**, 703–710.

Pillans, B. (1983) Upper Quaternary terrace chronology and deformation, South Taranaki, New Zealand. *Geology*, **11**, 292–297.

Plafker, G. (1972) Alaskan earthquake of 1964 and Chilean earthquake of 1960: implications for arc tectonics. *Journal of Geophysical Research*, **77**, 901–924.

Plafker, G., and Ward, S.N. (1992) Backarc thrust faulting and tectonic uplift along the Caribbean sea coast during the April 22, 1991 Costa Rica earthquake. *Tectonics*, **11**, 709–718.

Platt, J.P., Balanya, J.-C., Garcia-Duenas, V., Azanon, J.M., and Sanchez-Gomez, M. (1998) Alternating contractional and extensional events in the Alpujarride nappes of the Alboran Domain (Betics, Gibraltar Arc); discussion and reply. *Tectonics*, **17**, 973–981.

Poblet, J., Muñoz, J.A., Travé, A., and Serra-Kiel, J. (1998) Quantifying the kinematics of detachment folds using three-dimensional geometry: application to the Mediano anticline (Pyrenees, Spain). *Geological Society of America Bulletin*, **110**, 111–125.

Pollitz, F.F., and Sacks, S.I. (2002) Stress triggering of the 1999 Hector Mine earthquake by transient deformation following the 1992 Landers earthquake. *Bulletin of the Seismological Society of America*, **92**, 1487–149.

Pollitz, F.F., Bürgmann, R., and Romanowicz, B. (1998) Viscosity of oceanic asthenosphere inferred from remote triggering of earthquakes. *Science*, **280**, 1245–1249.

Porter, S.C. (1981) Lichenometric studies in the Cascade Range of Washington; establishment of *Rhizocarpon geographicum* growth curves at Mount Rainier. *Arctic and Alpine Research*, **13**, 11–23.

Porter, S.C. (1989) Some geological implication of average Quaternary glacial conditions. *Quaternary Research*, **32**, 245–261.

Porter, S.C., and Orombelli, G. (1981) Alpine rockfall hazards. *American Scientist*, **69**, 67–75.

Powell, J.W. (1875) *Exploration of the Colorado River of the West (1869–72)*. Smithsonian Institution, Washington, DC.

Pratt, B., Burbank, D.W., Heimsath, A., and Ojha, T. (2002) Impulsive alluviation during early Holocene strengthened monsoons, central Nepal Himalaya. *Geology*, **30**, 911–914.

Pratt-Sitaula, B., Garde, M., Burbank, D.W., Oskin, M., Heimsath, A., and Gabet, E. (2007) Bedload-to-suspended load ratio and rapid bedrock incision from Himalayan landslide-dam lake record. *Quaternary Research*, **68**, 111–120.

Quigley, M., Sandiford, M., Fifield, L.K., and Alimanovic, A. (2007) Bedrock erosion and relief production in the northern Flinders Ranges, Australia. *Earth Surface Processes and Landforms*, **32**, 929–944.

Ratschbacher, L., Frisch, W., Neubauer, F., Schmid, S.M., and Neugebauer, J. (1989) Extension in compressional orogenic belts: the eastern Alps. *Geology*, **17**, 404–407.

Raymo, M.E., and Ruddiman, W.F. (1992) Tectonic forcing of late Cenozoic climate change. *Nature*, **359**, 117–122.

Read, S.A.L. (1984) The Ostler Fault zone. In Wood, P.R. (ed.), *Guidebook to the South Island Scientific Excursions, International Symposium on Recent Crustal Movements of the Pacific Region*, Royal Society Miscellaneous Series. Royal Society of New Zealand, Wellington, New Zealand.

Reid, H.F. (1910) The mechanism of the earthquake. In *The California Earthquake of April 18, 1906. Report of the State Earthquake Investigation Commission*, vol. 2, pp. 1–192. Carnegie Institution, Washington, DC.

Reid, J.B., Jr. (1992) The Owens River as a tiltmeter for Long Valley caldera, California. *Journal of Geology*, **100**, 353–363.

Reilinger, R.E., Ergintav, S., Burgmann, R., McClusky, S., Lenk, O., Barka, A., Gurkan, O., Hearn, L., Feigl, K.L., Cakmak, R., Aktug, B., Ozener, H., and Toksoz, M.N. (2000) Coseismic and postseismic fault slip for the 17 August 1999, M=7.5, Izmit, Turkey earthquake. *Science*, **289**, 1519–1524.

Reiners, P.W. (2005) Thermochronologic approaches to paleotopography. *Reviews in Mineralogy and Geochemistry*, **66**, 243–267.

Reiners, P.W., and Brandon, M.T. (2006) Using thermochronology to understand orogenic erosion. *Annual Review of Earth and Planetary Sciences*, **34**, 419–466.

Reiners, P.W., Ehlers, T.A., Mitchell, S.G., and Montgomery, D. (2003) Coupled spatial variations in precipitation and long-term erosion rates across the Washington Cascades. *Nature*, **426**, 645–647.

Reiners, P.W., Spell, T.L., Nicolescu, S., and Zanetti, K.A. (2004) Zircon (U–Th)/He thermochronometry: He diffusion and comparisons with ^{40}Ar/^{39}Ar dating. *Geochimica et Cosmochimica Acta*, **68**, 1857–1887.

Repka, J.L., Anderson, R.S., and Finkel, R.C. (1997) Cosmogenic dating of fluvial terraces, Fremont River, Utah. *Earth and Planetary Science Letters*, **152**, 59–73.

Richter, C.F. (1935) An instrumental earthquake-magnitude scale. *Bulletin of the Seismological Society of America*, **25**, 1–32.

Riebe, C.S., Kirchner, J.W., Granger, D.E., and Finkel, R.C. (2001) Minimal climatic control on erosion rates in the Sierra Nevada, California. *Geology*, **29**, 447–450.

Riebe, C.S., Kirchner, J.W., and Finkel, R.C. (2003) Long-term rates of chemical weathering and physical erosion from cosmogenic nuclides and geochemical mass balance. *Geochimica et Cosmochimica Acta*, **67**, 4411–4427.

Ritter, J.B., Miller, J.R., Enzel, Y., Howes, S.D., Nadon, G., Grubb, M.D., Hoover, K.A., Olsen, T., Seneau, S.L., Sack, D., Summa, C.L., Taylor, I., Touysinhthiphonexay, K.C.N., Yodis, E.G., Schneider, N.P., Ritter, D.F., and Wells, S.G. (1993) Quaternary evolution of the Cedar Creek alluvial fan, Montana. *Geomorphology*, **8**, 287–304.

Rockwell, T.K., Keller, E.A., Clark, M.N., and Johnson, D.L. (1984) Chronology and rates of faulting of Ventura River terraces, California. *Geological Society of America Bulletin*, **95**, 1466–1474.

Rockwell, T.K., Lindvall, S., Herzberg, M., Murbach, D., Dawson, T., and Berger, G. (2000) Paleoseismology of the Johnson Valley, Kickapoo, and Homestead Valley faults: clustering of earthquakes in the Eastern California Shear Zone. *Bulletin of the Seismological Society of America*, **90**, 1200–1236.

Rodgers, D.W., and Little, T.A. (2006) World's largest coseismic strike-slip offset: the 1855 rupture of the Wairarapa Fault, New Zealand, and implications for displacement/length scaling of continental earthquakes. *Journal of Geophysical Research*, **111**, 12408, doi:10.1029/2005JB004065.

Roe, G.H. (2005) Orographic precipitation. *Annual Review of Earth and Planetary Sciences*, **33**, 645–671.

Roe, G.H., and Baker, M. (2006) Microphysical and geometrical controls on the pattern of orographic precipitation. *Journal of Atmospheric Science*, **63**, 861–880.

Roe, G.H., Montgomery, D.R., and Hallet, B. (2002) Effects of orographic precipitation variations on the concavity of steady-state river profiles. *Geology*, **30**, 143–146.

Roe, G.H., Montgomery, D.R., and Hallet, B. (2003) Orographic precipitation and the relief of mountain ranges. *Journal of Geophysical Research*, **108**, 2315, doi:10.1029/2001JB001521.

Roering, J.J. (2008) How well can hillslope evolution models "explain" topography? Simulating soil transport and production with high-resolution topographic data. *Geological Society of America Bulletin*, **120**, 1248–1262.

Roering, J.J., Kirchner, J.W., Sklar, L.S., and Dietrich, W.E. (2001) Experimental hillslope evolution by nonlinear creep and landsliding. *Geology*, **29**, 143–146.

Rosenbloom, N.A., and Anderson, R.S. (1994) Hillslope and channel evolution in a marine terraced landscape, Santa Cruz, California. *Journal of Geophysical Research*, **99**, 14013–14029.

Rowley, D.B., and Garzione, C.N. (2007) Stable isotope-based paleoaltimetry. *Annual Review of Earth and Planetary Sciences*, **35**, 463–508.

Rowley, D.B., Pierrehumbert, R.T., and Currie, B.S. (2001) A new approach to stable isotope-based paleoaltimetry: implications for paleoaltimetry and the paleohypsometry of the High Himalaya since the Late Miocene. *Earth and Planetary Science Letters*, **188**, 253–268.

Royden, L.H., Burchfiel, B.C., King, R.W., Wang, E., Chen, Z., Shen, F., and Liu, Y. (1997) Surface deformation and lower crustal flow in Eastern Tibet. *Science*, **276**, 788–790.

Rubin, C.M. (1996) Systematic underestimation of earthquake magnitudes from large intercontinental reverse faults: historical ruptures break across segment boundaries. *Geology*, **24**, 989–992.

Rubin, C.M., Lindvall, S.C., and Rockwell, T.K. (1998) Evidence for large earthquakes in metropolitan Los Angeles. *Science*, **281**, 398–402.

Ruhl, K.W., and Hodges, K.V. (2005) The use of detrital mineral cooling ages to evaluate steady-state assumptions in active orogens: an example from the central Nepalese Himalaya. *Tectonics*, **24**, TC4015, doi:10.1029/2004TC001712.

Ryan, J.W., and Ma, C. (1998) NASA-GSFC's geodetic VLBI program; a twenty-year retrospective. *Physics and Chemistry of the Earth*, **23**, 1041–1052.

Sadybakasov, I. (1990) *Neotectonics of High Asia* (in Russian). Nauka, Moscow.

Sahagian, D., and Proussevitch, A. (2007) Paleoelevation measurement on the basis of vesicular basalts. *Reviews in Mineralogy and Geochemistry*, **66**, 195–213.

Sahagian, D., Proussevitch, A., and Carlson, W. (2002a) Quantitative analysis of vesicular basalts for appli-

cation as a paleobarometer/paleoaltimeter. *Journal of Geology*, **110**, 671–685.

Sahagian, D., Proussevitch, A., and Carlson, W. (2002b) Timing of Colorado Plateau uplift: initial constraints from vesicular basalt-derived paleoelevations. *Geology*, **30**, 807–810.

Saleeby, J., and Foster, Z. (2004) Topographic response to mantle lithosphere removal in the southern Sierra Nevada region, California. *Geology*, **32**, 245–248.

Sandiford, M. (2007) The tilting continent: a new constraint on the dynamic topographic field from Australia. *Earth and Planetary Science Letters*, **261**, 152–163.

Sandiford, M., and Quigley, M. (2009) TOPO-OZ: insights into the various modes of intraplate deformation in the Australian continent. *Tectonophysics*, **474**, 405–416.

Santamaria Tovar, D., Shulmeister, J., and Davies, T.R. (2008) Evidence for a landslide origin of New Zealand's Waiho Loop moraine. *Nature Geoscience*, **1**, 524–526.

Sarna-Wojcicki, A.M., Lajoie, K.R., Meyer, C.E., Adam, D.P., and Rieck, H.J. (1991) Tephrochronology correlation of upper Neogene sediments along the Pacific margin, conterminous United States. In Morrison, R.B. (ed.), *Quaternary nonglacial geology: conterminous U.S.* (*The geology of North America*, vol. K-2), pp. 117–140. Geological Society of America, Boulder, CO.

Satake, K., and Atwater, B.F. (2007) Long-term perspectives on giant earthquakes and tsunamis at subduction zones. *Annual Review of Earth and Planetary Sciences*, **35**, 49–74.

Satake, K., Shimazaki, K., Tsuji, Y., and Ueda, K. (1996) Time and size of a giant earthquake in Cascadia inferred from Japanese tsunami records of January 1700. *Nature*, **379**, 246–249.

Savage, J.C., and Thatcher, W. (1992) Interseismic deformation at the Nankai Trough, Japan, subduction zone. *Journal of Geophysical Research*, **97**, 11117–11135.

Schaller, M., von Blanckenburg, F., Hovius, N., and Kubik, P.W. (2001) Large-scale erosion rates from in situ-produced cosmogenic nuclides in European river sediments. *Earth and Planetary Science Letters*, **188**, 441–458.

Schaller, M., von Blanckenburg, F., Veldkamp, A., Tebbens, L.A., Hovius, N., and Kubik, P.W. (2002) A 30,000 year record of erosion rates from cosmogenic ^{10}Be in Middle European river terraces. *Earth and Planetary Science Letters*, **204**, 307–320.

Schaller, M., von Blanckenburg, F., Veldkamp, A., van den Berg, M.W., Hovius, N., and Kubik, P.W.

(2004) Paleo-erosion rates from cosmogenic [10]Be in a 1.3 Ma terrace sequence: River Meuse, the Netherlands. *Journal of Geology*, **112**, 127–144.

Scharer, K., Burbank, D., Chen, J., and Weldon, R.J.I. (2006) Kinematic models of fluvial terraces over active detachment folds: constraints on the growth mechanism of the Kashi–Atushi fold system, Chinese Tian Shan. *Geological Society of America Bulletin*, **118**, 1006–1021.

Scherler, D., Bookhagen, B., and Strecker, M.R. (2011) Hillslope–glacier coupling: the interplay of topography and glacial dynamics in High Asia. *Journal of Geophysical*. doi:10.1029/2010JF001751.

Schildgen, T.F., Ehlers, T.A., Whipp, D.M., Jr., van Soest, M.C., Whipple, K.X., and Hodges, K.V. (2009) Quantifying canyon incision and Andean Plateau surface uplift, southwest Peru: a thermochronometer and numerical modeling approach. *Journal of Geophysical Research*, **114**, F04014, doi:10.1029/2009JF001305.

Schlische, R.W., Young, S.S., Ackermann, R.V., and Gupta, A. (1996) Geometry and scaling relations of a population of very small rift-related normal faults. *Geology*, **24**, 683–686.

Schmidt, K.M., and Montgomery, D.R. (1995) Limits to relief. *Science*, **270**, 617–620.

Schmidt, K.M., and Montgomery, D.R. (1996) Rock mass strength assessment for bedrock landsliding. *Environmental and Engineering Geoscience*, **2**, 325–338.

Scholz, C.H. (1990) *The Mechanics of Earthquakes and Faulting*. Cambridge University Press, Cambridge.

Scholz, C.H. (1998) A further note on earthquake size distributions. *Bulletin of the Seismological Society of America*, **88**, 1325–1326.

Scholz, C.H. (2002) *The Mechanics of Earthquakes and Faulting*, 2nd edn. Cambridge University Press, Cambridge.

Schumm, S.A. (1986) Alluvial river response to active tectonics. In *Active Tectonics*, pp. 80–94. National Academy Press, Washington, DC.

Schumm, S.A., and Khan, H.R. (1972) Experimental study of channel patterns. *Nature*, **233**, 407–409.

Schumm, S.A., Mosley, M.P., and Weaver, W.E. (1987) *Experimental Fluvial Geomorphology*. John Wiley and Sons, New York.

Schwartz, D.P., and Coppersmith, K.J. (1984) Fault behavior and characteristic earthquakes: examples from the Wasatch and San Andreas fault zones. *Journal of Geophysical Research*, **89**, 5681–5698.

Scott, K.M., and Williams, R.P. (1978) *Erosion and Sediment Yields in the Transverse Ranges, Southern California*, Professional Paper 1030. United States Geological Survey, Washington, DC.

Seeber, L., and Gornitz, V. (1983) River profiles along the Himalayan arc as indicators of active tectonics. *Tectonophysics*, **92**, 335–367.

Seidl, M.A., and Dietrich, W.E. (1992) The problem of channel erosion into bedrock. *Catena (Supplement)*, **23**, 101–124.

Seidl, M.A., Dietrich, W.E., and Kirchner, J.W. (1994) Longitudinal profile development into bedrock: an analysis of Hawaiian channels. *Journal of Geology*, **102**, 457–474.

Sempere, J.C., and Macdonald, K.C. (1986) Overlapping spreading centers: implications from crack growth simulation by the displacement discontinuity method. *Tectonics*, **5**, 151–163.

Shen, Z.-K., Wang, Q., Bürgmann, R., Wan, Y., and Ning, J. (2005) Pole-tide modulation of slow slip events at Circum-Pacific subduction zones. *Bulletin of the Seismological Society of America*, **95**, 2009–2015.

Sheppard, P.R., and White, L.O. (1995) Tree-ring responses to the 1978 earthquake at Stephens Pass, northeastern California. *Geology*, **23**, 109–112.

Shi, B., Anooshehpoor, A., Zeng, Y., and Brune, J.N. (1996) Rocking and overturning of precariously balanced rocks by earthquakes. *Bulletin of the Seismological Society of America*, **86**, 1364–1371.

Shimaki, K., and Nakata, T. (1980) Time-predictable recurrence model for large earthquakes. *Geophysical Research Letters*, **7**, 279–282.

Shuster, D.L., and Farley, K.A. (2005) $^4He/^3He$ thermochronometry: theory, practice, and potential complications. *Reviews in Mineralogy and Geochemistry*, **58**, 181–203.

Sieh, K. (1978) Prehistoric large earthquakes produced by slip on the San Andreas fault at Pallett Creek, California. *Journal of Geophysical Research*, **83**, 3907–3939.

Sieh, K.E. (1984) Lateral offsets and revised dates of large earthquakes at Pallett Creek, southern California. *Journal of Geophysical Research*, **89**, 7641–7670.

Sieh, K.E., and Jahns, R.H. (1984) Holocene activity of the San Andreas fault at Wallace Creek, California. *Geological Society of America Bulletin*, **95**, 883–896.

Sieh, K., Stuiver, M., and Brillinger, D. (1989) A more precise chronology of earthquakes produced by the San Andreas fault in Southern California. *Journal of Geophysical Research*, **94**, 603–623.

Sieh, K., Ward, S.N., Natawidjaja, D.H., and Suwargadi, B.W. (1999) Crustal deformation at the Sumatran subduction zone. *Geophysical Research Letters*, **26**, 3141–3144.

Sieh, K., Natawidjaja, D., Meltzner, A., Shen, C., Cheng, H., Kuei-Shu Li, K.-S., Suwargadi, B., Galetzka, J., Philibosian, B., and Edwards, R. (2008) Earthquake supercycles inferred from sea-level changes recorded in the corals of west Sumatra. *Science*, **322**, 1674–1678.

Simoes, M., and Avouac, J.P. (2006) Investigating the kinematics of mountain building in Taiwan from the spatiotemporal evolution of the foreland basin and western foothills. *Journal of Geophysical Research*, **111**, B10401, doi:10.1029/2005JB004209.

Simons, M., Fialko, Y., and Rivera, L. (2002) Coseismic deformation from the 1999 Mw 7.1 Hector Mine, California, earthquake as inferred from InSAR and GPS observations. *Bulletin of the Seismological Society of America*, **92**, 1390–1402.

Simpson, D.W., and Anders, M.H. (1992) Tectonics and topography of the western United States – an application of digital mapping. *GSA Today*, **2**, 117–121.

Simpson, R.W., and Reasenberg, P.A. (1994) Earthquake-induced static-stress changes on central California faults. In Simpson, R.W. (ed.), *The Loma Prieta, California, Earthquake of October 17, 1989 – Tectonic Processes and Models*, Professional Paper 1550-F, pp. F55–F89. United States Geological Survey, Washington, DC.

Simpson, R.W., Harris, R.A., and Reasenberg, P.A. (1994) Stress changes caused by the 1994 Northridge earthquake. [Proceedings of the 89th Annual Meeting of the Seismological Society of America; Program for Northridge abstracts.] *Seismological Research Letters*, **65**, 240.

Sklar, L., and Dietrich, W.E. (1998) River longitudinal profiles and bedrock incision models: stream power and the influence of sediment supply. In Tinkler, K.J., and Wohl, E.E. (eds), *Rivers Over Rock: Fluvial Processes in Bedrock Channels*, Geophysical Monograph Series, 107, pp. 237–260. American Geophysical Union, Washington, DC.

Sklar, L., and Dietrich, W.E. (2004) A mechanistic model for river incision into bedrock by saltating bed load. *Water Resources Research*, **40**, W06301, doi:10.1029/2003WR002496.

Sklar, L.S., Stock, J.D., Roering, J.J., Kirchner, J.W., Dietrich, W.E., Chi, W., Hsu, L., Hsieh, M., Tsao, S., and Chen, M. (2005) Evolution of fault scarp knickpoints following 1999 Chi-Chi earthquake in West-Central Taiwan. *Eos (American Geophysical Union Transactions)*, **86**, H34A-06.

Small, E.E., and Anderson, R.S. (1995) Geomorphically driven Late Cenozoic rock uplift in the Sierra Nevada, California. *Science*, **270**, 277–280.

Small, E.E., and Anderson, R.A. (1998) Pleistocene relief production in Laramide mountain ranges, western United States. *Geology*, **26**, 123–136.

Small, E.E., Anderson, R.S., Repka, J.L., and Finkel, R. (1997) Erosion rates of alpine bedrock summit surfaces deduced from in situ ^{10}Be and ^{26}Al. *Earth and Planetary Science Letters*, **150**, 413–425.

Small, E.E., Anderson, R.S., and Hancock, G.S. (1999) Estimates of the rate of regolith production using ^{10}Be and ^{26}Al from an alpine hillslope. *Geomorphology*, **27**, 131–150.

Smith, D.G., and Smith, N.D. (1980) Sedimentation in anastomosed river systems, examples from alluvial valleys near Banff, Alberta. *Journal of Sedimentary Petrology*, **50**, 157–164.

Smith, G.I., and Street-Perrott, F.A. (1983) Pluvial lakes of the western United States. In Porter, S.C. (ed.), *Late Quaternary Environments of the United States: the Late Pleistocene*, pp. 190–212. University of Minnesota Press, Minneapolis, MN.

Smith, G.I., Friedman, I., Gleason, J.D., and Warden, A. (1992) Stable isotope composition of waters in southeastern California: 2. Groundwaters and their relation to modern precipitation. *Journal of Geophysical Research*, **97**, 5813–5823.

Smith, R.B. (1979) The influence of mountains on the atmosphere. *Advances in Geophysics*, **21**, 87–230.

Smith, R.B., and Barstad, I. (2004) A linear theory of orographic precipitation. *Journal of the Atmospheric Sciences*, **61**, 1377–1391.

Smith, S.W., and Wyss, M. (1968) Displacement on the San Andreas fault subsequent to the 1966 Parkfield earthquake. *Bulletin of the Seismological Society of America*, **58**, 1955–1973.

Snow, R.S., and Slingerland, R.L. (1987) Mathematical modeling of graded river profiles. *Journal of Geology*, **95**, 15–33.

Snow, R.S., and Slingerland, R.L. (1990) Stream profile adjustment to crustal warping: nonlinear results from a simple model. *Journal of Geology*, **98**, 699–708.

Snyder, N.P., Whipple, K.X., Tucker, G.E., and Merritts, D.J. (2000) Landscape response to tectonic forcing: digital elevation model analysis of stream profiles in the Mendocino triple junction, northern California. *Geological Society of America Bulletin*, **112**, 1250–1263.

Snyder, N.P., Whipple, K., Tucker, G.E., and Merritts, D.J. (2003) Channel response to tectonic forcing: field analysis of stream morphology and hydrology in the Mendocino triple junction region, northern California. *Geomorphology*, **53**, 97–127.

Sobel, E.R., Oskin, M., Burbank, D.W., and Mikolaichuk, A. (2006) Exhumation of basement-cored uplifts:

example of the Kyrgyz Range quantified with apatite fission-track thermochronology. *Tectonics*, **25**, TC2008, doi:10.1029/2005TC001809.

Somerville, P., and Pitarka, A. (2006) Differences in earthquake source and ground motion characteristics between surface and buried crustal earthquakes. *Bulletin of the Earthquake Research Institute, University of Tokyo*, **81**, 259–266.

Spotila, J.A., and Sieh, K. (2000) Architecture of transpressional thrust faulting in the San Bernadino Mountains, southern California, from deformation of a deeply weathered surface. *Tectonics*, **19**, 589–615.

Spotila, J.A., Niemi, N., Brady, R., House, M., Buscher, J., and Oskin, M. (2007) Long-term continental deformation associated with transpressive plate motion: the San Andreas fault. *Geology*, **35**, 967–970.

Stark, C.P., Barbour, J.R., Hayakawa, Y.S., Hattanji, T., Hovius, N., Chen, H., Lin, C.-W., Horng, M.-J., Xu, K.-Q., and Fukahata, Y. (2010) The climatic signature of incised river meanders. *Science*, **327**, 1497–1501.

Steffensen, J.P., Andersen, K.K., Bigler, M., Clausen, H.B., Dahl-Jensen, D., Fischer, H., Goto-Azuma, K., Hansson, M., Johnsen, S.J., Jouzel, J., Masson-Delmotte, V., Popp, T., Rasmussen, S.O., Rothlisberger, R., Ruth, U., Stauffer, B., Siggaard-Andersen, M.-L., Sveinbjornsdottir, A.E., Svensson, A., and White, J.W.C. (2008) High-resolution Greenland ice core data show abrupt climate change happens in a few years. *Science*, **321**, 680–684.

Stein, R.S. (1999) The role of stress transfer in earthquake occurrence. *Nature*, **402**, 605–609.

Stein, R.S., and Ekstrom, G. (1992) Seismicity and geometry of a 110-km-long blind thrust fault 2. Synthesis of the 1982–1985 California earthquake sequence. *Journal of Geophysical Research*, **97**, 4865–4883.

Stein, R.S., King, G.C.P., and Rundel, J.B. (1988) The growth of geological structure by repeated earthquakes 2. Field examples of continental dip-slip faults. *Journal of Geophysical Research*, **93**, 13319–13331.

Stern, T.A., Baxter, A.K., and Barrett, P.J. (2005) Isostatic rebound due to glacial erosion within the Transantarctic Mountains. *Geology*, **33**, 221–224.

Stevens, G. (1974) *Rugged Landscape: The Geology of Central New Zealand*. Reed Ltd, Wellington, New Zealand.

Stewart, R.J., Hallet, B., Zeitler, P.K., Malloy, M.A., Allen, C.M., and Trippett, D. (2008) Brahmaputra sediment flux dominated by highly localized rapid erosion from the easternmost Himalaya. *Geology*, **36**, 711–714.

Stock, G.M., Anderson, R.S., and Finkel, R.C. (2004) Cave sediments reveal pace of landscape evolution in the Sierra Nevada, California. *Geology*, **32**, 193–196.

Stock, G.M., Anderson, R.S., and Finkel, R.C. (2005) Late Cenozoic topographic evolution of the Sierra Nevada, California, inferred from cosmogenic ^{26}Al and ^{10}Be concentrations. *Earth Surface Processes and Landforms*, **30**, 985–1006.

Stock, J.D., and Dietrich, W.E. (2006) Erosion of steepland valleys by debris flows. *Geological Society of America Bulletin*, **118**, 1125–1148.

Stock, J.D., and Montgomery, D.R. (1996) Estimating palaeorelief from detrital mineral age ranges. *Basin Research*, **8**, 317–328.

Stockli, D., Farley, K.A., and Dumitru, T.A. (2000) Calibration of the apatite (U–Th)/He thermochronometer on an exhumed fault block, White Mountains, California. *Geology*, **28**, 983–986.

Stolar, D., Roe, G., and Willett, S. (2007) Controls on the patterns of topography and erosion rate in a critical orogen. *Journal of Geophysical Research*, **112**, F04002.

Stone, J.O. (2000) Air pressure and cosmogenic isotope production. *Journal of Geophysical Research*, **105**, 23753–23759.

Stuiver, M. (1970) Tree ring, varve and carbon-14 chronologies. *Nature*, **228**, 454–455.

Stuiver, M., and Reimer, P.J. (1993) Extended ^{14}C data base and revised CALIB 3.0 ^{14}C age calibration program. *Radiocarbon*, **35**, 215–230.

Stuiver, M., Reimer, P., and Reimer, R. (2009) CALIB radiocarbon calibration, http://intcal.qub.ac.uk/calib.

Stüwe, K., White, L., and Brown, R. (1994) The influence of eroding topography on steady-state isotherms: application to fission track analysis. *Earth and Planetary Science Letters*, **124**, 63–74.

Subarya, C., Chlieh, M., Prawirodirdjo, L., Avouac, J.-P., Bock, Y., Sieh, K., Meltzner, A.J., Natawidjaja, D.H., and McCaffrey, R. (2006) Plate-boundary deformation associated with the great Sumatra–Andaman earthquake. *Nature*, **440**, 45–51, doi:10.1038/nature04522.

Summerfield, M.A. (1991) *Global Geomorphology. An Introduction to the Study of Landforms*. Longman, Harlow, Essex.

Suppe, J.S., Chou, G.T., and Hook, S.C. (1992) Rates of folding and faulting determined from growth strata. In McClay, K.R. (ed.), *Thrust Tectonics*, pp. 105–122. Chapman and Hall, London.

Suppe, J. (1983) Geometry and kinematics of fault bend folding. *American Journal of Science*, **283**, 648–721.

Suppe, J., and Medwedeff, D.A. (1990) Geometry and kinematics of fault-propagation folding. *Eclogae Geologicae Helvetiae*, **83**, 409–454.

Suppe, J., Connors, C.D., and Zhang, Y. (2004) Shear fault-bend folding. *American Association of Petroleum Geologists Memoir*, **82**, 303–323.

Suppe, J., Sabat, F., Muñoz, J.A., Poblet, J., Roca, E., and Vergés, J. (1997) Bed-by-bed fold growth by kink-band migration: Sant Llorenc de Morunys, Eastern Pyrenees. *Journal of Structural Geology*, **19**, 443–461.

Sylvester, A.G. (1986) Near-field tectonic geodesy. In *Active Tectonics*, pp. 164–180. National Academy Press, Washington, DC.

Sylvester, A.G. (1988) Strike-slip faults. *Geological Society of America Bulletin*, **100**, 1666–1703.

Talling, P.J., and Sowter, M.J. (1999) Drainage density on progressively tilted surfaces with different gradients, Wheeler Ridge, California. *Earth Surface Processes and Landforms*, **24**, 809–824.

Talling, P.J., Stewart, M.D., Stark, C.P., Gupta, S., and Vincent, S.J. (1997) Regular spacing of drainage outlets from linear fault-blocks. *Basin Research*, **9**, 275–302.

Taylor, F.W., Mann, P., Bevis, M.G., Edwards, R.L., Cheng, H., Cutler, K.B., Gray, S.C., Burr, G.S., Beck, J.W., Phillips, D.A., Cabioch, G., and Recy, J. (2005) Rapid forearc uplift and subsidence caused by impinging bathymetric features: examples from the New Hebrides and Solomon arcs. *Tectonics*, **24**, TC6005, doi:10.1029/2004TC001650.

Thatcher, W. (1986a) Geodetic measurement of active-tectonic processes. In *Active Tectonics*, pp. 155–163. National Academy Press, Washington, DC.

Thatcher, W. (1986b) Cyclic deformation related to great earthquakes at plate boundaries. *Royal Society of New Zealand Bulletin*, **24**, 245–272.

Thatcher, W. (2007) Microplate model for the present-day deformation of Tibet. *Journal of Geophysical Research*, **112**, B01401, doi:10.1029/2005JB004244.

Thiede, R.C., Bookhagen, B., Arrowsmith, J.R., Sobel, E.R., and Strecker, M.R. (2004) Climatic control on rapid exhumation along the Southern Himalayan Front. *Earth and Planetary Science Letters*, **222**, 791–806.

Thompson, S.C., Weldon, R., III, Rubin, C.M., Abdrakhmatov, K.E., Molnar, P., and Berger, G.W. (2002) Late Quaternary slip rates across the central Tien Shan, Kyrgyzstan, central Asia. *Journal of Geophysical Research – Solid Earth*, **107**, 2203, doi:10.1029/2001JB000596.

Thorson, R.M. (1989) Glacio-isostatic response of the Puget Sound area, Washington. *Geological Society of America Bulletin*, **101**, 1163–1174.

Tippett, J.M., and Hovius, N. (2000) Geodynamic processes in the Southern Alps, New Zealand. In Summerfield, M.A. (ed.), *Geomorphology and Global Tectonics*, pp. 109–134. John Wiley & Sons, Chichester.

Tippett, J.M., and Kamp, P.J.J. (1993) Fission track analysis of the Late Cenozoic vertical kinematics of continental Pacific crust, South Island, New Zealand. *Journal of Geophysical Research*, **98**, 16119–16148.

Tippett, J.M., and Kamp, P.J.J. (1995) Quantitative relationships between uplift and relief parameters for the Southern Alps, New Zealand, as determined by fission track analysis. *Earth Surface Processes and Landforms*, **20**, 153–175.

Toda, S., and Stein, R. (2003) Toggling of seismicity by the 1997 Kagoshima earthquake couplet: a demonstration of time-dependent stress transfer. *Journal of Geophysical Research*, **108**, 2567, doi:10.1029/2003JB002527.

Tomkin, J.H. (2007) Coupling glacial erosion and tectonics at active orogens: a numerical modeling study. *Journal of Geophysical Research*, **112**, F02015, doi:10.1029/2005JF000332.

Tomkin, J.H., and Roe, G.H. (2007) Climate and tectonic controls on glaciated critical-taper orogens. *Earth and Planetary Science Letters*, **262**(3–4), 385–397.

Tucker, G.E., and Hancock, G.R. (2010) Modelling landscape evolution. *Earth Surface Processes and Landforms*, **35**, 28–50.

Tucker, G.E., and Slingerland, R. (1996) Predicting sediment flux from fold and thrust belts. *Basin Research*, **8**, 329–349.

Tucker, G., Lancaster, S., Gasparini, N., and Bras, R. (2001) The Channel–Hillslope Integrated Landscape Development model (CHILD). In Harmon, R.S., and Doe, W.W., III (eds), *Landscape Erosion and Evolution Modeling*. Kluwer Academic/Plenum, New York.

Turcotte, D.L., and Schubert, G. (1982) *Geodynamics Applications of Continuum Physics to Geological Problems*. John Wiley and Sons, New York.

Turcotte, D.L., and Schubert, G. (2002) *Geodynamics*, 2nd edn, Cambridge University Press, Cambridge.

Turnbull, J.C., Lehman, S.J., Miller, J.B., Sparks, R.J., Southon, J.R., and Tans, P.P. (2007) A new high precision $^{14}CO_2$ time series for North American continental air. *Journal of Geophysical Research*, **112**, D11310, doi:10.1029/2006JD008184.

Turowski, J.M., Hovius, N., Hsieh, M.-L., Lague, D., and Chen, M.-C. (2008) Distribution of erosion across bedrock channels. *Earth Surface Processes and Landforms*, **33**, 353–363.

Unsworth, M.J., Egbert, G., and Booker, J.R. (1999) High-resolution electromagnetic imaging of the San

Andreas fault in central California. *Journal of Geophysical Research*, **104**, 1131–1150.

Van Arsdale, R.B., Stahle, D.W., Cleaveland, M.K., and Guccione, M.J. (1998) Earthquake signals in tree-ring data from the New Madrid seismic zone and implications for paleoseismicity. *Geology*, **26**, 515–518.

Van den Berg, J.H. (1995) Prediction of alluvial channel pattern of perennial rivers. *Geomorphology*, **12**, 259–279.

van der Woerd, J., Klinger, Y., Sieh, K., Tapponnier, P., Ryerson, F.J., and Mériaux, A.-S. (2006) Long-term slip rate of the southern San Andreas Fault from ^{10}Be–^{26}Al surface exposure dating of an offset alluvial fan. *Journal of Geophysical Research*, **111**, B04407, doi:10.1029/2004JB003559.

Van Dissen, R.J., Berryman, K.R., Pettinga, J.R., and Hill, N.L. (1992) Paleoseismicity of the Wellington–Hutt Valley segment of the Wellington Fault, North Island, New Zealand. *New Zealand Journal of Geology and Geophysics*, **35**, 165–176.

van Wijk, J.W., Baldridge, W.S., van Hunen, J., Goes, S., Aster, R., Coblentz, D.D., Grand, S.P., and Ni, J. (2010) Small-scale convection at the edge of the Colorado Plateau: implications for topography, magmatism, and evolution of Proterozoic lithosphere. *Geology*, **38**, 611–614.

Vance, D., Bickle, M., Ivy-Ochs, S., and Kubik, P.W. (2003) Erosion and exhumation in the Himalaya from cosmogenic isotope inventories of river sediments. *Earth and Planetary Science Letters*, **206**, 273–288.

Vergés, J., Burbank, D.W., and Meigs, A. (1996) Unfolding: an inverse approach to fold kinematics. *Geology*, **24**, 175–178.

Vermeesch, P. (2004) How many grains are needed for a provenance study? *Earth and Planetary Science Letters*, **224**, 441–451.

Vincent, K.R., Bull, W.B., and Chadwick, O.A. (1994) Construction of a soil chronosequence using the thickness of pedogenic carbonate coatings. *Journal of Geological Education*, **42**, 316–324.

von Blanckenburg, F. (2006) The control mechanisms of erosion and weathering at basin scale from cosmogenic nuclides in river sediment. *Earth and Planetary Science Letters*, **242**, 224–239.

Wallace, R.E. (1978) Geometry and rates of change of fault-generated range front, north-central Nevada. *Journal of Research of the United States Geological Survey*, **6**, 637–650.

Wallace, R.E. (1987) Grouping and migration of surface faulting and variations in slip rates on faults in the Great Basin province. *Bulletin of the Seismological Society of America*, **77**, 868–876.

Walsh, J.J., Nicol, A., and Childs, C. (2002) An alternative model for the growth of faults. *Journal of Structural Geology*, **24**, 1669–1675.

Ward, S.N. (1990) Pacific–North America plate motions: new results from very long baseline interferometry. *Journal of Geophysical Research*, **95**, 21965–21981.

Ward, S.N., and Valensise, G. (1994) The Palos Verdes terraces, California: bathtub rings from a buried reverse fault. *Journal of Geophysical Research*, **99**, 4485–4494.

Watts, A.B. (2001) *Isostasy and Flexure of the Lithosphere*. Cambridge University Press, Cambridge.

Weiland, R.J., and Cloos, M. (1996) Pliocene–Pleistocene asymmetric unroofing of the Irian fold belt, Irian Jaya, Indonesia: apatite fission-track thermochronology. *Geological Society of America Bulletin*, **108**, 1438–1449.

Weissel, J.K., and Karner, G.D. (1989) Flexural uplift of rift flanks due to mechanical unloading of the lithosphere during extension. *Journal of Geophysical Research*, **94**, 13919–13950.

Weldon, R.J. (1986) *Late Cenozoic geology of Cajon Pass; implications for tectonics and sedimentation along the San Andreas fault*, Ph.D. thesis, California Institute of Technology, Pasadena.

Weldon, R.J., and Sieh, K.E. (1985) Holocene rate of slip and tentative recurrence interval for large earthquakes on the San Andreas fault in Cajon Pass, southern California. *Geological Society of America Bulletin*, **96**, 793–812.

Weldon, R., Scharer, K., Fumal, T., and Biasi, G. (2004) Wrightwood and the earthquake cycle: what a long recurrence record tells us about how faults work. *GSA Today*, **14**, 4–10.

Wells, A., and Goff, J. (2006) Coastal dune ridge systems as chronological markers of palaeoseismic activity: a 650-yr record from southwest New Zealand. *The Holocene*, **16**, 543–550.

Wells, S.G., Bullard, T.F., Menges, C.M., Drake, P.A., Kelson, K.I., Ritter, J.B., and Wesling, J.R. (1988) Regional variations in tectonic geomorphology along a segmented convergent plate boundary, Pacific coast of Costa Rica. *Geomorphology*, **1**, 239–265.

Wesnousky, S.G. (1994) The Gutenberg–Richter or characteristic earthquake distribution, which is it? *Bulletin of the Seismological Society of America*, **84**, 1940–1959.

Wesnousky, S.G. (2006) Predicting the endpoints of earthquake ruptures. *Nature*, **444**, 358.

Wesson, R.L., Helley, E.J., Lajoie, K.R., and Wentworth, C.M. (1975) Faults and future earthquakes. In

Borchardt, R.D. (ed.), *Studies for Seismic Zonation of the San Francisco Bay Region*, Professional Paper 941A, pp. 5–30. United States Geological Survey, Washington, DC.

Westgate, J.A., and Gorton, M.P. (1981) Correlation techniques in tephra studies. In Self, S., and Sparks, R.S.J. (eds), *Tephra Studies*, NATO Advanced Studies Institute Series C, pp. 73–94. Reidel, Dordrecht.

Whipp, D.M., and Ehlers, T.A. (2007) Influence of groundwater flow on thermochronometer-derived exhumation rates in the central Nepalese Himalaya. *Geology*, **35**, 851–854.

Whipp, D.M., Ehlers, T.A., Blythe, A.E., Huntington, K.W., Hodges, K.V., and Burbank, D.W. (2007) Plio-Quaternary exhumation history of the central Nepalese Himalaya: 2. Thermokinematic and thermochronometer age prediction model. *Tectonics*, **26**, TC3003, doi:10.1029/2006TC001991.

Whipple, K.X. (2004) Bedrock rivers and the geomorphology of active orogens. *Annual Review of Earth and Planetary Sciences*, **32**, 151–185.

Whipple, K.X. (2009) The influence of climate on the tectonic evolution of mountain belts. *Nature Geoscience*, **2**, 97–104.

Whipple, K.X., and Dunne, T. (1992) Debris-flow fans in Owens Valley, California. *Geological Society of America Bulletin*, **104**, 887–900.

Whipple, K.X., and Meade, B.J. (2004) Controls on the strength of coupling among climate, erosion, and deformation in two-sided, frictional orogenic wedges at steady state. *Journal of Geophysical Research*, **109**, F01011, doi:10.1029/2003JF000019.

Whipple, K.X., and Meade, B. (2006) Orogen response to changes in climatic and tectonic forcing. *Earth and Planetary Science Letters*, **243**, 218–228.

Whipple, K.X., and Tucker, G.E. (1999) Dynamics of the stream-power river incision model: implications for height limits of mountain ranges, landscape response timescales, and research needs. *Journal of Geophysical Research*, **104**, 17661–17674.

Whipple, K., Kirby, E., and Brocklehurst, S. (1999) Geomorphic limits to climate-induced increases in topographic relief. *Nature*, **401**, 39–43.

Whipple, K.X., Hancock, G.S., and Anderson, R.S. (2000) River incision into bedrock: mechanics and relative efficacy of plucking, abrasion, and cavitation. *Geological Society of America Bulletin*, **112**, 490–503.

Whittaker, A.C., Cowie, P.A., Attal, M., Tucker, G.E., and Roberts, G.P. (2007) Bedrock channel adjustment to tectonic forcing: implications for predicting river incision rates. *Geology*, **35**, 103–106.

Wickham, J. (1995) Fault displacement-gradient folds and the structure at Lost Hills, California (USA). *Journal of Structural Geology*, **17**, 1293–1302.

Willemse, E.J.M., Pollard, D.D., and Aydin, A. (1996) Three-dimensional analyses of slip distributions on normal fault arrays with consequences for fault scaling. *Journal of Structural Geology*, **18**, 295–309.

Willenbring, J.K., and von Blanckenburg, F. (2010) Meteoric cosmogenic beryllium-10 adsorbed to river sediment and soil: applications for Earth-surface dynamics. *Earth Science Reviews*, **98**, 105–122.

Willett, S.D. (1999) Orogeny and orography: the effects of erosion on the structure of mountain belts. *Journal of Geophysical Research*, **104**, 28957–28982.

Willett, S.D., and Brandon, M.T. (2002) On steady states in mountain belts. *Geology*, **30**, 175–178.

Willett, S., Beaumont, C., and Fullsack, P. (1993) Mechanical model for the tectonics of doubly vergent compressional orogens. *Geology*, **21**, 371–374.

Willett, S.D., Slingerland, R., and Hovius, N. (2001) Uplift, shortening, and steady state topography in active mountain belts. *American Journal of Science*, **301**, 455–485.

Willett, S.D., Fisher, D., Fuller, C., Yeh, E.-C., and Lu, C.-Y. (2003) Erosion rates and orogenic-wedge kinematics in Taiwan inferred from fission-track thermochronometry. *Geology*, **31**, 945–948.

Wintle, A.G. (1993) Luminescence dating of aeolian sands: an overview. In Pye, K. (ed.), *The Dynamics and Environmental Context of Aeolian Sedimentary Systems*, Special Publication 72, pp. 49–58. Geological Society of London, London.

Wobus, C.W., Hodges, K.V., and Whipple, K.X. (2003) Has focused denudation sustained active thrusting at the Himalayan topographic front? *Geology*, **31**, 861–864.

Wobus, C., Heimsath, A., Whipple, K.X., and Hodges, K.V. (2005) Active surface thrust faulting in the central Nepalese Himalaya. *Nature*, **434**, 1008–1011.

Wobus, C., Crosby, B.T., and Whipple, K.X. (2006a) Hanging valleys in fluvial systems: controls on occurrence and implications for landscape evolution. *Journal of Geophysical Research*, **111**, F02017, doi:10.1029/2005JF000406.

Wobus, C.W., Whipple, K.X., and Hodges, K.V. (2006b) Neotectonics of the central Nepalese Himalaya: constraints from geomorphology, detrital $^{40}Ar/^{39}Ar$ thermochronology, and thermal modeling. *Tectonics*, **25**, TC4011, doi:10.1029/2005TC001935.

Wobus, C., Whipple, K.X., Kirby, E., Snyder, N., Johnson, J., Spyropolou, K., Crosby, B., and Sheehan,

D. (2006c) Tectonics from topography: procedures, promise, and pitfalls. In Willett, S.D., Hovius, N., Brandon, M.T., and Fisher, D.M. (eds), *Tectonics, Climate and Landscape Evolution*, Special Paper 398, pp. 55–74. Geological Society of America, Boulder, CO.

Wolf, R.A., Farley, K.A., and Silver, L.T. (1996) Helium diffusion and low temperature thermochronometry of apatite. *Geochimica et Cosmochimica Acta*, **60**, 4231–4240.

Wolfe, J.A. (1971) Tertiary climatic fluctuations and methods of analysis of Tertiary floras. *Palaeogeography, Palaeoclimatology, Palaeoecology*, **9**, 27–57.

Wolfe, J.A. (1990) Paleobotantical evidence for a major temperature increase following the Cretaceous/ Tertiary boundary. *Nature*, **343**, 153–156.

Wolfe, J.A. (1993) A method of obtaining climatic parameters from leaf assemblages, *U.S. Geological Survey Bulletin*, **2040**, 71p.

Wolfe, J.A., Forest, C.E., and Molnar, P. (1998) Paleobotanical evidence of Eocene and Oligocene paleoaltitudes in midlatitude western North America. *Geological Society of America Bulletin*, **110**, 664–678.

Wolfe, S.A., Huntley, D.J., and Ollerhead, J. (1995) Recent and late Holocene sand dune activity in southwestern Saskatchewan. *Current Research, Geological Survey of Canada*, **1995-B**, 131–140.

Working Group on California Earthquake Probabilities (1995) Seismic hazards in Southern California: probable earthquakes, 1994 to 2024. *Bulletin of the Seismological Society of America*, **85**, 379–439.

Wyss, M., and Brune, J.N. (1967) The Alaska earthquake of 28 March 1964; a complex multiple rupture. *Bulletin of the Seismological Society of America*, **57**, 1017–1023.

Xu, X., Chen, W., Ma, W., Yu, G., and Chen, G. (2002) Surface rupture of the Kunlunshan earthquake (Ms 8.1), northern Tibetan plateau, China. *Seismological Research Letters*, **73**, 884–892.

Yamaguchi, D.K., and Hoblitt, R.P. (1995) Tree-ring dating of pre-1980 volcanic flowage deposits at Mount St. Helens, Washington. *Geological Society of America Bulletin*, **107**, 1077–1093.

Yamaguchi, D.K., Atwater, B.F., Bunker, D.E., Benson, B.E., and Reid, M.S. (1997) Tree-ring dating the 1700 Cascadia earthquake. *Nature*, **389**, 922–923.

Yamanaka, H., and Iwata, S. (1982) River terraces along the middle Kali Gandaki and Marsyandi Khola, central Nepal. *Journal of Nepal Geological Society*, **2**, 95–112.

Yanites, B.J., Tucker, G.E., Mueller, K.J., Chen, Y.-G., Wilcox, T., Huang, S.-Y., and Shi, K.-W. (2010) Incision and channel morphology across active structures along the Peikang River, central Taiwan: implications for the importance of channel width. *Geological Society of America Bulletin*, **122**, 1192–1208.

Zachariasen, J., Sieh, K., Taylor, F.W., and Hantoro, W.S. (2000) Modern vertical deformation above the Sumatran subduction zone: paleogeodetic insights from coral microatolls. *Journal of Geophysical Research*, **90**, 897–913.

Zandt, G. (2003) The southern Sierra Nevada drip and the mantle wind direction beneath the southwestern United States. *International Geological Review*, **45**, 213–223.

Zeitler, P.K. (1985) Cooling history of the NW Himalaya, Pakistan. *Tectonics*, **4**, 127–151.

Zeitler, P.K., Koons, P., Bishop, M.P., Chamberlain, C.P., Edwards, M.A., Hamidullah, S., Qasim Jan, M., Asif Khan, M., Umar Khan Khattak, M., Kidd, W.S.F., Mackie, R.L., Meltzer, A.S., Park, S.K., Pecher, A., Poage, M.A., Sarker, G., Schneider, D.A., Seeber, L., and Shroder, J.F. (2001a) Crustal reworking at Nanga Parbat, Pakistan: metamorphic consequences of thermal–mechanical coupling facilitated by erosion. *Tectonics*, **20**, 712–728.

Zeitler, P.K., Meltzer, A.S., Koons, P.O., Craw, D., Hallet, B., Chamberlain, C.P., Kidd, W.S.F., Park, S.K., Seeber, L., Bishop, M., and Shroder, J. (2001b) Erosion, Himalayan geodynamics, and the geomorphology of metamorphism. *GSA Today*, **11**, 4–9.

Zepeda, R.L. (1993) *Active tectonics and soil chronology of Wheeler Ridge, Southern San Joaquin Valley, California*, Ph.D. thesis, University of California.

Zhang, P., Molnar, P., and Downs, W.R. (2001) Increased sedimentation rates and grain sizes 2–4 Myr ago due to the influence of climate change on erosion rates. *Nature*, **410**, 891–897.

Zhang, P.-Z., Shen, Z., Wang, M., Gan, W., Bürgmann, R., Molnar, P., Wang, Q., Niu, Z., Sun, J., Wu, J., Sun, H., and You, X. (2004) Continuous deformation of the Tibetan Plateau from global positioning system data. *Geology*, **32**, 809–812.

Zielke, O., and Arrowsmith, J.R. (2008) Depth variation of coseismic stress drop explains bimodal earthquake magnitude–frequency distribution. *Geophysical Research Letters*, **35**, L24301, doi:10.1029/ 2008GL036249.

Zielke, O., Arrowsmith, J.R., Ludwig, L.G., and Akciz, S.O. (2010) Slip in the 1857 and earlier large earthquakes along the Carrizo Plain, San Andreas fault. *Science*, **327**, 1119–1122.

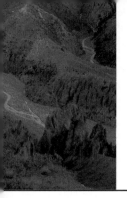

Index